**WITHDRAWN
UTSA LIBRARIES**

WORLD ATLAS
OF OIL AND GAS BASINS

WORLD ATLAS
OF OIL AND GAS BASINS

Li Guoyu

A John Wiley & Sons, Ltd., Publication

This edition first published 2011 © 2011 by John Wiley & Sons Ltd

Wiley-Blackwell is an imprint of John Wiley & Sons, formed by the merger of Wiley's global Scientific, Technical and Medical business with Blackwell Publishing.

Registered Office: John Wiley & Sons Ltd, The Atrium, Southern Gate, Chichester, West Sussex, PO19 8SQ, UK

Editorial Offices: 9600 Garsington Road, Oxford, OX4 2DQ, UK
The Atrium, Southern Gate, Chichester, West Sussex, PO19 8SQ, UK
111 River Street, Hoboken, NJ 07030-5774, USA

For details of our global editorial offices, for customer services and for information about how to apply for permission to reuse the copyright material in this book please see our website at www.wiley.com/wiley-blackwell.

The right of the author to be identified as the author of this work has been asserted in accordance with the UK Copyright, Designs and Patents Act 1988.

All rights reserved. No part of this publication may be reproduced, stored in a retrieval system, or transmitted, in any form or by any means, electronic, mechanical, photocopying, recording or otherwise, except as permitted by the UK Copyright, Designs and Patents Act 1988, without the prior permission of the publisher.

Designations used by companies to distinguish their products are often claimed as trademarks. All brand names and product names used in this book are trade names, service marks, trademarks or registered trademarks of their respective owners. The publisher is not associated with any product or vendor mentioned in this book. This publication is designed to provide accurate and authoritative information in regard to the subject matter covered. It is sold on the understanding that the publisher is not engaged in rendering professional services. If professional advice or other expert assistance is required, the services of a competent professional should be sought.

Library of Congress Cataloging-in-Publication Data

Guoyu, Li.
 World atlas of oil and gas basins / Li Guoyu.
     p. cm.
 Includes bibliographical references and index.
 ISBN 978-0-470-65661-7 (hardcover : alk. paper)
1. Oil fields–Maps. 2. Gas fields–Maps. 3. World atlas. I. Title.
 G1021.H8.L5 2011
 553.2′80223–dc22
                                             2010040514

A catalogue record for this book is available from the British Library.

This book is published in the following electronic formats: ePDF 9781444390049; Wiley Online Library 9781444390032; ePub 9781444390056

Set in 12/16pt Times by SPi Publisher Services, Pondicherry, India
Printed and bound in Malaysia by Vivar Printing Sdn Bhd

1 2011

Contents

Editorial Board
About the Author
Preface
Foreword
Introduction
The Geological Time Scale
Key to Maps

		Pages
Part I	**Overview**	**1**
1	World Topography	2
2	World Political Map	6
3	Geological Map of the Continents	10
4	World Tectonic Map	14
5	World Map of Oil and Gas Basins	18
6	Classification of Oil and Gas Basins by Geometry of Cross-Sections	22
7	Classification of World Oil and Gas Fields	26
8	World Map of Oil and Gas Resources	30
9	Graphs of World Oil and Gas Reserves, Production and Price	34
10	World Oil Trade	38
Part II	**Asian Oil and Gas Basins**	**42**
11	China	46
12	North and South Korea	52
13	Mongolia	56
14	Japan	60
15	Burma (Myanmar), Thailand, Laos, Cambodia and Vietnam	64
16	Malaysia, Singapore and Brunei	68
17	Indonesia and East Timor	72
18	Sumatra (Sumatera) Central Basin	76
19	Philippines	80
20	Pakistan, Afghanistan, Nepal, Bhutan and Bangladesh	84
21	India, Sri Lanka and The Maldives	88
22	Bombay Basin	92
23	Central Asia: Kazakhstan, Turkmenistan, Uzbekistan, Tajikistan and Kyrghyzstan	96
24	Caspian Sea Oil and Gas Region	100

Contents

Pages

25	Pre-Caspian Sea Basin	104
26	North Ustyurt Basin	110
27	Mangyshlak Basin	114
28	South Caspian Sea Basin	118
29	Amur (Kara-Kum) Basin	122
30	Chu-Sarysu Basin	128
31	Turgay Basin	132
32	Fergana Basin	136
33	Afghan-Tajik Basin	140
34	Azerbaijan, Georgia, Armenia and Kura Basin	144
35	Middle East	148
36	Iran and Iraq	152
37	Saudi Arabia, Kuwait, Bahrain, Qatar, United Arab Emirates, Oman and Yemen	156
38	Persian Gulf Oil and Gas Region	160
39	Geological Map of the Persian Gulf Oil and Gas Region	168
40	Syria, Lebanon, Jordan, Cyprus, Palestine and Israel	172

Part III African Oil and Gas Basins — **176**

41	Egypt	182
42	Libya	186
43	Sirte Basin	190
44	Algeria, Morocco and Tunisia	194
45	Mauritania, Western Sahara, Senegal, Gambia, Mali and Burkina Faso	198
46	Guinea, Guinea Bissau and Sierra Leone	202
47	Liberia, Cote D'ivoire, Ghana, Togo and Benin	206
48	Niger, Nigeria, Cameroon, Sao Tome and Principe and Equatorial Guinea	210
49	Niger Delta Basin	214
50	Congo, Democratic Republic of Congo and Gabon	218
51	Zambia, Angola and Malawi	222
52	Sudan, Central Africa and Chad	226
53	Muglad Basin	230
54	Ethiopia, Somalia, Djibouti and Eritrea	234
55	Kenya and Uganda	238
56	Tanzania, Rwanda and Burundi	242
57	Mozambique, Comoros, Madagascar, Seychelles, Mauritius and Reunion	246
58	Namibia, Zimbabwe, Botswana, South Africa, Swaziland and Lesotho	250

Contents

Pages

Part IV European Oil and Gas Basins — **254**

59	Romania, Serbia, Montenegro, Slovenia, Croatia, Bosnia and Herzegovina, Hungary, Bulgaria and Macedonia	258
60	Carpathian Basin	262
61	Poland, Czech Republic and Slovakia	266
62	Germany, Luxembourg, Switzerland and Liechtenstein	270
63	Austria, Italy, Albania, Greece, San Marino and Malta	274
64	Spain, Portugal and Andorra	278
65	France, Netherlands, Belgium and Monaco	282
66	United Kingdom and Ireland	286
67	Norway, Sweden, Finland, Denmark and Iceland	290
68	North Sea Oil and Gas Region	294
69	Estonia, Latvia, Lithuania, Belarus, Ukraine and Moldova	298
70	Dnept-Donets Basin	302
71	Russia	306
72	Oil and Gas Pipelines of Russia and Neighbouring Countries	314
73	Volga-Urals Basin	318
74	Timano-Pechora Basin	322
75	West Siberia Basin	326
76	East Siberia Basin	330
77	North Kavkaz Basin	334
78	Sakhalin Basin	338

Part V North American Oil and Gas Basins — **342**

79	Canada	346
80	Western Canada Oil and Gas Region	350
81	USA	354
82	Los Angeles Basin	362
83	Rocky Mountain Basins	366
84	Williston Basin	370
85	Michigan Basin	374
86	Mexico Gulf Oil and Gas Region	378
87	Appalachian Basin	382
88	Alaska Oil and Gas Region	386
89	Mexico	390
90	Tampico Basin	394
91	Caribbean Sea Region	398

Part VI South American Oil and Gas Basins — **404**

92	Colombia and Ecuador	408
93	Venezuela, Guyana and Surinam	412

		Pages
94	Maracaibo Basin	416
95	Venezuela East Basin	420
96	Peru, Bolivia and Paraguay	424
97	Putumayo Basin	428
98	Brazil and Uruguay	432
99	Campos Basin	436
100	Chile and Argentina	440

Part VII Australasia and the Poles — **445**

101	Australia and Papua New Guinea	446
102	Gippsland Basin	450
103	New Zealand, Samoa, Fiji and Tonga	454
104	Antarctica	458
105	Arctic Ocean	462

	References	466
	Index of Countries	468
	Index of Basins	470

The Editorial Board of World Atlas of Oil and Gas Basins

Chairman

Tang Ke   Former Minister of Petroleum Industry, People's Republic of China

Vice Chairmen

Zhou Jiping	Vice President, China National Petroleum Corporation (CNPC)
Mu Shuling	Former Vice President, China Petroleum and Chemical Corporation (SINOPEC); Academician, Russian Academy of Natural Sciences
Fu Chengyu	President, China National Offshore Oil Corporation (CNOOC)
Li Guoyu	Professor, Academician, Russian Academy of Natural Sciences; Director of the Oil and Gas Resources Bureau of Former China's Ministry of Energy

Advisors

Lord Browne of Madingley	President, Royal Academy of Engineering; Managing Partner, Riverstone Europe LLP
Elik. M. Khalimov	Former Deputy Minister of Oil, USSR; Academician, Russian Academy of Natural Sciences
A.E. Kontorovich	Academician, Russian Academy of Sciences; Director, Institute of Petroleum Geology of Russian Academy of Sciences, Siberia Branch
V.I. Vysotsky	Director, Oil and Gas Department, VNII Zarupezhe-geologia, Russia

Members

Li Desheng	Academician, Chinese Academy of Sciences; Former Chief Geologist, Research Institute of Exploration and Development, CNPC
Tian Zaiyi	Academician, Chinese Academy of Sciences; Former Vice President, Research Institute of Exploration and Development, CNPC
Luo Yingjun	Former Vice President, PetroChina; President Assistant, CNPC
Jin Zhijun	Deputy Chief Geologist, SINOPEC
Xavier Chen	President, Beijing Energy Club
Dapo Odesanya	Professor; Director, Andaz Global Solutions, Beijing

The Editorial Board of World Atlas of Oil and Gas Basins

Editorial Staff

Author: Li Guoyu

Maps: Li Guoyu

Text: Li Guoyu

Editorial consultant – English: Dapo Odesanya

Translators: Chen Jiahuang (place names); Liu Qian,
              Wei Haifeng (maps 1–10); Shi Jun (basin names);

English proofreaders: Liu Qian, Li Wei, Li Jun, Wang Haiyun, Ren Shuoyi

Index editor: Chen Jiahuang

About the Author: Professor Li Guoyu

Li Guoyu was born in Lanzhou City, the capital of Gansu Province in northwest China on 17August 1930. He finished his higher education in the Russian Language Department of Lanzhou University in 1952, and furthered his studies at the Beijing Petroleum Geology School in the next year.

Li Guoyu's professional and academic career has been varied and multifaceted. He was the Vice Director of the Exploration Department of China's Ministry of Petroleum for many years and has also held the position of Director of the Oil and Gas Resources Bureau of Former China's Ministry of Energy. His career in the petroleum industry spans over 57 years, during which he was directly involved in the exploration, discovery and development of some of China's most significant oil and gas fields, including Daqing and Shengli.

Professor Li participated in the efforts of the Chinese authorities to promote international cooperation regarding resources in the world's continental shelf regions, which was a strategically important objective in an increasingly globalized environment.

One of his most significant achievements, which has become a new and radical school of thought in the petroleum geology field worldwide, is in the field of theoretical research concerning the development of sedimentary basins. A fundamental tenet of his views is the contention that any sedimentary basin must have potential to be oil and gas bearing.

Professor Li's written output has been prodigious. He has written 84 scholarly works, papers and theses, many of which have been published and have received professional and academic accolades and acknowledgement in countries such as the USA, Russia, UK and France. Professor Li is an Honorary Academician of the Russian Academy of Sciences, and has travelled to 47 countries in pursuit of this work.

A pioneering achievement is an Atlas compendium of the world's oil and gas basins and fields. This compendium is divided into two parts, the first covering Professor Li's home country of China and the second dealing with the rest of the world. These two parts were further subdivided into two series of colour Atlases, each comprised of four sections.

China: Vol. I: China Oil and Gas Basins Atlas
        Vol. II: China Oil Fields Atlas (Part 1)
        Vol. III: China Oil Fields Atlas (Part 2)
        Vol. IV: China Gas Fields Atlas

World: Vol. V: World Atlas of Oil and Gas Basins
Vol. VI: World Atlas of Oil Fields (Part 1)
Vol. VII: World Atlas of Oil Fields (Part 2)
Vol. VIII: World Atlas of Gas Fields

Following this impressive effort, Professor Li created a further compendium, this time in the form of a collection of his major papers and thesis. This three-volume series comprise the following categories.

Vol. I: China Petroleum Geology
Vol. II: World Petroleum Geology
Vol. III: Collection of Inspection Reports on World Oil Regions and Areas

A third large-scale series is three volumes of a coloured Atlas.

Vol. I: Atlas of Oil and Gas Basins of China (Second edition)
Vol. II: New World Atlas of Oil and Gas Basins (Parts 1 and 2)
Vol. III: World Atlas of Oil and Gas Basins (English version)

Given his national and international academic stature, Professor Li has been invited as Visiting Professor at many of China's petroleum universities, as well as at the University of Texas in the USA and the West Siberian University of Russia. He continues to receive invitations to write and present papers on petroleum geology and on his views on sedimentary basin development at international conferences and forums, both in China and around the world.

Preface

When the manuscript of this English edition of *World Atlas of Oil and Gas Basins* was presented to me I was deeply impressed.

Data on the world's oil and gas resources do exist in the hands of oil and gas companies and academic institutions. But this is the first time that all the world's major basins – more than 500 – are presented to the general public in a succinct, reader-friendly format.

Oil and gas have played a critical role in fuelling human progress and will continue to do so in the 21st century. Through the colourful maps and detailed explanatory notes, Professor Li conveys a very important message: the world remains endowed with abundant oil and gas resources. His work shows, in detail, where these resources are located, in various geological structures across five continents and 190 countries.

Only a world-class expert with a long-term career in the oil industry could accomplish such a masterful job. I recently learned of Professor Li's 57 years of work as a geologist and that the Chinese edition of the book, published a few years ago, was widely acclaimed not only in China but elsewhere.

I know that Professor's Li's great wish is to share his work with the global community – hence this first translation of his Atlas into English. This is an admirable ambition, one for which Professor Li should be congratulated.

Lord Browne of Madingley
President, Royal Academy of Engineering
Managing Partner, Riverstone Europe LLP
London, 1st September 2009

Foreword

Professor Li Guoyu is a world famous petroleum geologist, whose many papers and theses have been published in different languages worldwide, including in countries such as the UK and Russia. Professor Li is a familiar and respected figure in the international community of petroleum geologists and oil and gas experts, and he has received justifiable praise for his vast scholarly and research work.

I am extremely pleased that Professor Li will soon publish another substantial work which will be a valuable addition to the world's store of written knowledge and expertise – the *World Atlas of Oil and Gas Basins*. This Atlas provides expert overviews and summaries of 507 petroliferous basins and 560 significant oil and gas fields in 190 countries and regions worldwide. In addition to geological and sedimentary data, the Atlas provides important information on the geography of the global oil and gas infrastructure, including major pipelines, refineries, terminals and port facilities.

There have been similar comprehensive works published in Russia in the past, such as the particularly noteworthy work of the Visotsky father and son team during the 1970s and 1980s (Visotsky, 1995; Visotsky *et al.*, 1995). However, Professor Li's masterly publication with its wide array of maps, diagrams and explanatory notes ushers in a new and important historical phase in the further development of a global collection of integrated atlases.

I have no doubt that this new Atlas will be of critical importance and value to experts, scholars, investors and professional practitioners in the global oil and gas industry. Not only will it exert great influence on the continuing international study of the distribution laws and theories of oil and gas fields, it will also play a significant role in predicting trends in global oil and gas in the 21st century.

I am proud to be associated with this valuable and important publication and I wish my friend and colleague, Professor Li Guoyu future success, more vigorous creativity and good health and a long life of achievement.

A.E. Kontorovich
Academician, Russian Academy of Sciences
July 2009

Introduction

I drew the maps and wrote the various notes contained in this atlas in order to provide a summary of my views on the past, present and future of the world's oil and gas industry. The *World Atlas of Oil and Gas Basins* is a culmination of my research and studies on world oil and geology carried out over the past 57 years. The compilation of this atlas has evolved over many years (Li Gouyu, 1982a–c, 1988a,b, 1990a,b, 1991a–d, 1997, 2000, 2009; Kewan Gan *et al.*, 1990; Yongxin Jiang and Yishan Dou, 2003). Although every effort has been made to update individual maps, this has not been possible in all cases. The reader is therefore urged to consult up-to-date international (such as the US Geological Survey World Energy Assessment – http://pubs.usgs.gov/dds/dds-060/) and local databases for more up-to-date information. I am pleased to say that I am optimistic, based on scientific research and theory, about the world's oil and gas resources for the future. The atlas covers an introduction to world petroleum geology, world oil and gas resources and the development of the global oil industry. This coverage includes the world's five continents, 190 countries and regions, 507 petroleum basins, as well as 560 large and significant oil and gas fields. The atlas provides what I consider to be a profound analysis and discussion of the past and present of the world oil and gas industry, and most importantly, its future.

Basic premise

According to my views on the development of sedimentary basins, which forms the basis of the atlas, any sedimentary basin must have the potential to be oil and gas bearing. However, each basin is distinct, particularly in respect of the commercial viability of developing and extracting the oil and gas within them. It is my scientific opinion that the world contains abundant oil and gas resources. There are one trillion tons of oil that are discoverable in the approximately 100 million km^2 of sedimentary rocks located on Earth. Approximately 70 million km^2 of which are distributed on continental landmasses, while the remainder are found under the oceans. In my view more and more large-scale oil and gas provinces will be discovered in the future and therefore the prospects of the industry for the 21st century are excellent. Furthermore, it is my belief that the industry will continue to play a key role in the global energy field well into the 22nd century.

The past

The history of the human race, particularly between 2000 BC and AD 1900, shows a remarkable transition from near-ignorance to a better appreciation of the Earth's resources and their functional value to human development. Three ancient civilizations – Egyptian, Babylonian and Indian – exploited oil asphalt from surface seepages for construction, lighting and rudimentary medicine production. It was only later that this rare fossil resource came to be exploited for oil production.

Total annual world production was merely 300 tons* in 1857, but then showed dramatic increases to 800,000 tons in 1870, 4 million tons in 1880, 10 million tons in 1890 and 20 million

---

* In the atlas SI units of measurement are used (e.g. ton, m, km), but units used for oil and gas reserves and production are often presented differently:
1 ton = 7.3 barrels of oil
1 barrel = 0.137 t
1 m^3 = 35.31 ft^3 (gas)
1 ft^3 = 0.02832 m^3

Introduction

tons in 1900. As of 1900, 8.5 million tons of oil have been produced in the USA, representing 42% of world total. At the same time, Russia produced 10.68 million tons, 52% of the world total. The interesting fact is that these two countries remain the largest oil producers in the world and this reinforces the premise that there are still many large oil and gas provinces to be discovered.

The present

'The Present' refers to the whole of the 20th century, which represents a critical historical phase of the development of the global petroleum industry, particularly the 55 years after the end of Second World War. Consider the interesting statistic that worldwide production was a mere 20 million tons in 1900, but reached 3.36 billion tons by 2000, a staggering 164-fold increase over the course of the century. This spectacular growth rate clearly indicates the importance of oil and gas in developing a perspective for future analysis of the global energy industry.

There are five key features of the stages of the development of the global petroleum industry, which can be summarized as follows:

1 Oil and gas fossil fuels and their use in high-efficiency internal combustion engines were important drivers for the rapid human development during the early stages of the 20th century. By the year 2000, oil and gas accounted for 61% of the world's energy mix.

2 There were multiple significant discoveries of oil and gas provinces in the 20th century including:

Mexico	1901
Iran	1904
Venezuela	1922
Kuwait	1938
Saudi Arabia	1940
Baku (Second)	1948
Daqing (China)	1959
West Siberia (Russia)	1960
North Sea Oil	1969

These and other discoveries of oil and gas provinces have laid a solid basis for the development of the global oil industry both for the present and for the future.

3 Economic statistics clearly confirm that the level of a country's aggregate prosperity is directly proportional to national fuel consumption. Consider the following figures of fuel consumption in some of the world's major economies in 2000:

USA	890 million tons (mt)
Japan	250 mt
China	230 mt
Germany	129 mt
Russia	123 mt
France	94.9 mt
Italy	93.5 mt
Britain	78.9 mt

In all of these countries, the strong long-term industrial and agricultural bases and rapid sustained economic development were driven by abundant oil and gas resources.

Introduction

4   Geopolitical competition for oil and gas resources continues unabated and indeed is intensifying. The USA, western Europe, Japan and China are four major oil consuming nations and regions. The competition among them for oil and gas is fierce, and in the case of the USA and western Europe, wars were waged in pursuit of access to these resources.

5   Oil exploration and extraction technologies have flourished. Geological conditions for oil and gas exploration and development are becoming more complicated, but humankind has tackled these challenges with vigorous advancement and enhancement of technologies such as geophysical techniques (e.g. three- and four-dimensional seismic mapping), horizontal drilling technologies and polymer flooding techniques, just to name a few.

The future

Here I refer to the third millennium, the period from the year 2001 to the year 3000. I must start by acknowledging that predicting oil and gas trends for a century is an extremely difficult task. Therefore, predictions for a millennium are unimaginably challenging. However, we may observe some trends from the analysis of oil reserves and production for this period, which are given in Table 1.

Table 1 demonstrates the production potential of the next millennium based on the best data available for reserves by the end of 2004. Based on scientific analysis of the trends of the global world industry I firmly believe the following:

- The quantity of global conventional and non-conventional oil resources that can be produced is in the range of 811.2 billion tons to 1112 billion tons. By the end of 2004, 129.9 billion tons had been produced cumulatively, representing between 12% and 16% of these reserves and resources. This suggests that 84% to 88% still remain unexplored and undeveloped.
- Seventy per cent of the Original Oil in Place (OOIP) figures excluded for 'recovery factor' reasons represent truly enormous reserves. The estimated recoverable oil reserves of 811.2 billion tons to 1112 billion tons mentioned above are calculated based on the 'recovery factor' basis, and merely represent 30% of global OOIP. This suggests that the excluded reserves are between 1622.4 billion tons and 2222.4 billion tons. I am not suggesting that all of this 70% can be produced. But we must not underestimate the ingenuity of humankind for further development of revolutionary oil and gas exploration, development and production technologies.

Table 1   World oil production forecast from 2001 to 3000

Period	Total number of years	Average oil production per annum (million)	Oil production of each period (billion)
2001–2100	100	3400	340
2101–2200	100	2000	200
2201–2300	100	1500	150
2301–2400	100	1000	100
2401–2500	100	500	50
2501–3000	500	300	150
Total			990

Introduction

The conclusion is simple and definite: potential oil and gas reserves that are yet to be exploited are massive in scale.

I am indebted to the Chairman, Vice Chairmen and Members of the Editorial Board of the Atlas, as well as to the Advisors for their suggestions for the development of this important publication. I would also like to place on record my gratitude to all the editors and proofreaders. I am particularly grateful to Lord Browne and Dr A.E. Kontorovich for their willingness to write the Preface and Foreword to this Atlas. I wish also to express my appreciation to Mr Wang Junqiao, President of SJ Petroleum Machinery Co., Mr Wang Xun, President of Beijing Conspase Trading Co., Ltd for his strong support and to Mrs Li Wen for her warm help.

Professor Li Guoyu
1st October 2009

The Geological Time Scale

Era	Period		Epoch	Ma
Cenozoic (Cz)	Quaternary (Q)		Holocene (Q_2)	0.01
			Pleistocene (Q_1)	1.6
	Neogene* (N)		Pliocene (N_2)	5.3
			Miocene (N_1)	23.7
	Paleogene* (E)		Oligocene (E_3)	36.6
			Eocene (E_2)	57.8
			Paleocene (E_1)	66.4
Mesozoic (Mz)	Cretaceous (K)			144
	Jurassic (J)			208
	Triassic (T)			245
Palaeozoic (Pz)	Permian (P)			286
	Carboniferous (C)	Pennsylvanian (C_2)		320
		Mississippian (C_1)		360
	Devonian (D)			408
	Silurian (S)			438
	Ordovician (O)			505
	Cambrian (ε)			570
Precambrian (Pε)	Proterozic (Pt)			2500
	Archaean (Ar)			4600

*Paleogene + Neogene = Tertiary (R). (Based on Tarbuck and Lutgens, 1991.)

KEY TO MAPS

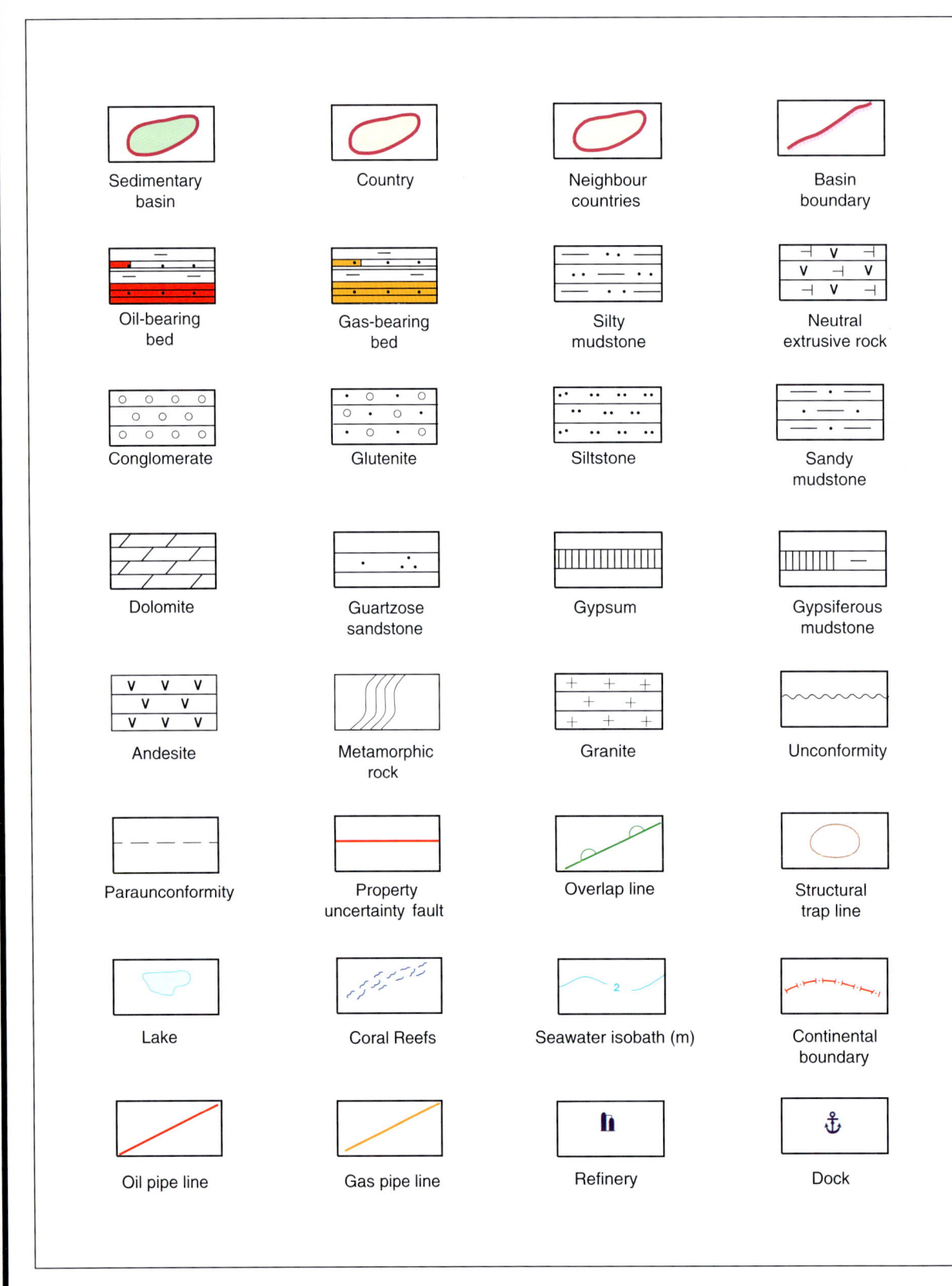

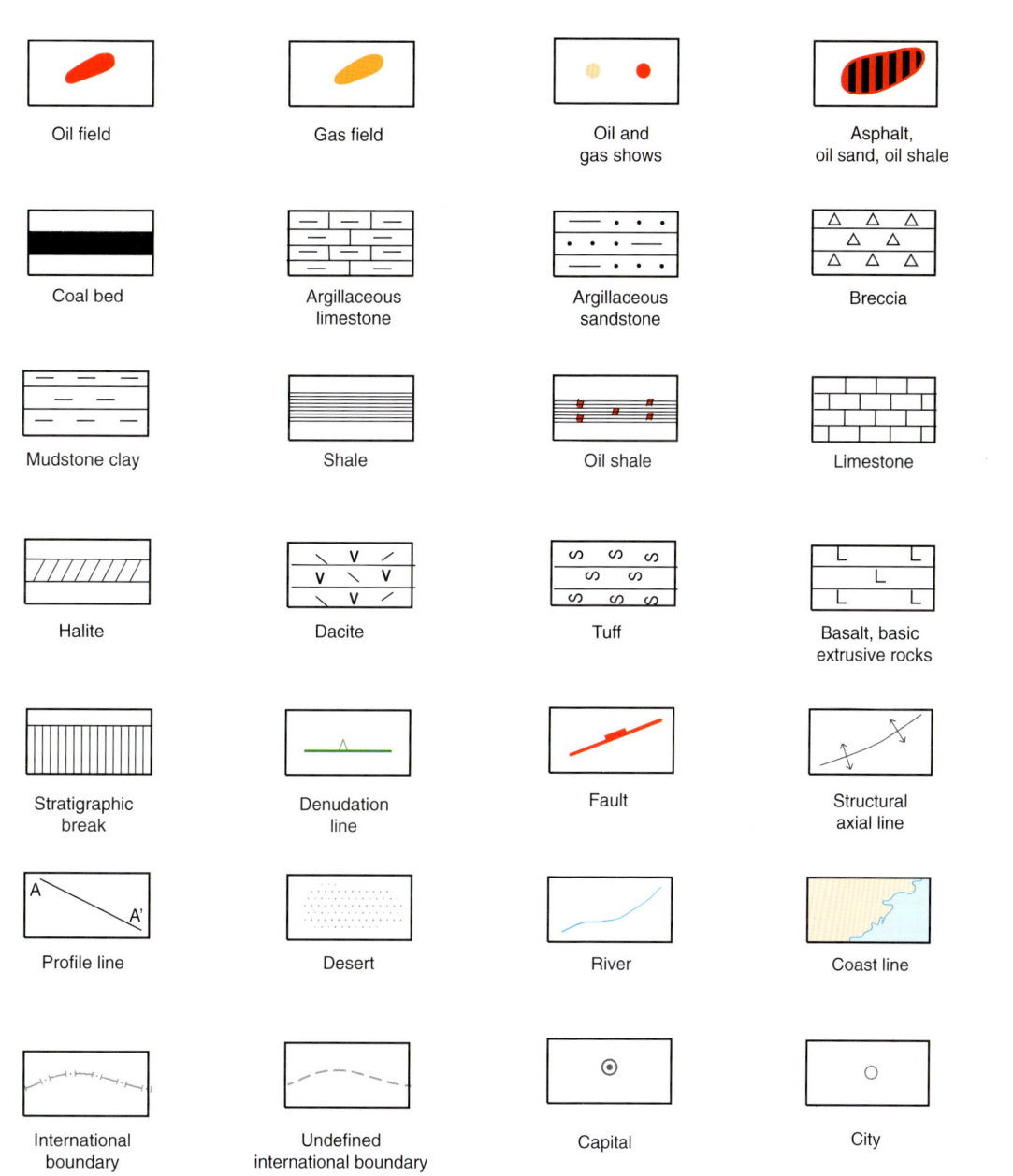

Part I
Overview

WORLD TOPOGRAPHY

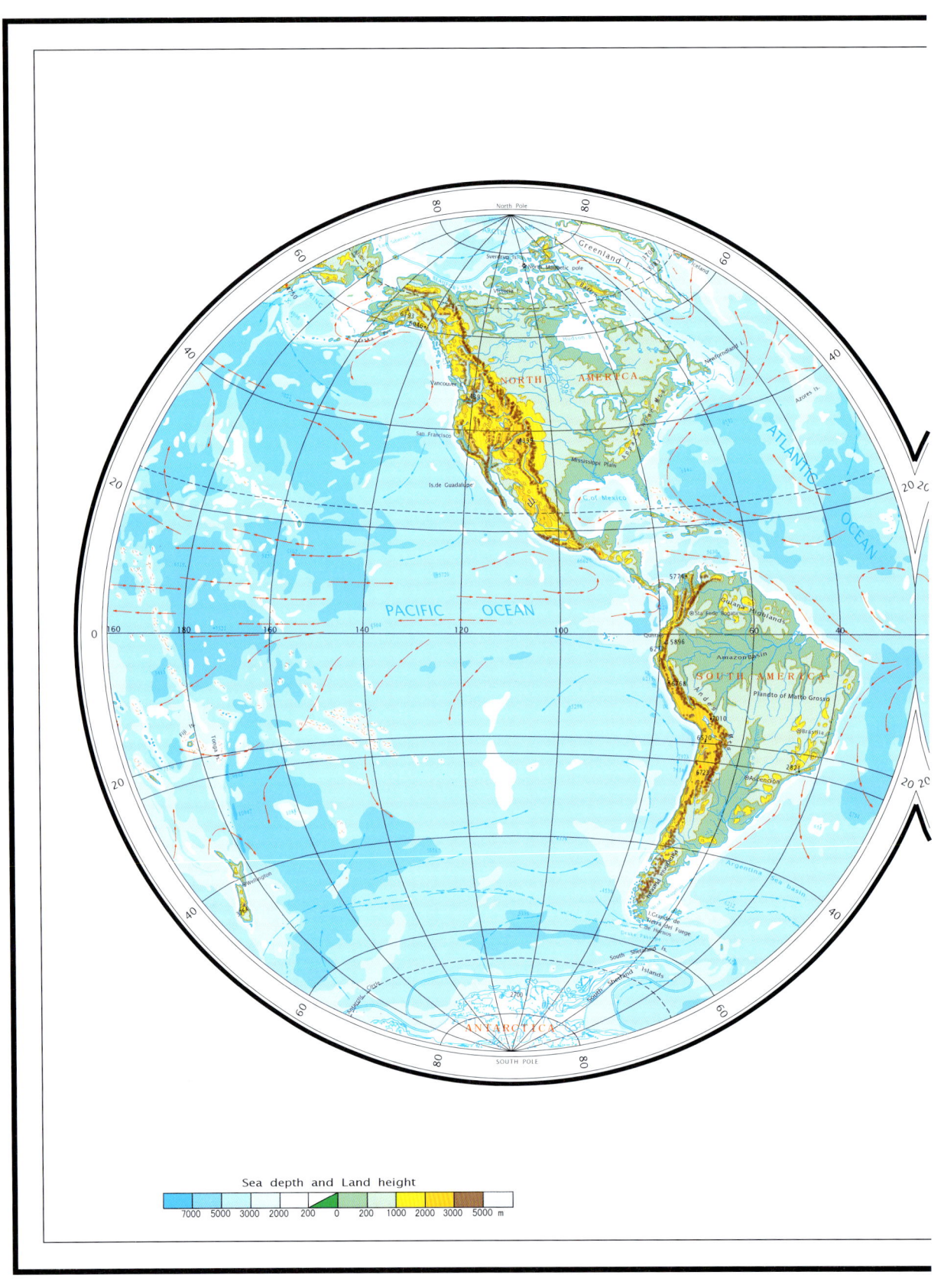

CHAPTER 1

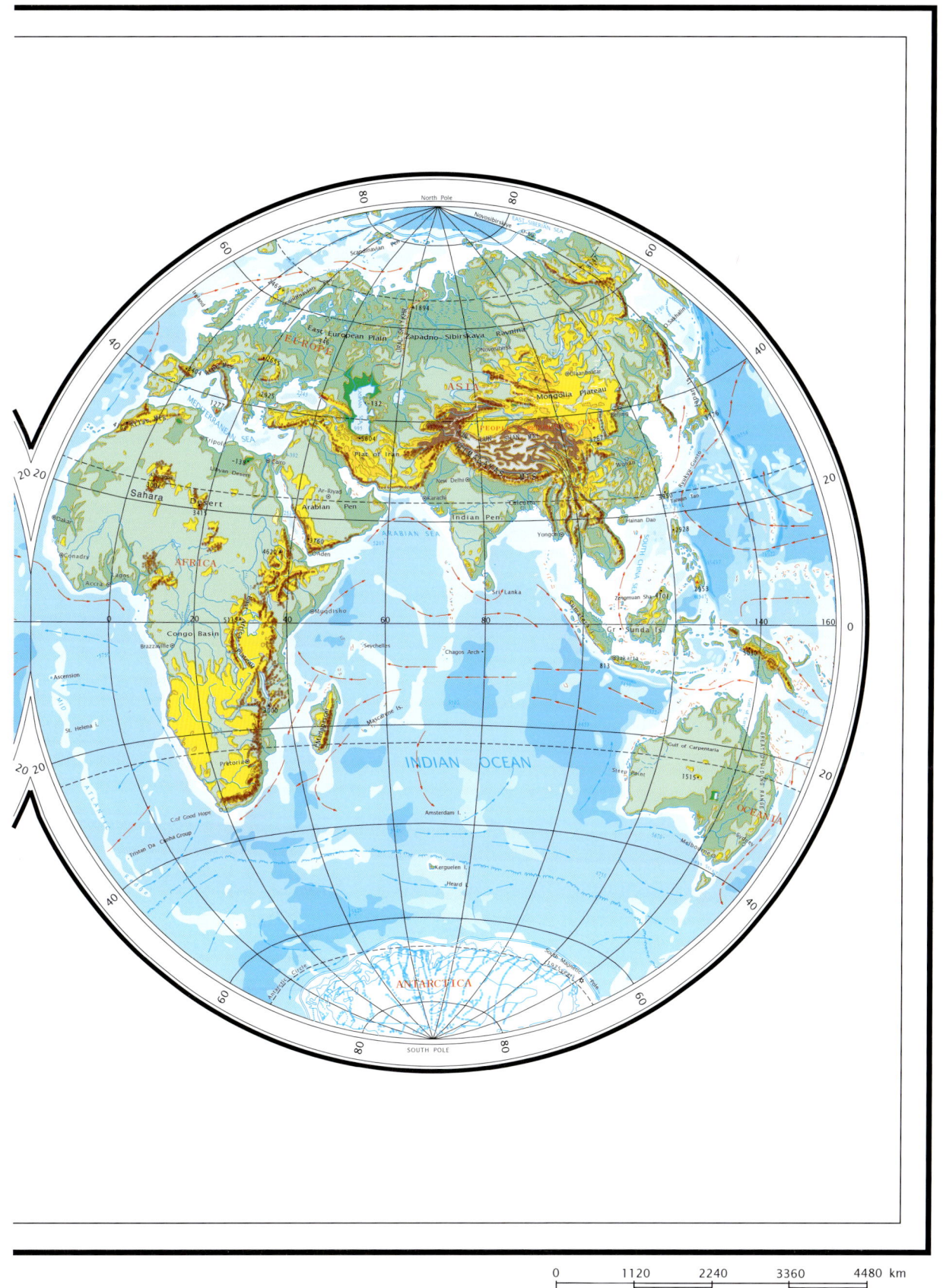

1 World Topography

Oceans and continents

Earth has an area of about 510 million km^2 (197 million square miles). Of this total, approximately 360 million km^2 (140 million square miles), or 71 per cent, are represented by oceans and marginal seas. The continents comprise the remaining 29 per cent, or 150 million km^2 (58 million square miles).

Land With an average altitude of about 875 m, land can be classified into continents, islands and peninsula. There are six mainland masses, namely: Eurasia, Africa, North America and South America, Antarctic, and Australia. Islands that are located near each other are called an archipelago.

Oceans Oceans refer to broad and continuous bodies of saline (salty) water on the Earth's surface. They are 3795 m deep on average. There are four oceans on Earth, namely, the Pacific, the Atlantic, the Indian and the Arctic. Seas are the smaller subdivisions of oceans. The largest sea in the world is the Coral Sea located off northeastern Australia with an area of 4.79 million km^2. Seas can be further divided into marginal seas, inland seas and intercontinental seas. Inland seas refer to those seas that extend onto mainland masses and which may connect with marginal seas or even with oceans by narrow waterways. The Bohai Sea and the Baltic Sea are illustrations of this type. A third common type of sea, the intercontinental, separates two or more continental land masses. The Mediterranean Sea is an example of this type.

Land and submarine topography

Land The surface of the Earth varies greatly in height and morphology. Using these two features as defining parameters, land presents itself in five forms: plains, mountains, plateaus, hills and basins.

A plain is a broad area of land with relatively low relief that has no cliffs at its edges. Plains are mostly less than 200 m in altitude and account for just under 35 per cent of the total land area. The largest plain in the world is the Amazon with an area of about 5.6 million km^2.

Mountains are often spectacular features that rise several hundred metres or more above the surrounding terrain. Mountainous areas have large altitudinal variations, steep slopes and great heights. Linearly extensive mountains are called mountain ranges. Adjacent mountain ranges that share similar genesis are called mountain systems. These ranges are mostly distributed in two main belts in the world. One belt comprises the south–north trending coastlines along both sides of the Pacific Ocean. It runs continuously from the tip of South America through Alaska, to ranges in Asia, the coastlines along Oceania as well as the Pacific Ocean, and islands outside marginal seas. The other is a belt that runs generally in an east–west direction, traversing Asia, southern Europe and northern Africa. This belt includes ranges in Java Island and Sumatra, the Himalayas, the Alps in southern Europe, and the Atlas in northwestern Africa.

World Atlas of Oil And Gas Basins, First Edition. Li Guoyu.
© 2011 John Wiley & Sons, Ltd. Published 2011 by John Wiley & Sons, Ltd.

Ranges in the above-mentioned belts are typically grand in scope and possess high peaks of above 4000–5000 m. There are 14 peaks with altitude of above 8000 m, most of them are distributed in the Karakorum and the Himalayas Ranges in Asia. Among these peaks, the Qumolangma (Everest) in the Himalayas at an altitude of 8848 m is the highest point in the world.

Plateaus refer to areas with moderately high elevations and relatively flat surfaces and edged by steep cliffs. The world's highest plateau is China's Tibetan Plateau with an area of 2.2 million km^2 and an average altitude of 4500 m. The world largest plateau in area is the Brazil Plateau (Mato Grosso) in South America. Its area is about 5 million km^2.

A basin is a depression in the landscape, typically below the surrounding area, such as Sichuan Basin in China and the Congo Basin in Africa.

Submarine landforms The Earth's surface waters tend to obscure the true nature of submarine landforms. It is known that the submarine topography fluctuates as much as the visible landforms above sea level. Submarine topography can be described as consisting of the continental shelf, the continental slope and the ocean floor.

The continental shelf accounts for approximately 7.5 per cent of the Earth's total sea area. The continental slope is defined as the transitional belt between the continental shelf and the ocean floor. This type of slope is the world's largest. It has gentle inclines and relatively shallow water depths which would typically be no more than 200 m. There are, however, exceptions of up to 500–600 m. The difference in submarine elevation from the continental shelf to the base of the continental slope is about 3,000 m. The continental slope makes up about 12 per cent of the Earth's total sea area.

The ocean floor (also known as the seabed) typically refers to the extension of the continental slope and other continental margin features, such as the continental rise below sea level. the ocean floor is the main physical feature of the Earth's oceans, with depths of between 3000 m and 6000 m. In area, the ocean floor accounts for approximately 80 per cent of the Earth's total sea area.

Submarine topographical features vary greatly, with several different physical features such as ocean ridges, marine basins, ocean trenches, sea knolls, seamounts, and submarine plateaus, to name just a few.

WORLD POLITICAL MAP

CHAPTER 2

2 World Political Map

There are 199 countries and regions in the world, but oil and gas is produced in only 90. Oil production exceeds 100 million t per annum in 13 of these countries (Table 2.1), but the majority of these countries produce low amounts of oil. By the end of 2008, only 12 of them played an important role in the world, with their annual production exceeding 100 million t. According to the statistics of 2008, the oil production of these 12 oil-producing countries was, in order of highest output: Russia, 488 million t; Saudi Arabia, 445 million t; USA, 245 million t; Iran, 195 million t; China, 190 million t; Mexico, 140 million t; Canada, 128 million t; United Arab Emirates (UAE), 122 million t; Iraq, 118 million t; Venezuela, 117 million t; Kuwait, 116 million t; and Norway, 108 million t (Grant and Middleton, 1987).

Another statistical method commonly used in the industry is to classify production on a 'per well per day basis'. This classification provides some insight into the commercial productivity of various geological basins and oil reservoirs. Using this method, oil-producing countries can be classified into three categories: the high production countries with oil production exceeding 100 t, the medium production countries with oil production ranging from 10 to 100 t, and the low production countries with oil production less than 10 t (Table 2.2).

These figures confirm the exceptionally high per-well-production features of Saudi Arabia, Norway and Iran in particular, who are among the leading producers in the world. In marked contrast, other major producers such as Venezuela and Russia rank only among the medium production countries. The most notable low production countries are China, Canada and the USA

These statistics have implications for long-term reservoir depletion and maintenance of reservoir integrity in many cases. However, drawing meaningful general conclusions would be a

Table 2.1 Classification of 90 oil producers in the world in 2007 (by production)

	Continents						Total
	Asia	Europe	Africa	South America	North America	Pacific region	
Total number of countries	48	45	56	13	23	14	199
Oil producers	31	23	17	9	7	3	90
>100 million (t)	6	2	1	1	3	—	13
50–100 million	—	1	3	1	—	—	5
30–50 million	6	1	1	1	—	—	9
10–30 million	5	1	4	2	—	1	13
5–10 million	4	1	1	1	1	—	7
1–5 million	4	7	6	1	1	2	22
0.1–1 million	4	8	1	2	1	—	16
<0.1 million	2	3	6	—	1	—	12

World Atlas of Oil And Gas Basins, First Edition. Li Guoyu.
© 2011 John Wiley & Sons, Ltd. Published 2011 by John Wiley & Sons, Ltd.

Table 2.2 Oil production per well per day in typical countries in 2007

Oil production grading	Countries	Oil wells	Oil production (t per well per day)
High production countries (more than 100 t per well per day)	Saudi Arabia	2310	520
	Norway	964	321
	Iran	1737	313
	Kuwait	1103	279
	Mexico	3153	137
	Algeria	1790	127
	Libya	1875	123
	Vietnam	340	122
Medium production countries (10 to 100 t per well per day)	Nigeria	2984	100
	Malaysia	1165	79.9
	Egypt	1850	50
	Australia	1280	44.8
	Brazil	8092	29.5
	India	4650	20
	Indonesia	7896	15.7
	Russia	131,343	10.2
Low production countries (less than 10 t per well per day)	Canada	56,891	6.6
	Argentina	19,900	4.6
	China	164,900	3.0
	Peru	5145	2.7
	Romania	8695	1.5
	USA	49,980	1.4
	Total world	970,689	10.5

very difficult task given the wide disparity in production completion technology, production drive mechanisms, production age distribution, geographical location and distribution of individual wells as well as prevailing national field development policies and regulations.

GEOLOGICAL MAP OF THE CONTINENTS

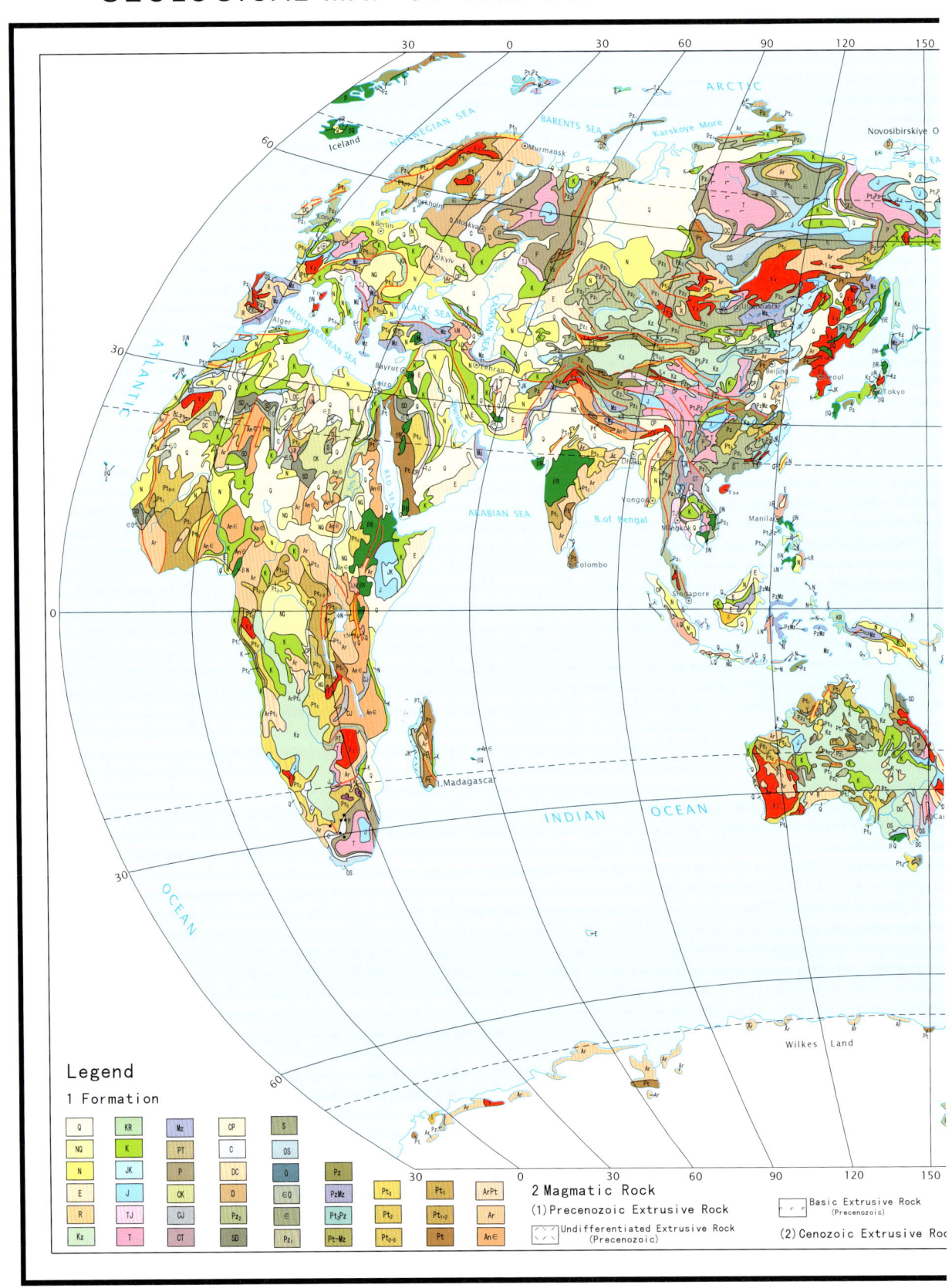

CHAPTER 3

3 Geological Map of the Continents

The geological map of the world continents is based on Tarbuck and Lutgens (1995). Sedimentary rocks account for only about of 5 per cent (by volume) of the Earth's outer 16 km, about 75 per cent of all rocks that crop out on the continents are sedimentary.

Dietz and Holden (1970) have carefully recorded the gross details of the migrations of individual continents over the past 500 million years. By extrapolating plate motion back in time using evidence such as the orientation of volcanic structures left behind on moving plates (e.g. Fig. 3.1), the distribution and movements of transform faults, and palaeomagnetism they were able to reconstruct Pangaea (see Fig. 3.2).

The fragmentation of Pangaea began about 180 million years ago. Figure 3.2 illustrates the breakup and subsequent paths taken by the landmasses involved. As we can see in Figure 3.2A, two major rifts initiated the breakup. The rift zone between North America and Africa generated numerous outpourings of Jurassic age basalts which are presently visible along the eastern seaboard of the USA. Radiometric dating of these basalts indicates that rifting occurred between 180 and 135 million years ago. This date can be used as the birth date of this section of the North Atlantic. The rift that formed in the southern landmass of Gondwanaland developed a 'Y'-shaped fracture which sent India on a northward journey and simultaneously separated South America–Africa from Australia–Antarctica.

Figure 3.2B illustrates the position of the continents 135 million years ago, about the time Africa and South America began splitting apart to form the South Atlantic. India can be seen halfway into its journey to Asia, and the southern portion of the North Atlantic has widened considerably. By the end of the Cretaceous Period, about 65 million years ago, Madagascar had separated from Africa, and the South Atlantic had emerged as a full-fledged ocean (Figure 3.2C).

The current map (Figure 3.2D) shows India' in contact with Asia, and the event that began about 45 million years ago created the highest mountains on Earth, the Himalayas, along with the Tibetan Plateau. By comparing Figures 3.2C and 3.2D, we can see that the separation of Greenland from

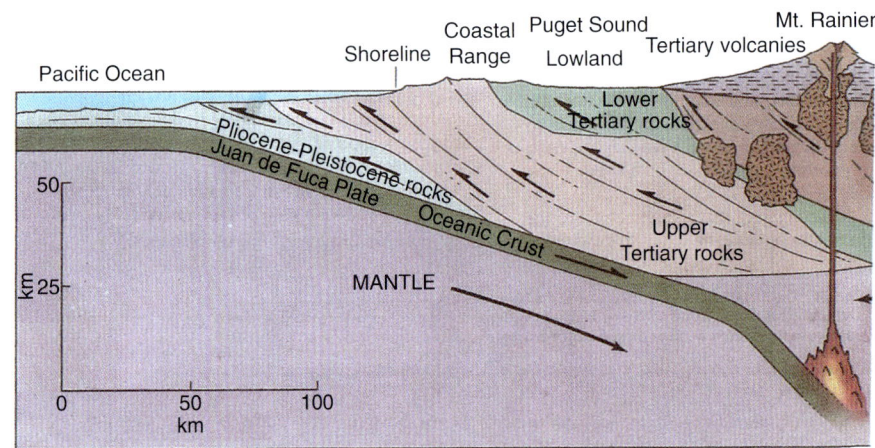

Fig. 3.1 An interpretation of the geology along the converging Juan do Fuca and American plates in the latitude of Mount Rainier, Washington. (After Cowan et al., 1986.)

Geological Map of the Continents

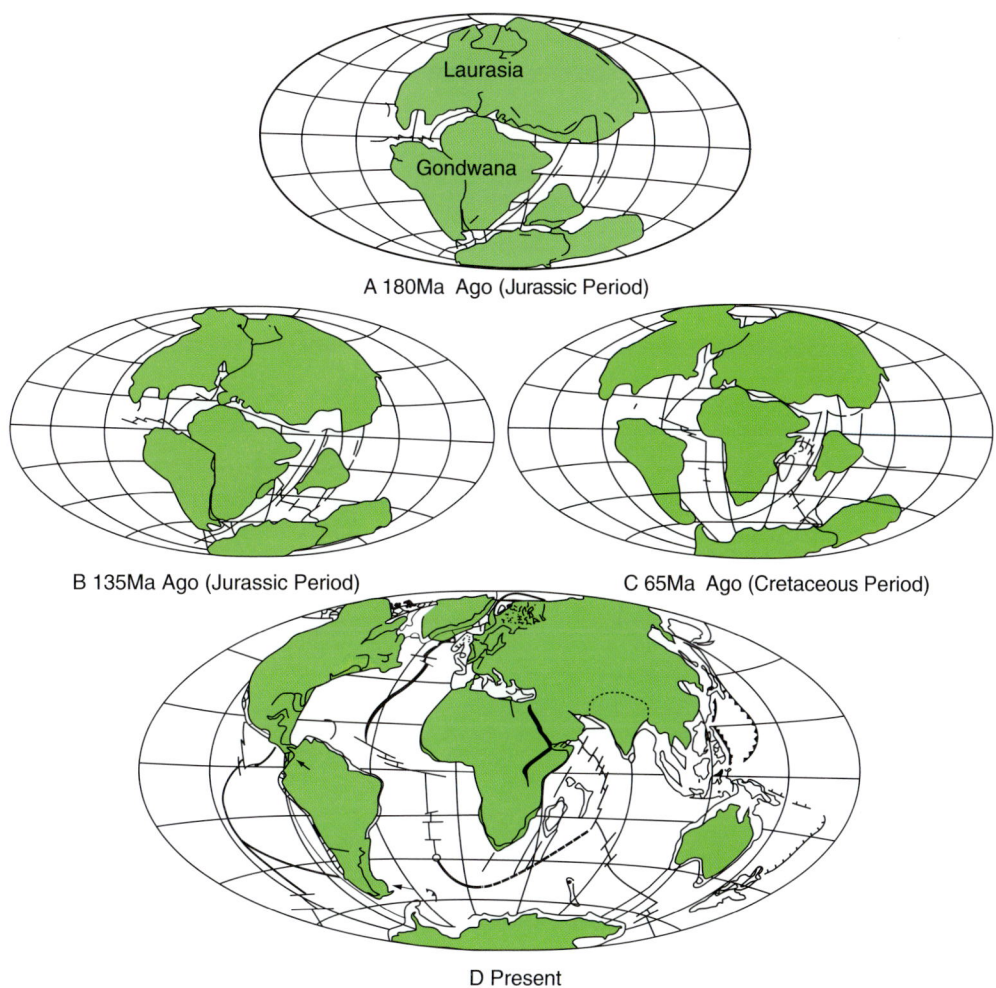

Fig. 3.2 Several views of the breakup of Pangaea over a period of 180 million years according to Dietz and Holden (1970). (Copyright by American Geophysical Union.)

Eurasia was a recent geological event. Also the recent formation of the Baja Peninsula and the Gulf of California is evident. This event is thought to have occurred less than 10 million years ago.

Prior to the formation of Pangaea, the landmasses had probably gone through several episodes of fragmentation similar to what we see happening today. Also like today, these ancient continents moved away from each other only to collide again at some other location. During the period between 500 and 225 million years ago, the fragments of an earlier dispersal began collecting to form the continent of Pangaea. These earlier continental collisions include the Ural Mountains of the Former Soviet Union (FSU) and the Appalachians of North America.

WORLD TECTONIC MAP

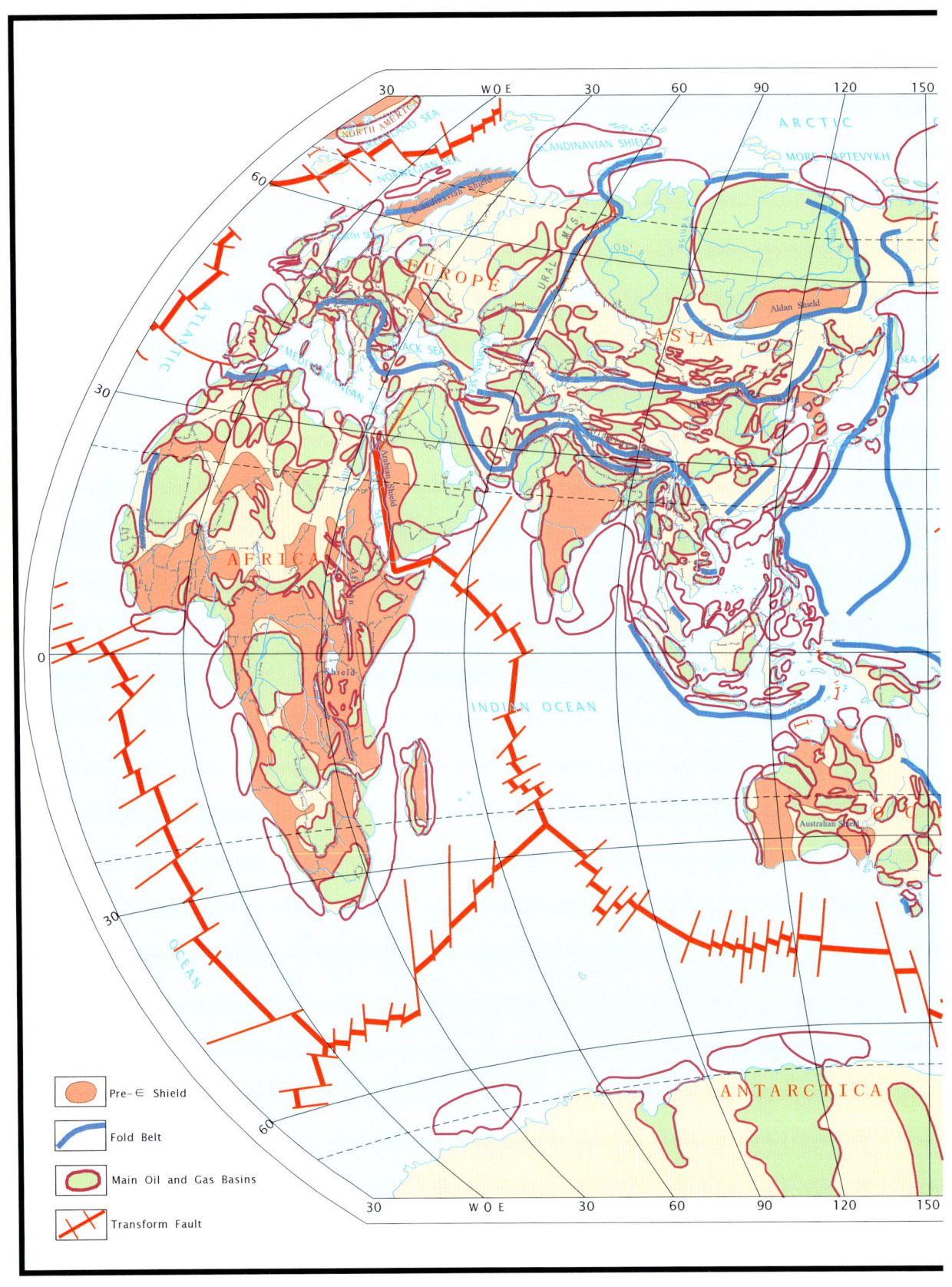

CHAPTER 4

4 World Tectonic Map

Observed crustal phenomena and the distribution and development of oil and gas basins can be understood in terms of plate-tectonic theory, and in the former geosynclinal approach that it replaced. The general application of plate-tectonic theory has been verified by modern geophysical studies as well as ocean floor investigations and geodetic surveys. However, caution is required because there remain theoretical uncertainties in horizontal plate movement mechanisms and those related to the sinking of continental crust during continental rifting and where plates collide, especially continental–continental plate collisions.

During the evolution of the crust, various states and processes can be recognized, i.e. stable and active areas, continental and oceanic plates, rifting and collision, horizontal and vertical movements, and compression and extension. The culmination of these states and processes is what we observe today on the surface of the Earth. Prevailing concepts recognize four types of regional structures in the world, which are summarized below.

Precambrian cratons

Cratons are large regions of continental crust that have remained tectonically stable over long periods of time, i.e. since the Proterozoic active (fold) belt formed stable parts of continental crust. The outcropping portions of the ancient Proterozoic fold belt are called Shields. On the Asian continent there are the Aldan Shield in eastern Siberia, the China-Korea Shield, the India Shield and the Arabia Shield (linked to the shield in eastern Africa). Africa can be regard as a shield except for its southern and northern ends. In Europe, there are the Baltic Shield, the Ukrainian Shield and the Scandinavian Shield. Oceania has the West Australian Shield. In North America there are the Canadian Shield and the Greenland Shield. South America has the Brazilian Shield and Antarctica has the Antarctica Shield.

Fold belts

The Palaeozoic fold belt is often divided into the Caledonian and the Hercynian tectonic cycles and the Meso-Cenozoic fold belt is often referred to as the Alps cycle.

The Palaeozoic fold belt The Caledonian fold belt is located towards the North Pole, generally around the Arctic Ocean. It winds through northern Greenland to the Franklin fold belt in the Arctic islands of Canada, disappears toward the modern Beaufort Sea and is buried by the coastal basin of the North Pole, only to emerge again in the northern margin of Siberia and along the coast line of northwestern Norway, and finally extends into Scotland and Ireland. The belt is an early Palaeozoic geosyncline affected by intense Caledonia folding, plutonism and metamorphism.

The east–west Hercynian fold belt that traverses Europe and Asia. It starts from northeast China, winding through Shayan in Mongolia, Yinshan, Tianshan, Kunlunshan and central Asia, then is buried by young sedimentation in the northern Caucasus. The belt in the western sphere is generally in a northeast–southwest direction and roughly coincides with the Appalachian Mountains, with its southwest end being buried by Meso-Cenozoic strata.

World Atlas of Oil And Gas Basins, First Edition. Li Guoyu.
© 2011 John Wiley & Sons, Ltd. Published 2011 by John Wiley & Sons, Ltd.

A third Palaeozoic fold belt trends east–west and is mainly distributed in the Cape Mountains at the southern end of Africa as well as on the Antarctic continent.

Meso-Cenozoic fold belts In the eastern hemisphere the Tethys fold belt (the name being derived from the former Tethys Sea) is represented by the Alpine–Himalayan fold belt, which starts from the Atlas Mountains in North Africa and Pyrenees Mountains in Europe at its western end, and then crosses the Alpine, Carpathian, Caucasus and Himalayan, mountain ranges. The circum-Pacific fold belt starts in the north from Alaska Bay to the southern end of South America along the Cordillera Central of the east Pacific coast. Its southern part is distributed along the Antarctic Peninsula.

Major petroleum provinces or basins

Examples of the most commonly known of these provinces or basins are listed below by continent (Nakicenovic, 1998).

Asia West Siberian Basin, East Siberian Basin, Karakum Basin, South Caspian Basin, Fergana Basin, Junggar Basin, Tarim Basin, Songliao Basin, Persian Gulf Basin, Central and South Sumatra

Africa Suez, Sirte Basin, Trias, Gefara Basin, Illizi Basin, Niger Delta

Europe North Sea, Dnieper-Donets, Volga-Urals Basin, Caspian

Oceania Cooper, Bowen, Surat, Gippsland, Taranaki

North America Alberta, Permian, Gulf of Mexico

Latin America Maracaibo, East Venezuela, Putumayo

Oceanic areas

These are tectonic units that differ from their continental counterparts. They feature thinner crusts dominated by basalts. The sediments are older with increasing distance from mid-oceanic ridges.

In conclusion, the numerous complicated structural phenomena presently observed at the Earth's surface can be understood within the basic framework of plate tectonics, especially the fact that petroleum basins are mainly distributed in continental plate settings.

WORLD MAP OF OIL AND GAS BASINS

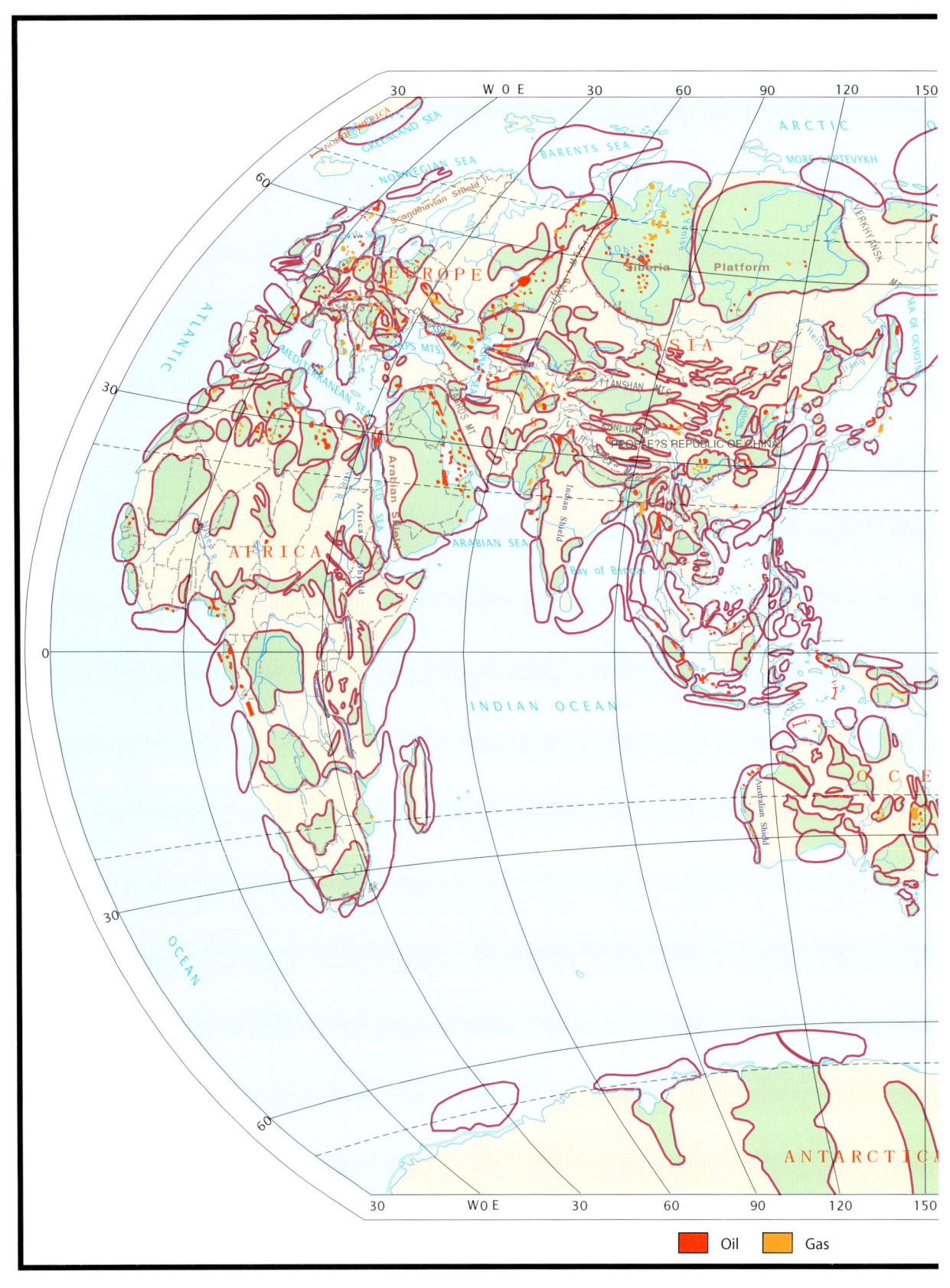

CHAPTER 5

5 World Map of Oil and Gas Basins

Statistical studies have shown that oil can be discovered in an area of approximately 100 million km^2 of sedimentary rocks globally, of which 70 million km^2 are distributed on present continental landmasses and 30 million km^2 in the present oceans. There is no agreement as to exactly how many petroleum-bearing sedimentary basins there are in the world. Several reasons may be adduced for this lack of consensus, amongst which is the relatively low degree of exploration.

At present, about 200 petroleum basins have been subject to large-scale exploration and development. Worldwide, approximately 67,000 oil and gas fields have been discovered comprising an estimated 41,000 oilfields and some 26,000 gas fields (Table 5.1).

Underpinning this Atlas are the observation-based concepts outlined in Li Guoyu (1996, 2002) regarding the development of sedimentary basins and their potential to be oil- and/or gas-bearing. These views can be summarized as follows:

1. all places on the Earth where downwarping occurs and which are filled with sedimentary rocks can be named sedimentary basins;
2. all sedimentary basins should contain oil and gas;
3. the conceptualization of the development of oil- and/or gas-bearing sedimentary basins depends on characterizing the generation and distribution patterns of hydrocarbons according to the nature of the sedimentary rocks, using the basin as a study unit;
4. within a sedimentary basin, all beds rich in organic matter can act as source rocks, all porous beds as reservoirs, all impermeable beds act as seals, all closed structures as traps for oil and gas, and all intense tectonism as forces to enhance hydrocarbon migration and/or to destroy oil and gas pools;
5. each basin unit comprises sedimentary rocks, organic matter, fluids (oil, gas and water) and geological structures.

Characterization of the development of sedimentary basins and their potential to be oil- and/or gas-bearing includes 14 interdependent lines of investigation deployed in a predetermined sequence.

1. Basin: All crust-downwarped areas with a complete system within itself can be called sedimentary basins, regardless of their size.
2. Sedimentary rocks: The research contents include the general thickness and volume of sedimentary rocks, geochronology, unconformity, lithology.
3. Source rocks: The research contents, based on the theory of organic origin.
4. Reservoir: The research contents lithology, thickness, pores, fractures, vugs, reservoir framework, continuity of occurrence, absence.
5. Seal: All compact rocks can act as seals, such as clay-stone, gypsum, halite and even compact carbonate rich in clay.
6. Fluids (oil, gas and water): Oil and gas are fluid resources, usually associated with water.
7. Migration: The research contents include primary migration and secondary migration.
8. Trap: Traps necessarily consist of 3 components, namely, reservoir, seal and barrier.

World Atlas of Oil And Gas Basins, First Edition. Li Guoyu.
© 2011 John Wiley & Sons, Ltd. Published 2011 by John Wiley & Sons, Ltd.

Table 5.1 Distribution of sedimentary basins in the world

Continents	Sedimentary basins	Marine basins (deep marine)	Oil and gas basins (marine)	Major oil and gas basins
Asia	375	136 (35)	86 (29)	24
Africa	95	42 (9)	25 (11)	3
Europe	84	50 (11)	39 (17)	13
Oceania	86	63 (12)	15 (9)	4
Latin America	123	72 (12)	45 (20)	11
North America	190	73 (20)	46 (9)	25
Antarctica	21	15 (9)	—	—
World Total	974	451 (108)	256 (95)	80

9. Oil and gas pools: The research contents include classification, geometry of oil and gas pools, main factors controlling the formation and preservation of oil and gas pools.
10. Destruction: The cause and degree of destruction of formed oil and gas pools.
11. Oil and gas occurrence pattern: This is mainly concerned about the study of the oil and gas occurrence patterns in a single, regional or world basin.
12. Resource appraisal: Resource appraisal and potential reserve calculations in a basin are carried out using the volumetric method, analogue method and concentration method.
13. Basement and edge mountains: Edge mountains are sources of sediment supply for basins.
14. Tectonics: Tectonics is a basic branch, involving the fundamental framework of petroleum geology. It cannot replace all branches in the theory of sedimentary basins.

The sedimentary basins described in the atlas can be considered at the scale of single basins, a country or a region and globally.

CLASSIFICATION OF OIL AND GAS BASINS BY

Symmetric basins	Asymmetric basins
Michigan basin (U.S.A)	Indian River basin (India)
Murzuk basin (Libya)	Alberta basin (Canada)
Gulf of Siam Malay basin (Malaysia)	Eastern Venezula basin (Venezuela)
Bass basin (Australia)	Arctic Slope basin (U.S.A)
West Sibiria basin (Russia)	Persian Gulf basin

GEOMETRY OF CROSS-SECTIONS CHAPTER 6

Platformal basins	Triangular basins
Tarim basin (China)	Beibu Gulf basin (China)
East Siberia basin (Russia)	Central Samatra basin (Indonesia)
Sichuan basin (China)	Dongying basin (China)
Kwanza basin (Angola)	Palawan basin (Philippines)
Essaouira basin (Morocco)	Erlian basin (China)

6 Classification of Oil and Gas Basins by Geometry of Cross-Sections

Generally oil and gas basins are classified by their sedimentary cover, plate-tectonic setting, stress history and subsidence characteristics. Based on subsidence characteristics basins can be divided into fault, depression and fault-depression types. When a section of crust subsides and filled in with sedimentary rocks it will develop its own unique characteristics and a classification method of oil and gas basins therefore can be devised on the basis of the geometry of basin cross-section.

In a sedimentary basin (or a sedimentary rock mass), oil and gas typically migrate from deep high-pressure areas to shallow low-pressure areas. The ultimate distribution is then characterized by the accumulation and settlement of gas, oil and water based on their differing densities. The geometry of a basin depends on tectonic movement, and the distribution pattern of gas, oil and water varies from basin to basin according their different cross-section geometry. Accordingly classification of sedimentary basins can be simplified into four types with the objective of identifying large oil and gas basins. Table 6.1 provides a useful summary of distribution of known sedimentary basins amongst these four types.

Symmetrical basins The West Siberia Basin (Russia) and the Michigan Basin (USA) are classified under this category, with their respective areas being 3,300,000 and 310,000 km^2. Regardless of the structural origin of the basin, oil and gas are mainly distributed in the central part of symmetrical basins, and the richness of the hydrocarbon deposits declines towards the sides as sedimentary rocks become thinner.

Asymmetrical basins The Persian Gulf Basin has an area of 3.2 million km^2, stretching from the Zagros piedmont zone, through the transition zone in Kuwait to the Saudi Arabian Platform. It holds substantial oil and gas reserves, several large oil and gas fields have been discovered in the basin. The Alberta Basin (Canada) ranges from the piedmont zone of the Rocky Mountains in the west to the Canadian Shield in the east. It is also a large and complete oil and gas zone. In addition to the discovery of multiple oilfields, several gas fields have been discovered in the east recently. Oil and gas deposits are distributed in both deep and low layers of the basin. Giant oil and gas fields are distributed in the slope zone of the basement uplift.

Platformal basins Tarim Basin (China) has an area of 560,000 km^2. It is shaped like a platform in the middle and there exist some sags at the periphery of the basin. It is composed of many sags and uplifts and the oil and gas distribution is controlled by giant sags. In addition, oil and gas fields are distributed on the slopes.

Triangular basins This type of basin is very prevalent in China. The extensional basins in East China are all grouped into this category, examples being Jiyang, Dagang, North China, Hailaer, and Beibu Gulf Basins. They feature substantial sedimentary thicknesses and the oil and gas distribution is determined by the extent of the basin.

World Atlas of Oil And Gas Basins, First Edition. Li Guoyu.
© 2011 John Wiley & Sons, Ltd. Published 2011 by John Wiley & Sons, Ltd.

Classification of Oil and Gas Basins by Geometry of Cross-Sections

Table 6.1 Classification of oil and gas basins

Basin types	Basins	Area (1000 km²)	Age	Thickness (m)
Symmetrical	Michigan (USA)	310	Cambrian–Carboniferous	4500
	Murzuk (Libya)	250	Palaeozoic–Cenozoic	6000
	Gulf of Siam–Malay (Malaysia)	414	Tertiary–Mesozoic	10,000
	Bass (Australia)	40	Meso-Cenozoic	7000
Asymmetrical	West Siberia (Russia)	3300	Meso-Cenozoic	10,000
	Indian River	360	Meso-Cenozoic	6000
	Alberta (Canada)	600	Cambrian-Cretaceous	6000
	Venezuela East	100	Tertiary	12,000
	Alaska Basin (USA)	330	Permian–Cenozoic	9000
Platformal	Persian Gulf	3280	Palaeozoic, Meso-Cenozoic	11,000
	Tarim (China)	560	Sinian–Quaternary	16,000
	East Siberia (Russia)	2000	Palaeozoic–Cenozoic	5000
	Sichuan (China)	200	Sinian–Quaternary	12,000
	Kwanza (Angola)	140	Meso-Cenozoic	4000
	Essaouira (Morocco)	40	Palaeozoic–Mesozoic	5000
Triangular	Beibu Gulf (China)	36	Palaeozoic Cenozoic	5500
	Central Samatra (Indonesia)	50	Cenozoic	2500
	Dongying (China)	6	Meso-Cenozoic	10,000
	Palawan (Philippines)	10	Cenozoic	8000
	Erlian (China)	109	Palaeozoic–Mesozoic	5000

CLASSIFICATION OF WORLD OIL AND GAS FIELDS

ANTICLINE	FAULTED	OVER THRUST	STRATIGRAPHIC	BURIED HILL	LITHOLOGICAL
DAQING (CHINA)	TEAK (TRINIDAD)	NORTH DOLINA (UKLAINA)	EAST TEXAS (USA)	INMALI 2 (CHINA)	BIG MUDDY (USA)
ZAKUM (ABU DABI)	CANOSURTA (MEXICO)	BOLISIAV (UKLAINA)	PRUDHOE BAY (USA)	LUNNAN (CHINA)	GUORICEMA (BRAZIL)
LAMAR (VENEZUELA)	VENTARA (USA)	KUNA (COLOMBIA)	PEMBINA (CANADA)	RENQIU (CHINA)	OIMASH (KAZAKHSTAN)
GACHSARAN (IRAN)	NEBITAG (TRUKMENSTAN)	NAME UNKNOWN (TAJIKISTAN)	STAFJORD (NORWAY)	DONGSHENGPU (CHINA)	PARENTI (FRANCE)
SHIKABOVO (RUSSIA)	GELOREBERG (AUSTRIA)	NAME UNKNOWN (TAJIKISTAN)	KARAMAI (CHINA)	COCAO (BRAZIL)	YANCHANG (CHINA)
BEATRICE (NORTH SEA)	SOUTH VIKING (NORTH SEA)	EAST PAINTER (USA)	KINGFISH (AUSTRALIA)	YOUROBXIN (RUSSIA)	MARTMIYA (RUSSIA)
URENGOY (RUSSIA)	KERN BLUFF (USA)	ACHISU (RUSSIA)	BACH HO (VIETNAM)	OFFSHORE (SPAIN)	MITSUKE (JAPAN)
LACQ (FRANCE)	CREOLE (USA)	MONESHITI (RUMANIA)	OKLAHOMA CITY (USA)	AUGILA (LIBYA)	MALIN (CHINA)

CHAPTER 7

FRACTURED	SALT AND MUD DOME	REEF	SYNCLINE	HYDRODYNAMIC	HEAVY OIL AND TAR SAND
NANCHONG (CHINA)	WEST BAY (USA)	ISHENBAI (RUSSIA)	NAME UNKNOWN (RUSSIA)	BURKETT (USA)	ATHABASKA (CANADA)
PAZANAN (IRAN)	SAND SEA (AZERBAIJAN)	ARUN (INDONESIA)	YUNYIJING (CHINA)	TORCHILIGHT (USA)	PEACE RIVER (CANADA)
AWALI (BAHRAIN)	BALLY LAKE (USA)	LEDUK D3A (CANADA)	GULONG (CHINA)	BILLINGS (USA)	SHUGUANG (CHINA)
KIRKUK (IRAQ)	SOUTH MARSH ISLAND (USA)	MICHIGAN (USA)	QIBEI (CHINA)	DANBEI (CHINA)	NINGJIN (CHINA)
SPRABERRY (USA)	HELDER (GERMAN)	KASIMUTARA (INDONESIA)	EAST SIBERIA (RUSSIA)	GAOTAIZI (CHINA)	GAOSHENG (CHINA)
YOUQUANZI (CHINA)	KINGIYAK (KAZAKHSTAN)	MICHIGAN (USA)	EAST BONE (USA)	NAME UNKNOWN (USA)	SUIZHUN36-1 (CHINA)
XINBAI (CHINA)	SERRA ISLAND (USA)	LIUHUA-11-1 (CHINA)	SESPE (CANADA)	LOS VILOS (USA)	RAGUSA (ITALY)
LUNNEUSI (CHINA)	SONGAI SONGAI (INDONESIA)	JIANNAN (CHINA)	CAP (MEXICO)	ELMWORTH (CANADA)	GUERE (VENEZUELA)

7 Classification of World Oil and Gas Fields

There are numerous oil and gas fields that have been discovered worldwide. They vary in terms of geological structure, mechanism of formation and morphology (Selley, 1985). According to currently available data, there are approximately 67,720 oil and gas fields discovered so far in the world. Despite the fact that many geological and petroleum institutions worldwide are engaged in long-term research on world oil and gas fields, there is still no universal acceptance regarding the exact number of oil and gas fields in the world (Levorsen, 1954).

The statistics shows that America holds the largest number of oil and gas fields (31,385 and 20,294), accounting for 70 per cent and 76 per cent of world total respectively. However, most of these fields are small. Apart from the USA, there are 9779 oil fields scattered in Latin America, Africa, East Asia, Canada, the Middle East, East Europe, China and Australia. There are 6262 gas fields in countries other than America: Canada with 1814, western Europe with 1201, Confederation of Independent States (CIS) with 756, eastern Europe with 698, East Asia with 547, Latin America with 470, Australia with 298, Africa 280 and China 119.

These oil and gas fields have various amounts of reserves. Petroleum basins with large areas and favourable geological conditions tend to contain numerous large-scale oil fields with large reserves, high daily production and high annual crude production. There are 12 fields whose original oil in place (OOIP) reserves are larger than 2 billion t (recoverable reserves of 6800 million t), of which 29 are in the Middle East, five in Latin America, four in the CIS, two in the USA, one in China.

Based on a comparative analysis of the substantial data available, Li Guoyu (1988) suggested a modification to the conventional ten types of oil and gas fields recognized. The modification has two implications. First is the addition of three new types of fields, namely overthrusts, reefs and tar sands. Second involves the re-categorization of diapiric fields and bitumen fields to the existing categories of salt-dome fields and heavy oil sand fields respectively. This results in a classification comprising the 12 types of oil and gas fields briefly summarized below by their key features:

Anticlinal field Anticlinal fields dominate world field types in terms of hydrocarbon reserves. They are usually found in anticlinal structures, and typically portray the shape of a dome. The Ghawar Field in Saudi Arabia, the world's largest field with original recoverable reserves of 11.4 billion t, belongs to this type. Other giant oil fields, such as the Burgan, the Safaniyah, the Kirkuk, the Daqing and the Samotlor, are all anticlinal fields.

Faulted field Fields formed in fault blocks are called fault block fields. This type of field is broadly distributed worldwide. Teak in Trinidad and Canosurta in Mexico are two examples of this type.

Overthrust field Overthrust fields are typically formed in fold belts and are associated with extensive horizontal plate movements. They have complicated structures and substantial resource potential, such as North Dolina in Ukraine, Kuna in Colombia and some oilfields in both Tajikistan and China.

Stratigraphic field Stratigraphic fields are related to a stratigraphic unconformity. The Oklahoma and the East Texas fields in the USA are excellent examples of stratigraphic fields (King, 1972).

World Atlas of Oil And Gas Basins, First Edition. Li Guoyu.
© 2011 John Wiley & Sons, Ltd. Published 2011 by John Wiley & Sons, Ltd.

Borehill field Borehill fields are formed by lengthy depositional hiatuses and weathering of all kinds of epicontinental sedimentary and chemical deposits, which are then buried by sedimentary caps. The Yourobxin in Russia, the Cocao in Brazil, the Renqiu in China and Augila in Libya are examples of this type of field.

Lithological field Lithological field refers to those formed by channel sandstone, partial recrystallization of limestone, sandstone lenses in mudstone and horizontal lithological change of sandstone. This kind of field is the most common around world. The most typical examples are the three petroleum fields in the Big Muddy in Wyoming, USA, Martmiya in Russia and Maling in China.

Fractured field Fractured fields formed in fractures in all kinds of strata, such as sandstone, mudstone and limestone. Examples include Spraperry in Texas, USA, Pazanan in Iran and Awali in Bahrain.

Salt and mud dome field Salt and mud dome fields refer to those formed in giant salt layers and mudstone, which are associated with diapirism or penetration of mud volcanos. Examples are the Helder in Germany, the Sand Sea in Azerbaijan, and the Beibu Gulf in China.

Reef field Reef field refers to those fields formed by reefs. Such fields can be found in Russia, Indonesia, Canada, England and China.

Syncline field Syncline field refers to those formed in synclines. The Bohai in China and the Cap in Mexico are good examples of this kind.

Hydrodynamic field Hydrodynamic field refers to suspended fields formed above water stored in the strata. They occur widely in Russia and in the USA. A most noteworthy example is the Elmworth field in Canada, which generated an abnormal amount of gas in the lower part with water above due to a bottleneck effect, which caused an inversion of the normal pattern.

Heavy oil and tar sands Heavy oil and tar sand fields are formed when heavy oil and bitumen is sealed. The Shuguang oilfield in China belongs to this type. It should be pointed out that while the previously listed 11 field types are of a generic classification, the heavy oil and tar sands refer to crude oil only. Now that heavy oil and tar sands occurrences are becoming more common, there is a need for an acceptable proper classification method to guide the practical exploration and development of these fields.

WORLD MAP OF OIL AND GAS RESOURCES

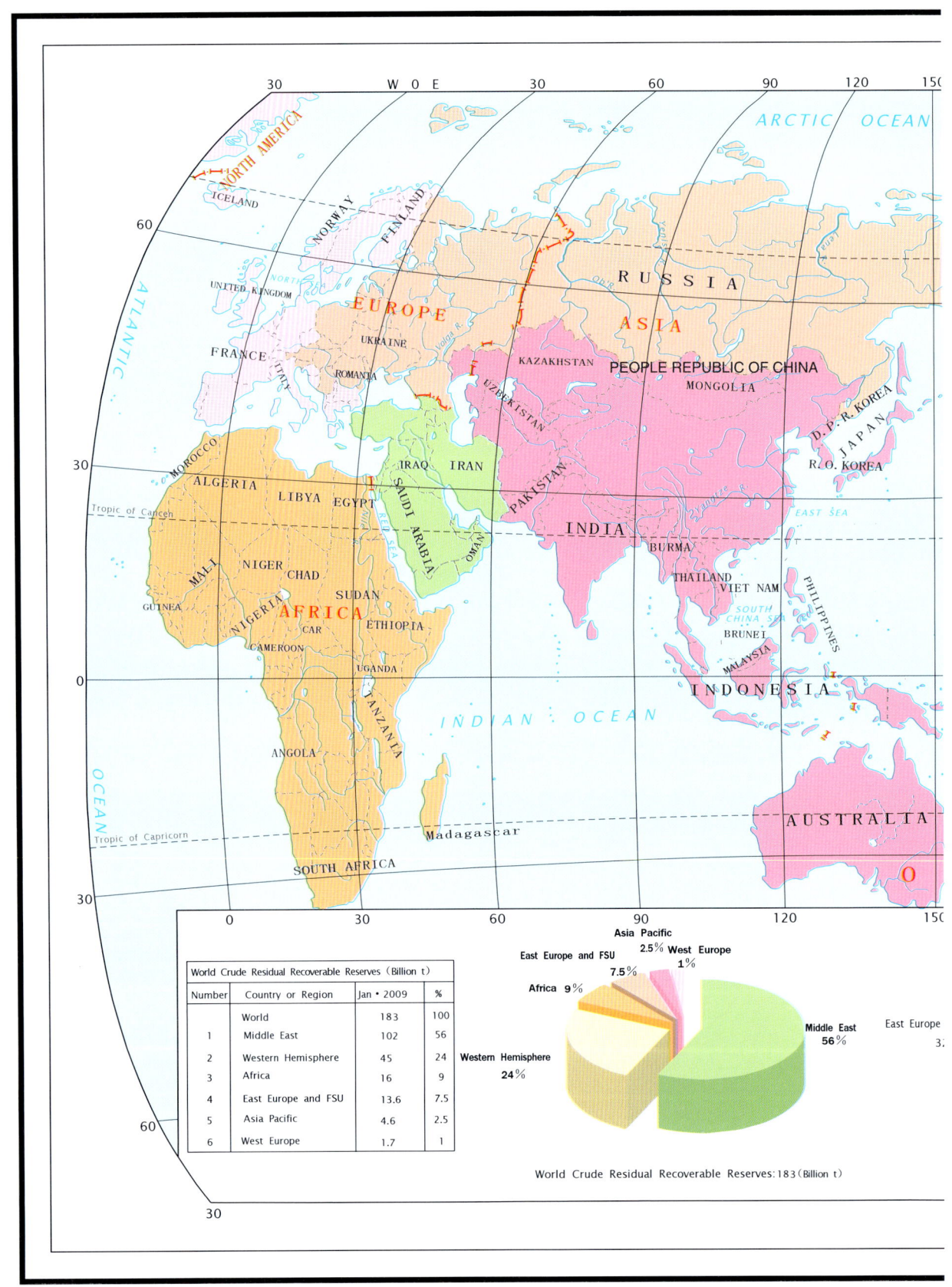

CHAPTER 8

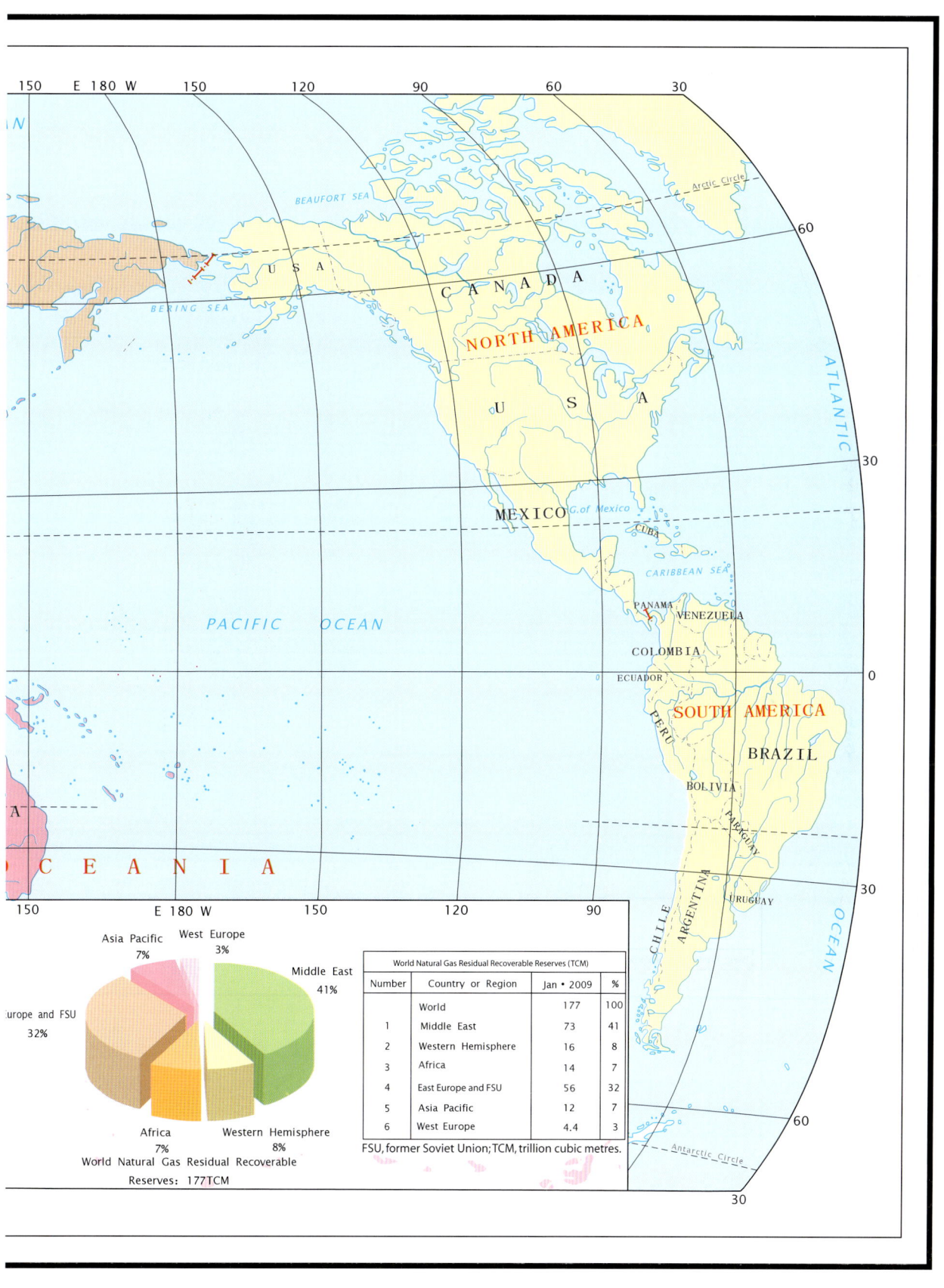

8 World Map of Oil and Gas Resources

World petroleum resources

For many years, numerous research centres have repeatedly evaluated and calculated the world's petroleum resources and produced different results. One of the most authoritative reports is the survey published by the United States Geological Survey (USGS) (Lapedas, 1976).

According to the data of petroleum resources published by USGS in 2000, the ultimate recoverable resources of the world's conventional petroleum were about 480 billion t, of which 112.8 billion t were recovered resources, leaving about 367.2 billion t. In addition, the ultimate recoverable unconventional oil resources (heavy oil, bitumen, and oil sands) were estimated to be 400–700 billion t, 30 per cent or twice as much as that of conventional resources. The conventional gas resources were 489 trillion m^3. The ultimate recoverable unconventional gas resources were 849 trillion m^3 (TCM), twice that of conventional gas resources (Table 8.1).

The conventional (480 billion t) and unconventional (400–700 billion t) oil totalled 880–1180 billion t and conventional and unconventional gas totalled to 1338 trillion m^3. It was a giant step in oil and gas resources evaluation, presenting an exciting prospect of rich oil and gas resources in the world.

After scrutinizing the relationship between the oil and gas production and reserves during the period between 1980 and 1998, it was found that crude production during these 18 years had experienced a dramatic rise, drop and rise again. Oil reserves had also increased but somehow less than that of production. Starting from 1989, oil production has remained stable but oil reserves have shown increasing momentum.

Table 8.1 World recoverable oil and gas reserves (USGS, 2000)

Country	Oil (billion t)	Gas (trillion m^3)	Country	Oil (billion t)	Gas (trillion m^3)
Saudi Arabia	76.8	33	Mexico	11.8	3
Russia	61.3	109	Kazakhstan	11.6	5.3
USA	49.5	57	Brazil	9.8	6
Iran	30.4	37	Libya	8.8	3
Iraq	26.4	6.7	Norway	8	10
UAE	21.1	9	Indonesia	7.5	13
China	20	22	Greenland	7	2.2
Kuwait	19.4	3	Algeria	6.2	9.4
Venezuela	18.2	9.8	UK	6.1	5
Nigeria	12.8	7.5	Canada	5.1	6
World total				480	489

World Atlas of Oil And Gas Basins, First Edition. Li Guoyu.
© 2011 John Wiley & Sons, Ltd. Published 2011 by John Wiley & Sons, Ltd.

World oil and gas resources appear more plentiful than evaluation results

Conservative estimation of the world's oil and gas resources, however, do not appear to agree with the facts. A review of all the evaluation results and comments issued by 16 companies and some analysts around world for the period 1969–1996 reveals that the lowest evaluation was proposed by the renowned petroleum geologist Dr Colin Campbell. He estimated that the ultimate recoverable reserves were 246.6 billion t. He pointed out that the world was entering into a period of oil crisis, its influence would be more severe than that of the crisis induced by political reasons in 1970s (Campbell, 1997). With stronger control of OPEC over more oil production (by the year of 2000, the Middle East countries would hold 30 per cent of world's total oil production) and further production decline, oil peak production and subsequent shortage would be unavoidable. He believed that 90 per cent of the world's conventional oil had already been discovered and the coming oil crisis could not be defused through putting proven undiscovered basins on stream. According to his opinion, almost all the potential basins had been found and explored. New discoveries that could provide us with cheap oil would be rare. Campbell's (1997) data showed that by the end of 1997, among the 246.6 billion t of the ultimate recoverable conventional oil reserves there were only 112.1 billion t left and 24 billion t waiting for discovery.

However, many geologists do not agree with Campbell's conclusion, as they believe that there are more than 24 billion t of undiscovered petroleum resources. Armed with deeper understanding of oil and gas field distribution patterns and new technologies, it is possible that undiscovered resources will be expanded. But petroleum is after all disposable energy and will run out some day in the future. The problem is when. Most experts believe that an oil shortage is doomed to come in the 21st century. However, Campbell's view, although one of petroleum depletion and possibly misguided, is still helpful for us to estimate the potential of the world's oil and gas resources (Odell and Rosing, 1983; Campbell, 1997).

GRAPHS OF WORLD OIL AND GAS RESERVES,

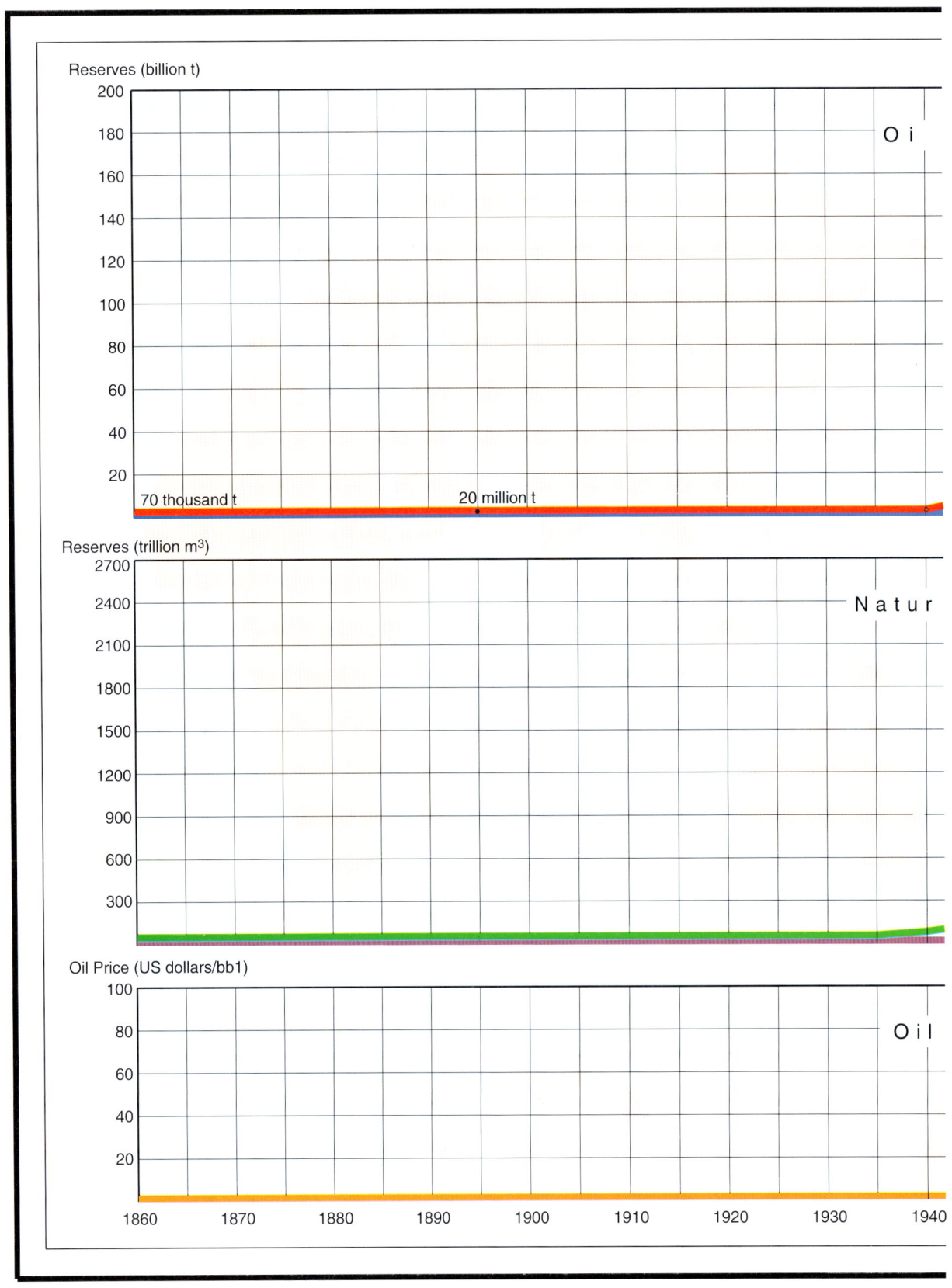

PRODUCTION AND PRICE CHAPTER 9

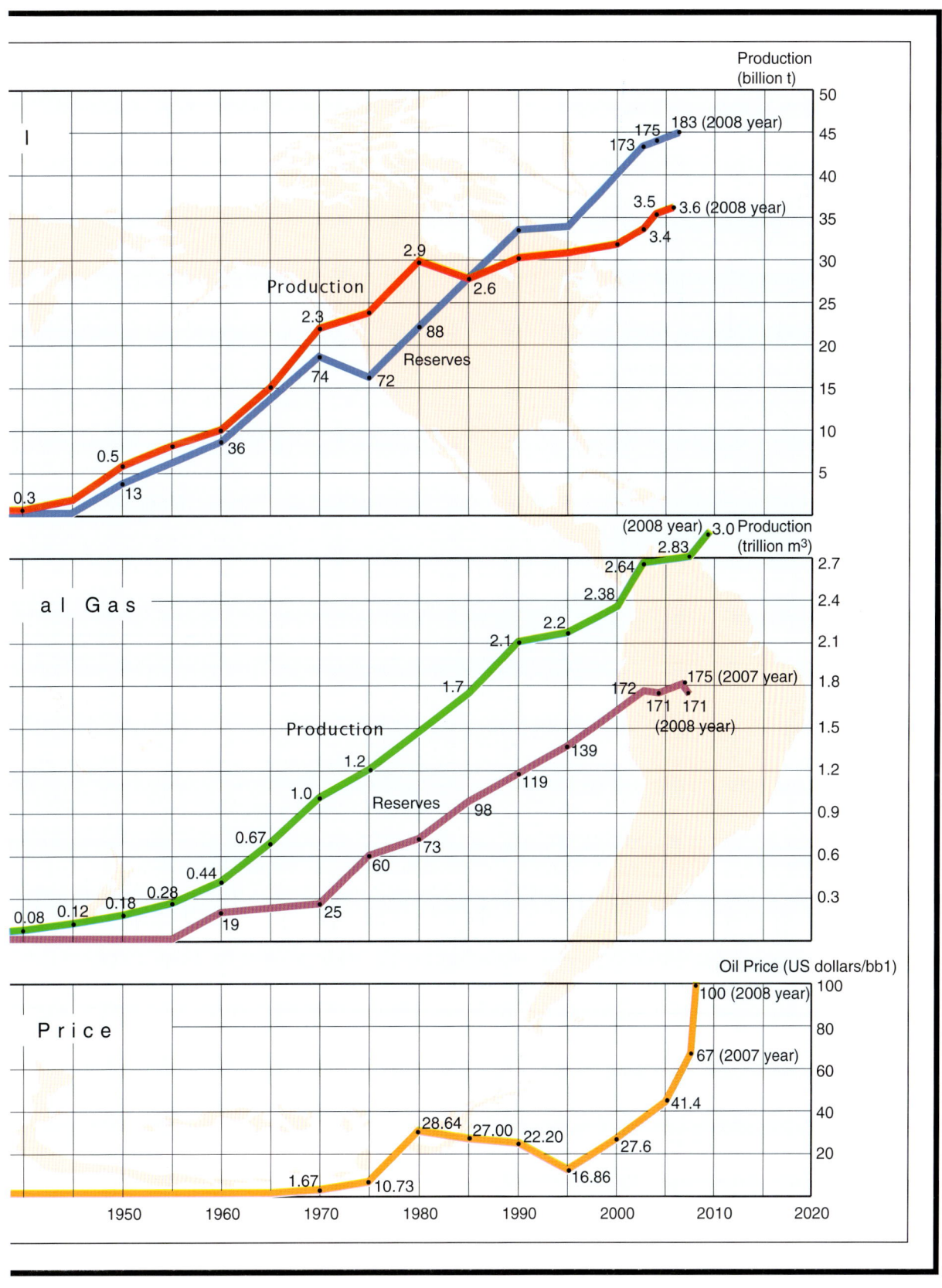

9 Graphs of World Oil and Gas Reserves, Production and Price

Computation and calculation of oil and gas reserves will be elaborated upon later in this atlas. The development and exploitation of remaining proven recoverable oil and gas reserves, however, will be discussed in this chapter (Riva, 1983; Howell, 1993).

For a long time, certainly up to 1945, calculation of oil reserves was conducted only on selected oil and gas fields in a few countries. For various political, commercial, technological and industrial reasons it was impossible to calculate the world's oil reserves on a truly global basis.

Crude oil reserves

The substantial jump in the estimation of crude oil reserves between the years 2000 and 2008 (Table 9.1) can be attributed to the inclusion of Athabasca Basin. Canadian oil-sand reserves in the computation of global oil reserves started from 2002. The simplistic conclusion that may be drawn from these figures is that in spite of models predicting relatively short-term peak oil discovery and production, and other pessimistic forecasts of word oil reserves, there is ample evidence to suggest that a more optimistic scenario is justifiable (Fig. 9.1).

Oil production

According to available historical data, total world annual crude oil production only reached a mere 300t in 1857, equivalent to the daily output of a high productivity well today. After nearly a half century, global annual production was still only 2043t in 1900. Even before the Second World War, oil production was a modest 29,000t annually.

However, propelled primarily by large-scale post-war industrialization in the USA, Europe, Russia and Japan, the development of the oil industry increased rapidly. Annual production reached

Table 9.1 World oil and gas reserves, production and price from 1945 to 2008

Year	Recoverable oil reserves (billion t)	Oil production (million t)	Recoverable gas reserves (trillion m^3)	Gas production (billion m^3)	Oil price (dollars/barrel)
1945	7.5	355	—	132	—
1950	13	538	—	185	—
1960	36	1081	—	448	—
1970	73.9	3236	41.6	1028	1.63
1980	88.8	2983	73.8	1543	28.64
1990	136.4	3015	119	2139	22.26
2000	140	3361	149	2388	27.6
2008	185	3648	177	3050	100

World Atlas of Oil And Gas Basins, First Edition. Li Guoyu.
© 2011 John Wiley & Sons, Ltd. Published 2011 by John Wiley & Sons, Ltd.

Graphs of World Oil and Gas Reserves, Production and Price

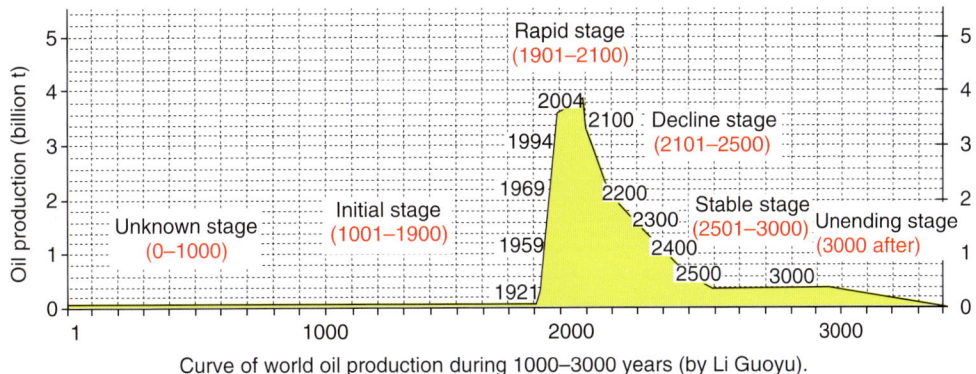

Curve of world oil production during 1000–3000 years (by Li Guoyu).

Fig. 9.1 Graph of world oil production, with postulated production up to the third millenium.

3.64 billion t in 2008. As a result, imbalances between supply and demand gradually disappeared to be replaced by market conditions characterized by close equilibrium between the two market forces.

Currently, most oil is produced in six distinct countries and regions, i.e. 1.1 billion t in the Middle East, 488 million t in Russia, 256 million t in the USA, 190 million t in China, 140 million t in Mexico, 178 million t in Norway and the UK combined.

Natural gas reserves

The initial neglect of the natural gas industry was reversed as more gas reserves were discovered (particularly in Russia, Iran, Qatar, Algeria, Saudi Arabia and Nigeria). This was coupled with rapid advancement in the technology of liquefying natural gas (LNG) for ocean transportation over huge distances in dedicated LNG tankers. Furthermore, gas utilization options were quickly commercialized resulting in even more demand for gas for manufacture of both petrochemicals and fertilizer and feedstock, and also for mega-electric power generating plants.

The remaining recoverable gas reserves were only 19.7 trillion m^3 in 1961, but they rose to 177 trillion m^3 in 2008. At present, the proven gas reserves are predominantly concentrated in two regions. These are estimated to be approximately 72 trillion m^3 in the Middle East and 47 trillion m^3 in Russia.

Gas production

Compilation of statistics for gas production started much later than oil production. In 1936, global oil production reached 245 million t, while gas production was 71 billion m^3. Worldwide gas production reached 1 trillion m^3, 2 trillion m^3, 2.38 trillion m^3, and 2.6 trillion m^3 in 1970, 1988, 2000, and 2004 respectively. Presently, gas production worldwide is concentrated in five countries and regions, i.e. 638.1 billion m^3 in Russia, 564.6 billion m^3 in the USA, 167.9 billion m^3 in Canada, 93.6 billion m^3 in Algeria, 89.9 billion m^3 in Norway, and 108.3 billion m^3 in the UK. The Middle East holds abundant quantities of gas reserves, but annual production is currently only 174.4 billion m^3. This is expected to increase rapidly in the region in the future (Frisch *et al.*, 1989).

WORLD OIL TRADE

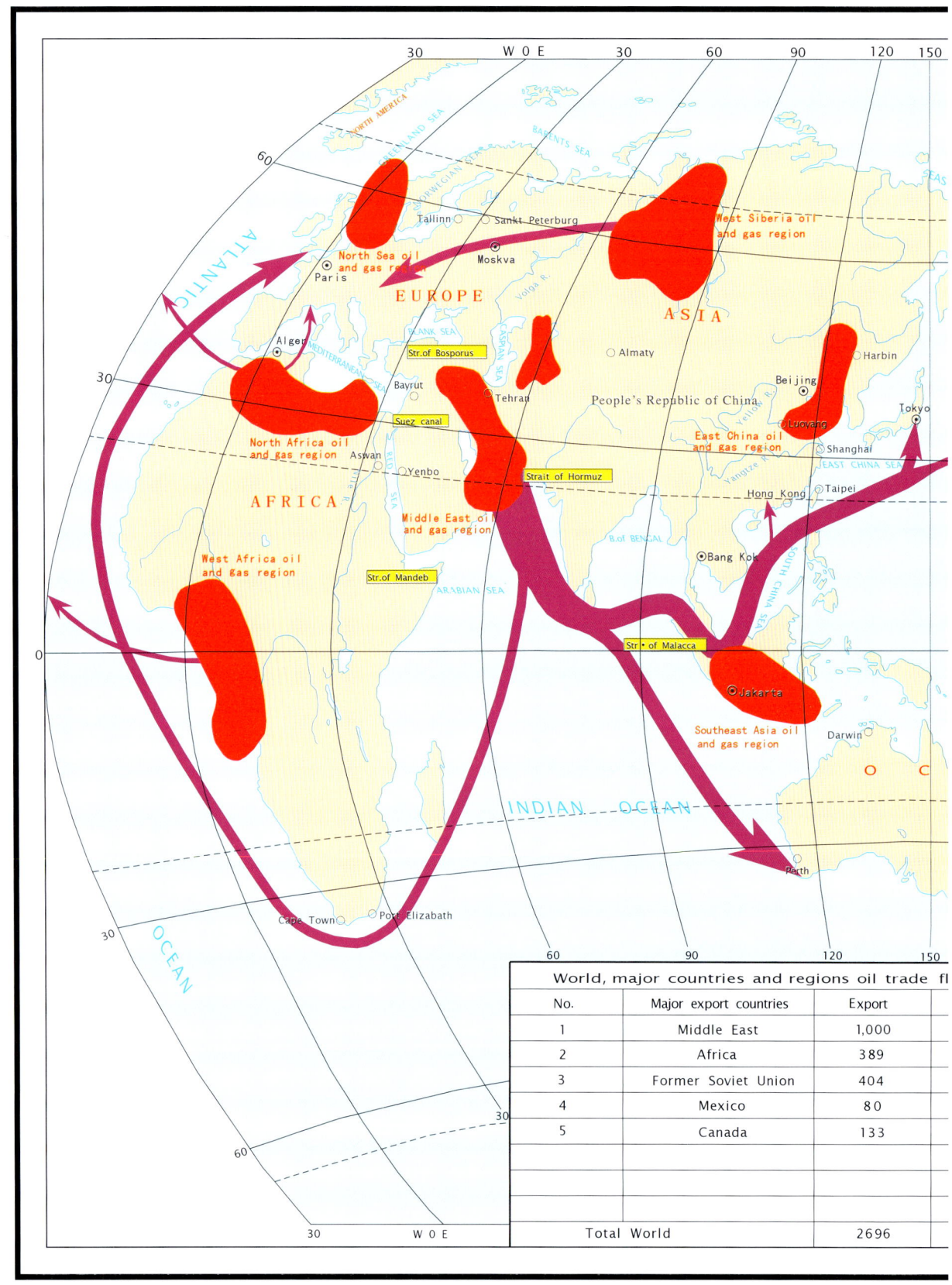

CHAPTER 10

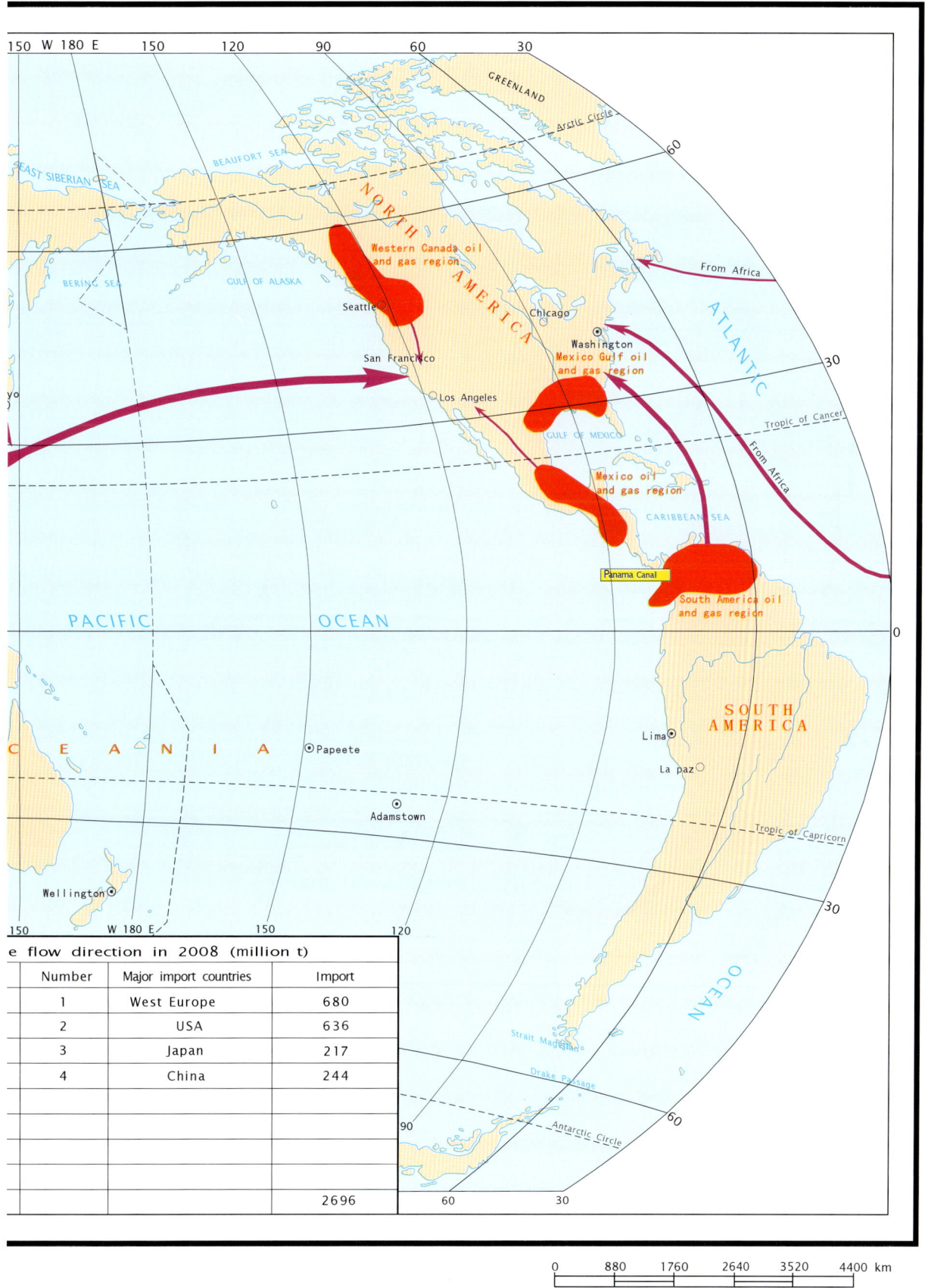

Number	Major import countries	Import
1	West Europe	680
2	USA	636
3	Japan	217
4	China	244
		2696

39

10 World Oil Trade

World oil trade movements (Table 10.1) are mainly effected by tanker shipment, pipeline and railway. Oil may be delivered through all the above-mentioned means, whereas natural gas, by virtue of the relatively large unit volumes of the commodity, is principally transported through pipelines only. Liquefied natural gas shipments by way of custom-built LNG tankers are an increasingly important component of global hydrocarbons trading.

In 2008, world oil production amounted to 3.6 billion t with oil trade totalling 2.6 billion t. Available figures for the same year showed that world natural gas production was at 3 trillion m^3, with global traded volume of 581.7 billion m^3. Currently, approximately 60 per cent of world oil trade is transported through crude oil tanker shipment and most of the balance is by overland pipeline systems. Railway transportation of crude oil is a very tiny proportion of global oil trade. Presently, more than 5 million t of oil flows daily worldwide mostly through carefully identified and dedicated narrow sea transportation pathways and pipeline systems.

At 2008 there were about 4800 oil tankers in the world. The six most important oil transportation routes include the Strait of Mandab, the Strait of Hormuz, the Strait of Bosporus, the Panama Canal, the Suez Canal and the Strait of Malacca.

Straight of Mandab The Mandab Straits, sometimes referred to as the 'corridor in the sea' that connects Europe, Asia and Africa, is located between the Arabian Peninsula and the Horn of Africa. The straits, about 25–32 km wide and 18 km long, consist of two channels divided by the island of Perim.

Strait of Hormuz The Strait of Hormuz is the only sea passage to the open ocean for Persian Gulf crude oil production. It is located between Iran and the Arabian Peninsula, connecting the Gulf of Oman and the Persian Gulf. It is about 150 km long from east to west, 64–97 km from south to north, and is up to 27.5 m deep. Presently, an estimated 650 million t of oil pass through the Straits annually. Due to geopolitical stand-offs between the USA and the Islamic Republic of

Table 10.1 World oil trade in 2008

Areas	Import (million t)			Export (million t)		
	Crude	Oil products	Total	Crude	Oil products	Total
USA	487	149	636	6.9	87.7	96.9
Europe	542	133	680	14	83	97
Former Soviet Union	1	7.1	8.1	311.34	93.5	404.8
Middle East	11	9.7	20.7	895	105	1000
Africa	17.3	21.7	39.1	353	36.6	389.6
China	178	39	217	3.7	15	18.7
Japan	203	41	244	—	17	17
World	1969	727	2696	1969	727	2696

World Atlas of Oil And Gas Basins, First Edition. Li Guoyu.
© 2011 John Wiley & Sons, Ltd. Published 2011 by John Wiley & Sons, Ltd.

Iran, the Strait of Hormuz is considered to be the most strategically important crude oil shipment pathway in the world.

Strait of Bosporus The Strait of Bosporus, also known as the Strait of Istanbul, is a strait that connects the Black Sea to the north and Sea of Marmara and Mediterranean Sea to the south, and forms the boundary between the European part of Turkey and its Asian counterpart. It is approximately 30.4 km long, with a maximum width of 7.5 km and a minimum width of 758 m. The depth varies from 27.5 m to 120 m. Currently, an estimated 100 million t of oil is transported through the Strait annually.

Panama Canal The Panama Canal is located in the Central American Republic of Panama. It joins the Pacific Ocean to the Caribbean Sea (Atlantic Ocean). The Canal is 81.3 km long, 150–304 m wide, and 13–15 m deep. Its water level is 26 m higher than that of the two oceans it connects. There are six ship locks over the length of the Canal. The maximum size of vessel that can pass through the Canal is referred to in maritime industry terms as Panamax-sized vessels. Currently, an estimated 30.65 million t of oil passes through the canal annually and heads to North America.

Suez Canal The Suez Canal is a canal in Egypt and an important transportation passage between Europe, Asia, and Africa. It links the Red Sea and the Mediterranean, allowing communication between the Atlantic Ocean, the Mediterranean and the Indian Ocean. The Canal is 175 km long, 135 m wide (average) and 22.5 m deep. The Suez Canal–Shamid oil pipeline system now has an estimated annual oil transportation volume of 190 million t. Suezmax is the maritime industry reference to the largest tankers that can pass through the Suez Canal fully loaded.

Strait of Malacca The Strait of Malacca is a narrow waterway that links the Indian Ocean and the South China Sea. It is 1080 km long and acts as an important sea transportation pathway between the Pacific Ocean and the India Ocean. Currently, an estimated 500 million t of oil is sent through the Strait. The Strait sees more than 50,000 ships each year. For China, as the second largest importer of crude oil, the Strait of Malacca is strategically vital for transporting imported crude from the Middle East, Africa and the Asia-Pacific region. In 2002, nearly 80 per cent of China's total imported crude oil volume of 70 million t was transported through the Strait.

ASIAN OIL AND GAS BASINS

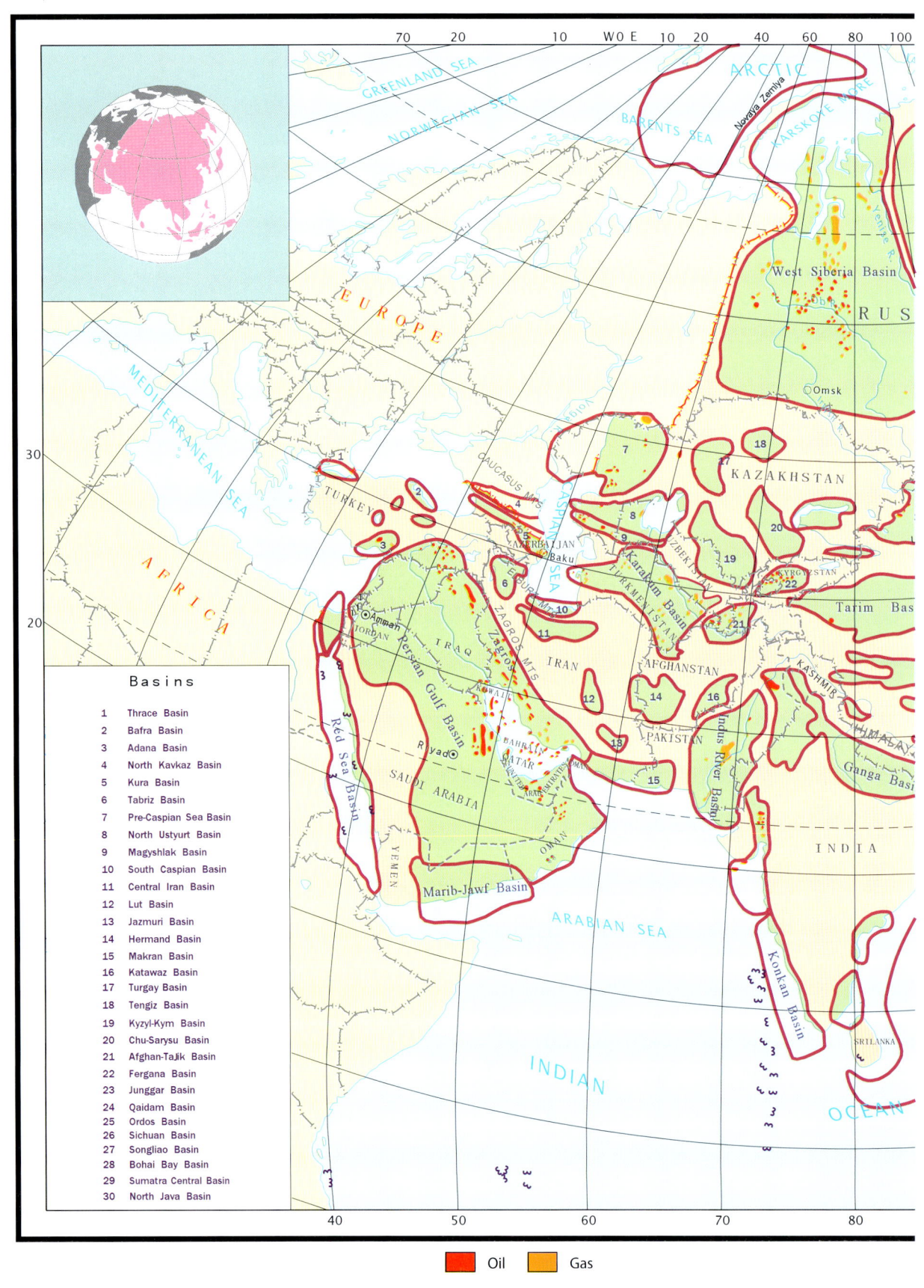

Basins

1. Thrace Basin
2. Bafra Basin
3. Adana Basin
4. North Kavkaz Basin
5. Kura Basin
6. Tabriz Basin
7. Pre-Caspian Sea Basin
8. North Ustyurt Basin
9. Magyshlak Basin
10. South Caspian Basin
11. Central Iran Basin
12. Lut Basin
13. Jazmuri Basin
14. Hermand Basin
15. Makran Basin
16. Katawaz Basin
17. Turgay Basin
18. Tengiz Basin
19. Kyzyl-Kym Basin
20. Chu-Sarysu Basin
21. Afghan-Tajik Basin
22. Fergana Basin
23. Junggar Basin
24. Qaidam Basin
25. Ordos Basin
26. Sichuan Basin
27. Songliao Basin
28. Bohai Bay Basin
29. Sumatra Central Basin
30. North Java Basin

Oil Gas

Part II
Asian Oil and Gas Basins

Geography

The continent of Asia is located in the northeastern part of the eastern hemisphere. It is the largest continent in the world with an area of 44 million km^2, which represents 29.4 per cent of the world's total. It has a population of 3.57 billion, which is 60 per cent of the world's total. In Asia there are 48 countries and regions, which are usually divided into Southeast Asia, South Asia, West Asia, Central Asia and North Asia.

History

As far back as 2000 years ago, oil seepages had occurred in China, Iraq and Baku. Before the Second World War, oil production of the continent was confined to only 12 countries, with an annual output of about 450 million t, similar to that of North America at the time. After the war, however, several giant oil and gas fields were discovered in the Middle East, Central Asia and in West Siberia, which is classified here as the Asian portion of Russia.

Reserves and production of oil and gas

At the end of 2008 the proven oil reserves in Asia stood at 149.5 billion t, 80 per cent of the world's total (182 billion t). About 56 per cent of the world's total oil reserves were found in the Middle East (102 billion t). Annual oil production of Asia in 2008 was 1.46 billion t, 40 per cent of the world's total (3.6 billion t). At the end of 2008 the proven gas reserves in Asia stood at 83.7 trillion m^3. Annual gas production was 0.6 trillion m^3. The 2008 statistics of oil and gas reserves for selected countries are shown in Table II.1.

Regional geology

This account of the regional geology is based on Vysotsky *et al.* (1995). The three largest petroleum basins in the world are found in Asia, i.e. Middle East, West Siberia and East Siberia, each with a size larger than 3 million km^2. They are the most prospective and prolific basins on Earth. In total, Asia has more than 60 petroleum basins. Most of the basins represent syneclises, which is evidenced by their basement and Palaeozoic infill. The distinctive feature of the basins is complicated by the presence of Meso-Cenozoic rifts, troughs and horst–graben dislocations.

The sedimentary cover comprises all systems of the Palaeozoic and attains up to 10–12 km in thickness, including deposits of marine origin. Mesozoic and Cenozoic systems are represented essentially by continental red-coloured sediments with widely developed salt and coal bearing formations.

Source rocks have been identified in both marine and continental sediments. Reservoir rocks are associated with sandy deposits, mostly in Mesozoic strata, and also with Mesozoic and Palaeozoic terrigenous and marine carbonate rocks. Particular reservoir types have been identified in the

World Atlas of Oil And Gas Basins, First Edition. Li Guoyu.
© 2011 John Wiley & Sons, Ltd. Published 2011 by John Wiley & Sons, Ltd.

Asian Oil and Gas Basins

Table II.1 Proven reserves and production of oil and gas (2008) in major countries of Asia (excluding the Middle East)

Country	Proven reserves		Production	
	Oil (million t)	Gas (billion m³)	Oil (million t)	Gas (billion m³)
Total Asia	10,600	19,870	510	675
Afghanistan	—	49	—	—
Bangladesh	3.8	141	0.2	—
Brunei	150	390	7.8	11.5
China	2191	2265	190	80
India	770	1074	33.4	29.1
Indonesia	546	3001	42.7	77.2
Japan	6	20	0.8	3.6
Malaysia	547	2350	37.5	48
Burma	6.8	283	0.95	—
Pakistan	46	885	2.3	41.4
Philippines	18	98	0.79	—
Thailand	60	317	11.4	14
Vietnam	821	192	13.7	5
Kazakhstan	4109	2408	69	28.8
Turkmenistan	82	2661	11	66
Uzbekistan	81	1840	5	62
Tajikistan	16	5.6	—	—
Kyrghyzstan	5.4	5.6	0.05	—

crust associated with the weathering of Archaean and Palaeozoic rocks of the basement. Coral seals and sealing strata associated with gypsiferous rocks and rock salt are widespread. Traps are represented by structural and non-anticlinal types. Oil and gas accumulations are multilayered and mainly confined to Meso-Cenozoic deposits. The petroleum potential of the relatively large basins is estimated as being high to very high, whereas the potential of the smaller basins is typically low (Shannon and Naylor, 1989).

Major oil and gas fields

More than 2000 oil and gas fields have been discovered in Asia, among which are 93 giant oil fields as well as 38 giant gas fields. Many giant oil fields are concentrated in the Persian Gulf. The Ghawar oil field of Saudi Arabia is the largest oil field in the world. This oil field was discovered in 1948 with an area of 2200 km² and holds proven oil reserves of 11.4 billion t. Many giant gas fields are located within the West Siberia Basin. The Urengoy gas field is the largest in the world. This gas field was discovered in 1966 with an area of 1260 km² and proven gas reserves of 8 trillion m³.

CHINA

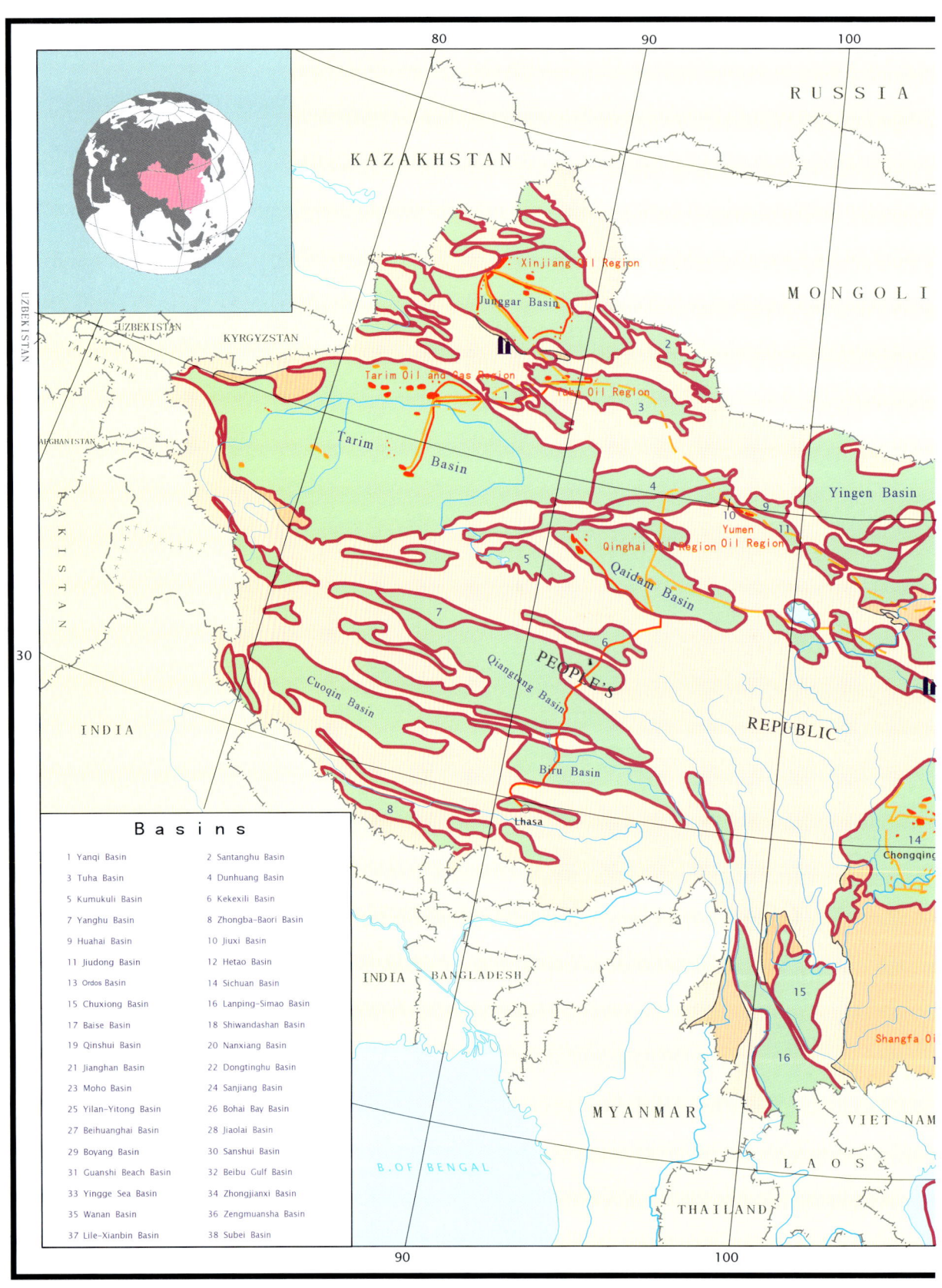

CHAPTER 11

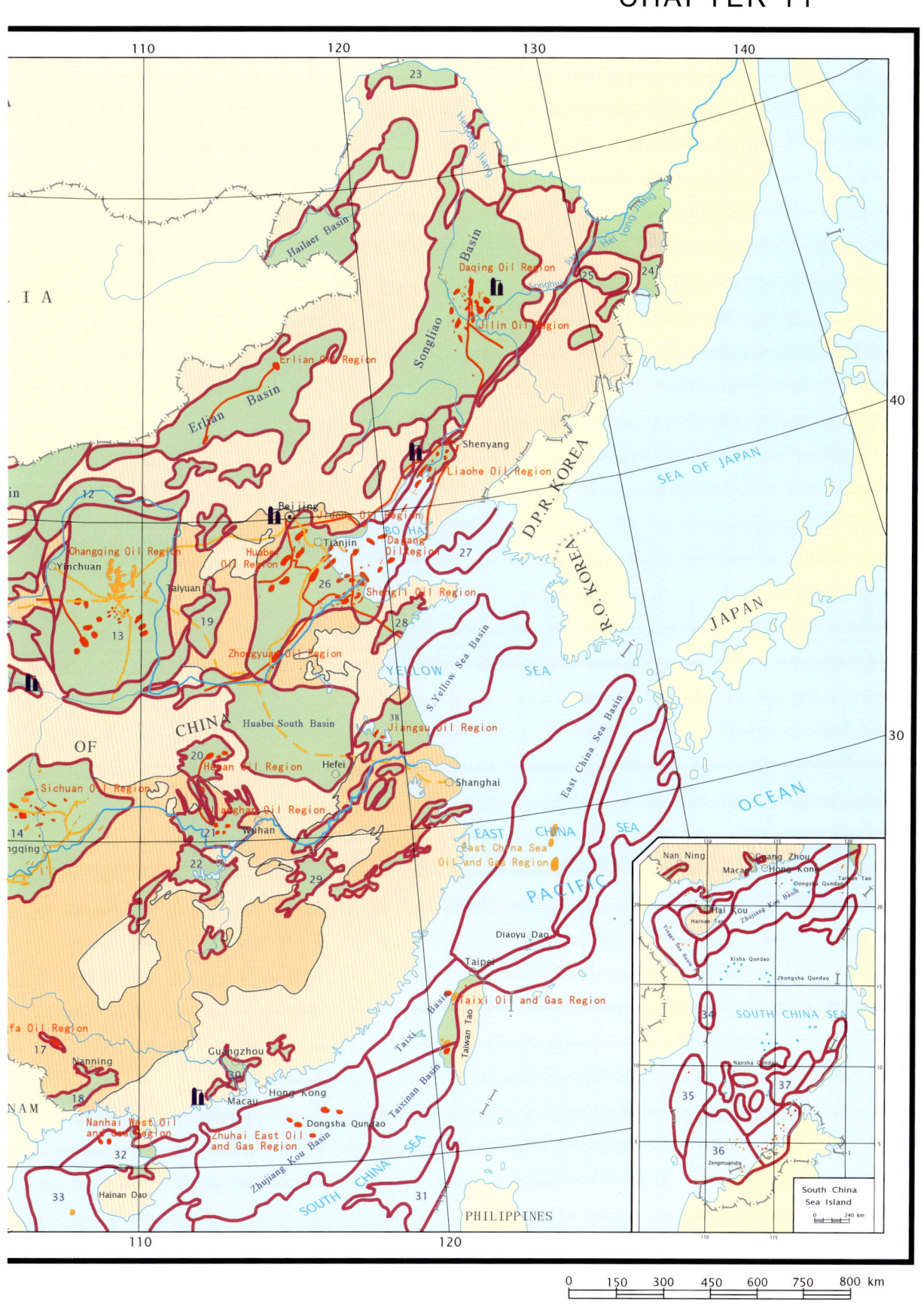

11 China

Geography

The country is situated in the eastern part of Asia, on the west coast of the Pacific Ocean. In area it is the third largest country in the world (after Russia and Canada).

History

China is an ancient oil- and gas-producing country. Indeed there are records of gas development going back as far as 211 BC in Sichuan Province. Recently China has become the fifth largest oil-producing country in the world and also ranks as the second, after the USA, in terms of consumption of petroleum and petroleum products. China's largest oil field, Daqing, has produced over 50 million t annually for the 26-year period, 1976–2002 (Fig. 11.1). In 2008 the production of the field dropped to 41.6 million t.

Regional geology

The Precambrian basement of China is widely distributed in the north as an east–west strip of Archaean and Proterozoic rocks. Proterozoic basement is also common in southern China. The Archaean rocks are not important for hydrocarbons and comprise volcano–sedimentary cycles that have suffered a high degree of metamorphism.

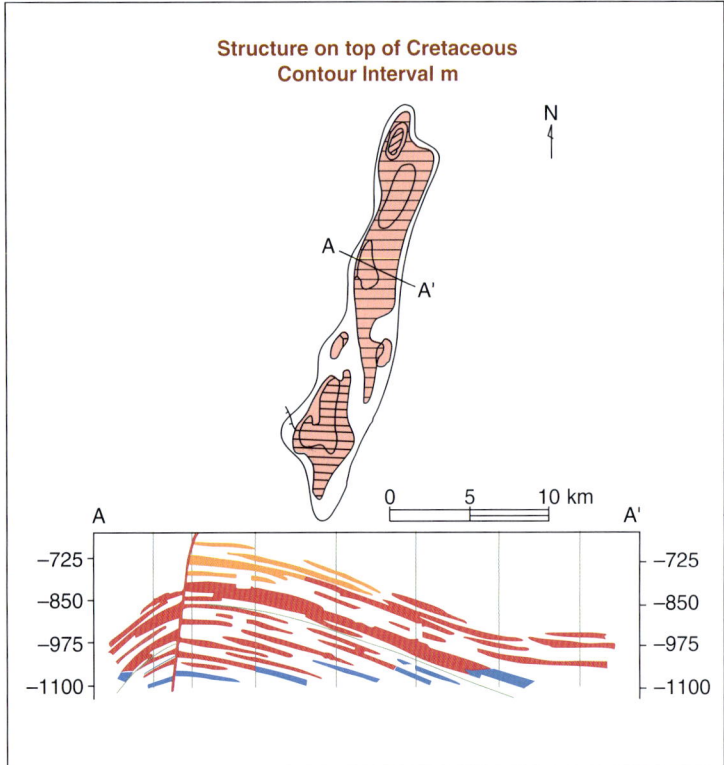

Daqing is the biggest non-marine oil field in China, and has a peak oil production of 56 million t.

Basic data

1. Name	Daqing
2. Discovery year	1959
3. Area (km²)	1890
4. Reservoir	Sandstone
5. Reservoir age	Cretaceous
6. Depth (m)	750–1,770
7. Thickness (m)	60
8. Porosity (%)	21–30
9. Permeability (mD)	500
10. Recoverable reserves (billion t)	2.5
11. Peak oil production (million t)	56 (1996)

Fig. 11.1 Details of the Daqing oil field.

World Atlas of Oil And Gas Basins, First Edition. Li Guoyu.
© 2011 John Wiley & Sons, Ltd. Published 2011 by John Wiley & Sons, Ltd.

The various productive and prospective basins run as distinct swathes across the Chinese landmass. The first of these, including the Tarim, Jungger, Qaidam and Jiuquan basins, runs in an approximately east–west trend to the south of Mongolia. The other groups lie in the eastern and central regions and have a northeast–southwest trend. An outer broad band of basins lies offshore and includes the East China Sea, South China Sea and Pearl River Mouth basins, while an inner belt includes the Songliao, Huabei and Sichuan basins.

The basins occur in three major tectono-sedimentary settings. Those in the west, displaying the east–west trend, developed in an area of compression-dominated tectonics. The eastern coast consists of an outer fore-arc belt and an inner zone of retro-arc (back-arc) basins. Further inland, a broad northeast–southwest belt of basins consists of a series of extension-dominated rift basins.

There are more than 236 separate Mesozoic and Cenozoic sedimentary basins or sub-basins (details of 16 of the major basins are provided in Table 11.1), which comprise more than one-third of the vast Chinese landmass of 9.6 million km². China nevertheless accounts for less than 5 per cent of the world's petroleum production.

Table 11.1 Major oil and gas basins of China

Basin	Area (thousand km²)	Major sedimentary rock		Reservoir	
		Age	Thickness (m)	Age	Lithology
Beibu Gulf	36.4	Mz, Kz	9000	R*	Sandstone
Bohai Bay	187	Pz, Mz	5000	R, Pz, Pt	Sandstone, carbonate rock
East China Sea	241	Kz	4500	E_2–N_1	Sandstone,
Erlian	109	Mz	6000	K	Sandstone, basalt, conglomerate, tuff, limestone
Hubei South	140	Pz–Kz	6000–9000	R	Sandstone
Junggar	134	C–R	13,000	P, T, R	Sandstone
Qaidam	104	J–Kz	10,000	R, Q_1	Sandstone
Ordos	250	Pz, C–K	7000	T, J	Sandstone
Sichuan	200	Pt–K	10,000	P, T, J	Carbonate rock, sandstone
Songliao	260	Mz, Kz	6800	K	Sandstone
South Yellow Sea	84	Pz–R	16,000	R	Glutenite
Taixi	65	Kz	5000	N	Sandstone
Tarim	560	Pz–Kz	10,000	J, R	Sandstone
Tuha	55	T–R	7500	J	Sandstone
Yingge Sea	98	Mz, Kz	>10,000	R	Reef, carbonate rock, sandstone
Zhujiang Kou	202	Mz, Kz	5800	R	Sandstone, reef

*R = Tertiary. This applies to all later tables and figures where this abbreviation is used.

Asian Oil and Gas Basins

Production in China has risen significantly during the past decades. In 1976, annual production reached 64 million t. By 1996 it had risen to 157 million t and now stands at approximately 184 million t. Most of the production comes from a relatively small number of fields located in a few, mostly onshore, basins. However, the offshore basins are becoming increasingly important due to a number of recent significant discoveries and the increased pace of exploration in these areas. By the end of 2008 oil production was 190 million t and gas production was 80 billion m^3.

The petroleum exploration of the Palaeozoic marine strata is an optimistic target for future replacement of other domains in China which are being gradually depleted. It is expected that these Palaeozoic marine strata will be one of the main exploration domains in the 21st century. It is suggested that the Tarim, Ordos and Sichuan craton basins are strategic exploration regions where Palaeozoic marine strata are well preserved. The slope facies mudstones are the main source rocks for the marine. Transitional facies source rocks of the Natal Group are the important exploration targets.

NORTH AND SOUTH KOREA

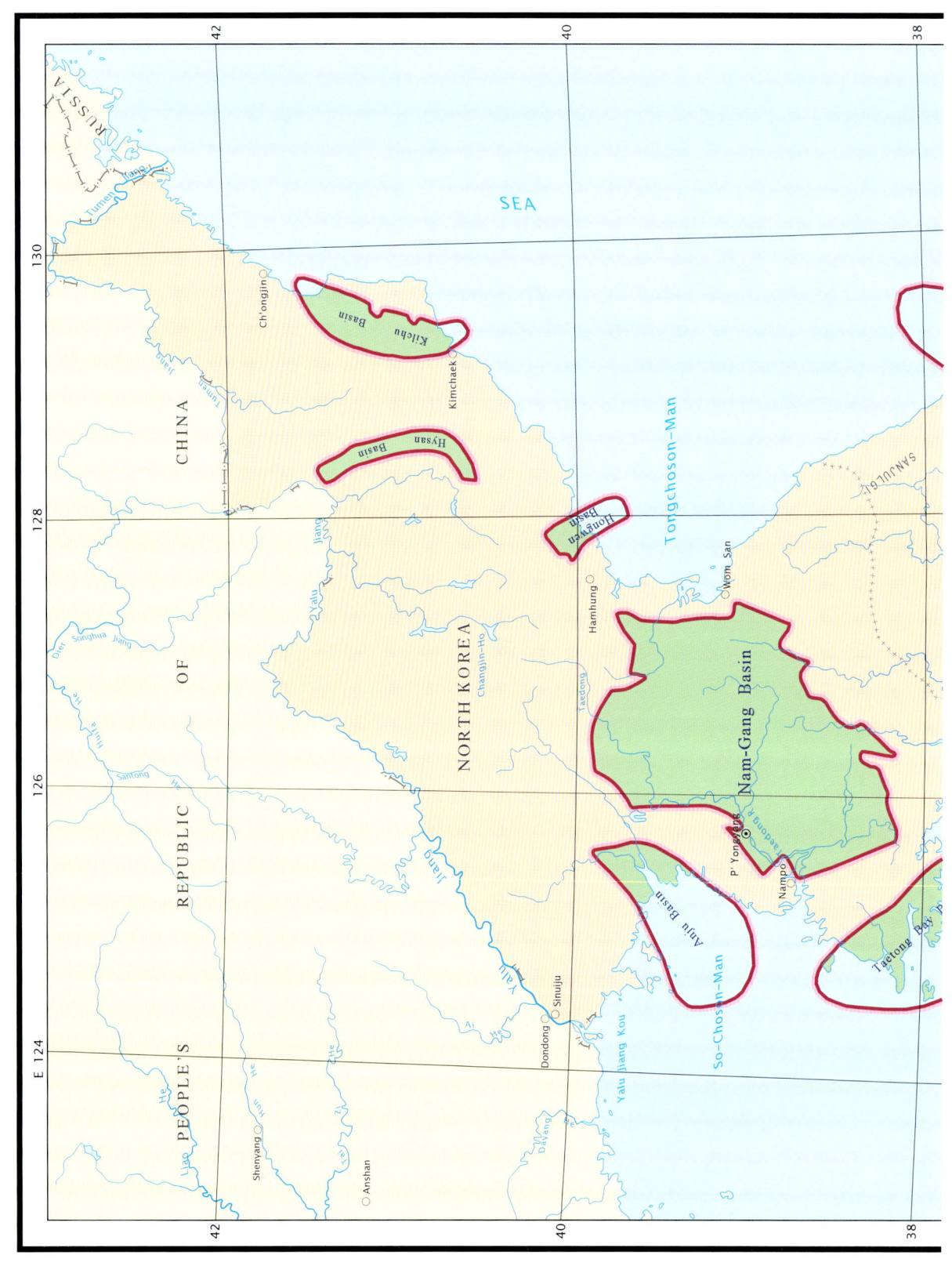

CHAPTER 12

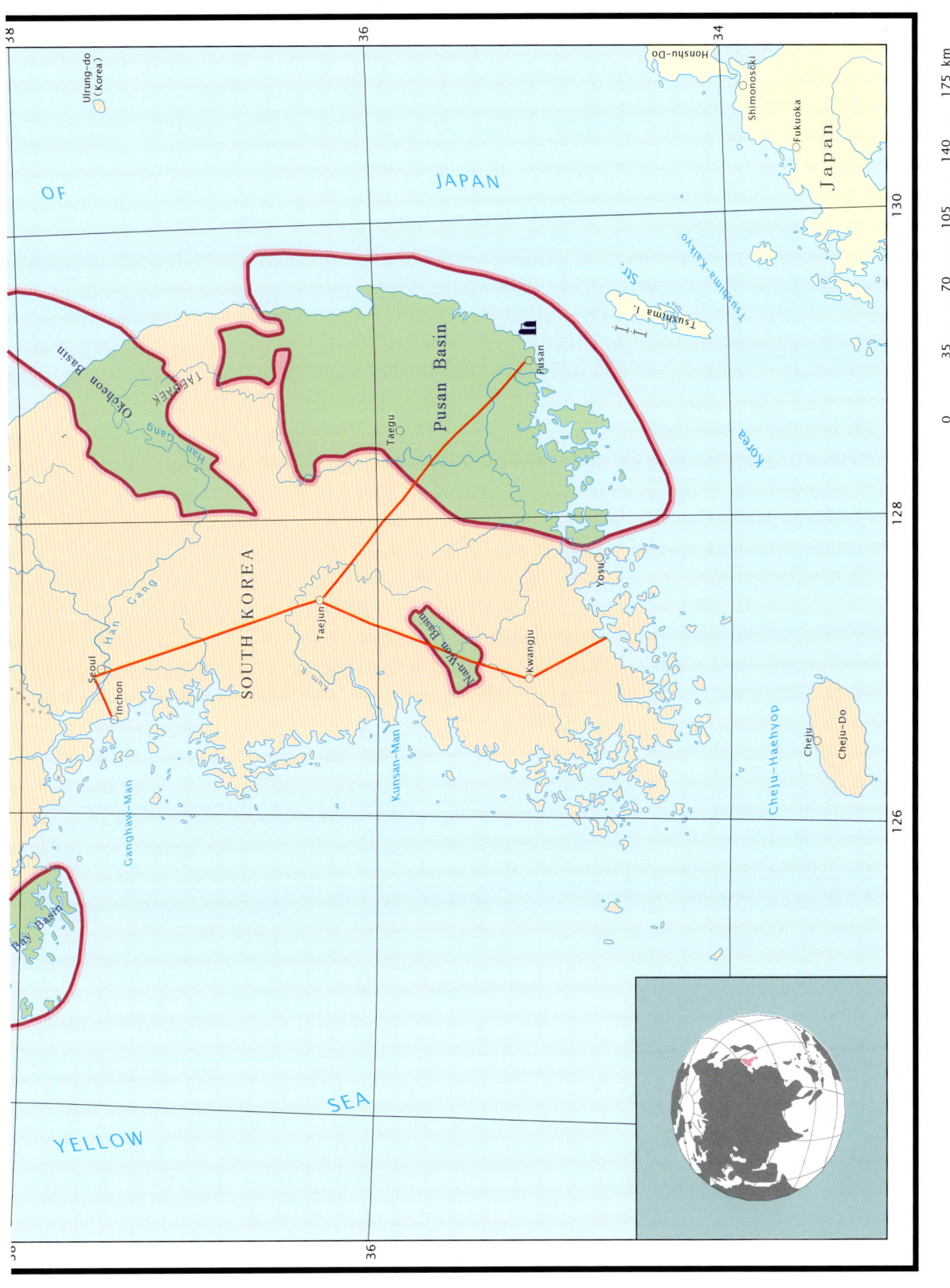

12 North and South Korea

Geography

Both countries occupy the Korean Peninsula in East Asia, bordered to the north by China and occupying an area of 221,400 km^2 in total with a combined population of 69.1 million.

History

Oil and gas exploration activity began some time ago, but up till now no major oil and gas fields have been discovered.

Regional geology

The Korean Peninsula is a single geological entity a basement of Archaean rocks and a sedimentary cover of Protozoic, Palaeozoic and Meso-Cenozoic rocks, with thicknesses in the range of 2000–7000 m. There are nine sedimentary basins, each of which is small in area.

The Sino-Korean Platform comprises Early Precambrian basement rocks cropping out within the area of the Sino-Korean Shield. The North China platform cover comprises Upper Proterozoic and Lower Palaeozoic marine deposits, Upper Palaeozoic paralic coal-bearing rocks and Mesozoic and Cenozoic continental sediments. The absence of Devonian–Carboniferous deposits is typical of the Palaeozoic succession. During the Mesozoic, the platform underwent strong orogenesis, especially in the east, which resulted in the formation of large superimposed structures such as grabens, troughs and depressions that complicated the general platform-like structural pattern. The different geological conditions between China and Korea has led to different oil and gas prospects.

North Korea In total, six basins are located in North Korea, but most are small in area (Table 12.1). The two most notable are Nam-Gang (210,000 km^2) and Taetong Bay (80,000 km^2). A consideration of the geology suggests that this country has good potential for oil and gas exploration.

Table 12.1 Sedimentary basins of North Korea

Basin	Area (thousand km^2)	Major sedimentary rock	
		Age	Thickness (m)
Anju	6	Mz, Kz	7000
Hongwen	1.2	—	—
Hysan	1	—	—
Kilchu	2.7	Mz, Kz	—
Nam-Gang	21	Z, Pz	—
Taetong Bay	8	Z, Pz	—

World Atlas of Oil And Gas Basins, First Edition. Li Guoyu.
© 2011 John Wiley & Sons, Ltd. Published 2011 by John Wiley & Sons, Ltd.

Table 12.2 Sedimentary basins of South Korea

Basin	Area (thousand km²)	Major sedimentary rock	
		Age	Thickness (m)
Nan-Won	1.2	—	—
Okcheon	12.5	Z, Pz	—
Pusan	29	R	—

Table 12.3 Oil consumption, import and refinery capacity of South Korea

Year	Oil consumption (million t)	Oil import (million t)	Refinery capacity (million t)
2000	103	122	128
2001	103	117	128
2002	104	107	128
2003	105	108	127
2004	104	114	128
2005	105	120	128
2006	105	121	131
2007	107	120	133
2008	103	123	135

South Korea South Korea has an area of 99,600 km² and a population of 48.29 million. No commercial discovery has been made in spite of sporadic and concerted exploration efforts. In spite of its status as one of the most developed countries in Asia, South Korea is wholly dependent on imported oil.

South Korea has three basins on land and some on the shelf (Table 12.2). There is therefore a good prospect of oil and gas exploration in the future. South Korea is rising rapidly in the ranks of world petroleum consumers. It is one of the world's largest petrochemical producers. By the end of 2008 oil consumption was 103 million t, oil production was 123 million t and refinery capacity was 135 million t (Table 12.3).

MONGOLIA

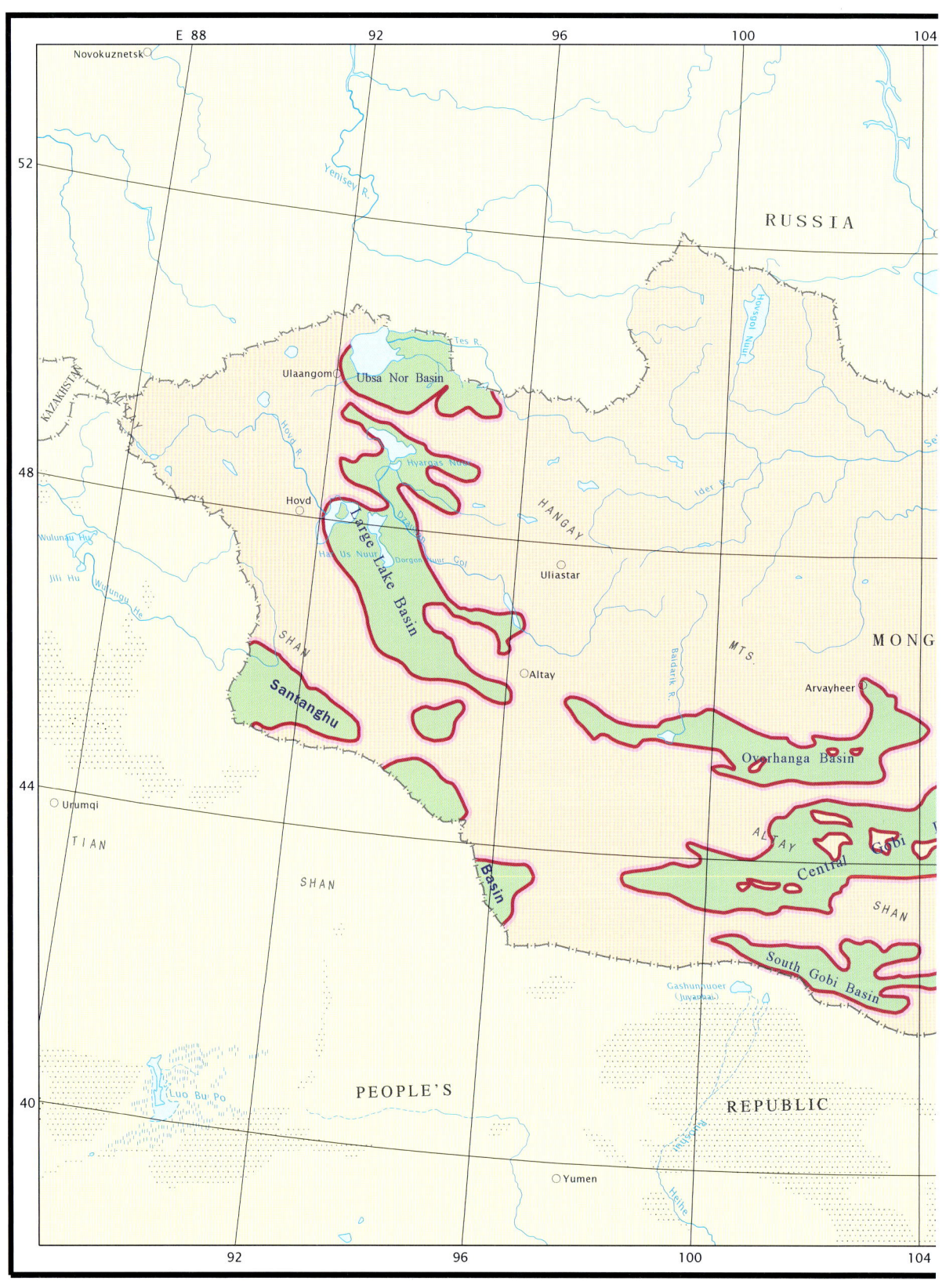

CHAPTER 13

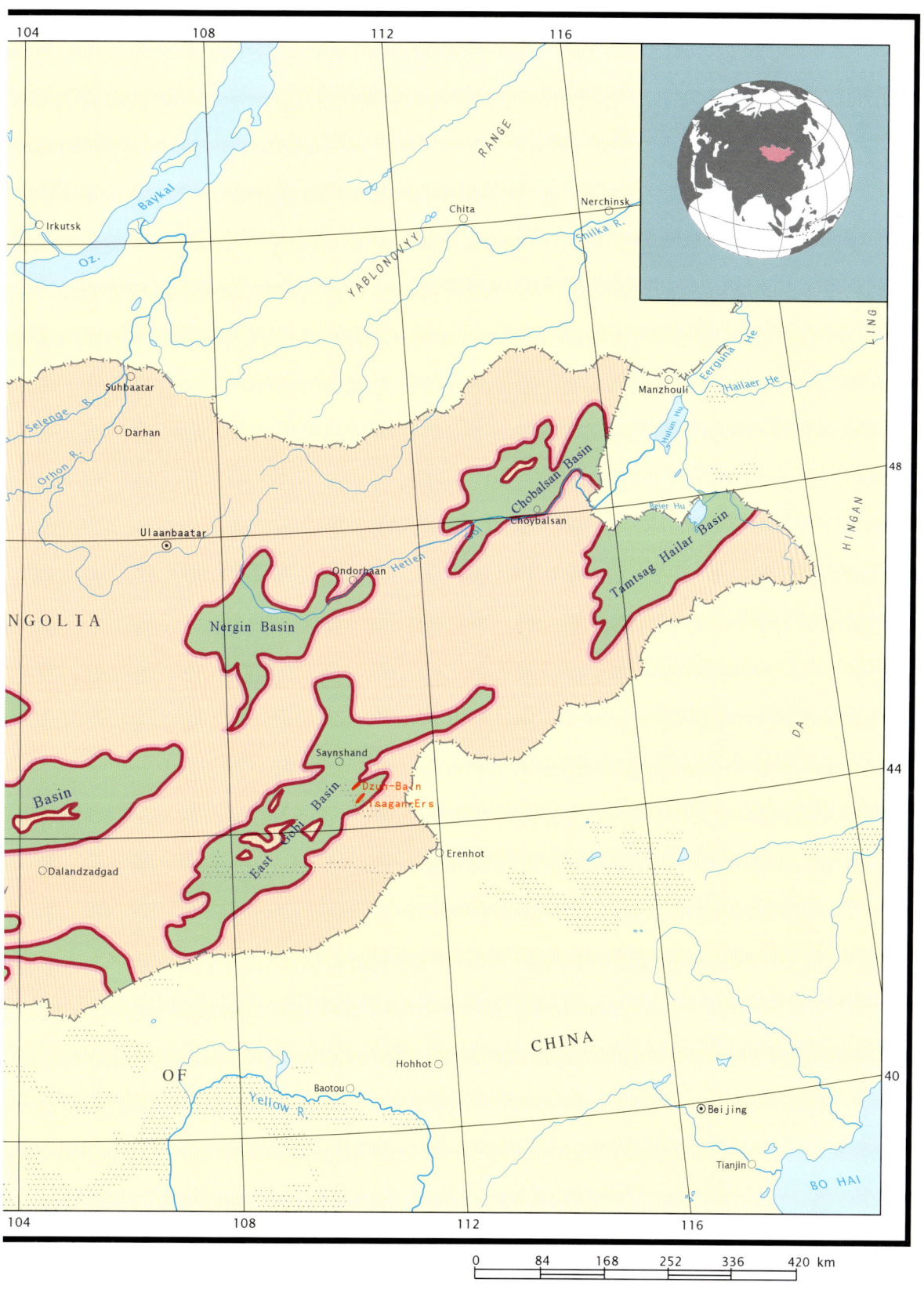

13 Mongolia

Geography

This landlocked desert country is located in eastern Asia, lying between China and Russia. It has a total area of 1,566,500 km^2 and a population of 2.56 million.

History

Oil exploration began in the early 1920s. The Zunmbayan oil field was discovered in 1947. Oil production of the field was 54,000 t in 1954. From 1969 to 1990 no oil exploration was carried out. New oil fields were discovered starting from 1995, with proven oil reserves of 9.24 million t. Export of oil to China was 16,900 t in 2002.

Recently, exploration work has surged. In the Tsagan-Ers Basin in Mongolia, wells 19–13 and 19-14 were completed and fracture-stimulated in the Tsagaantsav zone The wells tested at initial rates of approximately 350 barrels/day of oil. These two wells are expected to be put into production. Recent estimates of proven oil reserves have been very modest, in spite of these discoveries. Oil consumption was 51,700 t in 2003, mainly from imported sources.

Regional geology

An arc-shaped structure belt, part of the Palaeozoic Orogenic Belt, is the basic element of regional geological structure in Mongolia, formed by the Mesozoic–Cenozoic collision of the Indian continental plate with the Asian continental plate, which resulted in Palaeozoic continental crustal material in the collision zone being extruded to the east and consequent interaction with the Pacific oceanic plate. As a result of multistage structural movements over the history of this collision, most of the sedimentary basins today are remnant basins, foreland basins and refit-subsidence basins. An example of a fault-controlled oil field is shown in Fig. 13.1. Both western and eastern basin groups bear the features of intermontane basins. Upper Jurassic to Cretaceous strata developed in most sedimentary basins of Mongolia, with minor Neozoic deposits. The sedimentary cover comprises Cambrian, Ordovician, Silurian, Devonian–Permian, Triassic–Jurassic and Cretaceous–Tertiary rocks.

Source rocks have been identified in deposits of continental origin. Reservoirs are associated with Jurassic and Cretaceous sandstone and volcanic rocks. Traps are represented by both structural and non-anticlinal types. The 10 sedimentary basins (Table 13.1) have an area of 21–70,000 km^2 with sediment thicknesses of 2700–7570 m. The Central Gobi Basin is the largest with an area of 70,000 km^2. The petroleum potential of Mongolia has been estimated as very high (Vysotsky, 1995).

World Atlas of Oil And Gas Basins, First Edition. Li Guoyu.
© 2011 John Wiley & Sons, Ltd. Published 2011 by John Wiley & Sons, Ltd.

Mongolia

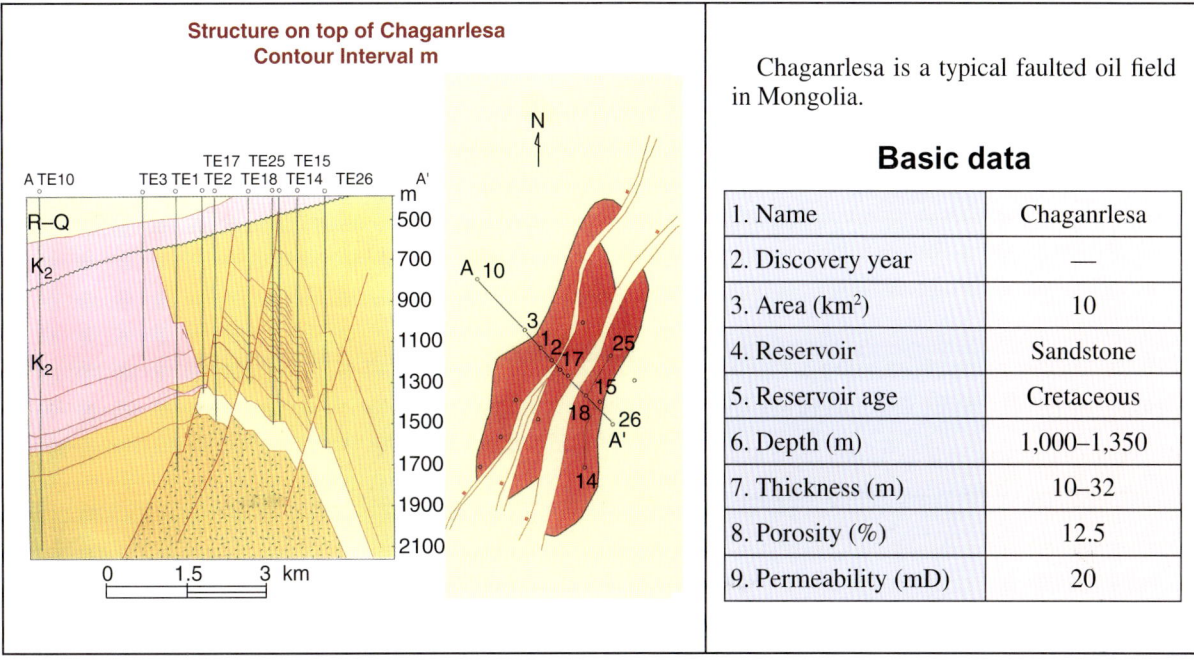

Chaganrlesa is a typical faulted oil field in Mongolia.

Basic data

1. Name	Chaganrlesa
2. Discovery year	—
3. Area (km²)	10
4. Reservoir	Sandstone
5. Reservoir age	Cretaceous
6. Depth (m)	1,000–1,350
7. Thickness (m)	10–32
8. Porosity (%)	12.5
9. Permeability (mD)	20

Fig. 13.1 Details of the Chaganrlesa oil field.

Table 13.1 Major oil and gas basins of Mongolia

Basin	Area (thousand km²)	Major sedimentary rock		Reservoir		Remarks
		Age	Thickness (m)	Age	Lithology	
Central Gobi	70	Mz				
Chobalsan	25					
East Gobi	56	Mz–R	5000	K	Sandstone	Extends into China
Tamtsag Hailar	28	J–R	75,700	J	Sandstone	
Large Lake	56					
Nergin	23					
Ovorhanga	28					
Santanghu	35	C–R	2700–3200			
South Gobi	25	Mz				
Ubsa Nor	21					

JAPAN

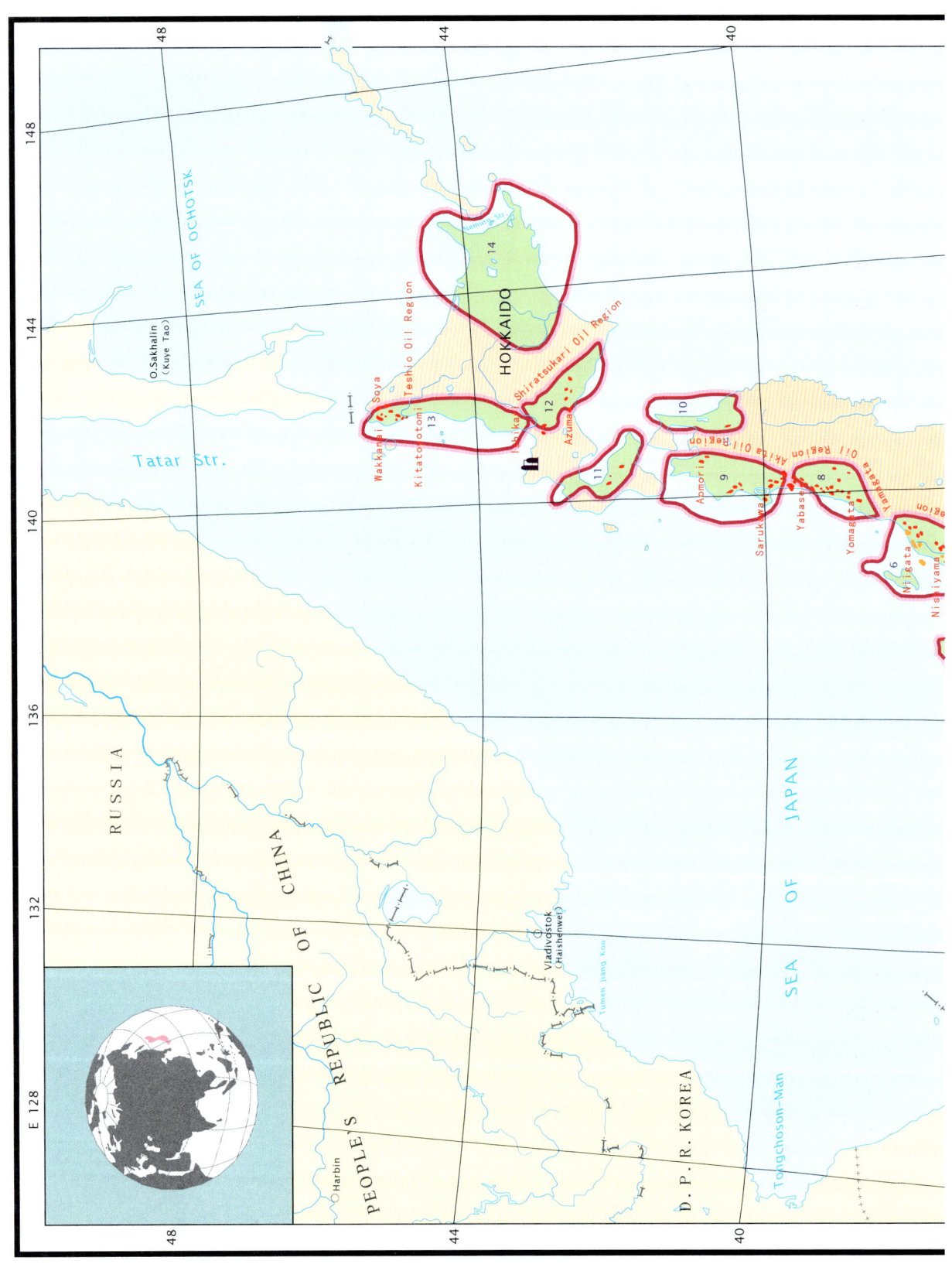

CHAPTER 14

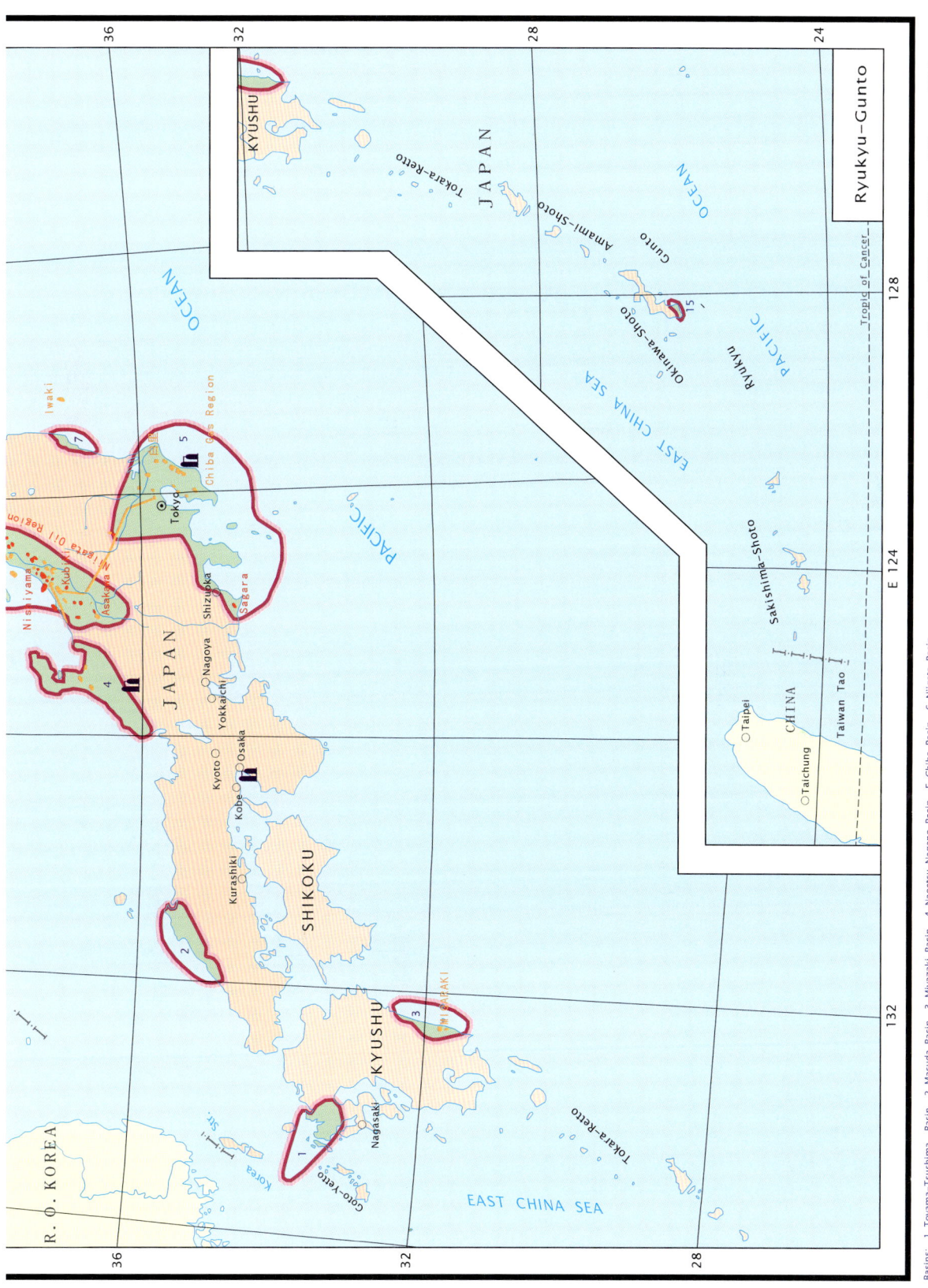

Basins: 1 Toyama-Tsushima Basin 2 Masuda Basin 3 Miyazaki Basin 4 Naoetsu-Nagano Basin 5 Chiba Basin 6 Niigata Basin
7 Abukuma-oki Basin 8 Yamagata Basin 9 Akita Basin 10 Towada Basin 11 Hakodate Basin 12 Shiratsukari Basin
13 Teshio Basin 14 Hokkaido Basin 15 Okinawa Basin

14 Japan

Geography

The country is located in East Asia. There are four main islands in Japan, i.e. Hokkaido, Honshu, Shikoku and Kyushu. It has a total area of 377,800 km² and a population of 127 million.

History

As far back as AD 615 oil was discovered in Niigata. The first oil field was discovered in Akita in 1874. In 1857 oil production was 700 t. The first offshore well was drilled in September 1958. In 2008 oil production was 820,000 t, while gas production was 3.5 billion m³.

Regional geology

Japan is located on the east side of the Euro-Asia continental plate. The sedimentary cover is composed of Tertiary rocks. All the source rocks, reservoirs and traps are associated with those strata. The area of sedimentary cover in Japan is 84,000,000 km² representing 22 per cent of the whole area of the country. There are 15 sedimentary basins (Table 14.1) and the largest one has

Table 14.1 Major oil and gas basins of Japan

Basin	Area (thousand km²)	Major sedimentary rock		Reservoir	
		Age	Thickness (m)	Age	Lithology
Toyama-Tsushima	8	Kz			
Masuda	7.2	Kz			
Miyazaki	6	Kz			
Naoetsu-Nagano	12.2	Kz		R	Tuffaceous sandstone
Chiba	50	Kz, Q	3500	N, Q	Sandstone
Niigata	23	Kz		N	Sandstone, pyroclastic rock
Abukuma-Oki	18	Kz	5000	R	Sandstone
Yamagata	9	Kz		N	Sandstone, tuff
Akita	18	Kz		N	Sandstone, volcanic rock
Towada	9	Kz			
Hakodate	10	Kz		N	Sandstone, volcanic rock
Shiratsukari	8.6	Kz		R	Sandstone
Teshio	20	Kz	>10,000	R	Sandstone
Hokkaido	50	Kz	>10,000	N	Sandstone, tuff
Okinawa	0.7	Kz			

World Atlas of Oil And Gas Basins, First Edition. Li Guoyu.
© 2011 John Wiley & Sons, Ltd. Published 2011 by John Wiley & Sons, Ltd.

Japan

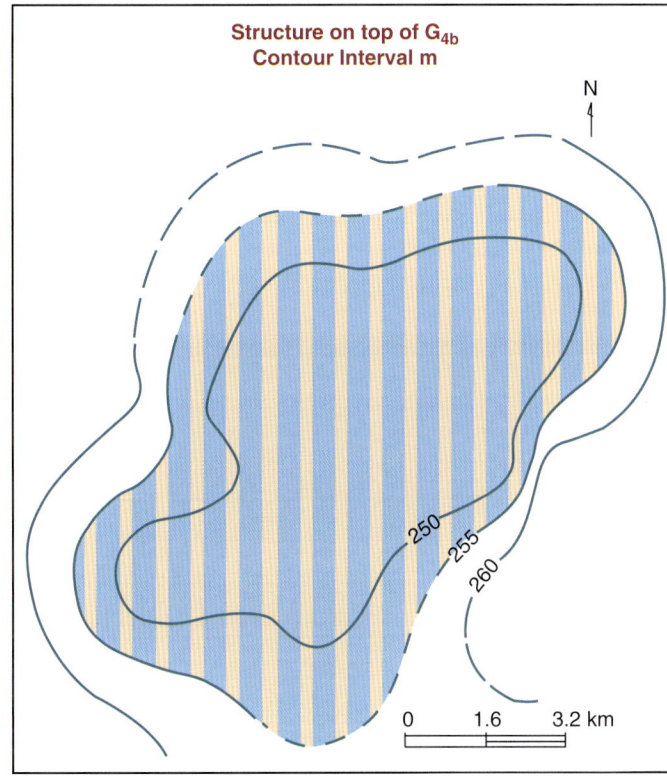

East Nigata is a typical water solution gas field in Japan.

Basic data

1. Name	East Nigata
2. Discovery year	1955
3. Area (km²)	70
4. Reservoir	Sandstone
5. Reservoir age	Tertiary
6. Depth (m)	100–1,820
7. Thickness (m)	25–100
8. Porosity (%)	25–30
9. Permeability (mD)	—
10. Recoverable reserves (billion m³)	6.8

Fig. 14.1 Details of the East Nigata gas field.

Table 14.2 Oil import of Japan

Year	Oil production (thousand t)	Gas production (billion m³)	Oil import (million t)
1950	293	0.07	1.2
1960	526	0.82	27
1970	784	2.57	170
1980	500	2.54	211
1990	500	1.60	240
2000	640	2.30	266
2007	800	3.5	257
2008	800	3.6	244

area of 98,000 km². All the basins share common features of small area, small reserves and low productivity. Details of a typical water-solution gas field are shown in Fig. 14.1.

Oil and gas importation

Japan, the world's third largest energy importer after the USA and China, remains heavily dependent on oil imports from the Middle East (Table 14.2). Japan is a highly developed industrial country, but its relative scarcity of natural resources has not hindered it in its post-war industrial development, which now makes the country's economy the second largest in the world. All of the country's requirements of crude petroleum, iron ore and bauxite, and most of its requirements of coking coal and copper ore, have to be met by imports. Japan's post-war industrial development was based on three main industries, namely steel, shipbuilding and petrochemicals.

BURMA (MYANMAR), THAILAND, LAOS, CAMBODIA

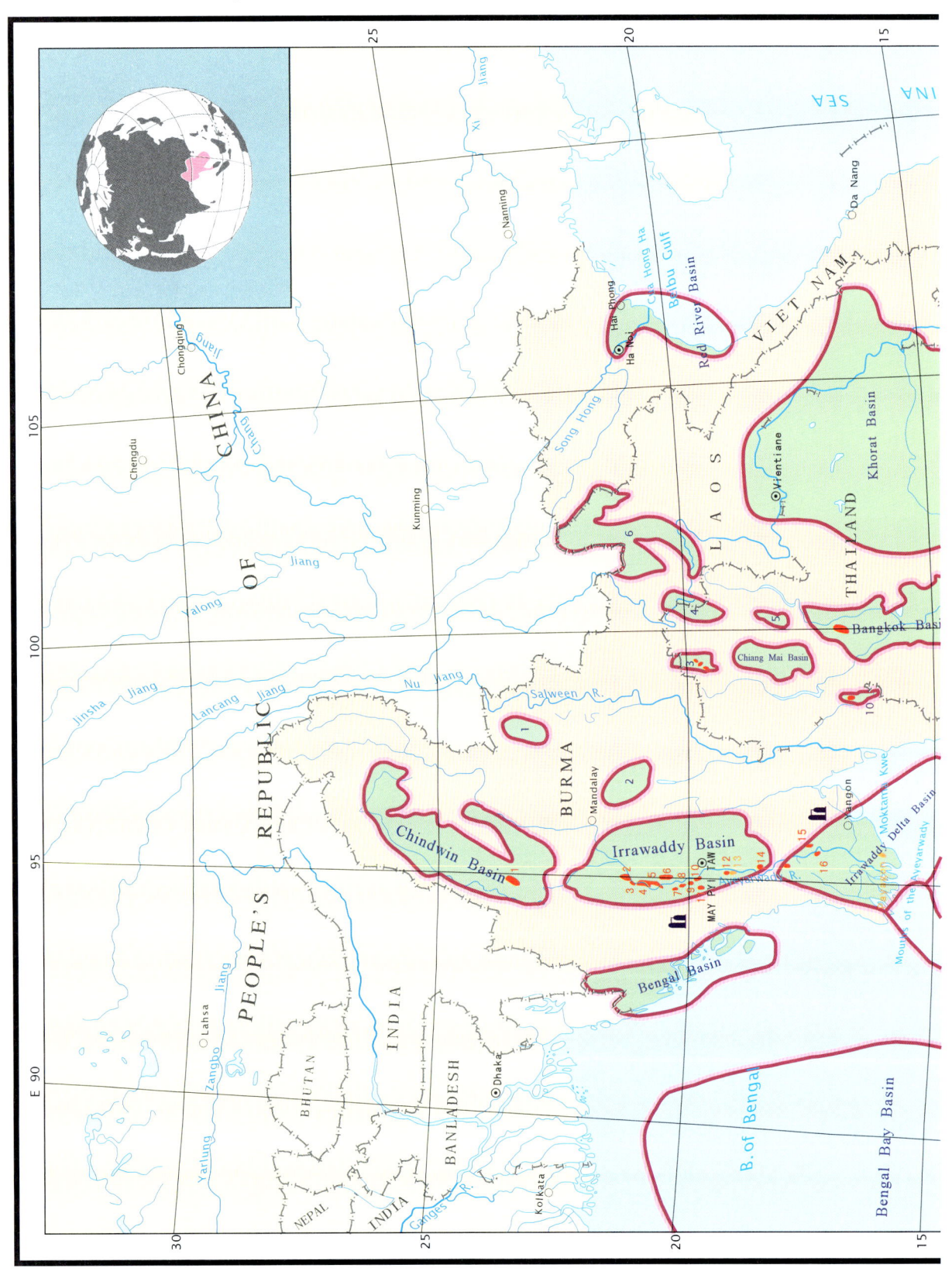

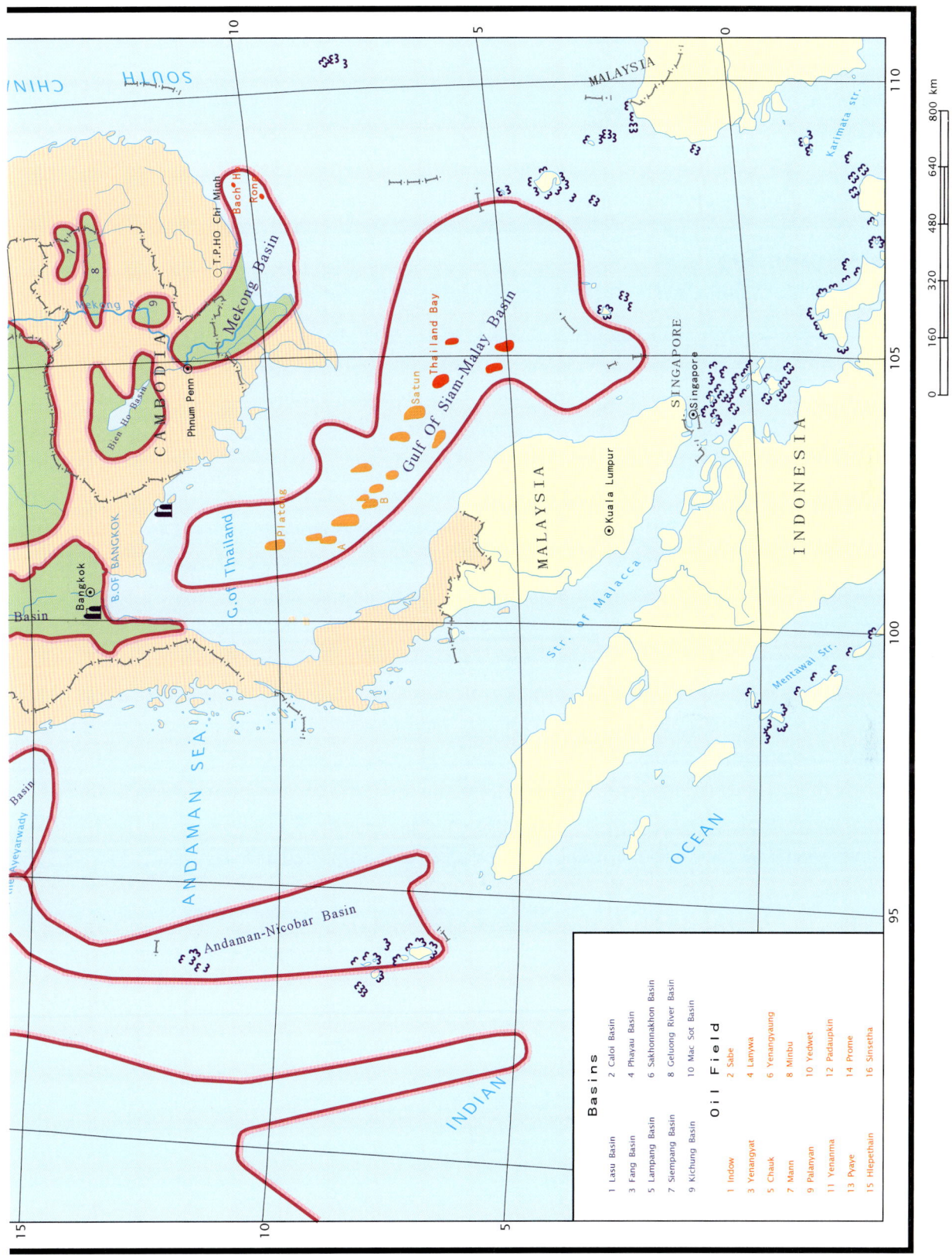

15 Burma (Myanmar), Thailand, Laos, Cambodia and Vietnam

Geography

Burma (Myanmar), Thailand, Laos, Cambodia and Vietnam are neighbouring countries located in Southeast Asia, contiguous with each other.

Regional geology

In practice, all the regions of Southeast Asia represent a fragment of the Eurasian Plate. Western and southern portions of it occupy the area of the plate (land, shelf, continental slope) with continental crust, while eastern parts of the region are located predominantly in the area where oceanic crust is developed (deep-water basins of marginal seas, deep-water trenches and island arcs).

Burma The total area of Burma (Myanmar) is 676,581 km² with a population of 55.4 million. Oil production dates back as far as 1759. Seven basins (Table 15.1) are located within the country, the largest being Andaman-Nicobar with a surface area of 256,000 km². In 2008 oil production totalled 950,000 t and gas production was 13.4 billion m³.

Thailand Thailand's total area is 513,115 km², with a population of 64.76 million. Geological surveys for oil and gas started in 1921. Four basins are major prospects for oil and gas exploration with a total sedimentary cover area of 500,000 km² on land and offshore. Khorat Basin is the largest with an area of 350,000 km². Two gas fields were discovered in this basin in Permian and Carboniferous strata. Oil consumption was 44.3 million t in 2008. Imported crude oil and petroleum products provide domestic energy requirements. Thailand is one of the largest net importers of petroleum among the non-oil less developed countries (LDCs). The most recent records show that gas production is 14.7 billion m³, while domestic consumption is 30.6 billion m³. The discovery of the large gas field 'B' with proven reserves of 200 billion m³ in the Gulf of Thailand represents the best hope of reducing dependence on imported energy.

Table 15.1 Major oil and gas basins of Burma

Basin	Area (thousand km²)	Major sedimentary rock		Reservoir	
		Age	Thickness (m)	Age	Lithology
Andaman-Nicobar	256	Kz	4000	E	Sandstone
Bengak	76	C–Kz	16,000	R	Sandstone
Caloi	13	Mz–Pz			
Chindwin	115	Kz		N	Sandstone
Irrawaddy	112	Kz	12,000	R	Sandstone, limestone
Irrawaddy Delta	140	Kz		N	Sandstone
Lasu	7	Mz–Pz			

World Atlas of Oil And Gas Basins, First Edition. Li Guoyu.
© 2011 John Wiley & Sons, Ltd. Published 2011 by John Wiley & Sons, Ltd.

Table 15.2 Sedimentary basins of Cambodia

Basin	Area (thousand km²)	Age of major sedimentary rock
Bien Ho	51	Pz–Kz
Geluong River	17	Pz–Mz
Kichung	7	Pz–Kz
Siempang	6	Pz–Mz

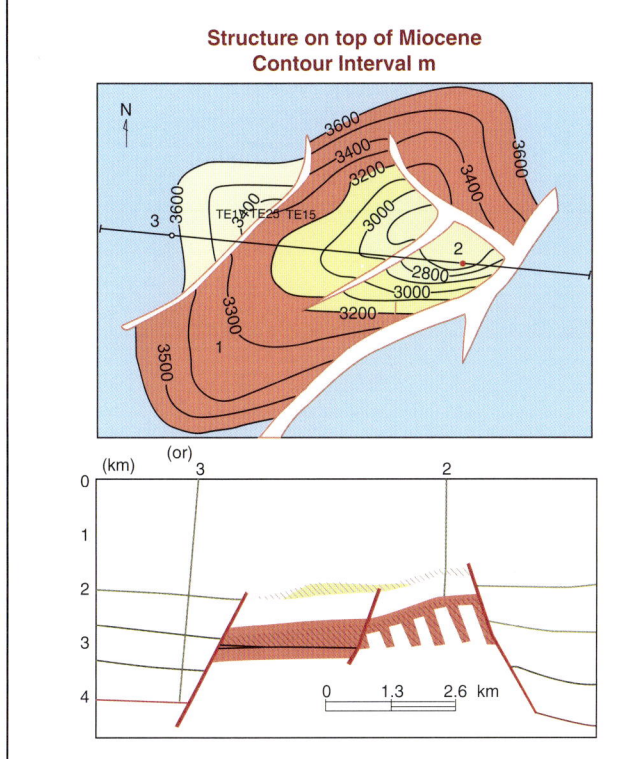

Dai Hung is a giant oil field in Vietnam.

Basic data

1. Name	Dai Hung
2. Discovery year	1974
3. Area (km²)	40
4. Reservoir	Sandstone
5. Reservoir age	Tertiary
6. Depth (m)	—
7. Thickness (m)	—
8. Porosity (%)	—
9. Permeability (mD)	—
10. Recoverable reserves (million t)	1.5
11. Water depth (m)	115

Fig. 15.1 Details of the Dai Hung giant oil field.

Laos The total area of Laos is 236,800 km² with a population of 5.84 million. Oil and gas production is virtually non-existent. Oil consumption was 150,000 t in 2004.

Cambodia Cambodia's total area is 181,035 km² with a population of 13.66 million. Cambodia is a non-oil-producing country. This country has four basins. Exploration for oil and gas is proceeding slowly. As at the present time, no important discovery has been made. Annual oil consumption is 190,000 t, met entirely by imports.

Vietnam The total area of Vietnam is 329,556 km² with a population of 84.11 million. Oil exploration began in the late 1960s. The first successful offshore oil well was finished in September 1974. After this, the petroleum industry of Vietnam started to develop very rapidly. Two large basins as well as other relatively smaller basins are located on land and on the continental shelf. Mekong Basin is a large oil-producing basin with a total area of 100,000 km², in which oil and gas accumulations are associated with Tertiary strata. Big Bear, White Tiger and Dragon are major oil and gas fields in the country. Details of the giant Dai Hung giant oil field are shown in Fig. 15.1. In 2008 oil production was 18.7 million t, gas production was 4.3 billion m³ and oil consumption was 11.5 million t.

Generally, the five countries are thought to have good prospects for oil and gas exploration.

MALAYSIA, SINGAPORE AND BRUNEI

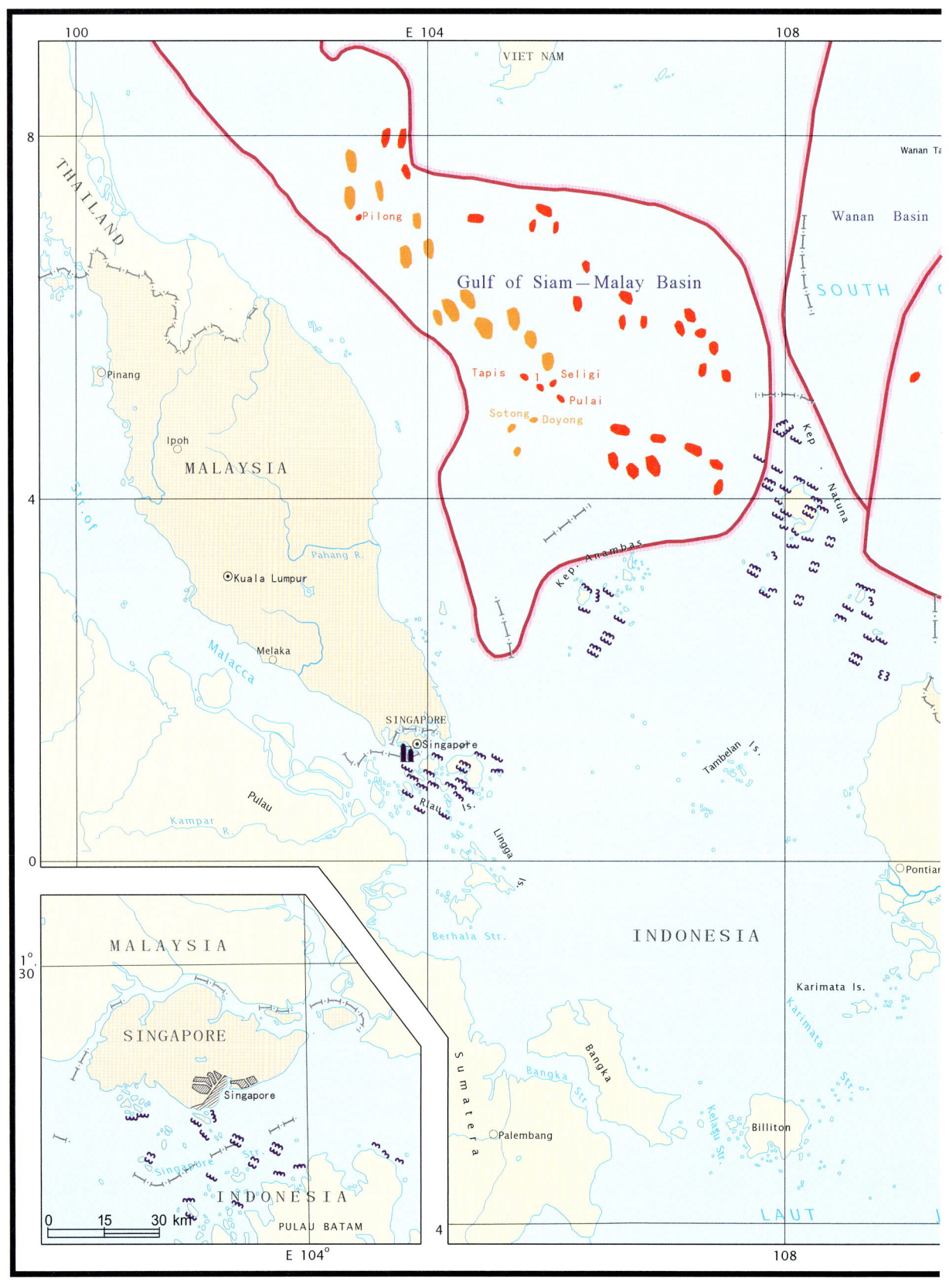

Oil Field: 1 Bekori 2 Seria 3 Southwest Ampa 4 West Lutong 5 Miri

CHAPTER 16

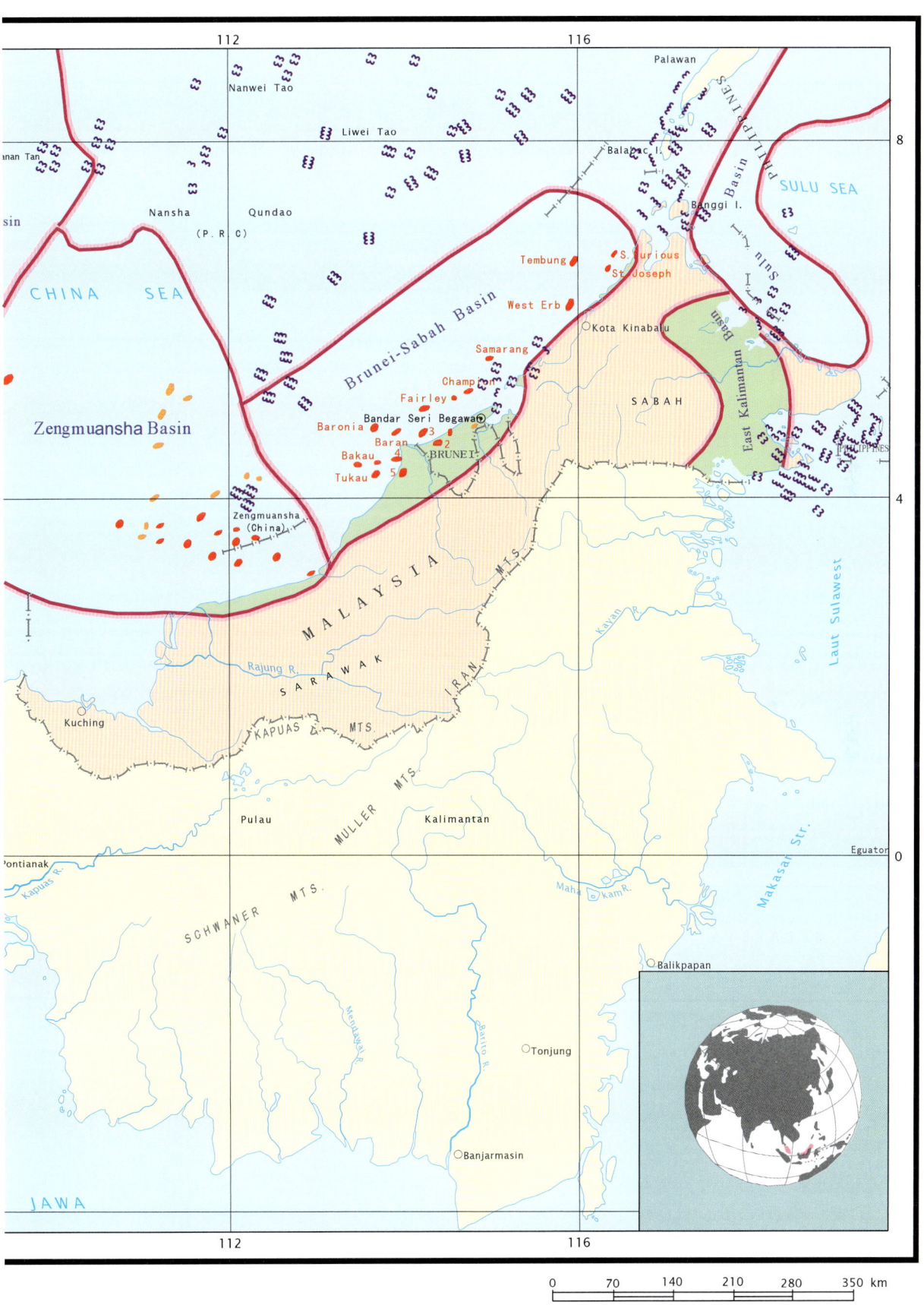

16 Malaysia, Singapore and Brunei

Geography

The three countries are located in Southeast Asia, in the southern portion of the South China Sea.

Regional geology

The Sunda plate and Sunda syncline are the major tectonic features of the region. The basement consists of Palaeozoic granite. Sedimentary cover is made up of Jurassic, Cretaceous and Tertiary rocks. Most basins are located within Malaysia on land and offshore.

Malaysia Malaysia is divided by the South China Sea into eastern and western parts. It has a total area of 330,257 km^2 and a population of 26.64 million. Oil was first discovered in Malaysia in 1866, but the petroleum industry of the country developed rapidly in the 1970s. Several large and medium oil and gas fields have been discovered. Because of continuously improving figures of proven reserves, oil and gas production has been kept stable from 1984 to 2008. Six basins are located in the country, three of which are major oil producing ones. A typical Malaysian oil field is shown in Fig. 16.1. In the Gulf of Siam-Malay Basin, the thickness of Tertiary Formation reaches 10 km. Oil-bearing strata are sandstone. Traps are related with fault anticlines.

Oil production increased significantly from 1970 to 2008. Malaysia possesses huge reserves of natural gas, which will play an increasingly important role in the country's economic development. There is little doubt that Malaysia will become one of the world's largest producers of liquefied natural gas. Reserves of natural gas are estimated as being the equivalent of twice the official estimate of oil reserves. Domestic consumption of energy is growing steadily and will eventually take up most if not all of total domestic production of oil.

In 2007 the proven reserves in oil were 547 million t, while proven gas reserves were 2.75 trillion m^3. Oil production was 37.75 million t and gas production was 48.4 billion m^3. Domestic oil consumption was 22.95 million t, while gas was 40.3 billion m^3.

Singapore Singapore is an island of 699 km^2 with a population of 4.88 million. It is located at one of the crossroads of the world and is considered a major air and sea hub of Asia. Due to its limited area, which constrains oil and gas exploration, it is essentially a non-oil-producing country. Because of its location on the Straits of Malacca it is also becoming a major petroleum transportation centre. Demand for oil and gas is very high because of Singapore's rapid pace of economic development. Oil consumption reached 44.04 million t recently, and consumption of gas amounted to 6.6 billion m^3. The principal exports of the country are petroleum products. Despite the lack of domestic oil and gas production, Singapore has leveraged on its strengths in infrastructure, economic development, key geographical location and excellent educational facilities to become one of the major petroleum centres of the world.

Brunei Brunei lies on the northwest coast of Borneo island, facing the South China Sea. The total area is 5,765 km^2 and the population is 370,000. The Brunei–Sabah Basin is a large and

World Atlas of Oil And Gas Basins, First Edition. Li Guoyu.
© 2011 John Wiley & Sons, Ltd. Published 2011 by John Wiley & Sons, Ltd.

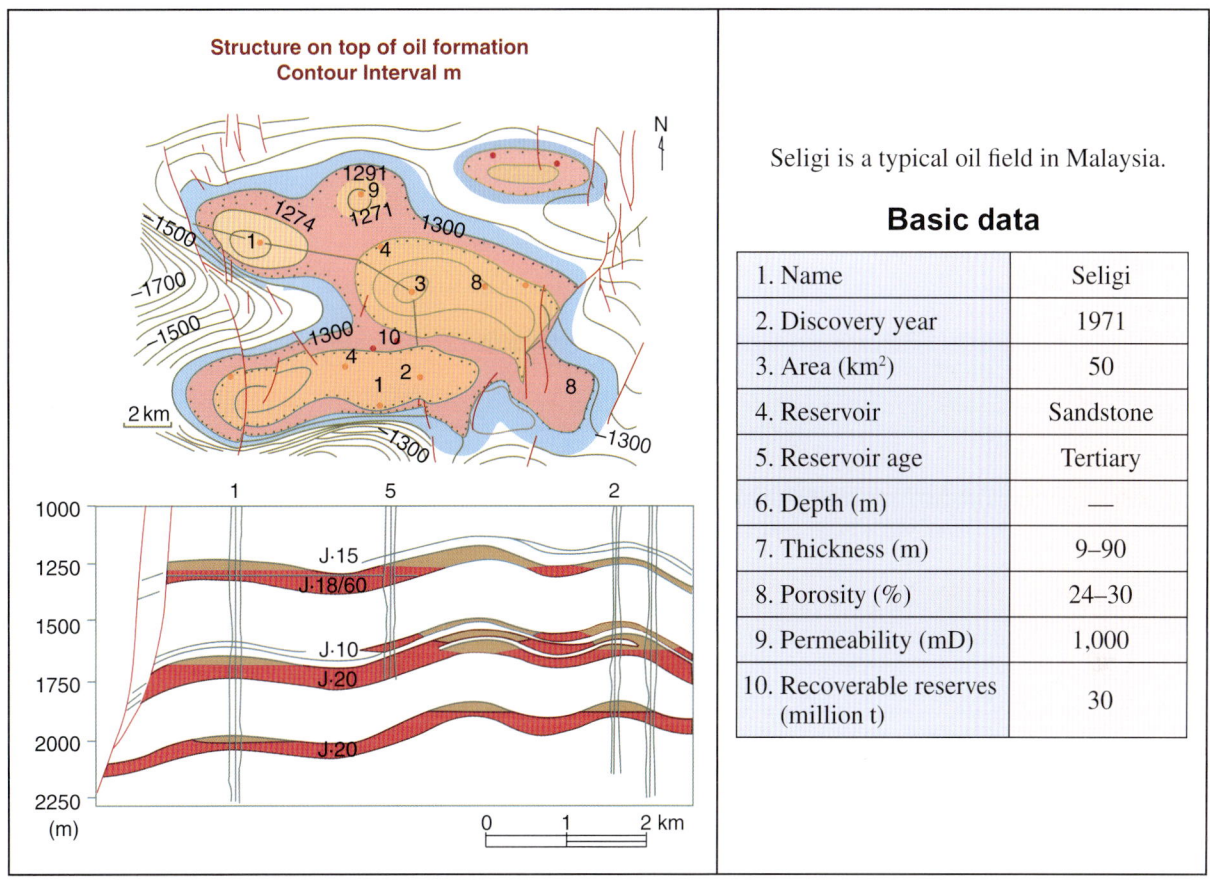

Fig. 16.1 Details of the Seligi oil field.

highly prolific petroleum basin, and is the reason why Brunei has become a major oil and gas producing country. Brunei is the fourth largest liquefied natural gas (LNG) producer of the world and it has excellent long-term prospects for further development in the natural gas and LNG sectors. Production of natural gas rose steadily from the 1970s to 2008. In 2008 proven oil reserves were 150 million t and proven gas reserves were 390 billion m^3. Oil production was 7.8 million t and gas production was 11.9 billion m^3. It is a major oil and gas exporting country.

INDONESIA AND EAST TIMOR

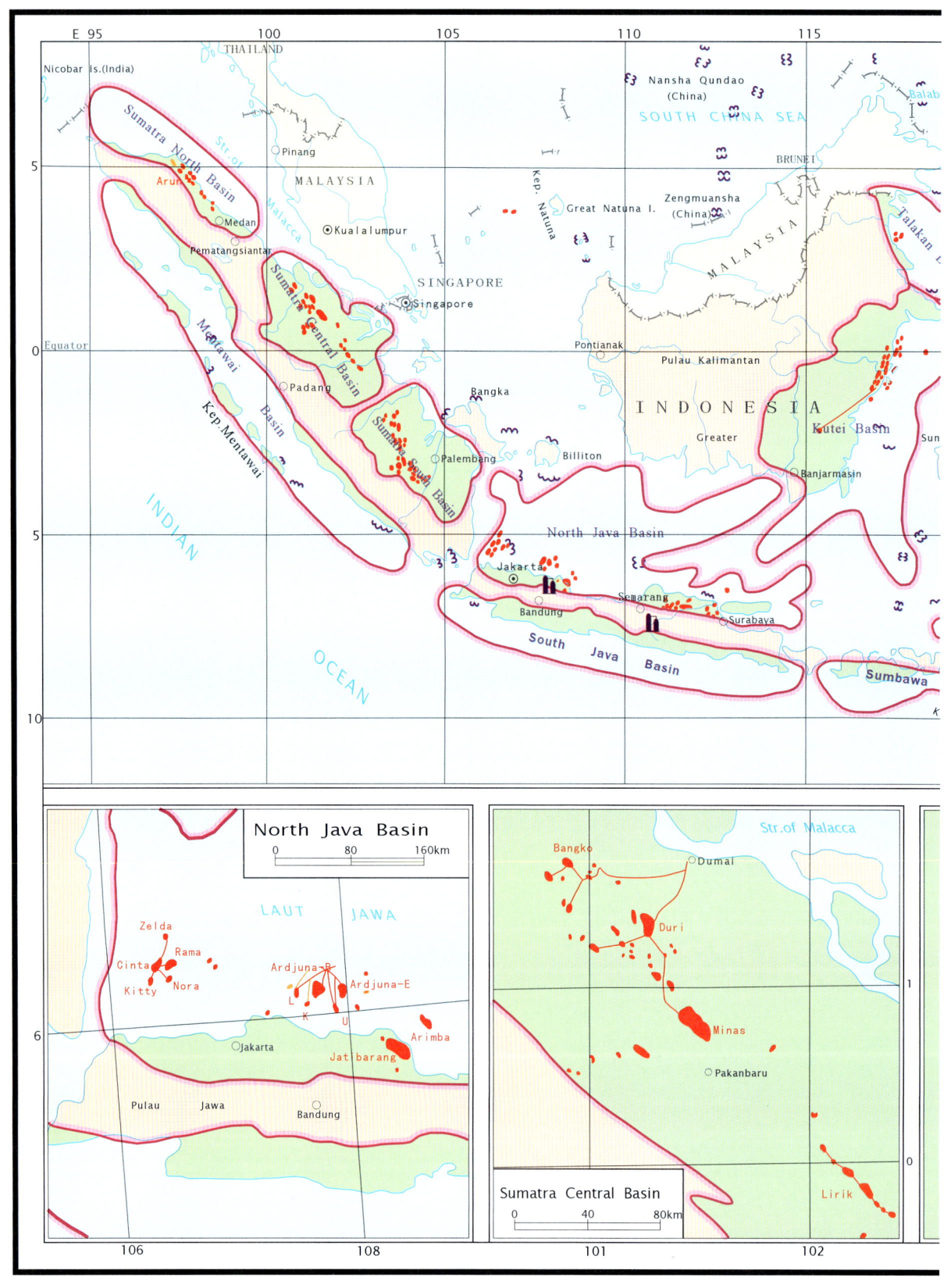

CHAPTER 17

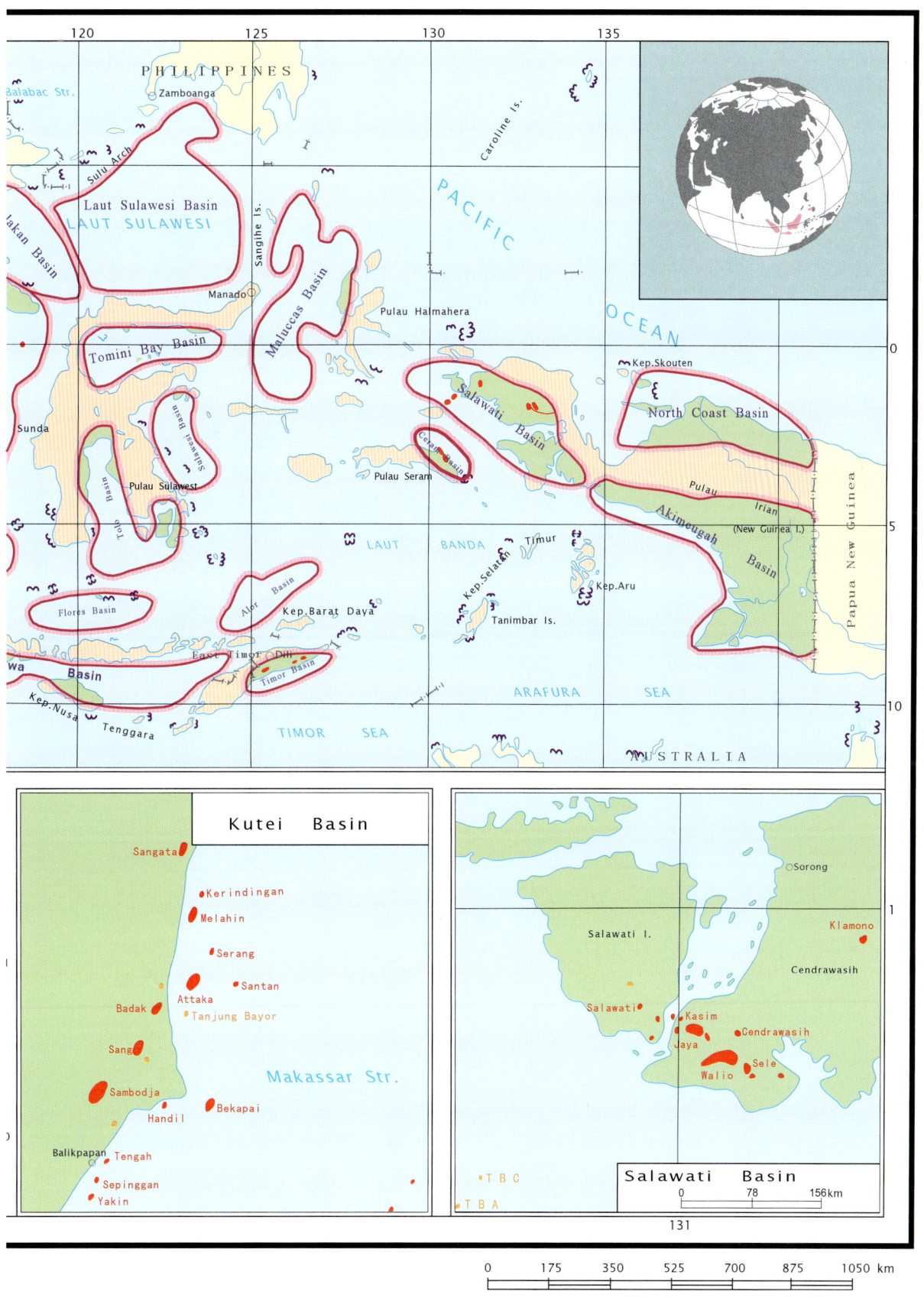

17 Indonesia and East Timor

Indonesia
Geography

The archipelagos of Indonesia comprise some 13,667 islands. The total area is 1.9 million km^2 with a population of 210 million.

History

Indonesia has a long history of oil exploration and production began in the 8th century AD. Modern oil and gas exploration commenced in earnest in 1993 after prominent proven reserves were discovered. Peak oil production reached 82.94 million t in 1977. During 1982–2001 annual oil production was maintained at 60 million t. It dropped to 40 million t thereafter. Gas production peaked at 74.1 billion m^3 in 2000. By the end of 2008 oil reserves were 546 million t, gas reserves were 3 trillion m^3, oil production was 42 million t and gas production was 77.2 billion m^3.

Although oil reserves and production in Indonesia have been declining, the gas reserves are plentiful and the country's output of liquefied natural gas (LNG) grew rapidly in the 1970s. Output of LNG increased from 312 trillion cubic feet in 1976 to 998 trillion cubic feet in 1979. Expansion plans for LNG exports are very ambitious, with the most recent being based on the giant Tangguh gas field. The LNG export destinations for this field include China whose CNOOC is an investor in the project.

Growth in domestic consumption of petroleum products, which is heavily subsidized, will progressively reduce Indonesia's ability to export.

Regional geology

Indonesia is located at the triple junction of the Euro-Asia, Indo-Australia and Pacific Plates. Owing to long-term orogeny and complex subsidence history many sedimentary basins developed. The peri-continental basins are characterized by continental crust, increased heat flow and consolidated Palaeozoic–Mesozoic basement with wide development of granite intrusions. The sedimentary cover succession includes continental, coastal-marine and shallow-water marine facies. The high values of geothermal gradient in the mostly oil producing basins confine hydrocarbon accumulations largely to Neocene and Paleocene sequences. Petroleum potential is estimated to be very high in these distinctive basins as exploration indicates unique or very large concentrations of hydrocarbon reserves.

The Tertiary sedimentary cover provides the source rocks, reservoir rocks and traps. About 60 basins are confirmed in the country. Most important are Sumatra, Java, Laut Sulawesi, Kutei, North Coast and Sumbawa. basins. In the Central Sumatra oil- and gas-bearing basin the unique Minas oil field has been discovered with original proven reserves of 993 million t. Significant reserves are contained in some other fields such as Bekasap, Duri and Bangkok, where oil reserves are estimated ay 75 million t, 65 million t and 65 million t, respectively. In the South Sumatra petroleum basin there are two large oil fields: Talang-Akar with original proven oil reserves of 82.4 million t; Talang-Jimar with 46 million t. In the North Java petroleum basin the Arjuna oil field with original

World Atlas of Oil And Gas Basins, First Edition. Li Guoyu.
© 2011 John Wiley & Sons, Ltd. Published 2011 by John Wiley & Sons, Ltd.

Table 17.1 Major oil and gas basins in Indonesia

Basin	Area (thousand km²)	Major sedimentary rock		Reservoir	
		Age	Thickness (m)	Age	Lithology
Akimeugah	214	Kz			
Alor	36				
Ceram	18	R	3000	N_2	Sandstone
Flores	45				
Kutei	390	R	11,000	N	Sandstone
Laut Sulawesi	240	R			
Mentawai	240	R	N_1	Reef	
Maluccas	120				
North Coast	125	R		N_2	Sandstone
North Java	270	R	6000	N	Reef, sandstone, volcanic rock
Sulawasi	120	R	6000	N_1	Reef, clastic rock
South Java	180	R	7600		
Sumatra Central	120	R,Q	5000	N_1	Sandstone
Sumatra North	100	R,Q	5000	R	Sandstone, reef
Sumatra South	120	R,Q	5000	E_3–N_2	Sandstone, carbonate rock
Sumbawa	120				
Talakan	76	Kz	9000	N_2	Sandstone, conglomerate
Tolo	90				
Tomini Bay	60				

reserves of 122 million t and the Pagerangan oil field with 84 million t are located. Future prospects for undiscovered resources of oil and gas have been estimated as very high.

East Timor Three oil fields with Triassic oil-producing rocks have been discovered in this mountainous country in Southeast Asia, bordering Indonesia to the west. It is located in the East Timor basin which has an area of 3000 km². The total area of East Timor is 14,874 km² and the population is 97,000.

SUMATRA (SUMATERA) CENTRAL BASIN

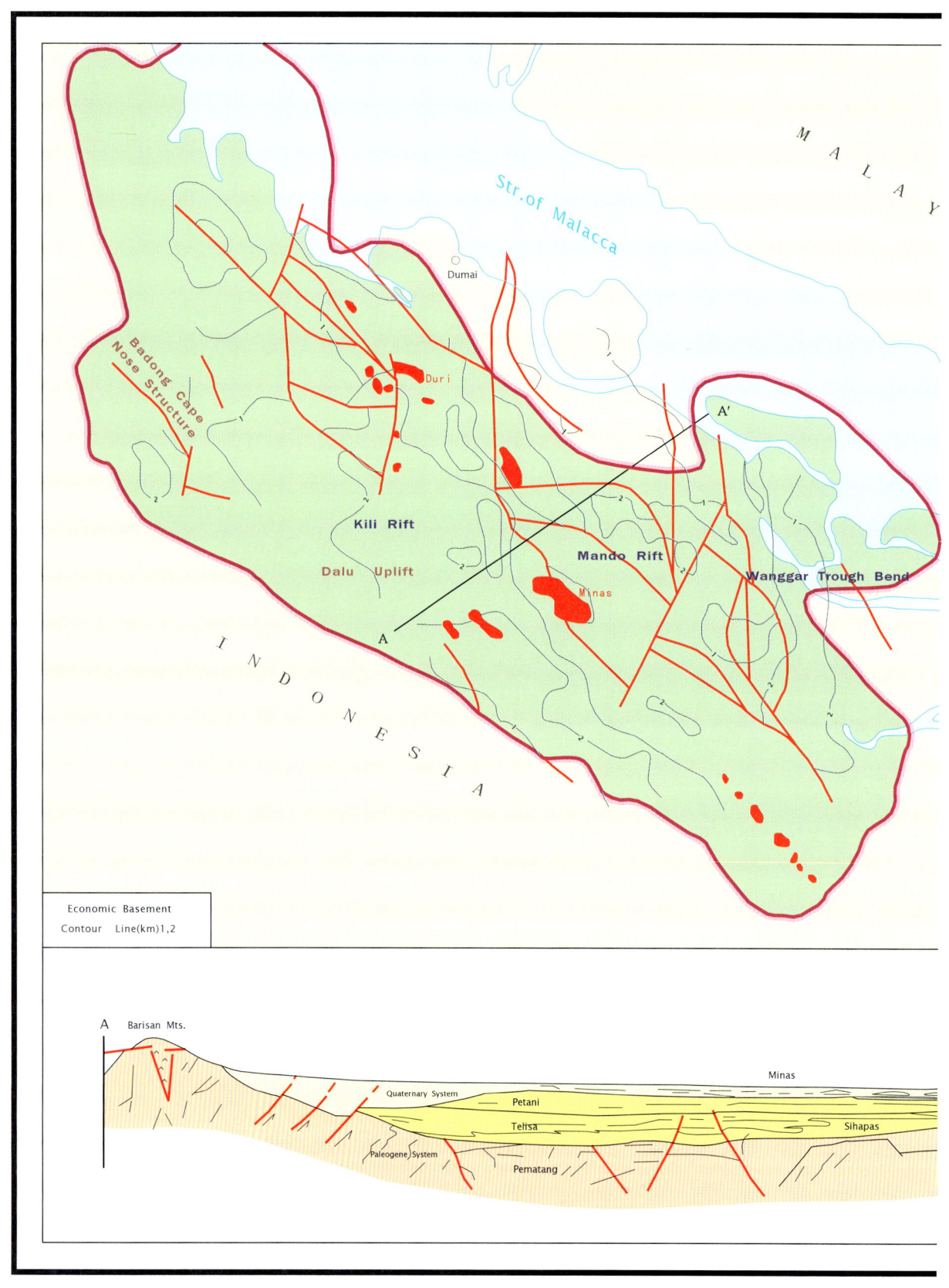

CHAPTER 18

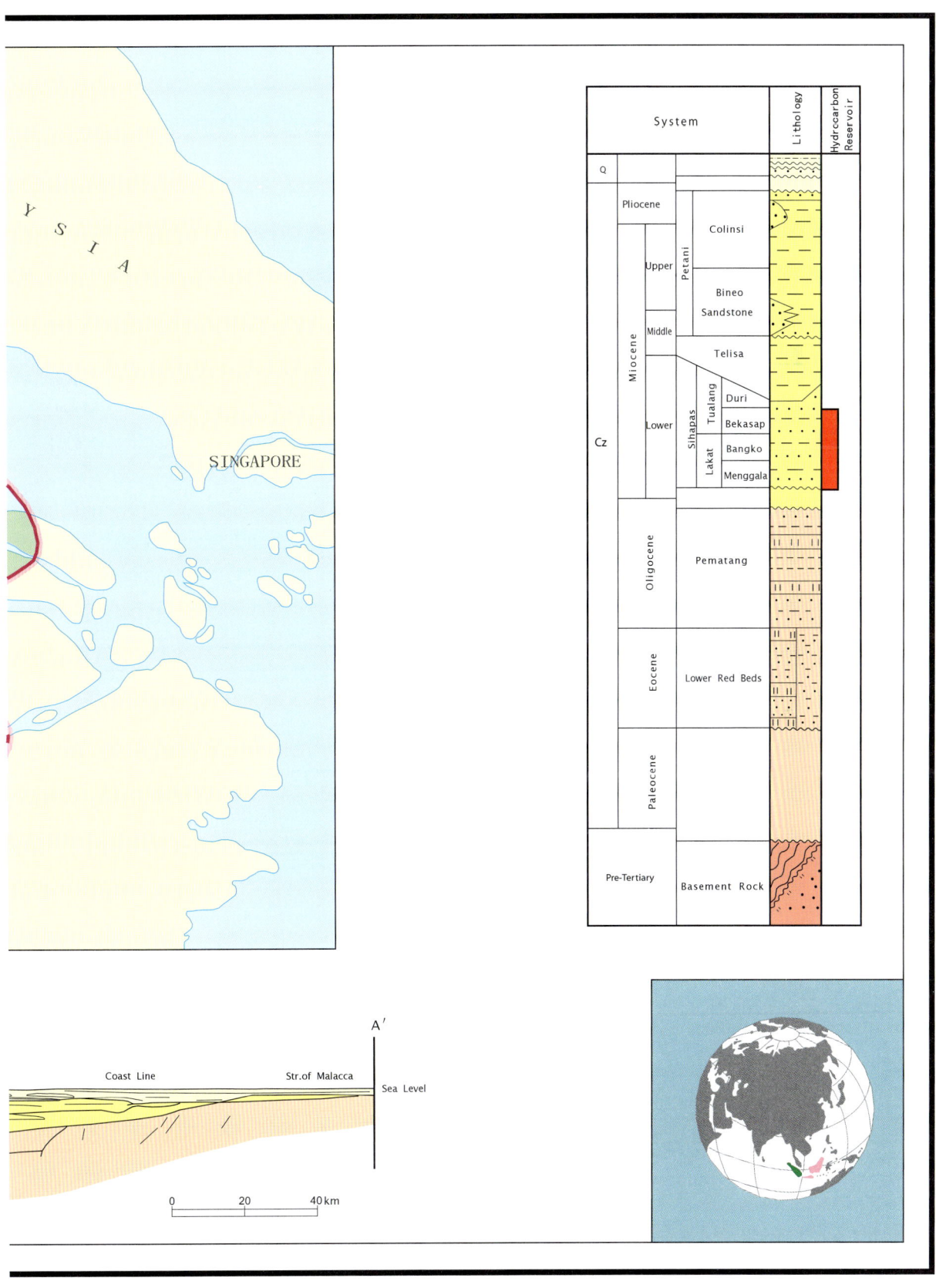

18 Sumatra (Sumatera) Central Basin

Geography

The Sumatra Central Basin is located in the southeastern portion of the Barisan Mountains in central Sumatra. The total area of the oil basin is 113,400 km², 76 per cent of which is located on land with the remainder offshore.

History

The Sumatra Central Basin is a very important oil producing area in the country. The first oil field, Lirik, was discovered in 1939. The giant Minas oil field was discovered in 1944, and together with the Duri oilfield (Table 18.1) accounts for 60 per cent of the total production of the country as far back as 1973. There are 40 fields discovered so far in the Sumatra Central Basin.

Regional geology

Sedimentary cover comprises Tertiary and Quaternary strata with thicknesses up to 5000 m. Tertiary source rocks have been identified and the sandstones of Sihapas Group form a particularly noteworthy reservoir. Traps are formed by anticlines and faulted anticlines. Many factors contribute to the accumulation of oil and gas in the Sumatra Central Basin : (i) fine-grained source rocks with high of organic matter content; (ii) the reservoir potential of the Sihapas Group with high porosity (27 per cent) and permeability (1500 mD); (iii) the Telisa and Petani Shales are highly effective sealing rocks; (iv) high temperature gradient (44.9°C/km); (v) free migration path; (vi) rifting structure.

The tectonic history of the Central Sumatra Basin can be divided into four periods. The first was rifting of basement with the rift filled by early Permian to early Cretaceous deposits. The second was extensional rifting during Eocene to Oligocene times. The third was post-rift subsidence. The fourth was localized structures formed by extensive Miocene to recent faulting.

Table 18.1 Major oil and gas fields of Indonesia

Name	Discovery year	Recoverable reserves		Basin	Depth (m)	Trap	Reservoir		Remarks
		Oil (million t)	Gas (billion m³)				Age	Lithology	
Minas	1944	560.0	50	Sumatra Central	700	Anticline	N	Sandstone	
Duri	1941	420.0	—	Sumatra Central	200	Anticline	E	Sandstone	
Handil	1974	112.0	—	Kutei	1900	Anticline	N	Sandstone	
Ardjuna B	1969	84.0	1.7	North Java	800	Anticline	N	Sandstone	
Attaka	1970	77.0	—	Kutei	2200	Anticline	N	Sandstone	
Arun	1971	—	38	Sumatra North	3000	Anticline	N	Limestone	Gas field
Badak	1972	—	19	Kutei	1300	Anticline	N	Sandstone	Gas field

World Atlas of Oil And Gas Basins, First Edition. Li Guoyu.
© 2011 John Wiley & Sons, Ltd. Published 2011 by John Wiley & Sons, Ltd.

Sumatra (Sumatera) Central Basin

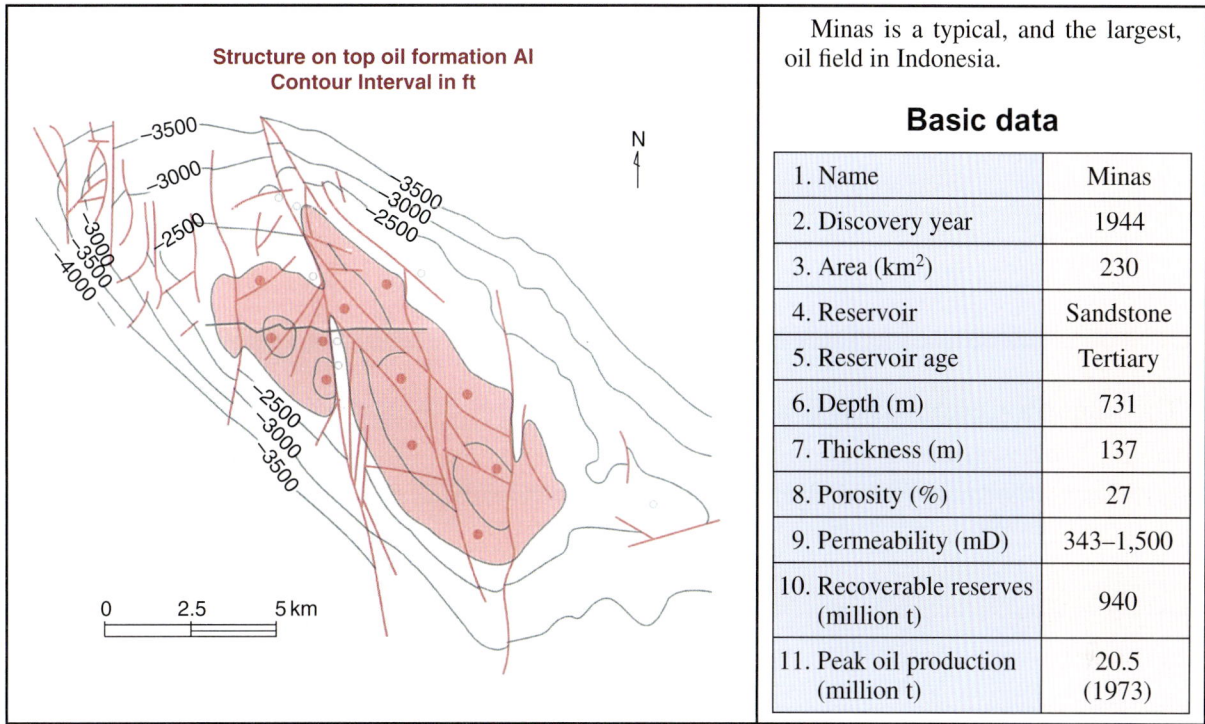

Fig. 18.1 Details of the Minas oil field.

Figure 18.1 provides details of the largest oil field Minas. This oil field is typical of the Central Sumatra Basin and was discovered in 1944, with an area of 23,240 km². The geological reserves were 940 million t, the recovery ratio was 57 per cent, the recoverable reserves were 560 million t, and the depth of oil sandstones was 731 m. There are five oil layers in the oil field, with an effective efficient thickness of 137 m, porosity of 27 per cent, permeability of 434–1,500 mD and pressure 6.54 MPa.

PHILIPPINES

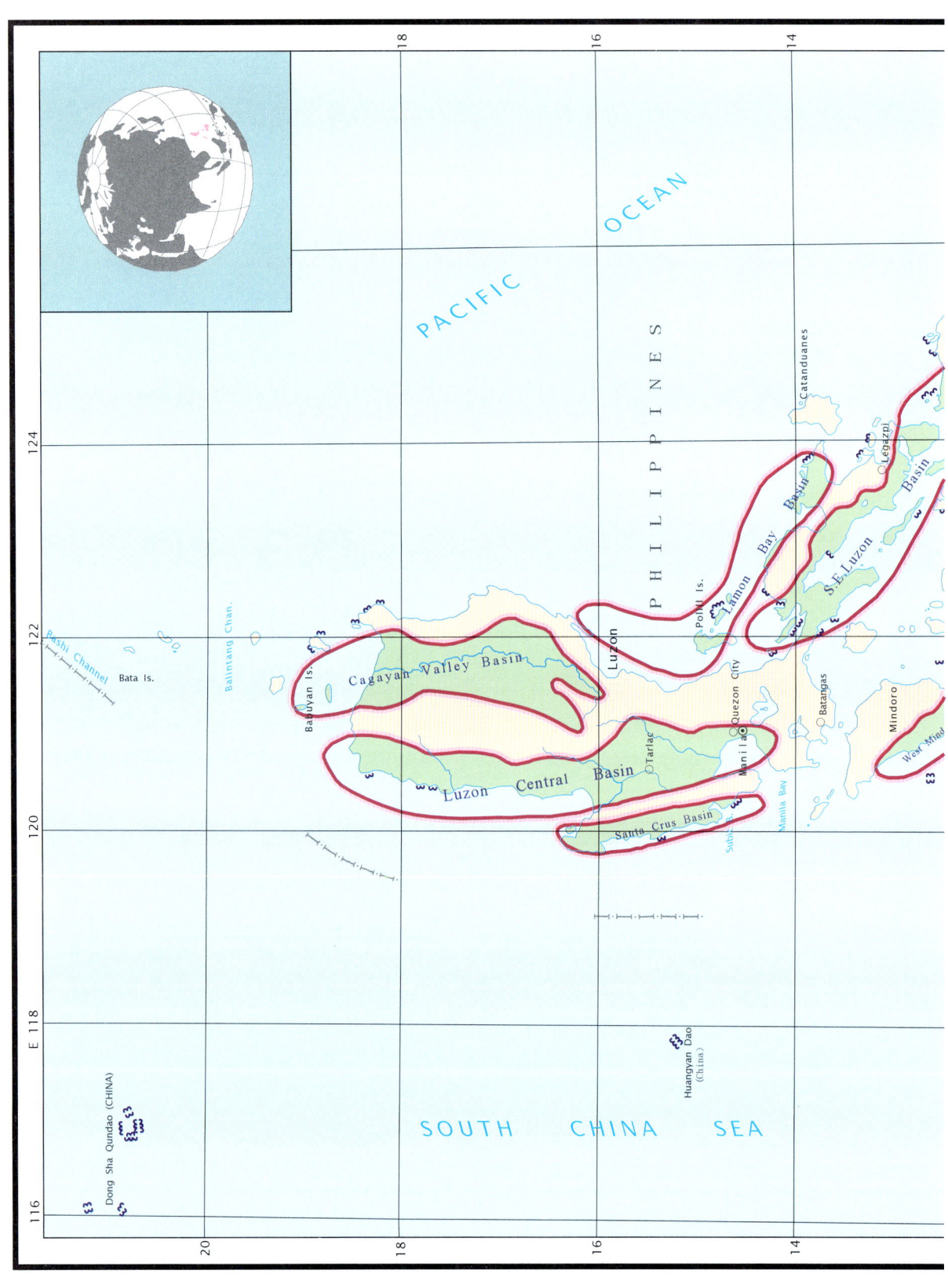

CHAPTER 19

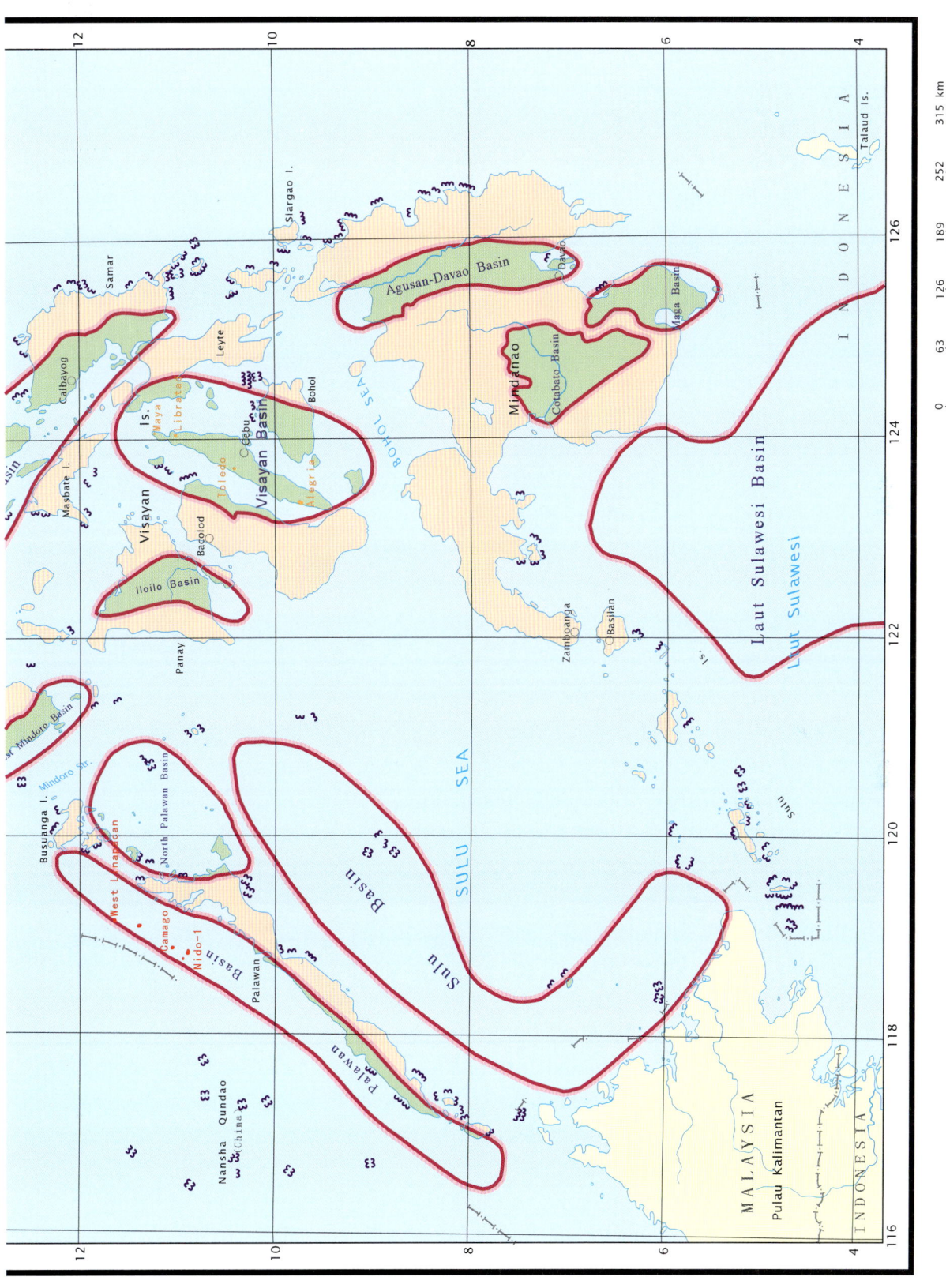

19　Philippines

Geography

The Philippines are located in Southeast Asia. They form an archipelago between the Philippine Sea and the South China Sea, comprising mostly mountainous islands with narrow and extensive coastal lowlands. The Philippines are one of the largest archipelagos in the world with more than 7000 islands. The total area is 299,000 km^2 and the population is 84.87 million.

History

Oil exploration began in 1896. During 1962–1972, 112 exploration wells were drilled in Luzon. In 1977, the Nido A oil field was discovered on the continental shelf in the Palawan Basin. This is commonly referred to as the modern-day beginning of the petroleum industry of the Philippines. In 1980, 93 per cent of energy requirements were met from petroleum, 80 per cent of which had to be imported. Petroleum had thus become an increasingly costly proportion of the country's import bill over the 1970s. Commercial petroleum production began offshore in 1979, with some 1500 t per day. By 1980 production had risen to 5500 t per day, but prospects of increasing this output or of new discoveries were uncertain. In 2007, oil was still imported to meet domestic demand. In 2008 proven oil reserves were 18.97 million t and proven gas reserves were 98.5 billion m^3. Oil production was 790,000 t, while oil consumption was 13.4 million t.

Regional geology

The Philippine Islands form part of the Pacific Orogenic Belt on the Euro-Asia continental plate. The sedimentary cover comprises Tertiary deposits with thicknesses ranging from 3000 to 8000 m that have undergone intensive Ceno-Mesozoic magmatic intrusion and deformation. There is frequent volcanic activity even at the present time.

There are 14 major oil- and gas-bearing sedimentary basins located in the Philippines (Table 19.1). Nido A (Fig. 19.1) is a typical oil field in the Philippines. Basins of this type are located in the zones where island arcs and associated deep-water depressions and trenches have developed.

Fore-arc settings provide a subtype comprising 10 sedimentary basins (3 petroleum-bearing and 7 possibly petroleum-bearing). They are confined to oceanic portions of island arcs and are located between volcanic arcs and the outer side of deep-water trenches. They correspond to early stages of evolution of petroleum basins. Fore-arc basins are characterized by the following features: predominant development of subcontinental crust and presence of suboceanic crust; confinement to continental (island) slope and deep waters; prevailing role of Cenozoic strata in the basement composition; decreased values of heat flow; essentially Cenozoic sedimentary cover formed principally under marine conditions; development of volcanic varieties; presence of turbidites; predominant gas-proneness; confinement of gas accumulations to island-faced side of basins.

Inter-arc settings provide another subtype of sedimentary basin that is widespread in the region. There are 12 basins of this type (6 petroleum-bearing and 6 possibly petroleum-bearing). Uplifts terminate the basins. Original petroleum potential is commonly classed as medium and sometimes estimated as high, but the sheer number of basins suggests that the prospects for oil and gas exploration are good.

World Atlas of Oil And Gas Basins, First Edition. Li Guoyu.
© 2011 John Wiley & Sons, Ltd. Published 2011 by John Wiley & Sons, Ltd.

Table 19.1 Major oil and gas basins of the Philippines

Basin	Area (thousand km²)	Major sedimentary rock		Reservoir	
		Age	Thickness (m)	Age	Lithology
Agusan-Davao	15	Kz		N	Reef
Cagayan Valley	28	Kz, Q	8000	R	Terrigenous deposit, carbonate rock
Cotabato	11	Kz			
Iloilo	8	Kz	5500		
Lamon Bay	15				
Luzon Centre	31	Kz	7500		
Maga	8				
North Palawan	19				
Palawan	36	Kz		R	Sandstone, carbonate rock
Santa Crus	8				
Southeast Luzon	35	Kz, Q	4000	R	Sandstone, mudstone
Sulu	67	R			
Visayan	33	Kz	7750	N_1	Sandstone, reef
West Mindoro	8	Kz	3000	E_2	Sandstone

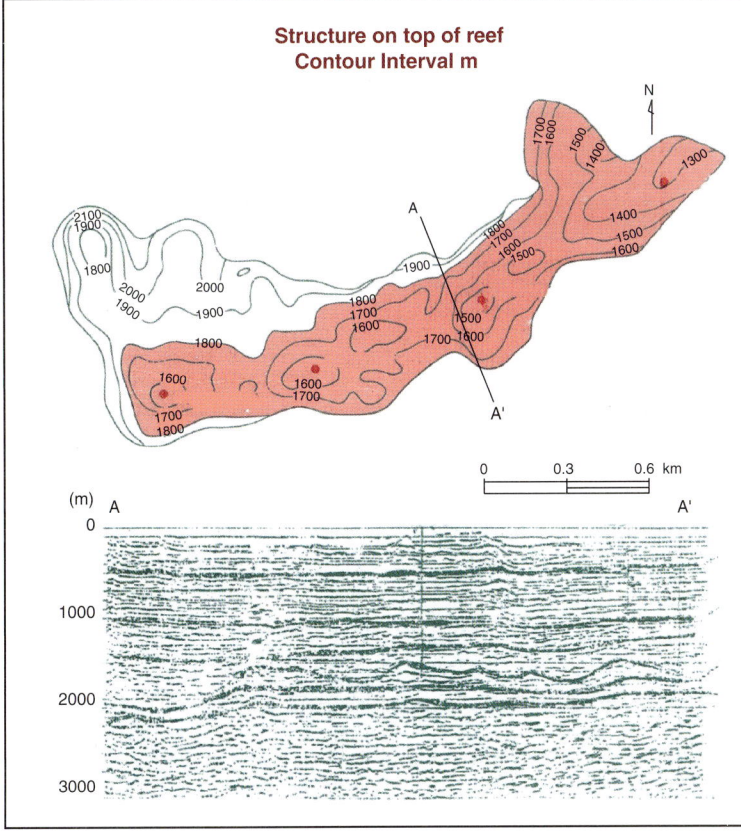

Nido A is a small and typical oil field in Philippines.

Basic data

1. Name	Nido A
2. Discovery year	1976
3. Area (km²)	2.5
4. Reservoir	Reef
5. Reservoir age	Tertiary
6. Depth (m)	2,071
7. Thickness (m)	52
8. Porosity (%)	5.9
9. Permeability (mD)	0.1–170
10. Recoverable reserves (million t)	2.1
11. Peak oil production (million t)	1.8
12. Water depth (m)	3

Fig. 19.1 Details of the Nido A oil field.

PAKISTAN, AFGHANISTAN, NEPAL, BHUTAN AND

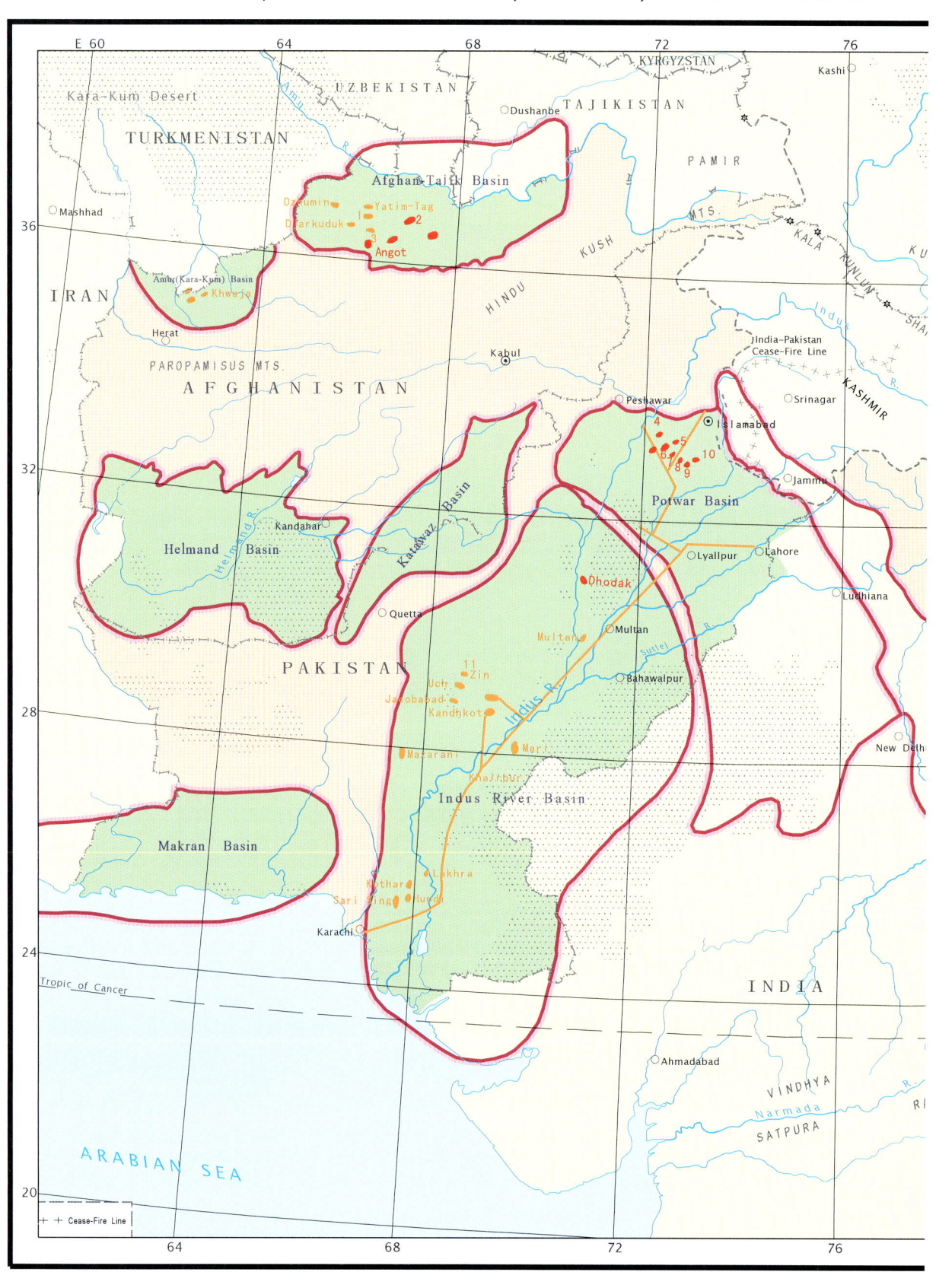

BANGLADESH CHAPTER 20

20 Pakistan, Afghanistan, Nepal, Bhutan and Bangladesh

Geography

These five countries are located in South Asia, being bounded by the Hindu Kush, the Himalayas and the Kunlun mountain belt, and bordered by the Arabian Sea and the Bay of Bengal to the south.

History

Oil and gas exploration began in the mid-19th century in Pakistan, but in the other countries it began much later.

Regional geology

Two basic geological factors have affected the development of this area: first is the Himalayan Orogenic Belt, and second is the India platform. Multiple plate-tectonic movements have led to the sedimentary cover being composed of Palaeozoic to Tertiary rocks. In addition, the intensive folding associated with the Himalayan Orogenic Belt has led to many faults, resulting in difficult conditions for oil exploration. Precambrian crystalline rocks form the basement. Most of the sedimentary basins are represented by syneclises, formed in basement and Meso-Cenozoic strata. The sedimentary cover comprises Mesozoic and Cenozoic strata. Source rocks have been identified in shale sediments and reservoir rocks are associated with mostly Mesozoic sandy deposits, and also with Palaeozoic marine-carbonate to terrigenous rocks. Caprock is formed by shales and salt. Traps are represented by structural and non-anticlinal types. Oil and gas accumulations are multilayered and mainly confined to Meso-Cenozoic deposits. Petroleum potential of the relatively large basins is estimated as high.

In 2007 oil production occurred only in Pakistan and Bangladesh. Domestic energy requirements for this region will continue to be met by oil imports.

Pakistan Pakistan has a total area is 796,095 km^2 with a population of 150.4 million. Oil seepages in some areas of Pakistan were well-known in the 1860s but exploration began only in 1893 when the first well was drilled in Sindh. There are three major basins in Pakistan (Table 20.1), with a total area of 800,000 km^2, including 600,000 km^2 on land and 200,000 km^2 offshore.

Table 20.1 Major oil and gas basins of Pakistan

Basin	Area (thousand km^2)	Major sedimentary rock		Reservoir		Remarks
		Age	Thickness (m)	Age	Lithology	
Indus River	360	Mz–Kz	6000	E	Carbonate rock	Extends into India
Makran	100	Kz				Extends into Iran
Potwar	110	Pz–Kz	6000	E	Sandstone, limestone	Extends into India

World Atlas of Oil And Gas Basins, First Edition. Li Guoyu.
© 2011 John Wiley & Sons, Ltd. Published 2011 by John Wiley & Sons, Ltd.

Table 20.2 Major sedimentary basins of Afghanistan

Basin	Area (thousand km²)	Major sedimentary rock		Reservoir		Remarks
		Age	Thickness (m)	Age	Lithology	
Afghan-Tadzhik	100	J–Kz	8000–12,000	K, R	Carbonate rock	Extends into Tajikistan and Uzbekistan
Helmand	120	Kz				Extends into Iran
Katawaz	60	R				Extends into Pakistan

Table 20.3 Major oil and gas fields of Bangladesh

Name	Discovery year	Reservoir	Depth (m)	Oil ultimate recoverable reserves (proven + control reserves) (million t of oil equivalent)	Gas production in 1995 (billion m³)
Bakhrabad	1996	Sandstone	1862	19	0.14
Titas	1962	Sandstone	2617	48	2,790.27
Habiganj	1963	Sandstone	1399	43	0.16
Rashidpur	1960	Sandstone	1387	21	0.07

The Indus River Basin has an area of 360,000 km², the Makran Basin 100,000 km² and Potwar 110,000 km². By the end of 2007 proven oil reserves were 39.61 million t, with annual oil production being 3.45 million t. At the same time proven gas reserves were 792.8 billion m³ and annual gas production was 40 billion m³.

Afghanistan   Afghanistan is completely landlocked, bordered by Iran to the west, with which it shares a 925 km-long border. It is mostly composed of rugged mountains and plains in the north and southwest. Afghanistan has an area of 647,500 km² with a population 28.5 million. As mentioned above there are very complex geological conditions, making oil and gas exploration very difficult. Details of major sedimentary basins with portions in Afghanistan are provided in Table 20.2. Oil exploration began in 1960 when one small oil field was discovered, with oil production more than 9000 t and gas 290 million m³. It is classified as a non-oil-producing country, but with good prospects for the future.

Nepal   Nepal has an area of 144,181 km² with a population of 24.79 million. The northern part consists of the oldest metamorphosed rocks in the Himalaya Orogenic Belt. The southern part is located on the northern edge of the Ganga Basin of India. The thickness of Tertiary sedimentary cover is more than 6000 m with overthrust faults and folds. There are many oils seepages at the surface but no oil and gas fields have been discovered at the present time.

Bhutan   Bhutan has an area of 38,000 km² with a population of 730,000. It is located in the Himalaya Mountains. The whole area consists of Precambrian and Palaeozoic metamorphosed rocks. Hydrocarbon prospects are not clear at present.

Bangladesh   This country has an area of 1,475,570 km², with a population of 140 million. The Bangladesh Basin is the major oil and gas producing area of the country. This basin is filled by Upper Cretaceous and Tertiary sediments. Oil exploration started in 1910. During 1977–1990, eight gas fields were discovered. The Titas gas field (Table 20.3) was discovered in 1962, with recoverable oil and gas reserves of 4.83 million t (oil equivalent). By the end of 2008 oil production was 200,000 t, the proven gas reserves were 141.5 billion m³ and gas production was 15.3 billion m³.

INDIA, SRI LANKA AND THE MALDIVES

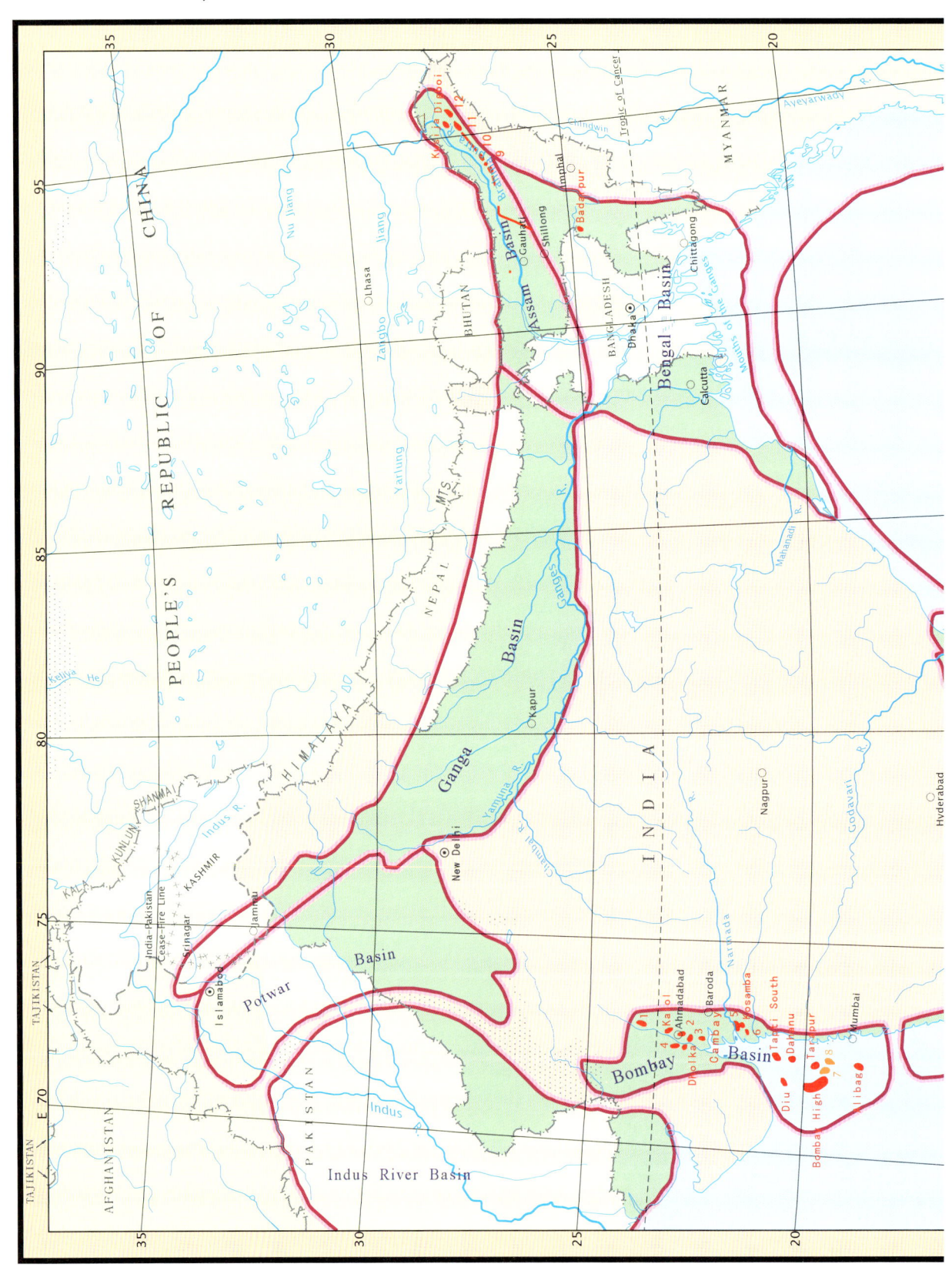

CHAPTER 21

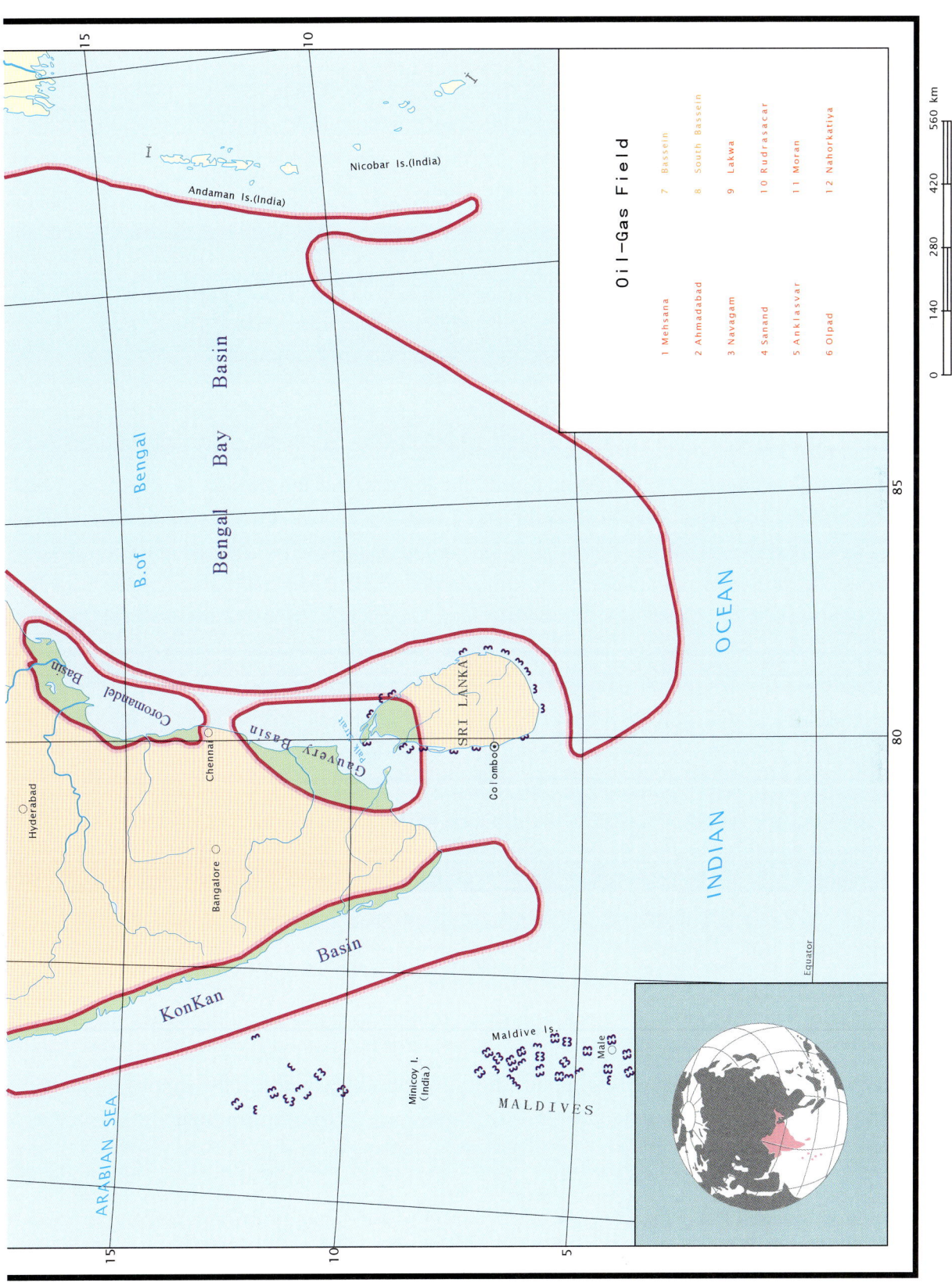

21 India, Sri Lanka and The Maldives

Geography

India, Sri Lanka and the Maldives are located in South Asia. India itself is often referred to as the largest country of the Indian sub-continent. All three countries are part of the sub-Himalayan group of countries of South Asia.

History

In India oil exploration began in the 19th century. During 1959–1974 oil exploration was mainly focused on land and offshore Bombay Basin. Oil production was 1.08 million t in 1962, and during 1994–2007 it was maintained at approximately 34 million t per year.

Regional geology

The Indian platform comprising Precambrian metamorphosed rocks is surrounded by four offshore sedimentary basins, namely, Bombay, KonKan, Gauvery and Coromandel (Table 21.1). The first two are primarily offshore, whereas Gauvery and Coromandel are distributed both offshore and onshore. The four known petroleum basins are filled predominantly with Cenozoic marine and deltaic sediments, being 3–4 km in thickness (sometimes up to 6–8 km). In some basins these sediments are in places underlain by Upper Cretaceous–Paleocene trappean rocks. Mesozoic and Palaeozoic complexes are composed of marine and continental terrigenous rocks from 1 to 2 km thisck.

Reservoir beds are spread throughout the succession being composed of terrigenous and carbonate rocks. Regional seals are confined mostly to Cenozoic deposits. In terms of area continuity, these are often disrupted by abrupt facies variations. Source rocks are clayey and carbonate deposits. The majority of source rock beds are concentrated in the Cenozoic complex and to a lesser extent-in the Mesozoic complex. Traps are of structural and combined types. Hydrocarbon accumulations are mostly confined to Cenozoic formations. Only sporadic accumulations can be found in the Mesozoic and Palaeozoic complexes. Original in-place resources are estimated as medium in these basins, except for the Bombay Basin where the resources are estimated as high.

Oil exploration and development on the west coast of India is concentrated on the broad shelf of the Arabian Sea near Bombay. One exploratory well has also been drilled in each of the Kutch and Kerala areas. The width of the continental shelf is 300 km near the coast off Bombay and decreases towards the north to about 80 km near Porbander and 120 km near the Kutch Peninsula. The width of the shelf narrows rapidly to the south to about 100 km near Ratnagiri and to 60 km near Cochin.

Sedimentary rocks are exposed along the coast of the Kutch Peninsula, Saurashtra Peninsula, and in a thin strip along the coast near Cochin. The sedimentary basins are half-graben type and the stratigraphy of the offshore basins is inferred principally from seismic data coupled with subsurface data from the wells.

World Atlas of Oil And Gas Basins, First Edition. Li Guoyu.
© 2011 John Wiley & Sons, Ltd. Published 2011 by John Wiley & Sons, Ltd.

Table 21.1 Major oil and gas basins of India

Basin	Area (thousand km²)	Major sedimentary rock		Reservoir		Remarks
		Age	Thickness (m)	Age	Lithology	
Assam	100	Mz, Kz	6000	R	Sandstone	Extends into Bangladesh
Bombay	160	Kz	5000	R	Sandstone	
Coromandel	70	J–R	4000	K, R	Sandstone	
Ganga	520	Kz, Q				Extends into Nepal and Pakistan
Gauvery	100	Mz, Kz	3000–4000	K, R	Sandstone, limestone	Extends into Sri Lanka
KonKan	280	K, R				

India India has a total area of 2.98 million km² and a population of 1.09 billion. India is commonly described as a subcontinent, occupying most of the peninsula of the same name. The peninsula is separated from mainland Asia by the Himalayas. As mentioned above, India is a major oil producing country, but as a result of rapid economic development, the country is increasingly reliant on imported oil. Imports currently account for about 72 per cent of total country consumption. The volume of oil imports is expected to continue rising by around 10 per cent annually for the short to medium term. The longer-term oil outlook is, however, more optimistic and the discovery of more oil fields on both the eastern and western continental shelves has boosted prospects of significantly increasing output. This is necessary so as to curtail the growth of imports and gradually move towards a desired national goal of self-sufficiency in oil. At the end of 2008 proven oil reserves were 770 million t with oil production of 33.4 million t. At the same time proven gas reserves totalled 1 trillion m³ with gas production of 29.1 billion m³. Total oil consumption was 136 million t and oil imports accounted 102 million t. At present India is a significant oil importing country.

Sri Lanka Sri Lanka is an island in the Indian Ocean, having an area of 65,600 km² and a population of 19.88 million. During Late Jurassic and Early Cretaceous times, Sri Lanka was separated from the Indian Platform. Between the two land masses, new basins developed, filled by Jurassic, Cretaceous and Tertiary sediments. Oil exploration began in 1966 and one well was completed at the same time. During 1973 and 1974 offshore seismic surveys and a few more exploration wells were completed.

The Maldives The Maldives comprise 19 coral islands in the India Ocean, lying to the southwest of the southern tip of the Indian peninsula. It has an area of 298 km² and a population of 299,000. Joint venture oil and gas exploration efforts started in 1968 with the French Elf Aquitaine Company. However, the three exploration wells drilled between 1976 and 1978 showed little commercial attractiveness. A second joint venture collaboration with Shell in the early 1990s was also unsuccessful. There have been recent plans announced to start another phase of active exploration.

BOMBAY BASIN

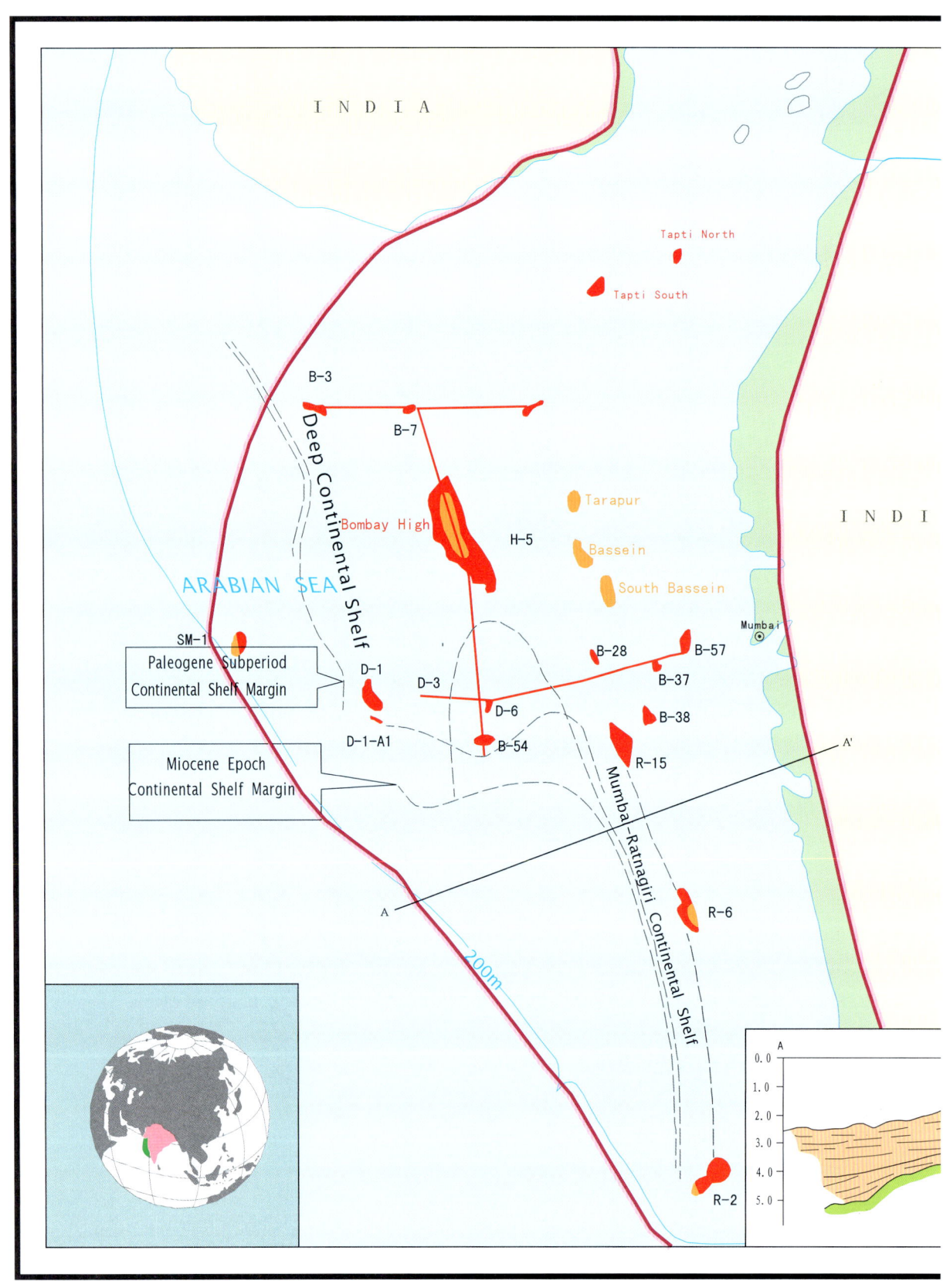

CHAPTER 22

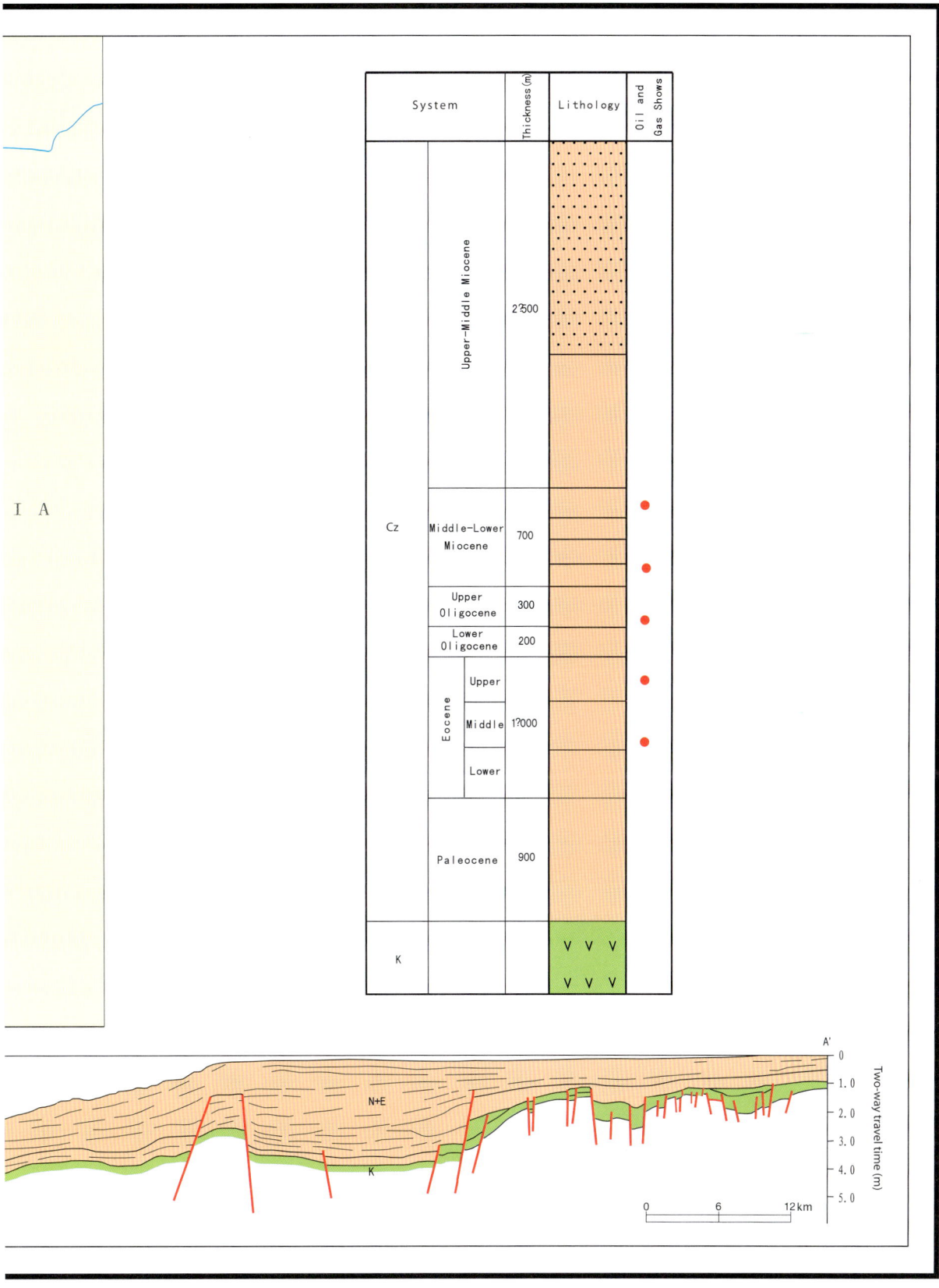

22 Bombay Basin

Geography

Bombay Basin is located on the western shelf of India with a total area of 16,000 km².

History

Oil exploration was initiated in 1960. The first well was finished in 1973. The second exploration well discovered the large Bombay High oil field (Fig. 22.1) in 1974. Exploration in the Bombay Offshore Basin for oil and gas intensified after the discovery of the Bombay High Oilfield in February 1974. Details of the major oil and gas fields in the Bombay Offshore Basin are provided in Table 22.1. The infilling Tertiary sediments have thicknesses of up to 5,000 m. The Bombay Offshore Basin is a continuation of that on land, to the north of the Gulf of Bombay. The tectonics and the origin of the structures are the same for both the onland and offshore basins.

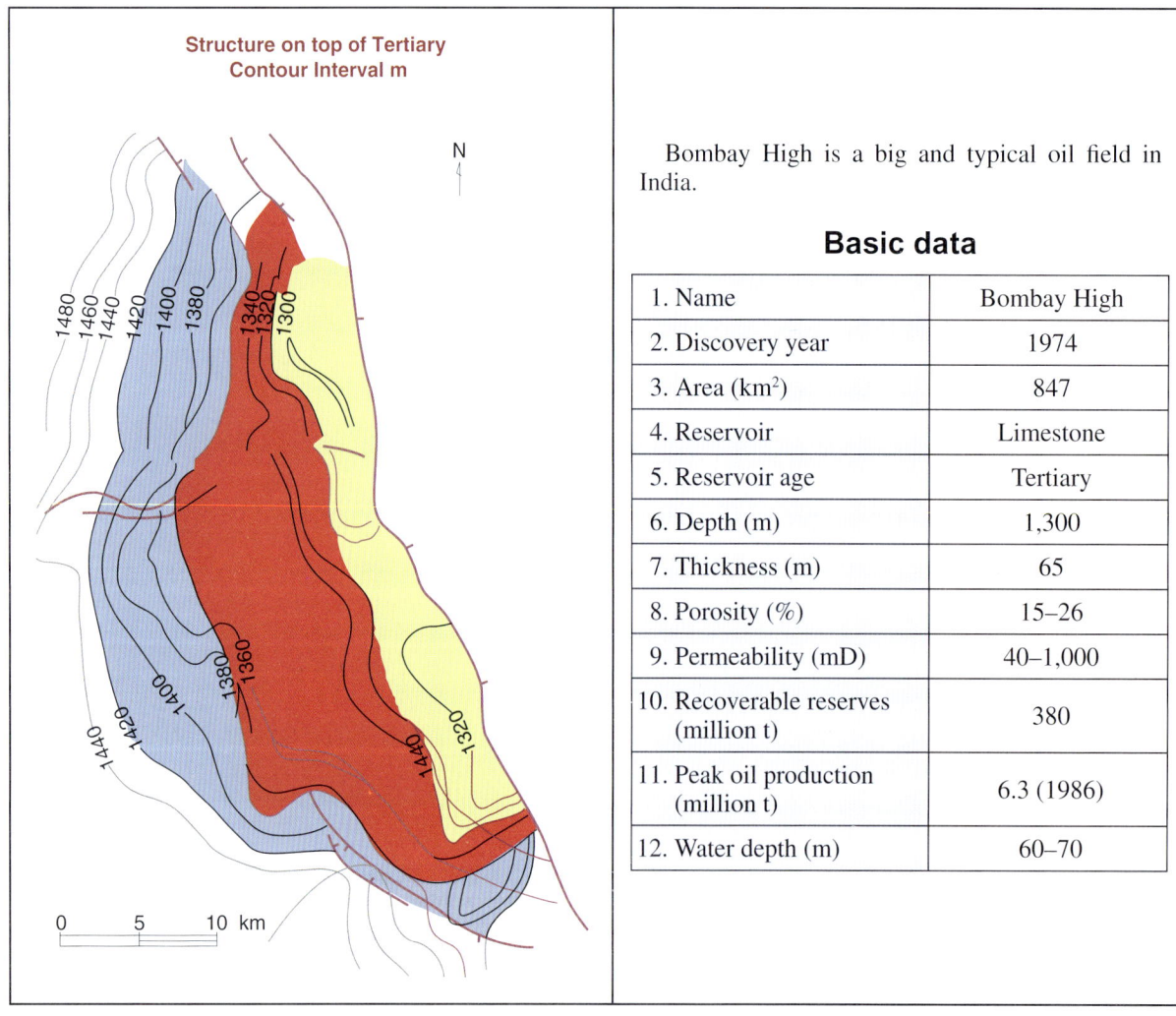

Bombay High is a big and typical oil field in India.

Basic data

1. Name	Bombay High
2. Discovery year	1974
3. Area (km²)	847
4. Reservoir	Limestone
5. Reservoir age	Tertiary
6. Depth (m)	1,300
7. Thickness (m)	65
8. Porosity (%)	15–26
9. Permeability (mD)	40–1,000
10. Recoverable reserves (million t)	380
11. Peak oil production (million t)	6.3 (1986)
12. Water depth (m)	60–70

Fig. 22.1 Details of the Bombay High oil field.

World Atlas of Oil And Gas Basins, First Edition. Li Guoyu.
© 2011 John Wiley & Sons, Ltd. Published 2011 by John Wiley & Sons, Ltd.

Table 22.1 Major oil and gas fields of the Bombay Offshore Basin

Name	Discovery year	Recoverable reserves		Depth (m)	Trap	Reservoir	
		Oil (million t)	Gas (billion m³)			Age	Lithology
Bombay	1974	380.0	2.2	1300	Anticline	E, N	Limestone
Ankleshvar	1960	73.0	—	1200	Anticline	E	Sandstone
Basein	1976	77.0	—	1700	Anticline	E	Limestone
Basein South	1977	—	23.4	1800	Anticline	E	Limestone

Regional geology

Development of the Bombay Basin is associated with rifting caused by the movement of the Indian and African continental plates. The sedimentary cover comprises Ceno-Mesozoic sediments, mainly Tertiary in age. Since the Paleocene, when sedimentation started in the basin, a few prominent sinks within the palaeoshelf provided accommodation potential for thick sedimentary sequences. The large accumulation of oil in the Bombay High oil field does not seem to have originated from the poor source rocks in the basin, but the underlying Middle Eocene shale has good source-rock potential.

The Bombay High oil field was discovered in 1974 with an area of 840 km² beneath water depths of 60–70 m. Recoverable oil reserves were estimated at 380 million t. The oil field is located approximately 160 km northwest of Bombay. The structure is a north-northwest trending doubly-plunging anticline with a gently dipping western flank and a faulted eastern flank. It has two structural culminations separated by a saddle. Fifteen exploratory wells were drilled on the structure.

Bombay High oil field has two oil plays, Limestone II and Limestone III and one gas play, Sand 1. Limestone III is the deeper of the two oil plays and is found over the entire structure. Limestones III and II are separated by a 300 m shale interval. Sand 1 is lenticular gas-bearing sand in the shale and is developed along the crestal part of the structure. Limestone III is a poorly consolidated biomicrite. Vugs, varying in diameter from a few millimetres to 5 cm, are seen in many parts of the field. Although the porosity of the play is mainly intergranular, secondary porosity forms vugs and solution channels also form an impartment part of the pore structure. The matrix permeability obtained from laboratory data is four to five times less than the reservoir permeability obtained from build-up tests during production testing. Limestone III limestone has a few shale or compact limestone beds, some of which can be traced all over the field. Based on these shales, the play is divided into layers and sublayers. Although the gas–oil contact is the same for all the layers, the oil–water contact changes level by over 35 m. The cause for the variation of this oil–water level and its effect on the production development is being studied (Shannon and Naylor, 1989).

The northern part of Bombay High oil field was developed first in view of the better reservoir properties of the main oil play. Five (4 well) platforms were drilled. The distance between the platforms is 3 km. Development wells were drilled separately for two zones, having different reservoir properties. Curved conductors are used in platforms to achieve horizontal drifts of 1:1 (with a maximum deviation angle of 58°) at a 1400 m depth, the level of the main oil play.

Commercial oil production from the field started on 21 May 1976, about two years after the discovery of the field. The initial gas production was 1112 m³/day from two wells on one platform. The oil was pumped from the platform to a storage tanker through a storage single buoy mooring connected by subsea-line to the platform. The production reached 12,700 m³/day by December 1977.

CENTRAL ASIA: KAZAKHSTAN, TURKMENISTAN,

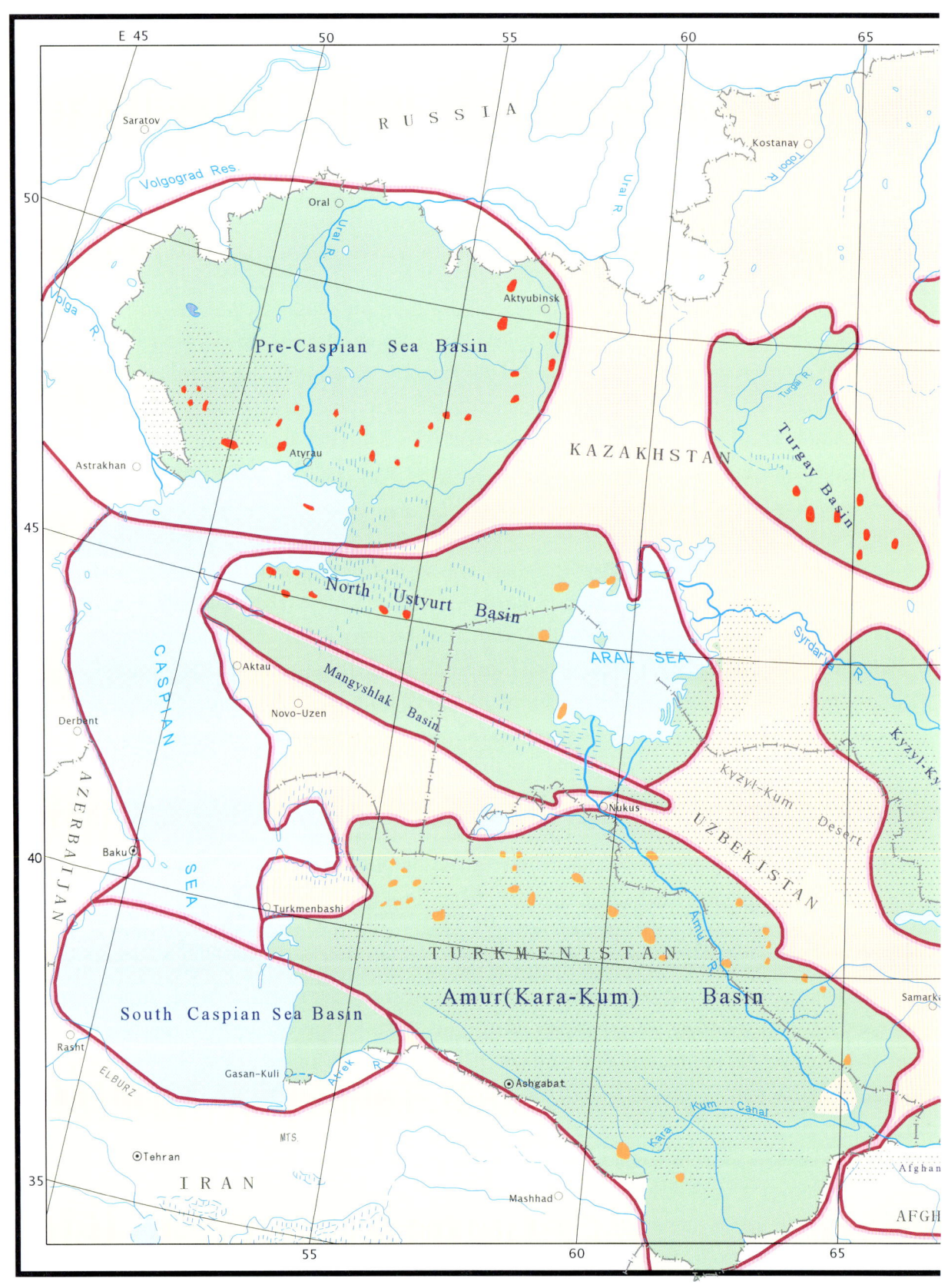

UZBEKISTAN, TAJIKISTAN AND KYRGHYZSTAN   CHAPTER 23

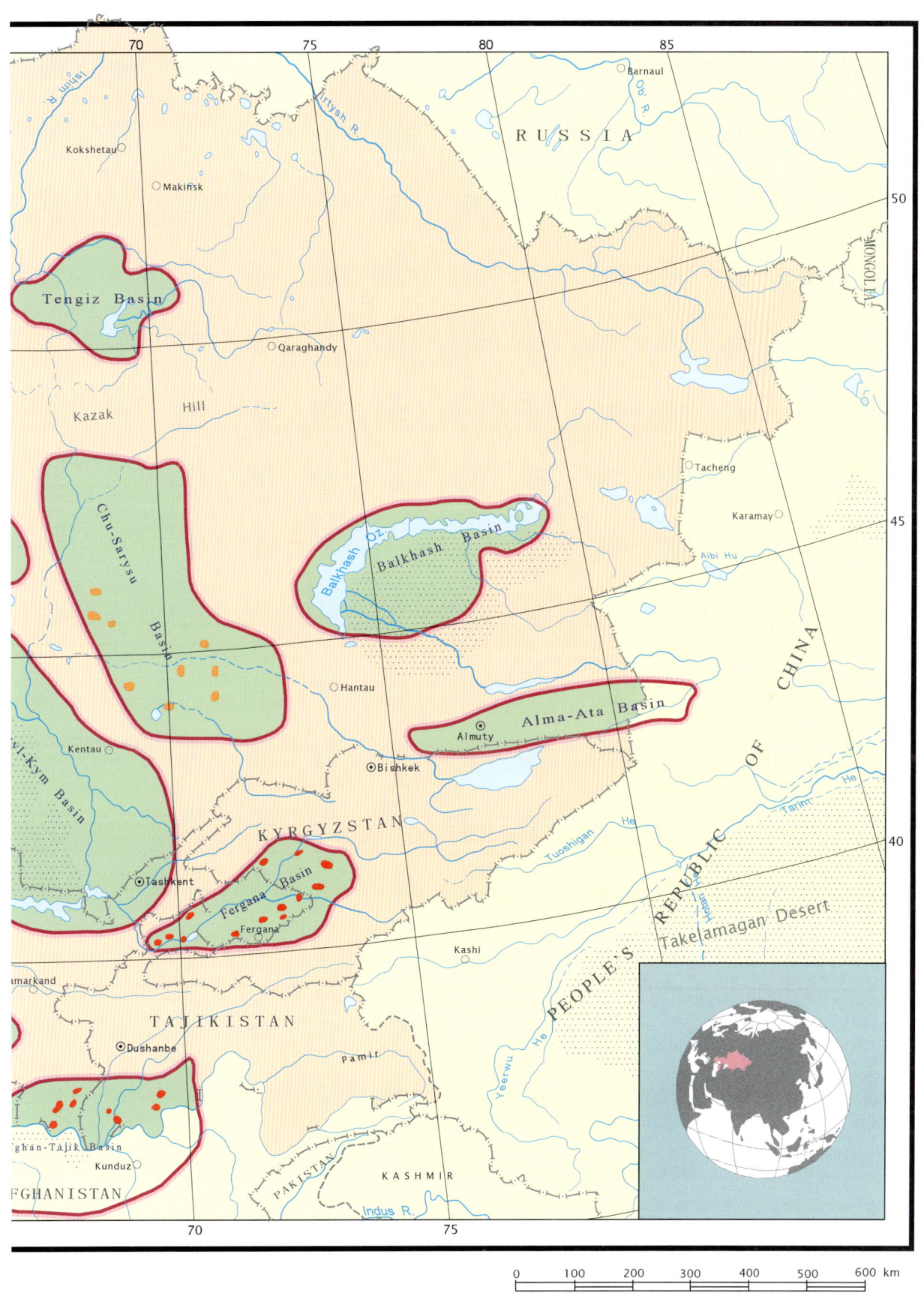

23 Central Asia: Kazakhstan, Turkmenistan, Uzbekistan, Tajikistan and Kyrghyzstan

Geography

The area referred to as Central Asia lies between the Caspian Sea in the west and China in the east. In the north it includes Kazakhstan, Turkmenistan, Uzbekistan, Tajikistan and Kyrghyztan, and in the south it includes Iran and Afghanistan. The area of the region is about 3.7 million km^2.

History

This is a huge oil and gas region covering an area of more than 3 million km^2. It is an attractive area for oil exploration. Oil seepages have been found in many places all over the region. Oil exploration began 120 years ago in the Fergana Basin in Uzbekistan. Details of the major oil and gas basins of Central Asia are provided in Table 23.1. Today this area is becoming one of the largest oil-producing areas worldwide with good prospects for future rapid development.

Regional geology

This region is a part of the Eurasian continental plate. Two main types of tectonic elements comprise the geological structure of the region, i.e. a young platform and various orogenic belts of different ages. The platform is represented by the Turanian epi-Palaeozoic Platform which is a part of the vast Central Eurasian Platform. The Turanian Platform basement is made up on the whole of Hercynian rocks. It also comprises blocks of Caledonian and Precambrian basement. Two main orogenic belts can be identified, being, the Uralo-Okhotsky and the Mediterranean. The Uralo-Okhotsky Orogenic Belt is represented by Palaeozoic folding of Kazakhstan, North and South Tianshan. Within the belt, the West Tian Shan epiplatform orogen is located in the area that had experienced the activation of the Turanian epi-Palaeozoic Platform (Shannon and Naylor, 1989).

Kazakhstan Kazakhstan ranks first among the oil-producing countries of Central Asia. Ten oil and gas basins are located here, infilled with Palaeozoic and Mesozoic rocks with thicknesses of between 4000 and 10,000 m. Kazakhstan is a very prospective oil and gas region. Many giant oil fields have been discovered, good examples being the Tengiz and Kashagan oil fields with recoverable reserves of more than 1.3 billion t each. A total of 210 oil and gas fields have been discovered in the country as at the end of 2008, with proven oil reserves of 4.1 billion t and annual oil production of 69 million t. Proven gas reserves were 2.4 trillion m^3 and gas production 26.6 billion m^3.

Turkmenistan The giant gas basin Amur (Kara-Kum) is located in Turkmenistan, with an area of 530,000 km^2. This basin comprises mainly Jurassic and Cretaceous rocks with thicknesses reaching 5000 m. Turkmenistan is rich in oil and gas resources. By the end of 2008 the

World Atlas of Oil And Gas Basins, First Edition. Li Guoyu.
© 2011 John Wiley & Sons, Ltd. Published 2011 by John Wiley & Sons, Ltd.

Table 23.1 Major oil and gas basins of Central Asia

Country	Basin	Area (thousand km²)	Major sedimentary rock Age	Major sedimentary rock Thickness (m)	Reservoir Age	Reservoir Lithology	Remarks
Kazakhstan	Chu-Sarysu	150	Pz, K, R	4000–5000	D–P	Sandstone, limestone	
	Kyzyl-Kym	180	Kz				Extends into Uzbekistan
	North Ustyurt	250	Mz, Kz	2000–3000	J, K	Clastic rock	Extends into Uzbekistan
	Pre-Caspian	500	Pz–Kz	10,000	D–J, R	Clastic rock, carbonate rock, limestone	Extends into Russia
	Turgai	110	Pz, Mz, Kz	5000	J, K	Clastic rock	
	Caspian	370	Pt–Q				
Uzbekistan	Fergana	45	T–Kz	8000	R, K	Carbonate rock, sandstone	Extends into Kirghizstan and Tajikistan
Turkmenistan	Amur (Kara-Kum)	530	J–Kz	5000	J, K	Carbonate rock, sandstone	
Afghanistan	Afghan-Tadzhik	100	J–Kz	8000–12,000	K, R	Carbonate rock	Extends into Tajikistan and Uzbekistan

proven oil reserves were 82 million t with annual oil production of 11 million t. Proven gas reserves stood at 2661 billion m³ and gas production was 67 billion m³.

Uzbekistan   Uzbekistan is the third most important oil and gas region among the five countries. Fergana Basin is a major basin, infilled by Triassic–Tertiary sediments with thicknesses of up to 8000 m and an area of 45,000 km². Oil exploration began in 1880. The first oil field was discovered in 1964. By the end of 2007 proven oil reserves were 82.19 million t with annual oil production 5 million t. Proven gas reserves were 1840 billion m³ with gas production of 58 billion m³.

Tajikistan   Tajikistan is a landlocked country with an area of 143,100 km² and a population of 6.92 million. The western part of the country is occupied by the Fergana Basin. The first oil field was discovered in the basin in 1908. More than 21 small oil and gas fields have been discovered in this country. Oil production reached 410,000 t in 1979. After that time oil production declined gradually reaching only 20,000 t annually between 1995 and 1999.

Kyrghyzstan   Kyrghyzstan is a landlocked country with an area of 198,500 km² and a population of 5.17 million. The first exploration well was finished in 1898. Oil production began in 1903. More than 14 sedimentary basins are distributed within this country. Forty-seven per cent of the largest basin in the region, the Fergana Basin, lies within Kyrghyzstan, with a total area of 45,000 km². By the end of 2008 proven oil reserves were 5.47 million t and oil production 50,000 t. Proven gas reserves were 5.6 billion m³.

CASPIAN SEA OIL AND GAS REGION

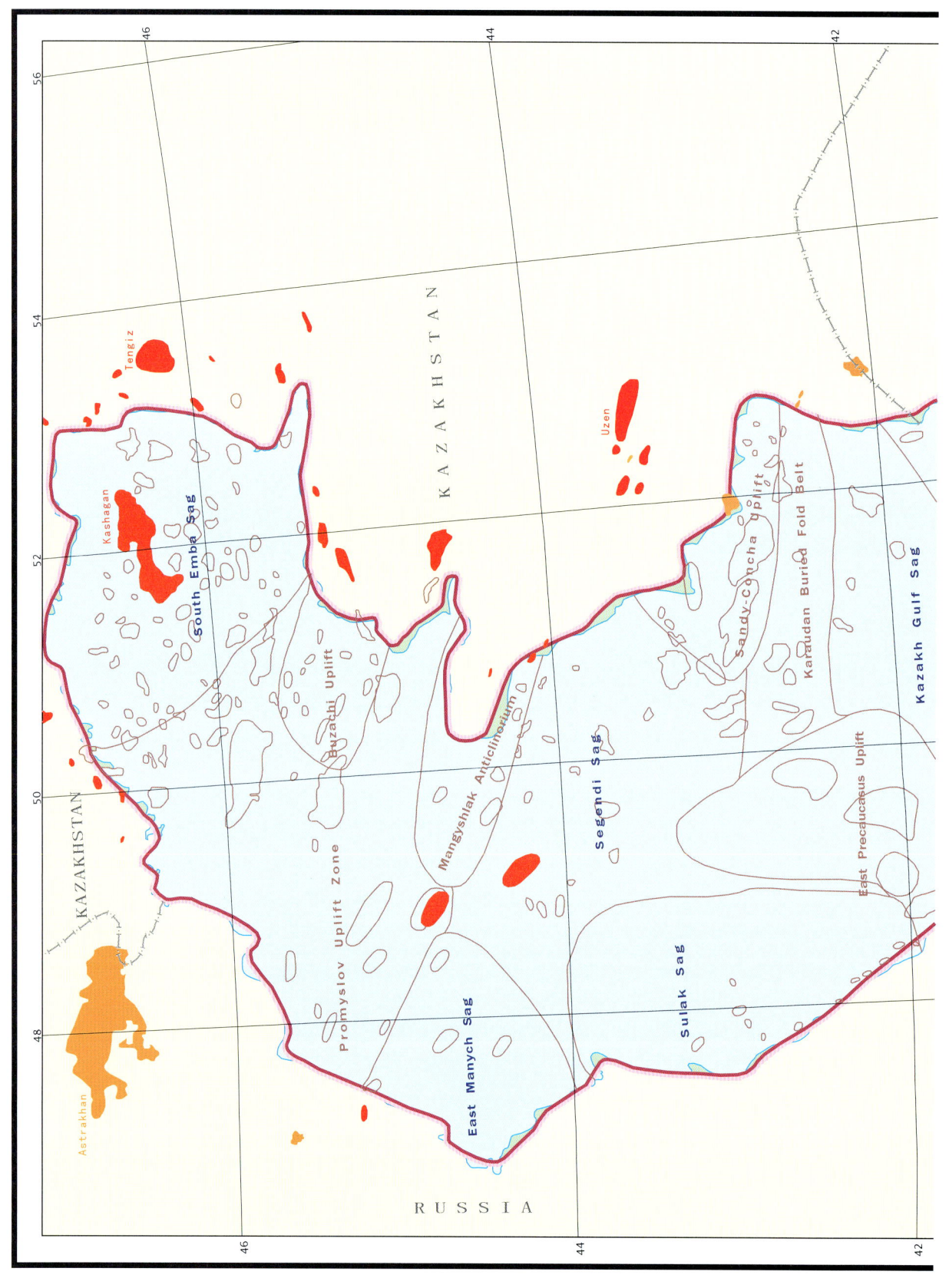

CHAPTER 24

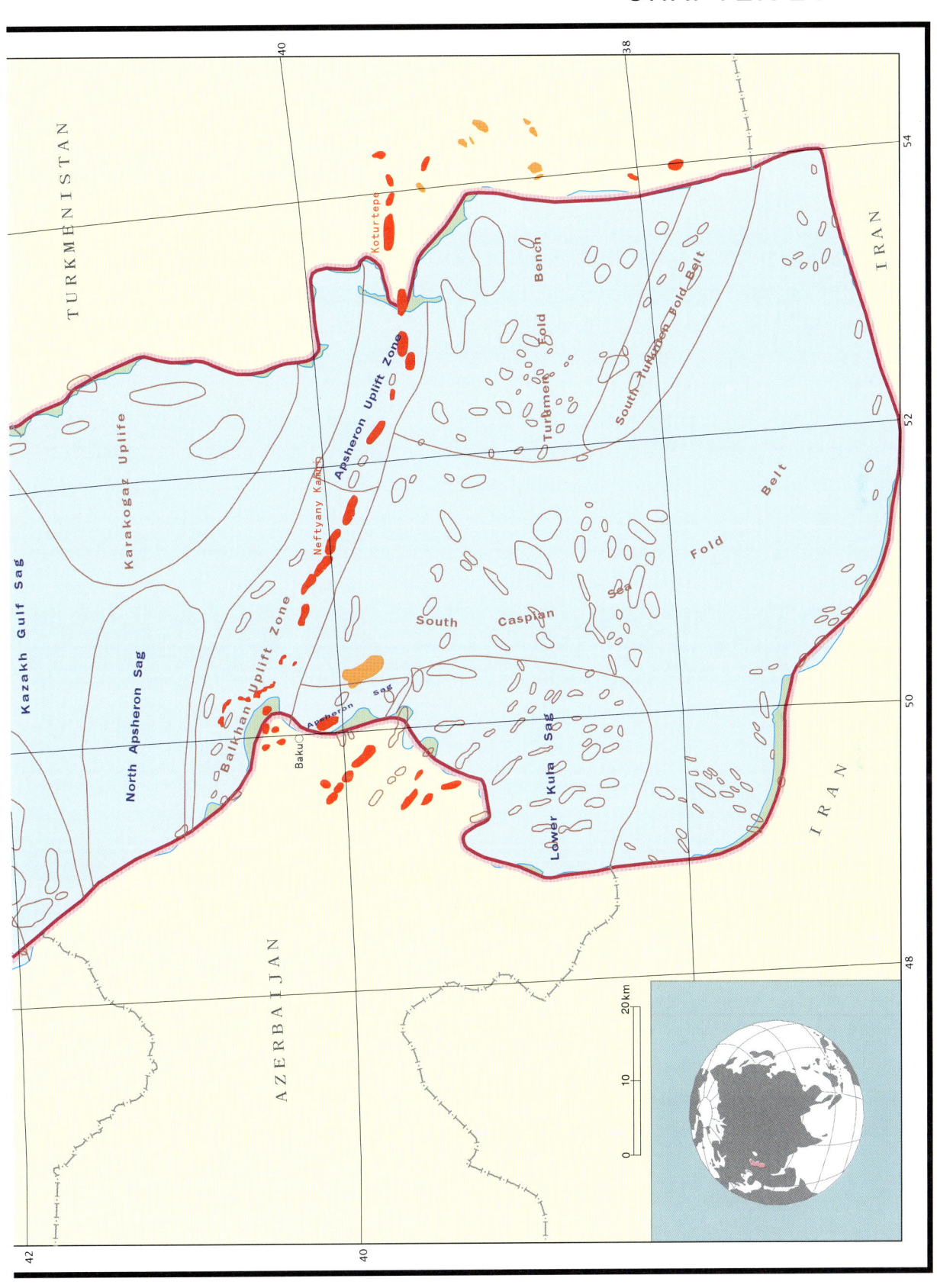

24 Caspian Sea Oil and Gas Region

Geography

The area of this region includes Kazakhstan, Turkmenistan, Russia, Azerbaijan and Iran. The Caspian Sea is located within three of the world's largest oil and gas regions: Volga–Urals, Persian Gulf and West Siberia. Due to the unique geological conditions of the region, the Caspian Sea and Central Asia have become a focus of oil and gas exploration and production interest. The Caspian Sea has an area of 371,000 km^2, with an average water depth of 180 m, reaching a maximum depth of 250 m. Its total coastline is 1204 km, and it is the largest inland body of water in the world and is classified as endorheic because it has no outflow.

History

Oil and gas exploration commenced in the last quarter of the 19th century and by 1900 the Baku area of Azerbaijan had over 2000 oil-producing wells out of 3000 wells that had been drilled there. The first large oil field, Nift Kamin, was discovered in 1949. For a relatively long period of time there was limited exploration work carried out in the region primarily because of the complex geological conditions, typified by the Pre-Caspian Basin with its many salt domes. It was only after the discovery of the giant onshore Tengiz oil field in Kazakhstan in 1979 that serious reconsideration was given to the overall potential of the region. The Caspian Sea has now become one of the largest and most important oil- and gas-producing regions of the world. For a long time geological conditions hampered oil exploration. Permian salt-bearing beds of 7000–10,000 thickness were the main problem, but after the discovery of the giant Tengiz oil field it was possible to salt movement and the important conditions it created for oil accumulation in large fields.

Regional geology

The Caspian Sea oil region can be divided into two main areas. One is the water-covered portion of the Caspian Sea with an area of 371,000 km^2. only. The other comprises the Caspian Sea onshore sedimentary basins with a combined area of more than 1.28 million km^2 (Pre-Caspian Basin 500,000 km^2, Mangyshlak Basin 70,000 km^2, South Caspian Sea Basin 90,000 km^2, North Ustyurt 250,000 km^2 – see Map 24). The surrounding onshore basins extend into the Caspian Sea and consist of three main tectonic elements. The northern portion is the Pre-Caspian Basin, filled by Tertiary to Cambrian strata with a thickness of over 22 km, and comprising more than 1600 salt domes. The middle portion comprises of the North Ustyurt and Mangyshlak basins, filled by Tertiary to Cambria strata with a thickness of 10–12 km. The southern portion consists primarily of the Caspian Basin, filled by Mesozoic and Cenozoic sediments with a thickness of 10–15 km. Clearly, the reason why the Caspian Basin has such a high oil and gas potential can be ascribed to the wide area and substantial thickness of its sedimentary rocks.

In the northern portion, oil and gas reservoirs are mainly in carbonaceous massive reef limestones below salt with a thickness of 1000–3000 m. Source rocks are Lower Carboniferous and

Caspian Sea Oil and Gas Region

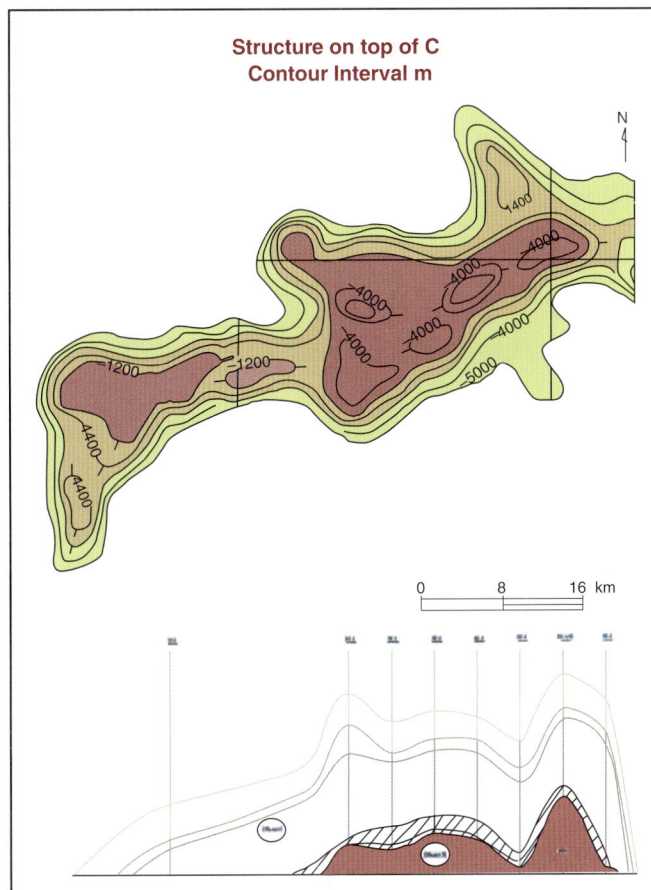

Kashagan is a giant oil and gas field in Caspian Sea Basin.

Basic data

1. Name	Kashagan
2. Discovery year	
3. Area (km²)	1,000
4. Reservoir	Reef
5. Reservoir age	C, D
6. Depth (m)	4,200–4,600
7. Thickness (m)	900–1,300
8. Porosiity (%)	12
9. Permeability (mD)	2–8
10. Recoverable reserves (billion t)	2.7
11. Water depth (m)	5

Fig. 24.1 Details of the Kashagan giant oil and gas field.

Lower Permian shale that contain organic matter in the range of 1–1.4 per cent, lying below salt. The giant Kashagan oil field (Fig. 24.1) was discovered in this area, with recoverable oil reserves of 2.7 billion t and gas reserves of 3 trillion m³.

In the southern portion the large Nift Kamin oil field was discovered offshore in 1949 with recoverable oil reserves estimated at 1.94 million t. The reservoir is Tertiary sandstone with depths of 200–680 m. Ninety per cent of the oil production of Azerbaijan (15 million t in 2001) comes from the Caspian Sea area.

The Caspian Sea area contains one large depression as well as nine distinct fold belts, which locally can be further subdivided into several other structural features in the different parts of the region, each with high potential for oil and gas reserves.

PRE-CASPIAN SEA BASIN

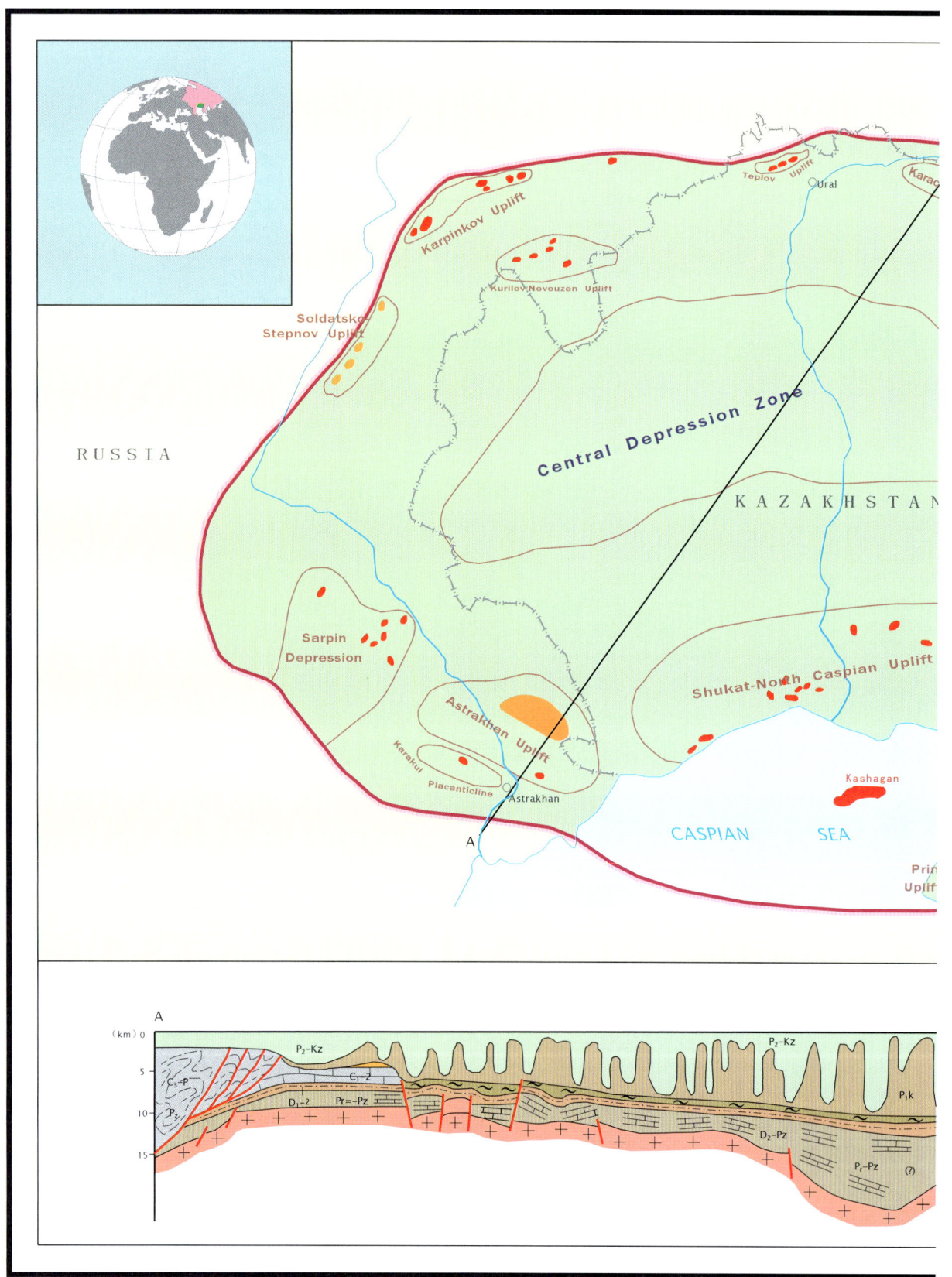

CHAPTER 25

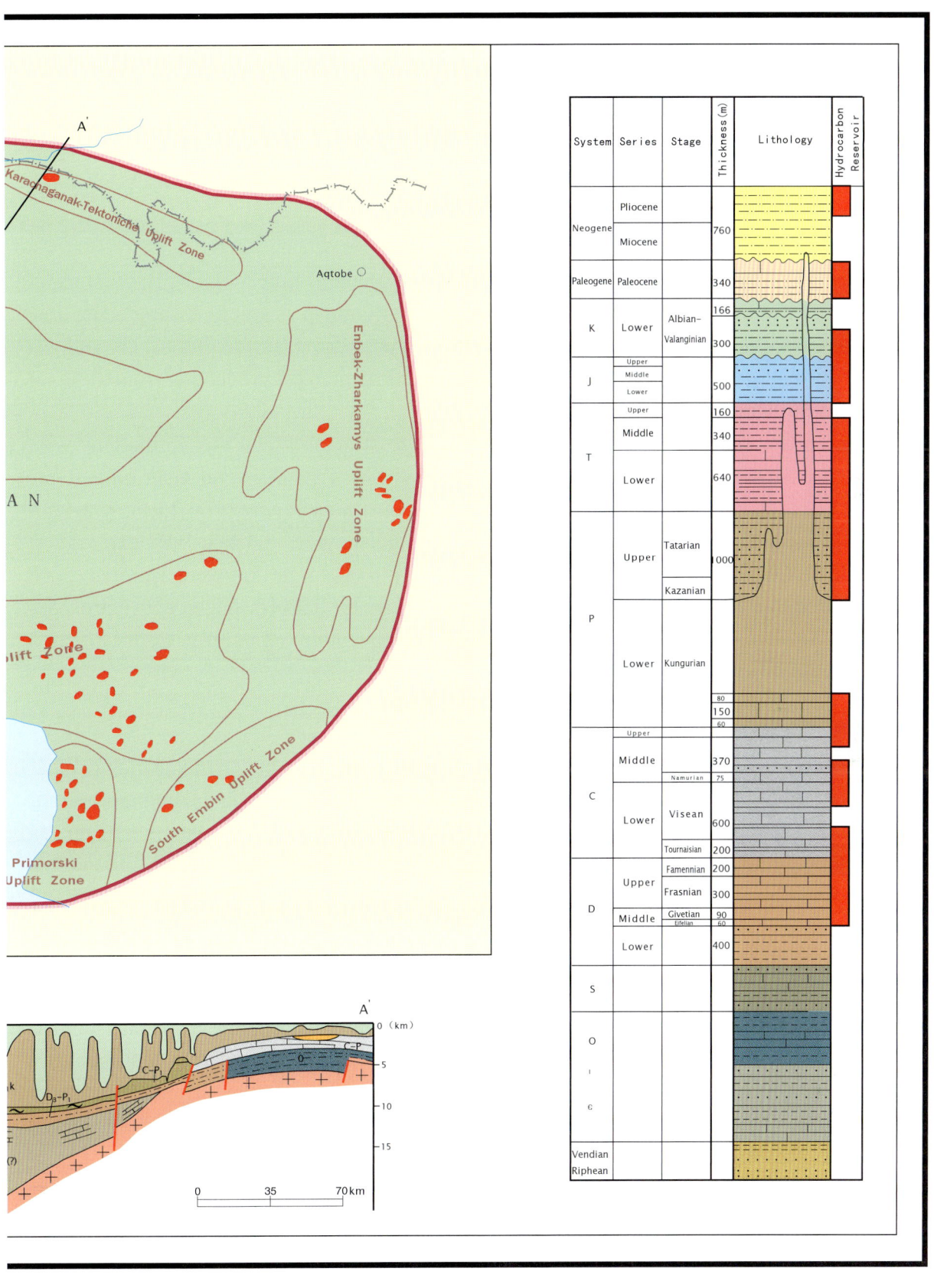

25 Pre-Caspian Sea Basin

Geography

The Pre-Caspian Sea Basin is the oldest oil exploration region associated with the salt domes. It is located in Kazakhstan on the northern onshore area of the Caspian Sea, with an area of 500,000 km^2.

History

Beforethe end of the 19th century exploration started in southern Emba region. The first small oil field was discovered in 1911. The region Emba was well-known as an old oil-producing area with oil plays associated with the more than 1000 salt domes above salt formations. Oil and gas accumulate in shallow, small, multifaulted types of trap. Maximum oil production from these oil fields was 1.5 million t in 1959.

By using seismic and drilling technology, many large structures and salt formations with depths ranging between 4000 and 5000 m were revealed. Since 1960 several giant oil and gas fields have been discovered among these large structures, such as Tengiz (Fig. 25.1), Kingiyak (Fig. 25.2), Zananor, Kalachaganak and Astlahan. During the period 1994–1997 exploration began in the Caspian Sea itself. Seismic lines measuring a total 26,180 km were acquired, from which 96 favourable structures were mapped. The supergiant oil field Kashagan with a prospective area of 5180 km^2 was discovered and 6 billion t of recoverable oil reserves were estimated as existing there.

Regional geology

The thickness of sedimentary rocks in this basin reaches 1.2 km. An interesting geological feature of the region is the thick Early Permian salt formation, which divides the oil and gas bearing column into two parts: above the salt formation and below the salt formation. Petroleum and gas basin development above salt is characterized by very complex shallow and small oil and gas fields, but development below the salt formation is characterized by simple, deep, giant oil and gas fields.

There are different types of salt diapirs: piercement forming caprock; pinnacle dome; truncation; other non-piercement. Very unique examples are provided from this basin of the role of salt domes in oil accumulation. In the Kingiyak oil field, oil traps are found above, alongside and beneath the salt. Initially the small oil traps above and alongside the salt showed poor prospect, but this changed with the new discovery of a huge field beneath the salt. The combination of faulting through structural deformation, syndepositional faulting, sedimentary facies variations, occurring simultaneously with basin subsidence and development of salt diapirs has created an array of hydrocarbon traps and mechanisms for the development oil and gas plays. Salt diapirism effects therefore range from simple seals to modification of sediment accommodation space. The salt diapirs of the Pre-Caspian Sea Basin undoubtedly had a greater volume of salt deposits upon

Pre-Caspian Sea Basin

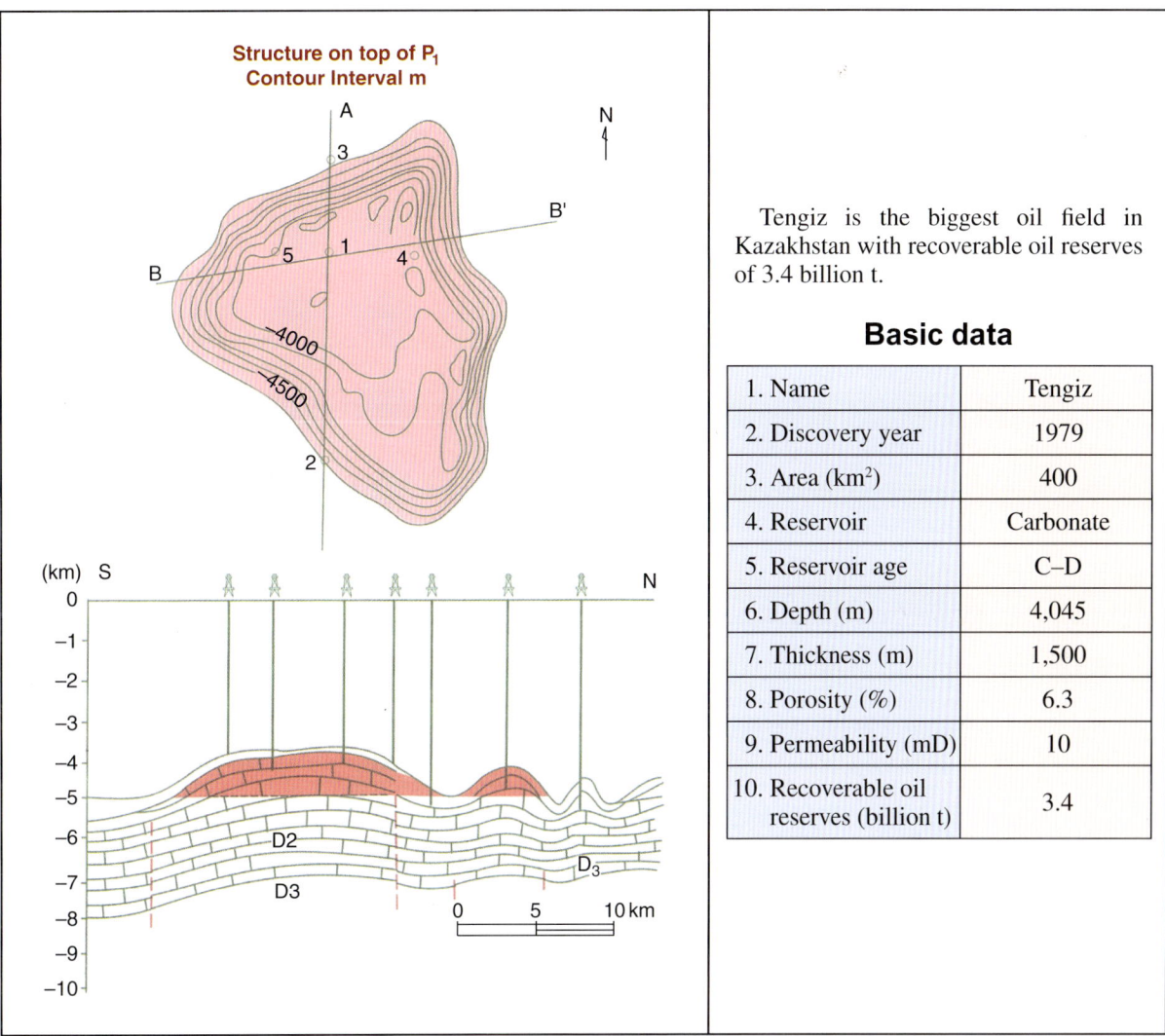

Fig. 25.1 Details of the Tenzig oil field.

which to feed than the diapirs of the interior basins. Sedimentation above the salt deposits was also greater in coastal areas.

The basin is asymmetrical, with the deeper portion in the south having a thickness of 15–26 km of sedimentary rocks. The sedimentary cover consists of Cambrian to Tertiary strata. Large uplifts affecting areas of several thousand square kilometres resulted in giant oil and gas fields beneath the salt deposits. Above the salt deposits a number of small and complex salt domes were found and many small oil and gas fields were discovered. The total number of salt domes reaches 1000 within an area of between 130,000 and 150,000 km². Beneath the salt deposits the Lower Carboniferous and Lower Permian shales are source rocks with an organic matter content in the 1–1.4 per cent range. Above the salt deposits the source rocks have been identified as Jurassic and Cretaceous sediments. Reservoirs beneath the salt deposits are in Carboniferous and Permian carbonate rocks. Above the salt deposits reservoirs are in Jurassic and Cretaceous sandstone. The salt deposits form caprocks to reservoirs beneath them, but above the salt deposits shales form the caprock. Traps are represented by mainly anticlinal types. Petroleum potential is estimated as very high.

Asian Oil and Gas Basins

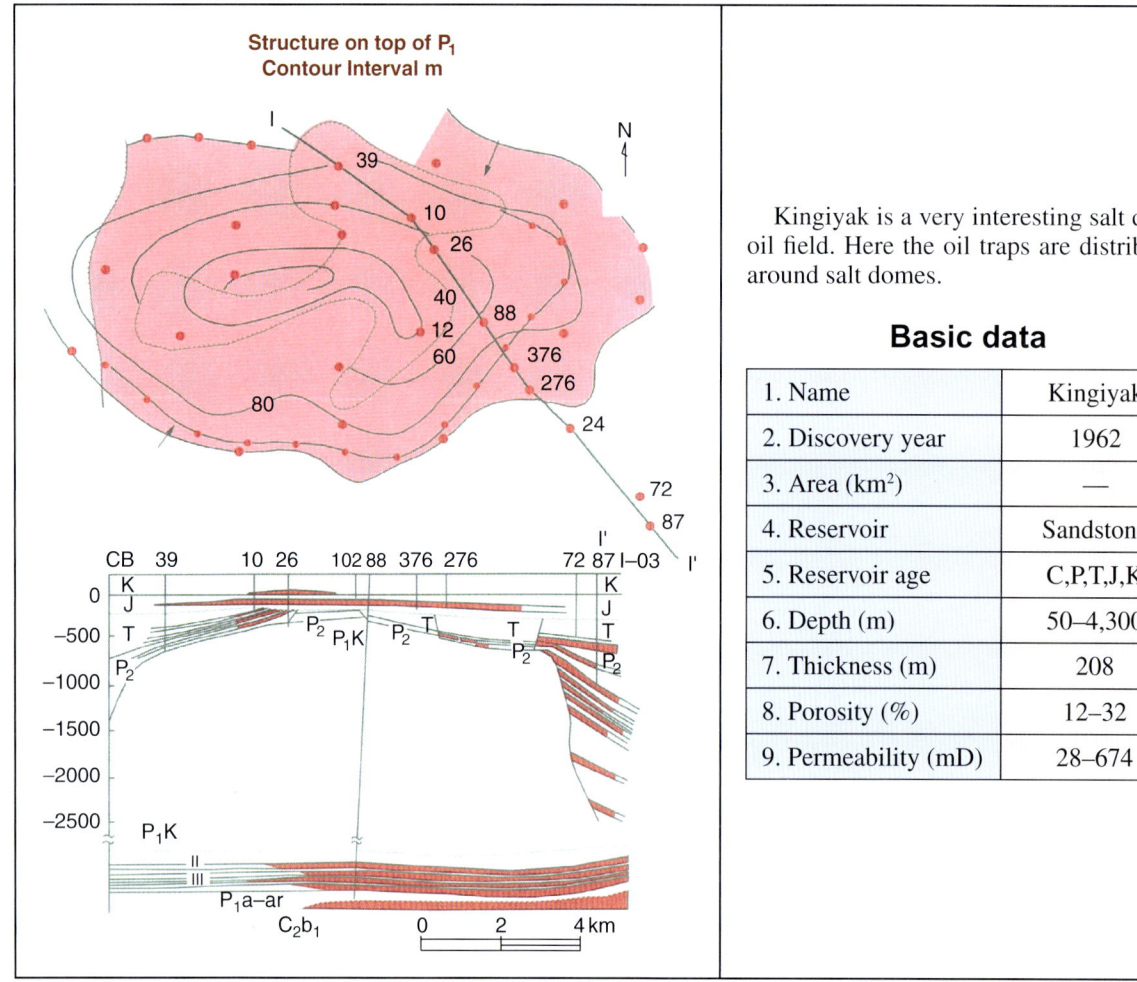

Kingiyak is a very interesting salt dome oil field. Here the oil traps are distributed around salt domes.

Basic data

1. Name	Kingiyak
2. Discovery year	1962
3. Area (km^2)	—
4. Reservoir	Sandstone
5. Reservoir age	C,P,T,J,K
6. Depth (m)	50–4,300
7. Thickness (m)	208
8. Porosity (%)	12–32
9. Permeability (mD)	28–674

Fig. 25.2 Details of the Kingiyak oil field.

NORTH USTYURT BASIN

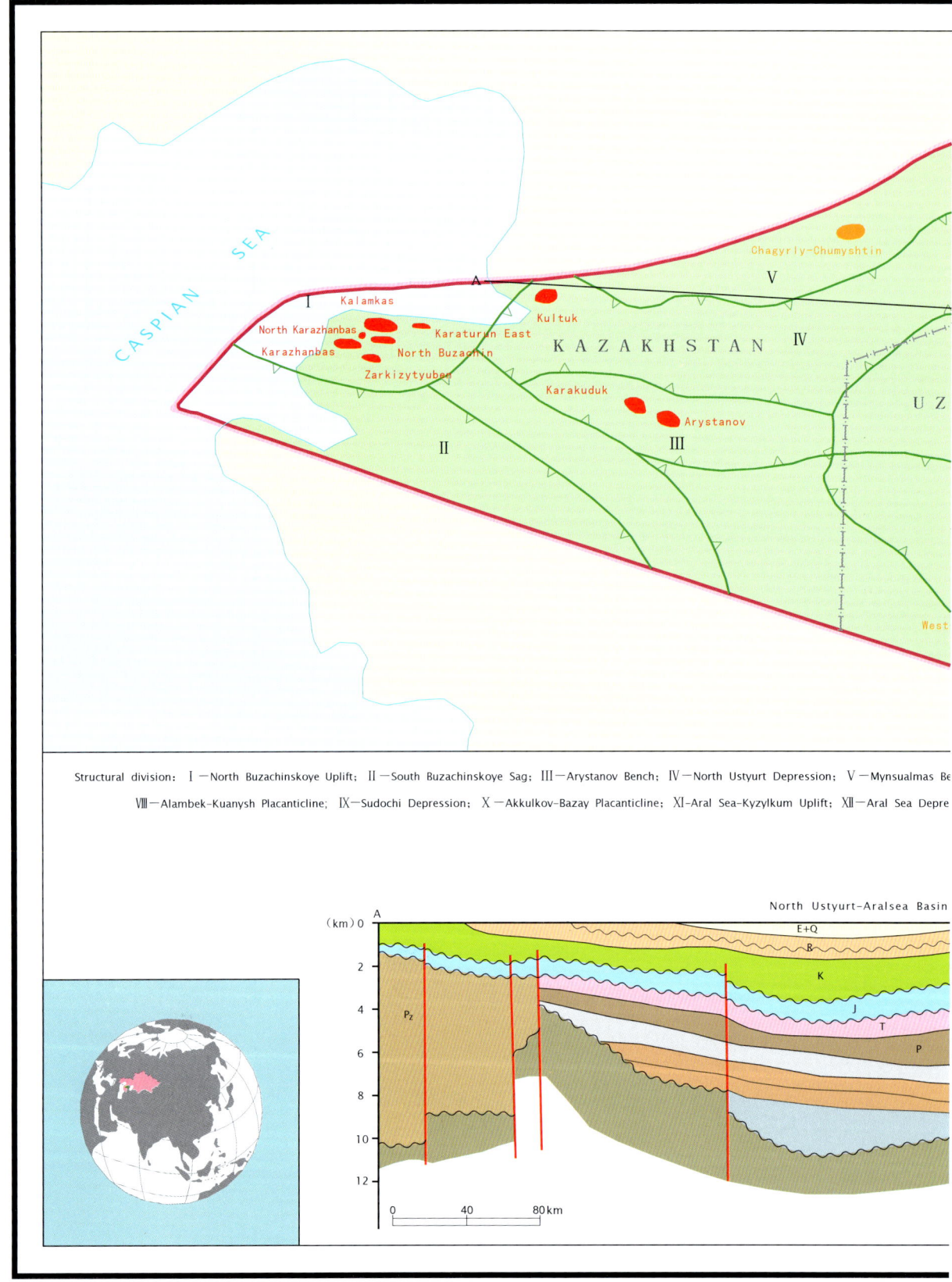

Structural division: Ⅰ—North Buzachinskoye Uplift; Ⅱ—South Buzachinskoye Sag; Ⅲ—Arystanov Bench; Ⅳ—North Ustyurt Depression; Ⅴ—Mynsualmas Be... Ⅷ—Alambek-Kuanysh Placanticline; Ⅸ—Sudochi Depression; Ⅹ—Akkulkov-Bazay Placanticline; Ⅺ-Aral Sea-Kyzylkum Uplift; Ⅻ-Aral Sea Depre...

CHAPTER 26

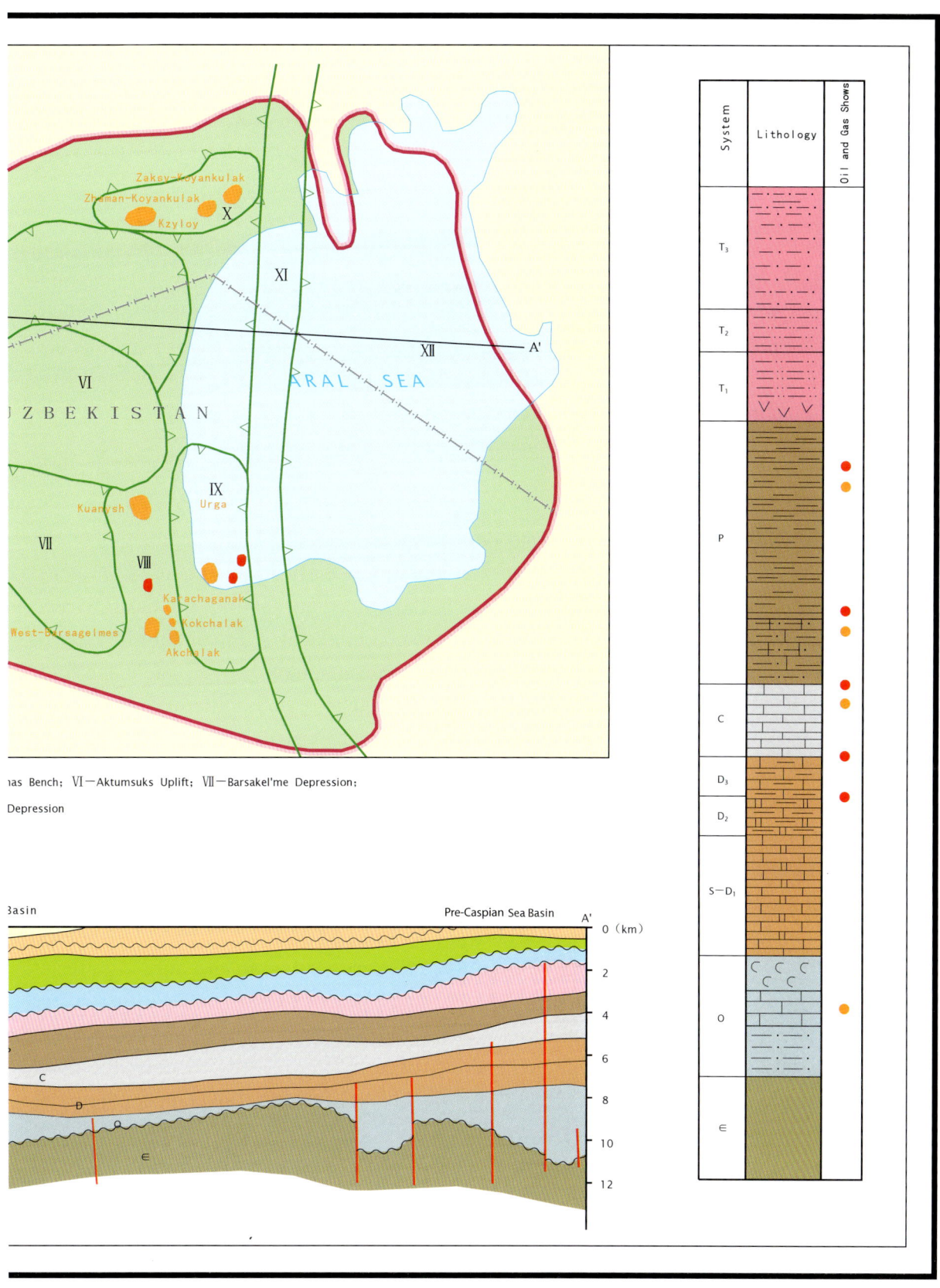

26 North Ustyurt Basin

Geography

The North Ustyurt Basin lies on the northwestern shore of the Caspian Sea, largely within Kazakhstan (70 per cent), but also extending eastwards into Uzbekistan and Turkmenistan.

History

Initial exploration started in the basin during the 1950s. The first gas field was discovered in Eocene sandstones. At the end of 1960s, two medium sized oil fields were discovered. In the mid-1970s a few large oil fields were discovered, including Kalamakas (Fig. 26.1) and Kalazhanbas, which were developed by steam injection because of heavy oil. The Kalamkas oil field was discovered in 1976 at a depth of 505–936 m in Jurassic and Cretaceous sandstone reservoirs 33 m thick. The gas pipeline of Central Asia passes through the basin to supply gas to Russia. Two abandoned gas fields in Kazakhstan are now used for gas storage.

Regional geology

Upper Palaeozoic passive margin carbonate deposits were laid down on the margin of the Euramerican Plate, but were deformed in the Late Carboniferous–Early Permian Uralian Orogeny. During Late Permian to Early Jurassic times, rifting to the immediate south of the basin formed a major graben. To the north, in the North Ustyurt area, continental sedimentation prevailed. Suturing of microcontinents at the end of the Triassic led to inversion of the rift system and folding and erosion of the North Ustyurt Basin area. In the subsequent period of thermal subsidence, fluvial, lacustrine and shallow marine clastics were deposited in Early to Middle Jurassic times. These are overlain by Late Jurassic marine carbonates. Localized uplift in Neocomian–Aptian times formed a widespread unconformity which is overlain by shallow marine clastics and Late Cretaceous carbonates. Shallow marine deposition continued until the Early Tertiary.

The organic matter content of Jurassic rocks reached 6.5 per cent, and they are thought to be mature for hydrocarbon generation in basinal deeps. Triassic shales and marls and Devonian–Permian volcanic clastics and carbonates are thought to have generated gas, especially in the east of the basin. Reservoirs occur in Carboniferous to Eocene strata, but the most significant are shallow marine to transitional sandstones and silty sandstones of Middle Jurassic age. Lower Cretaceous shoal sandstones are also significant oil and gas reservoirs. Upper Jurassic reservoirs have high porosity of 24 per cent and high permeability of 57–67 mD. Local seal units are found throughout the sedimentary sequence, but Upper Jurassic, Lower Cretaceous and Upper Oligocene intervals provide semi-regional and regional seals.

Structures in the basin are related to five main phases of tectonic activity: Uralian compression; Permian–Triassic rifting; Cimmerian inversion; Jurassic–Eocene subsidence (with a pulse of transpression in the Early Cretaceous); and Tertiary hinterland compression.

World Atlas of Oil And Gas Basins, First Edition. Li Guoyu.
© 2011 John Wiley & Sons, Ltd. Published 2011 by John Wiley & Sons, Ltd.

North Ustyurt Basin

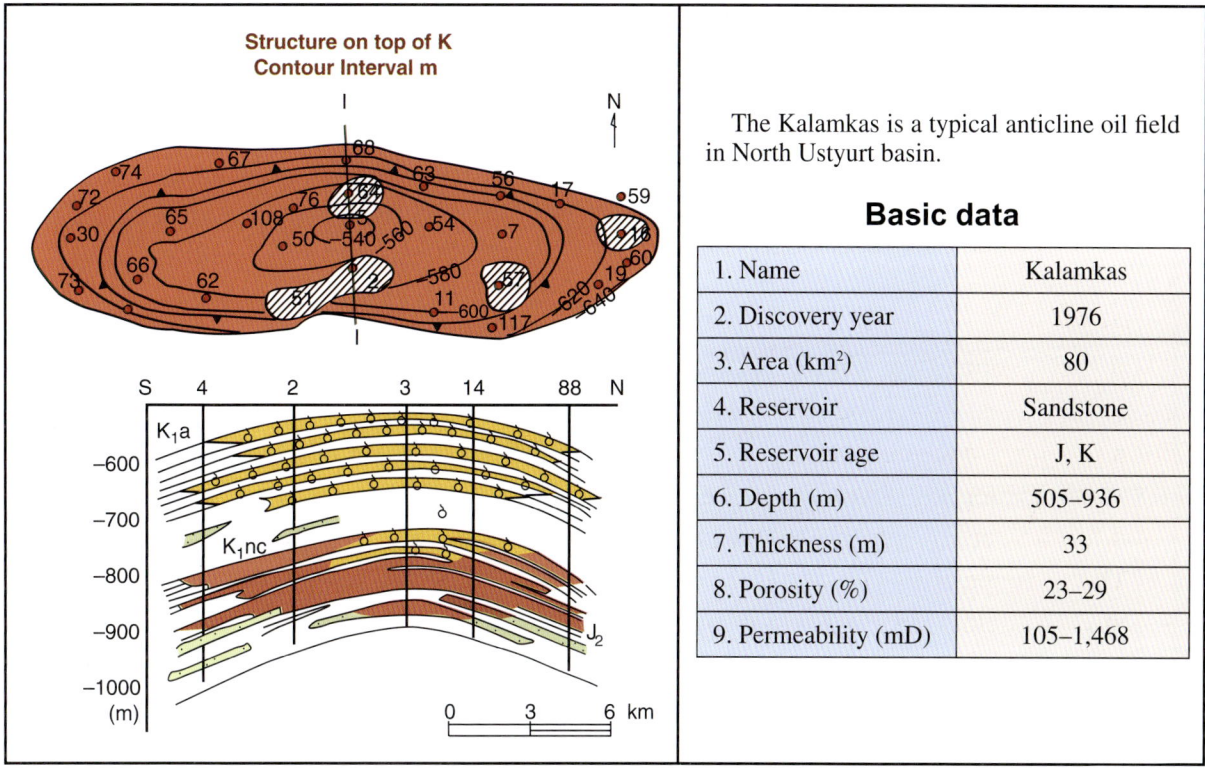

Fig. 26.1 Details of the Kalamkas oil field.

Carboniferous, Jurassic, Cretaceous and Eocene strata are recognized as play zones. Jurassic plays account for almost 70 per cent of the basin's hydrocarbons, which are largely hosted by the Middle Jurassic structural-stratigraphic play. The Lower Cretaceous structural-stratigraphic play contains about 25 per cent of the basin's hydrocarbon reserves. The Jurassic and the Late Palaeozoic Systems have been recognized as two petroleum systems in the basin. The Late Palaeozoic Petroleum System is relatively minor, containing only a small proportion of the basin's discovered hydrocarbons. Middle Jurassic plays are still likely to be a focus for exploration, especially in poorly explored parts of the basin, including the offshore and areas to the east. Cretaceous potential is limited to areas of Late Jurassic–Neocomian erosion (probably restricted to the North Buzachi Uplift), but it is likely that gas will be found in Eocene sands in the north and east of the basin. in The Palaeozoic sediments in many areas of the basin have been identified as stratigraphic prospects, and in the northwest of the basin analogies have been drawn with the Tengiz field in the Pre-Caspian Sea Basin to the north.

MANGYSHLAK BASIN

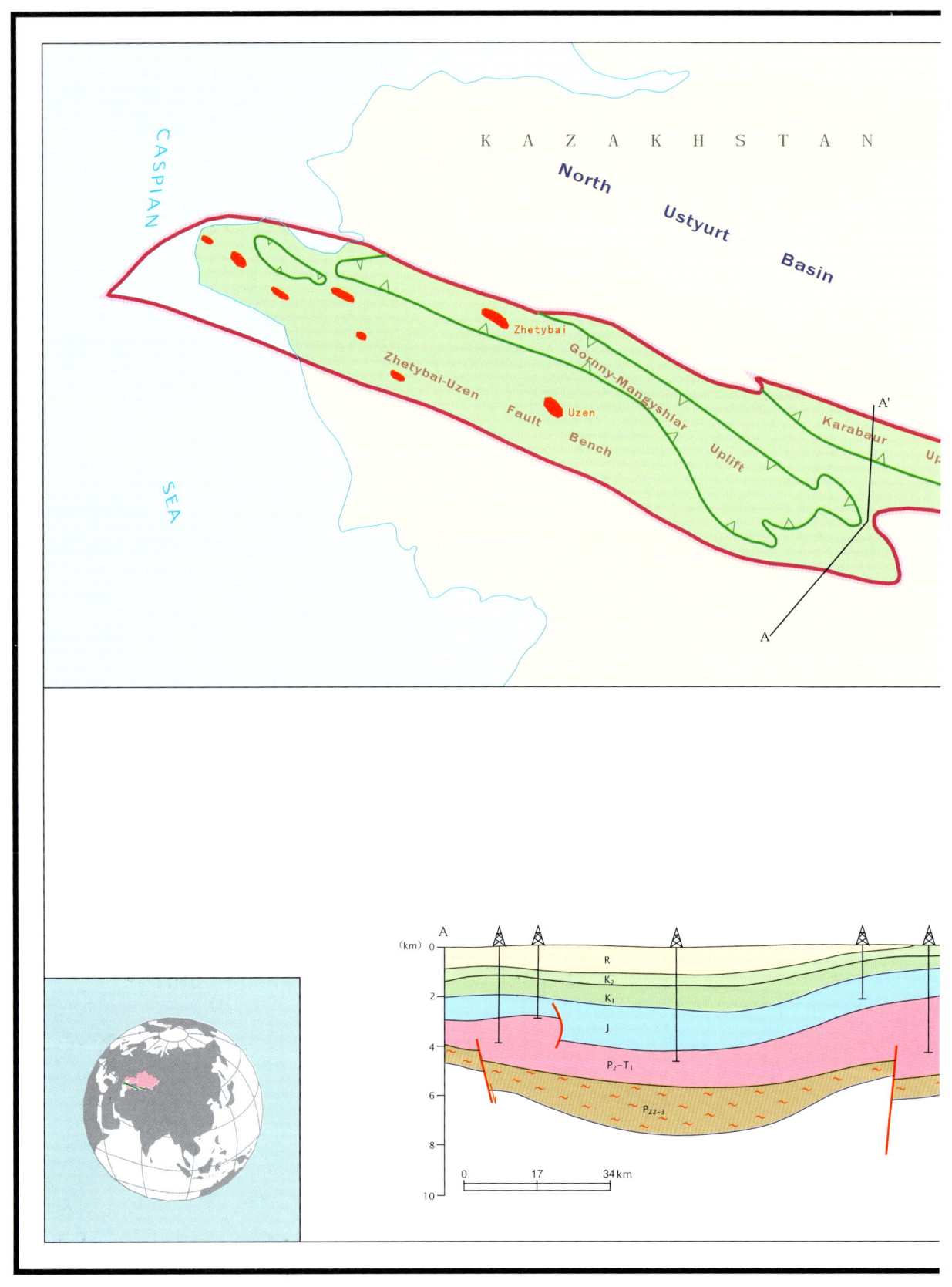

CHAPTER 27

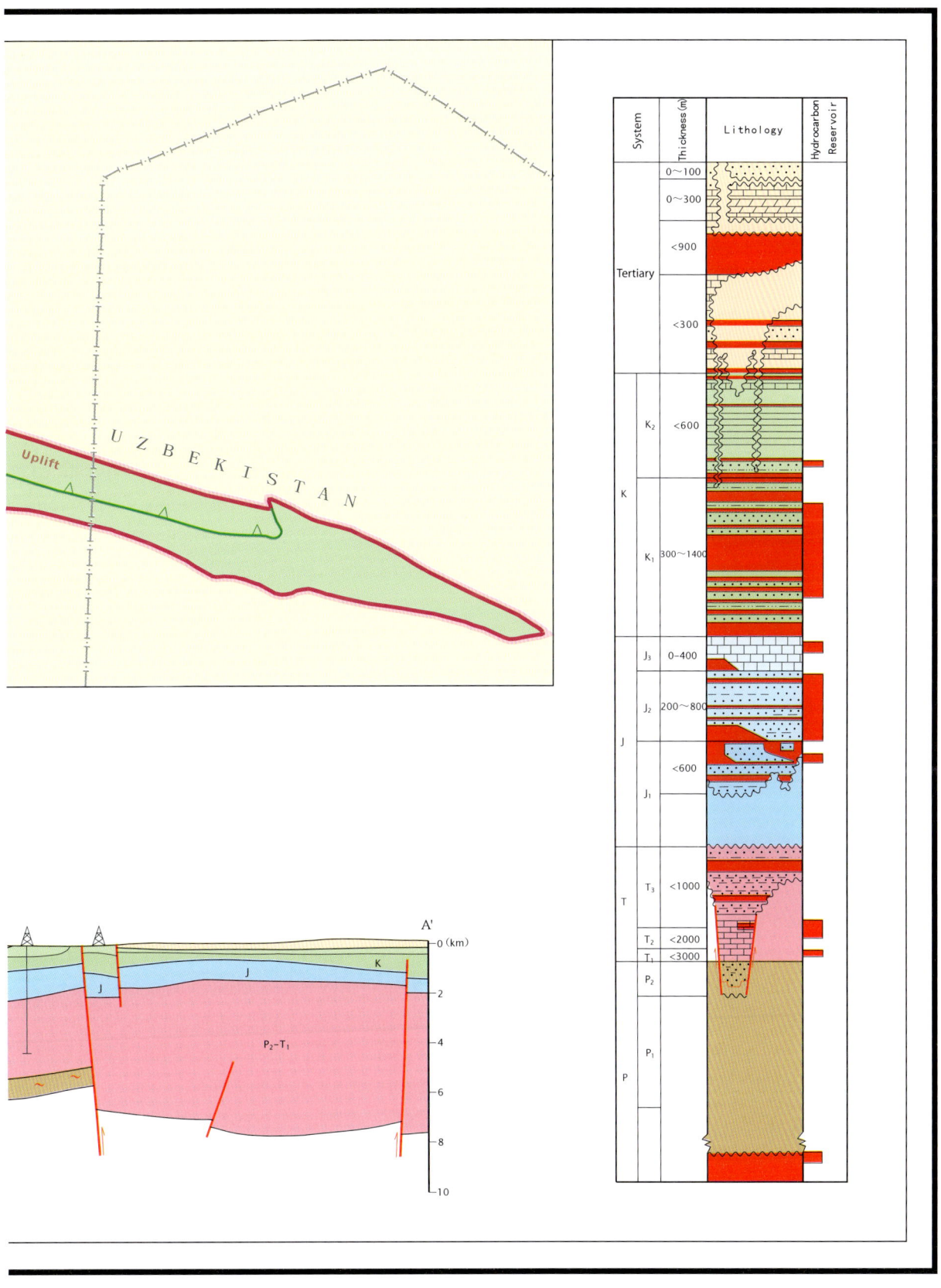

27 Mangyshlak Basin

Geography

The Mangyshlak Basin is located on the peninsula Mangyshlak of Kazakhstan on the eastern margin of the Caspian Sea and it extends into Uzbekistan and Turkmenistan. It extends offshore into the Caspian Sea where a small sector lies in Russian territorial waters.

History

Exploration began in the mid-1950s. The two giant oil fields Zhetybay and Uzen (Fig. 27.1) were discovered in 1960 and 1961, respectively. The Uzen oil field has an area of 405 km^2. Reservoirs are Jurassic and Cretaceous sandstones with depths ranging from 1200 to 1500 m, thickness from 11 to 33 m, porosity 21–24 per cent and permeability 1–24 mD. Recoverable reserves are 454 million t. The crude oil has a high wax content of 22–26 per cent, so development of the basin needs to be conducted using stream techniques. Offshore surveying began in the early 1960s and the first wells were drilled in 1976. Oil production began in 1965 and crude output peaked in 1975 at a rate of over 20 million t but declined to 9.5 million t in 2001, with the Uzen oil field producing 75 per cent of this. Gas production began in 1965 from Uzen and Kazakh output peaked in 1977 at over 50,046 billion m^3. Production has since been declining. In Uzbekistan, the Shakhpakhty hydrocarbon field was brought on stream in 1971. Another interesting situation is the discovery in 1980 of the Yimashay oil field in the granitic basement. This oil field is in an anticlinal trap and has an area of 30 km^2. The oil column reached 190 m (3580–3778 m) in fractured granite reservoirs with porosity of 5.2 per cent, permeability of 2.4 mD and a high productive rate per well.

Regional geology

The sedimentary cover comprises Palaeo–Mesozoic strata with thicknesses over 10 km. Rifting in Late Permian times led to the creation of a large west-northwest trending rift valley, within which up to 9 km of Permian and Triassic sediments accumulated. Due to these unique geological conditions there many source rocks, seals and traps. Intense compression at the end of the Triassic inverted the graben and caused folding and erosion. In the ensuing thermal subsidence phase, Early Jurassic continental clastics were overlain by increasingly marine sediments, with limestone deposition in the latest Jurassic. A short period of uplift in the Early Cretaceous was succeeded by renewed thermal subsidence and limestone deposition was again predominant in Late Cretaceous times. Effects of the collision of the Arabian and Indian plates with Eurasia were felt from Late Eocene times onwards and produced several phases of deformation, uplift and erosion.

Two major source rock units have been identified: Early–Middle Triassic and Early–Middle Jurassic. The Triassic shales were thought to have begun generating oil in the Late Jurassic; generation from Jurassic siltstones and shales began in the Late Cretaceous. Reservoir units are in Palaeozoic basement to Late Cretaceous strata, but the most important are Middle Jurassic sands and silts of units Yu1-XII. Sealing units are Triassic to Late Cretaceous strata, but the lack of a regional seal between the Triassic and Jurassic strata leads to vertical migration into the Jurassic

Mangyshlak Basin

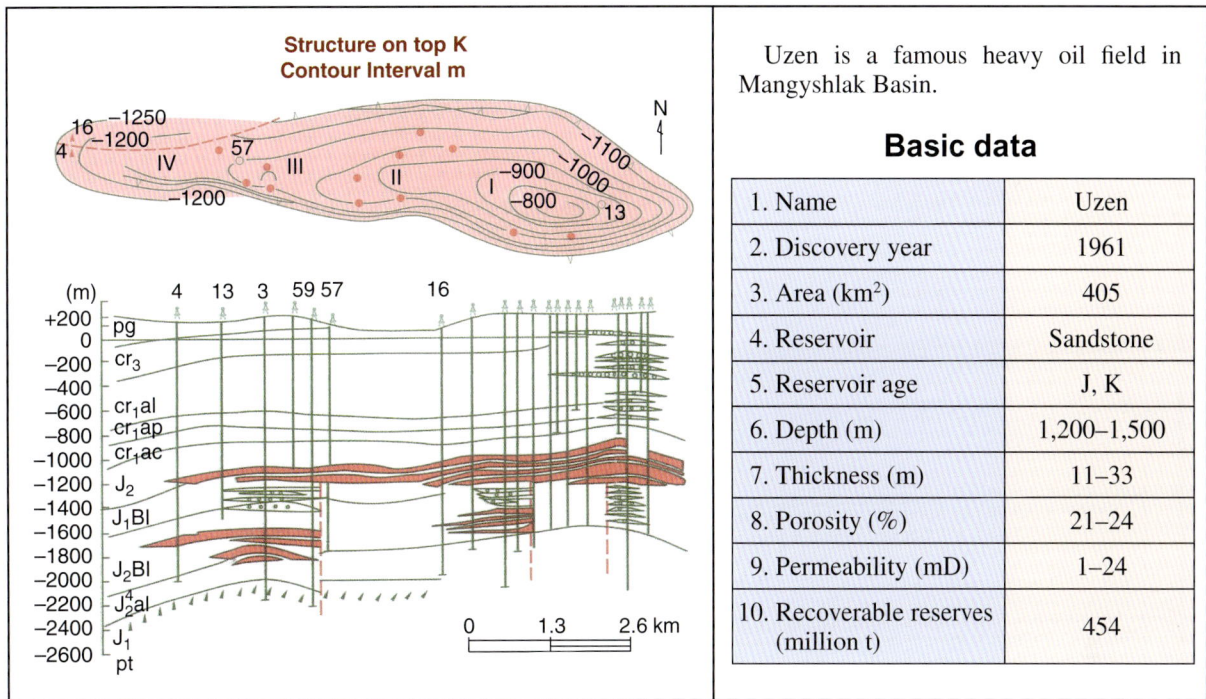

Uzen is a famous heavy oil field in Mangyshlak Basin.

Basic data

1. Name	Uzen
2. Discovery year	1961
3. Area (km^2)	405
4. Reservoir	Sandstone
5. Reservoir age	J, K
6. Depth (m)	1,200–1,500
7. Thickness (m)	11–33
8. Porosity (%)	21–24
9. Permeability (mD)	1–24
10. Recoverable reserves (million t)	454

Fig. 27.1 Details of the Uzen oil field.

reservoirs. In addition, the seal at the top of the Jurassic sequence is semi-permeable and has allowed significant degassing of trapped hydrocarbons.

Structures in the basin were formed in five main periods of tectonic activity: Permian–Triassic rifting; Late Triassic–Early Jurassic inversion; Late Mesozoic thermal subsidence; Early Cretaceous transpression and Mid–Late Tertiary hinterland compression. Eleven plays have been recognized in the basin, but the most significant are the four Jurassic plays, which together contain over 90 per cent of the basin's hydrocarbon reserves. Of these, the Middle Jurassic structural-stratigraphic play is found in 23 fields. One petroleum Triassic–Jurassic system is recognized in the basin.

SOUTH CASPIAN SEA BASIN

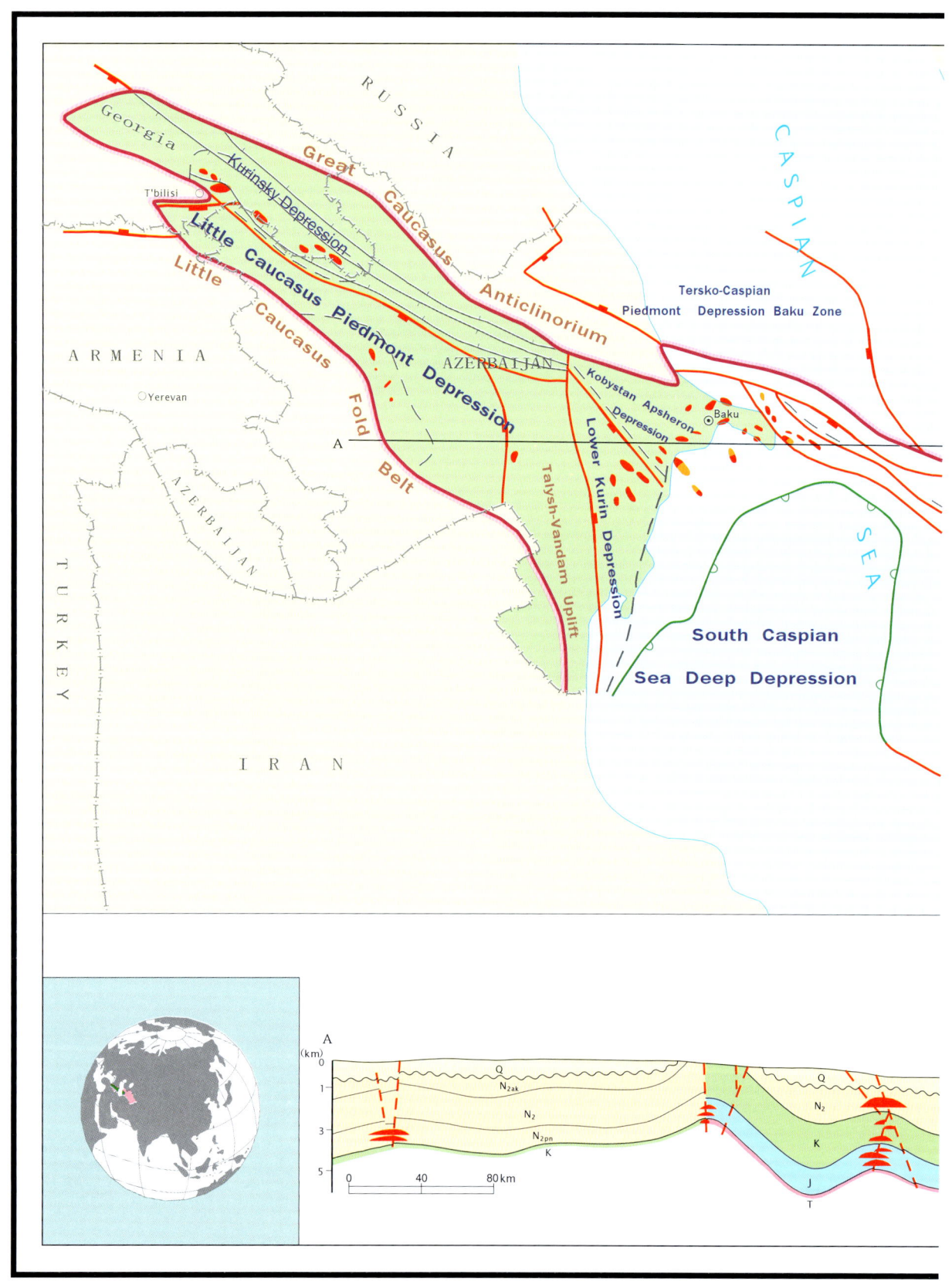

CHAPTER 28

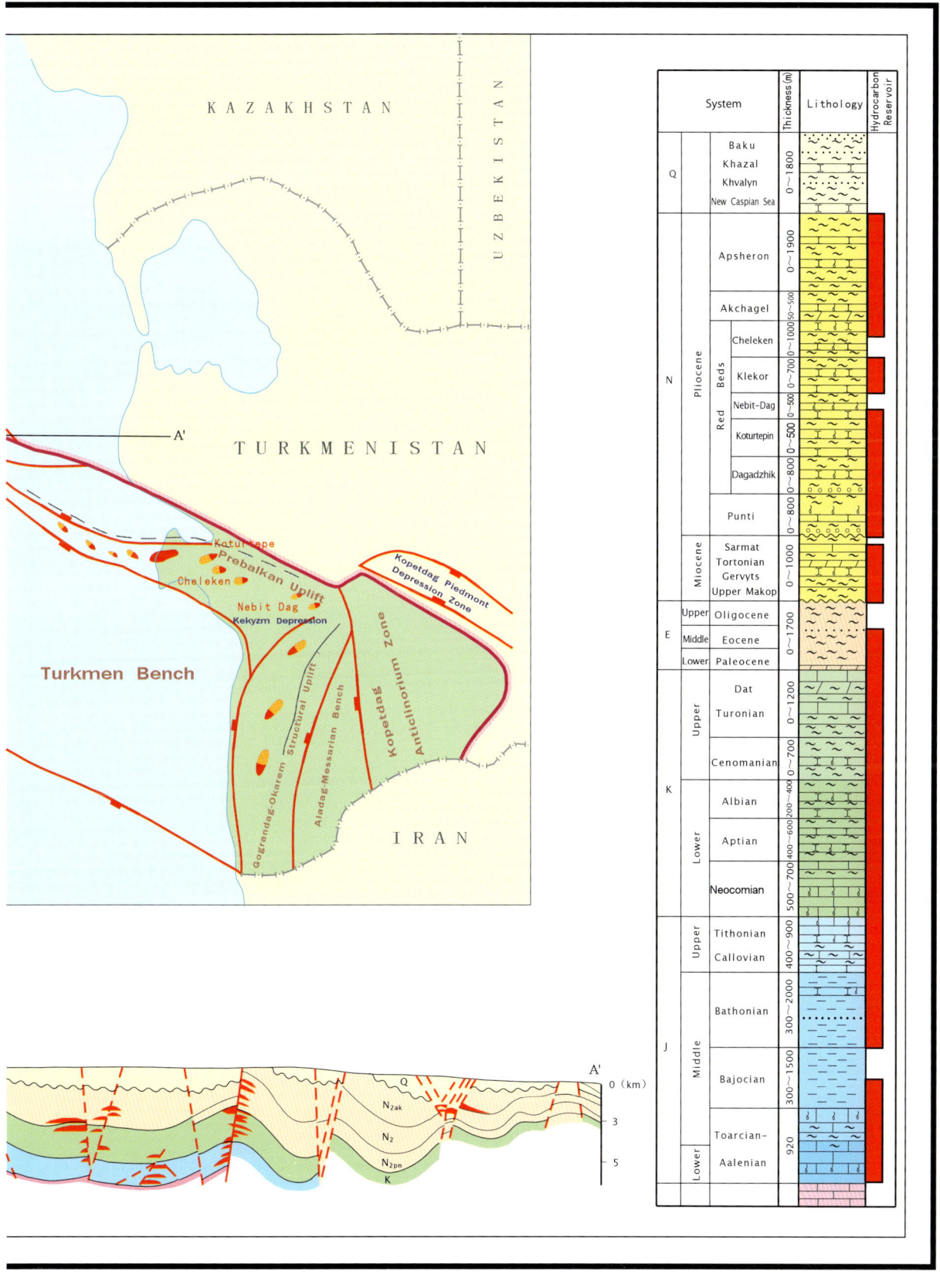

28 South Caspian Sea Basin

Geography

The South Caspian Sea Basin occupies Iran, Turkmenistan, Azerbaijan, Armenia and Georgia. It is an elongated intermontane basin with an area of 90,000 km². The central portion of the basin is covered by the southern Caspian Sea.

History

The South Caspian Basin has an extremely long exploration history, and more than 130 fields have been discovered. Significant discoveries are still being made and future potential is deemed to be very good. The presence of oil has been recognized in the Baku area of Azerbaijan for many centuries. The first well was finished in 1848. The giant Surakhany oil field was discovered in 1904. Exploration began in Georgia and Turkmenistan in the 1930s and the large Nebitdag oil field was discovered in 1933. The first offshore well was drilled adjacent to small islands in the shallow waters of the Caspian Sea near Baku and fields extend on and offshore (BibiEybat 1873; Pirallakhi 1901). The giant Neft Dashlary oil field was the first offshore discovery in 1950. The Neft-Kamin (Fig. 28.1) is another well known offshore oil field also discovered at about this time. Few details are known about development drilling activity, but it is known that over 35,000 development wells have

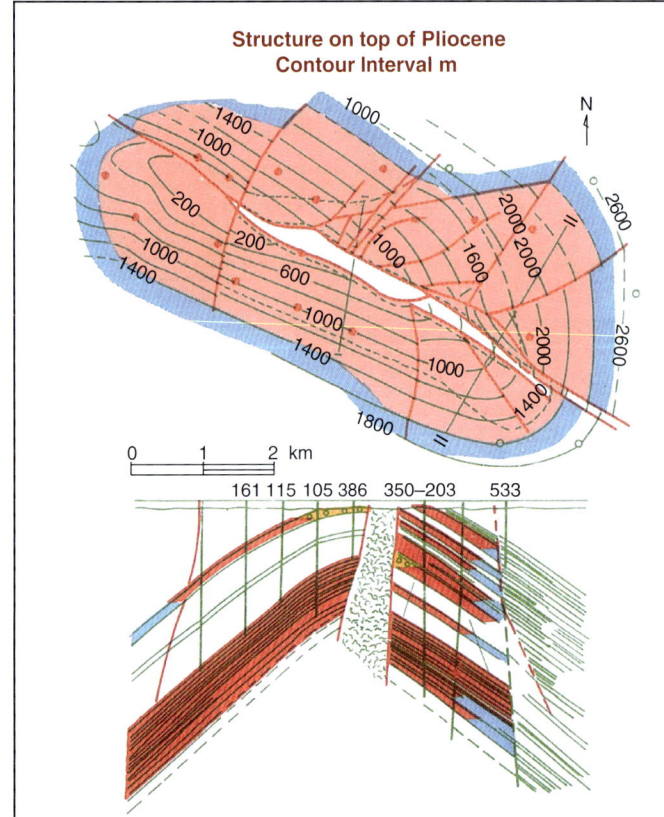

Neft Kamin is a large mud volcano oil field in the South Caspian Sea Basin.

Basic data

1. Name	Neft Kamin
2. Discovery year	1949
3. Area (km²)	32
4. Reservoir	Sandstone
5. Reservoir age	Tertiary
6. Depth (m)	200–680
7. Thickness (m)	120
8. Porosity (%)	21–26
9. Permeability (mD)	3–1,200
10. Recoverable reserves (million t)	190
11. Peak oil production (million t)	0.63
12. Water depth (m)	18.3

Fig. 28.1 Details of the Neft Kamin oil field.

World Atlas of Oil And Gas Basins, First Edition. Li Guoyu.
© 2011 John Wiley & Sons, Ltd. Published 2011 by John Wiley & Sons, Ltd.

been drilled throughout the basin, largely in Azerbaijan. The development of Neft Dashlary was particularly significant in that a whole town was built on trestles, linked by miles of walkways.

Regional geology

The South Caspian Basin correlates with the intermontane sag areas of the Alpine fold belt systems of the Caucasus and West Turkmenia, which open to the South Caspian Depression where oceanic crust is developed. The basin occupies an area of about 2,000,000 km^2. The largest onshore tectonic elements of the basin are the Lower Kurin Depression in the Caucasus and the Aladag-Messarian Bench in west Turkmenistan (Institute of Microeconomics, 2004). The basement of the basin comprises pre-Palaeozoic-Alpine rocks and occurs at a depth ranging from 5 to 10 km in the Lower Kurin Depression, and up to 20 km in the central portion of the South Caspian Sea Deep Depression. The sedimentary cover consists of a Jurassic–Eocene terrigenous–marine carbonate complex up to 8000 m thick and a Oligocene–Pliocene sandy–clayey complex up to 7000 m thick.

Oil and gas accumulations are concentrated mainly in the Pliocene rock sequence. Reservoirs are represented by sandstones with porosity of up to 22 per cent. Regional seals are clayey sequences of the upper part of Pliocene succession. Source rocks are Oligocene–Miocene clays. Anticlinal structures complicated by faults and mud volcanism serve as traps. The basins of this type are characterized by very high petroleum potential.

Following intense deformation in the Triassic, a period of extension led to the opening of a shallow marine basin. This back-arc basin development continued throughout the Jurassic and Cretaceous, interrupted by periods of compression related to the suturing of various continental terranes. A major period of regional compression associated with the final closure of the Tethys Ocean lasted throughout the Tertiary, and Eocene tuffs were overlain by Oligocene–Miocene clastic molassess. Uplift of the Greater Caucasus and Kopet-Dag fold belts in the Late Miocene led to the creation of a rapidly subsiding intermontane basin, which was filled with a thick sequence of Pliocene and Pleistocene clastic molasses of.

Source rocks in the basin are Oligocene–Miocene Maykop Group shales, although Jurassic, Lower Cretaceous, Eocene and Pliocene shales and carbonates have also been invoked as mature source rocks. Reservoir intervals in the basin are Cretaceous to Pliocene strata, with the bulk of the reserves being in the mid-Pliocene fluviodeltaic clastics of the Productive/Variegated Series. Eocene tuffs and clastics and Oligocene–Miocene clastics are minor reservoirs in the west of the basin. Seals are Paleocene to Upper Pliocene strata. The Mid-Pliocene intraformational shale members of the Productive/Variegated Series are the most important but the Upper Pliocene shales of the Akchagyl and Absheron Formations are also significant.

Structures are related to five main periods of deformation: basin opening in the Early Jurassic; Bathonian compression; Callovian–Maastrichtian extension; Danian–Middle Eocene compression; and Late Eocene to recent compression and transpression.

Five groups of plays have been recognized within the basin, but over 90 per cent of the basin's reserves are found in the Productive/Variegated Series group of plays. The uppermost Pliocene Absheron–Akchagyl group of plays is also significant, accounting for 6.5 per cent of the basin's recoverable hydrocarbons. The petroleum systems within the basin are the Jurassic–Cretaceous, Upper Cretaceous–Eocene and Maykop–Productive Systems. Significant levels of hydrocarbon loss have occurred due to late-stage (Late Pliocene) deformation.

AMUR (KARA-KUM) BASIN

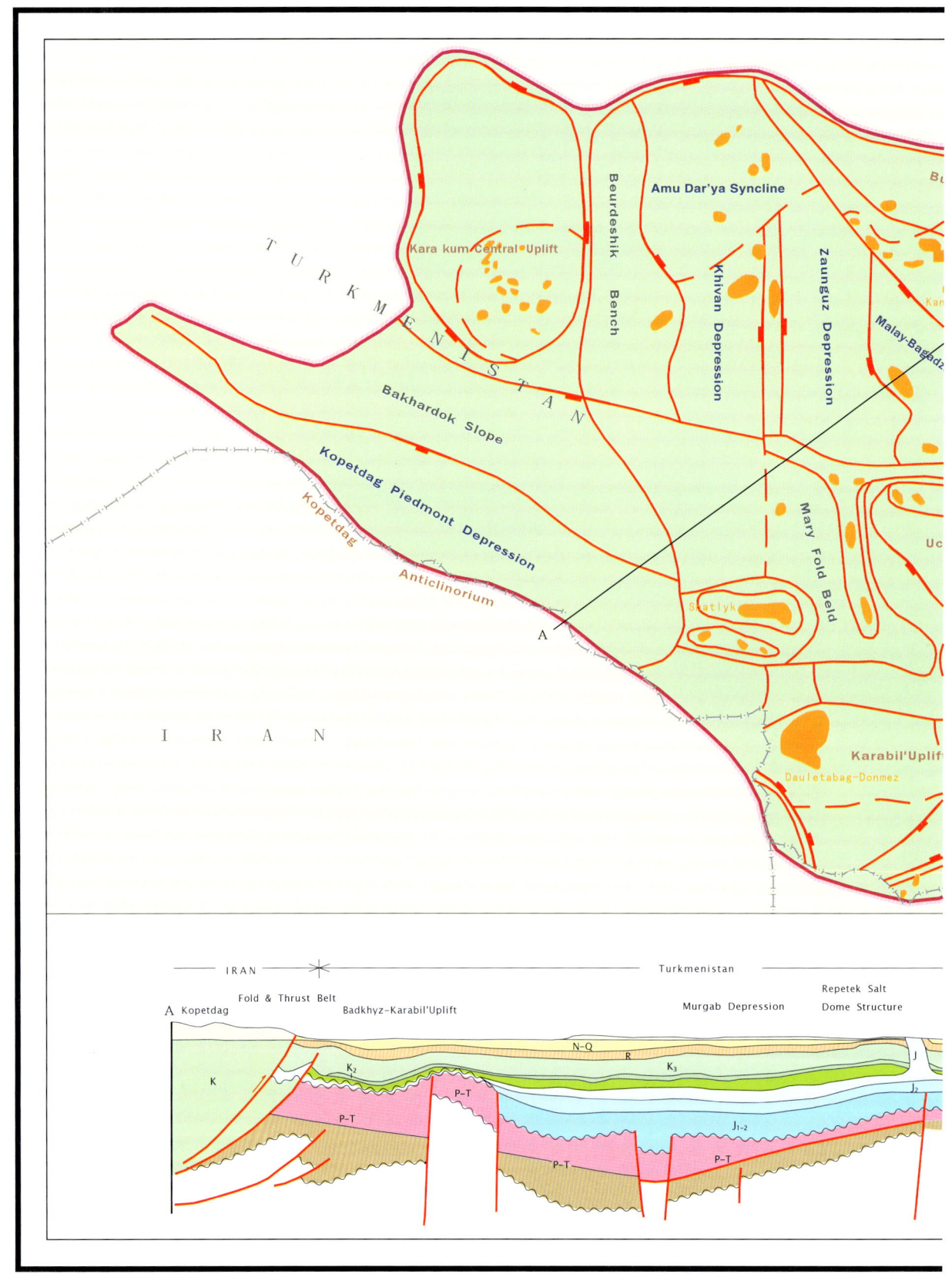

CHAPTER 29

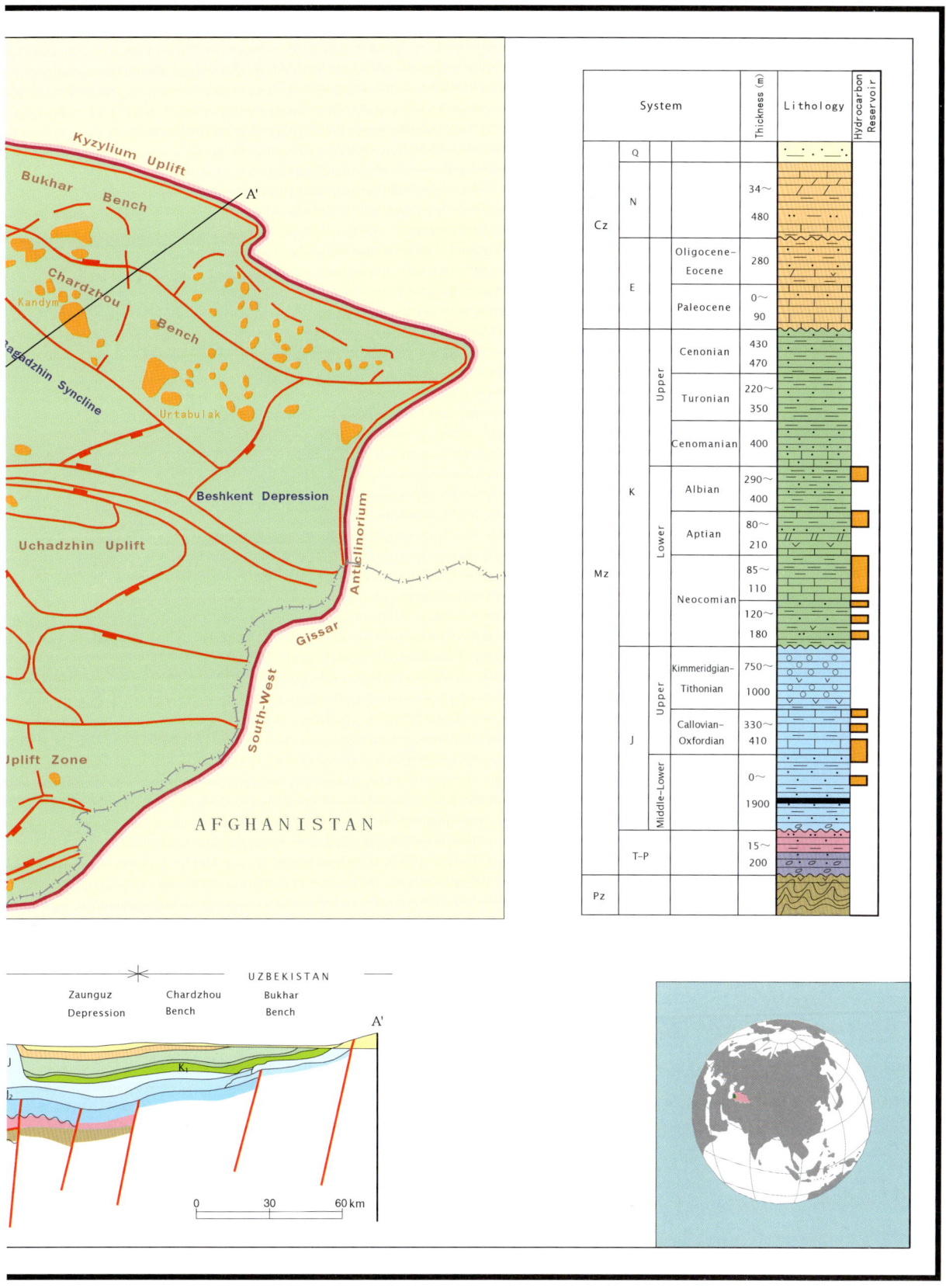

29 Amur (Kara-Kum) Basin

Geography

Amur Basin is mostly located in western Turkmenistan, facing the Caspian Sea to the west, with an area of 530,000 km². It is the largest among the basins of Central Asia and is a gas producing region.

History

During the 1930s–1950s, several field geological surveys were completed. In 1953 the first gas field was discovered in Uzbekistan. The giant gas field Gazli was discovered in 1956. Gas production reached a maximum of 13 billion m³ in the late 1960s. The major oil and gas fields of Turkmenistan and Uzbekistan, and their year of discovery, are listed in Tables 29.1 and 29.2.

Regional geology

The large Amu Dar'ya petroleum basin is the only one in the region, with an area of 360,000 km². It is located in the southeastern portion of the Turanian Plate and consists the Amu Dar'ya Syneclise, the Kara-Kum Central Uplift and the Kopetdag Piedmont Depression. The heterogeneous Hercynian basement of the basin includes individual blocks of Baikalian basement reworked during the Hercynian cycle of tectogenesis. The basement occurs at depths ranging from 2–3 km to 15 km. The sedimentary cover comprises Permo-Triassic terrigenous continental and volcano-sedimentary deposits with thicknesses up to 3000 m. Higher in the succession, Lower and Middle Jurassic terrigenous rocks up to 1200 m thick and Upper Jurassic carbonate and salt-bearing deposits up to 1,600 m thick occur. Cretaceous–Cenozoic terrigenous and carbonate rocks up to 1600 m thick occur at the top of the succession.

Reservoir rocks are Lower–Middle Jurassic sandstones with porosity of 5–17 per cent; Upper Jurassic carbonates with reef reservoirs of good quality where porosity reaches up to 20 per cent; Lower Cretaceous terrigenous and carbonate rocks where porosity of sandstones is up to 20 per cent and those of limestone reach 5–11 per cent. Regional seals are Upper Jurassic salt-bearing sequences and clayey beds of the upper part of the Cretaceous succession. Source rocks are thought to be clayey-carbonate and clayey deposits of Upper Jurassic-Cretaceous.

There are 97 gas and 23 oil and oil/gas fields discovered in the basin. The most important of them are Gazlin, Shatlyk and Dawletabad. The main type of trap is anticlinal (Fig. 29.1), although traps associated with individual reef build-ups are widely developed. Traps associated with reservoir pinching are observed on the slopes of the uplifts. The petroleum potential of the basin is considered to be high. The Gazli gas field was discovered in 1956 in an anticlinal trap. It had recoverable gas reserves of 501 billion m³ in a Cretaceous sandstone reservoir

World Atlas of Oil And Gas Basins, First Edition. Li Guoyu.
© 2011 John Wiley & Sons, Ltd. Published 2011 by John Wiley & Sons, Ltd.

Table 29.1 Major oil and gas fields of Turkmenistan

Name	Discovery year	Recoverable reserves		Basin	Depth (m)	Trap	Reservoir		Remarks
		Oil (million t)	Gas (billion m³)				Age	Lithology	
Kotur Tepe	1956	240.0	48	South Caspian	2600	Anticline	N	Sandstone	
Cheleken	1965	90.0	—	South Caspian	2500	Anticline	E	Sandstone	
Dauletabad	1976	—	1500	Amur	2900	Stratum formation	K	Sandstone	Gas field
Shatlyk	1968	—	975.5	Amur	3500	Anticline	K	Sandstone	Gas field
Naip	1970	—	174.4	Amur	1900	Anticline	K	Sandstone	Gas field
Kirpichli	1972	—	168.8	Amur	3000	Anticline	K	Carbonate rock	Gas field
Achakchli	1966	—	155.3	Amur	1600	Anticline	K	Sandstone	Gas field
Gugurtli	1965	—	113	Amur	1000	Anticline	K	Limestone	Gas field
Samantepe	1964	—	101.3	Amur	2300	Anticline	J	Carbonate rock	Gas field

Table 29.2 Major oil and gas fields of Uzbekistan

Name	Discovery year	Recoverable reserves		Basin	Depth (m)	Trap	Reservoir	
		Oil (million t)	Gas (billion m³)				Age	Lithology
Shyrtan	1976		548	Amur	2800	Anticline–reef	K	Carbonate rock
Gazli	1956		501	Amur	700	Anticline	K	Sandstone
Zevard	1968		188	Amur	2600	Reef	J	Carbonate rock
Kandym	1967		158	Amur	2500	Anticline	J	Carbonate rock
Dendizkul Khauzak	1966		124	Amur	2300	Anticline	J	Carbonate rock
Urtabulak	1963		102	Amur	2500	Anticline	J	Carbonate rock

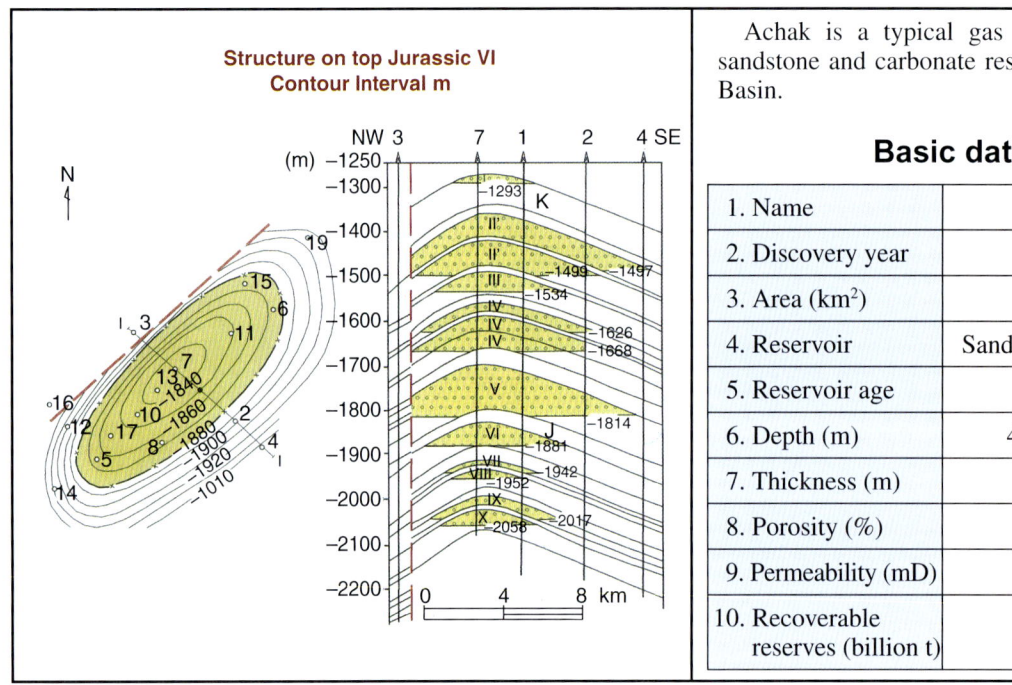

Achak is a typical gas field with both sandstone and carbonate reservoirs in Amur Basin.

Basic data

1. Name	Achak
2. Discovery year	1966
3. Area (km²)	160
4. Reservoir	Sandstone Carbonate
5. Reservoir age	J, K
6. Depth (m)	4,120–2,155
7. Thickness (m)	175
8. Porosity (%)	14.5
9. Permeability (mD)	270
10. Recoverable reserves (billion t)	170

Fig. 29.1 Details of the Achak gas field.

CHU-SARYSU BASIN

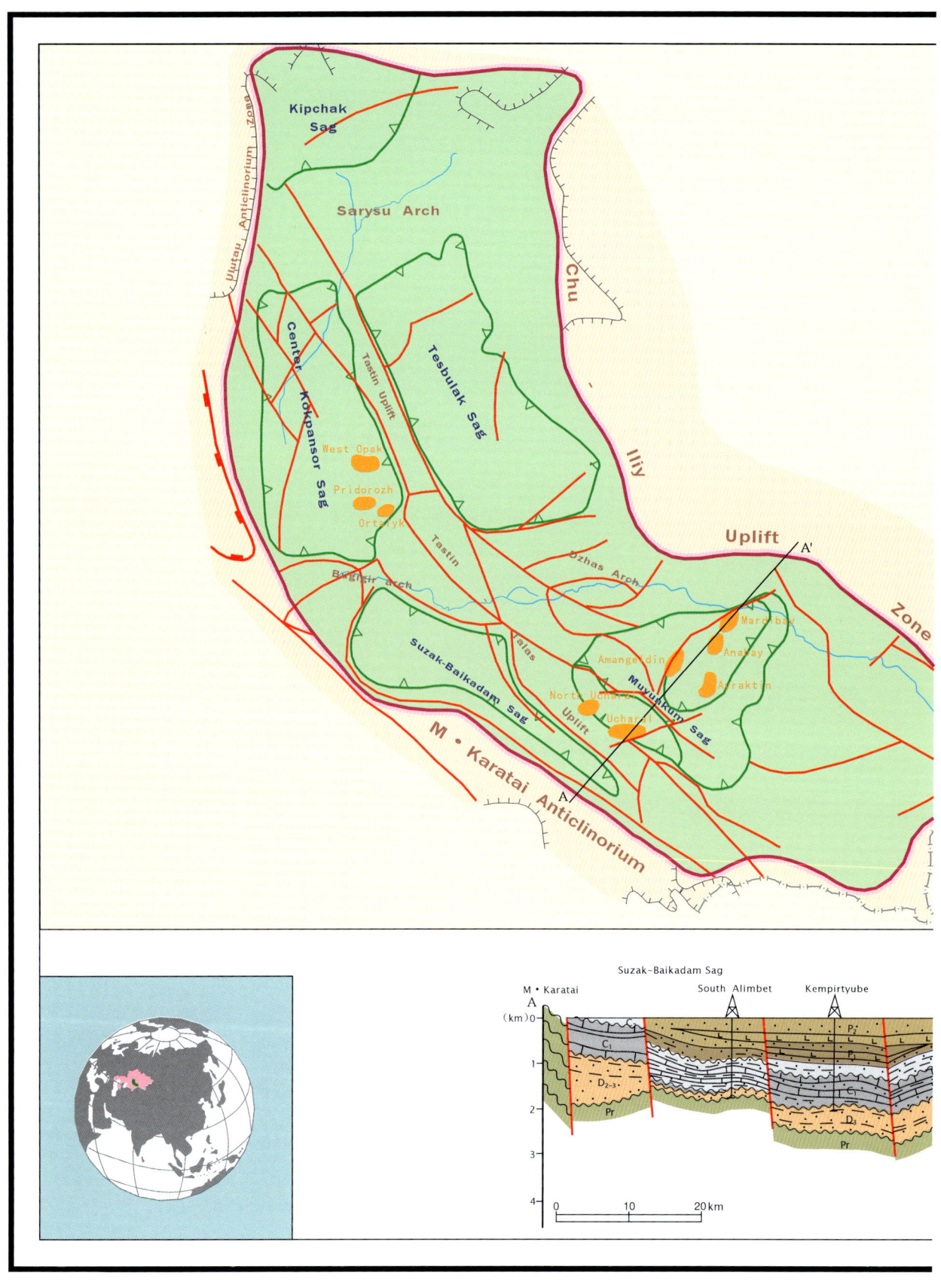

CHAPTER 30

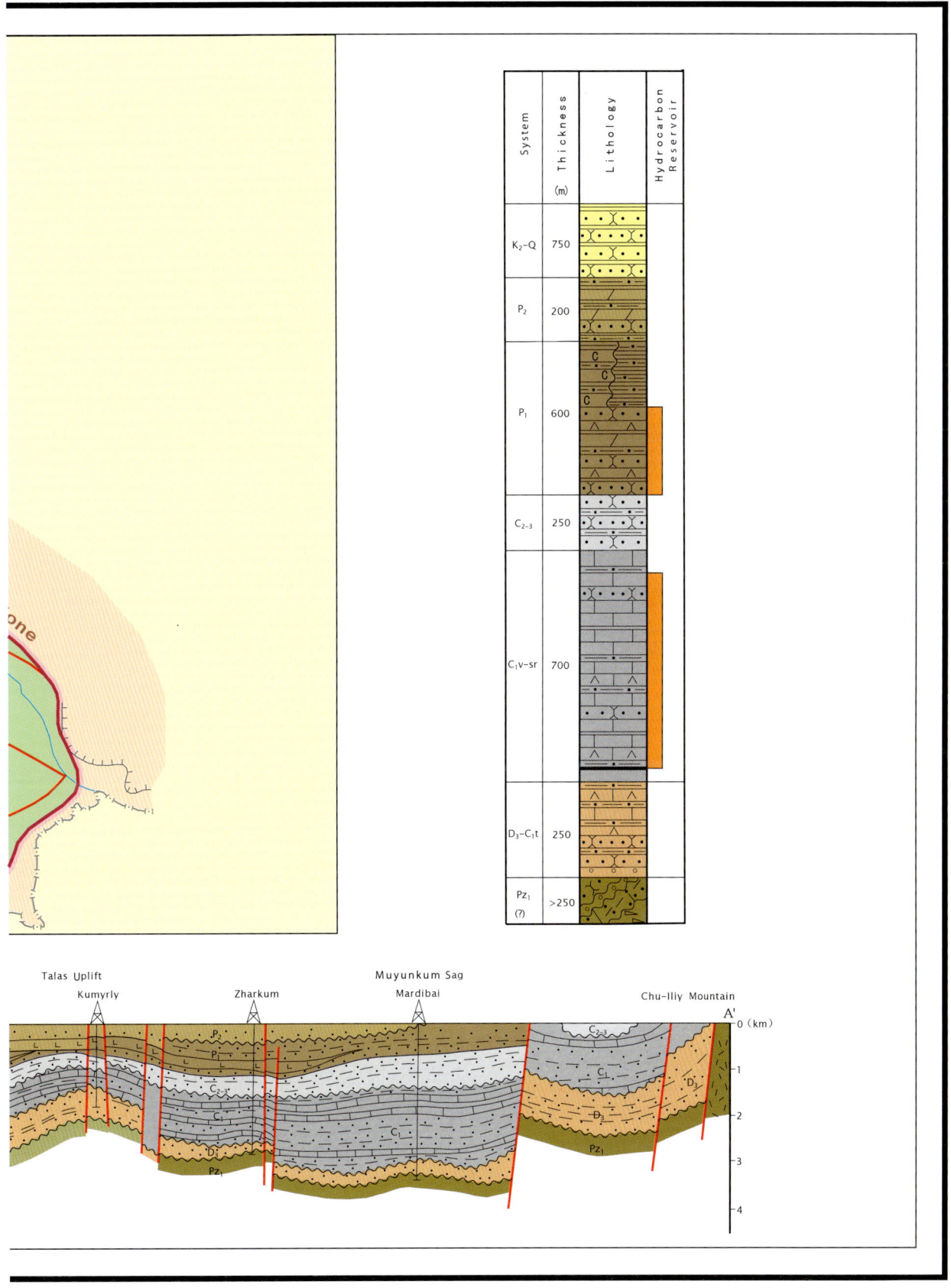

30 Chu-Sarysu Basin

Geography

The Chu-Sarysu Basin is located in central Kazakhstan, having an area of 150,000 km^2. It is an intermontane basin filled by Devonian to Tertiary sediments with a thickness of up to 8000 m.

History

Oil exploration started in 1960. The local structures at the bottom of the Permian gypsum and at the bottom of the Lower Carboniferous carbonate are mapped from the results of seismic exploration. Since 1962 a total of 136 parametric and exploration wells have been drilled. During 1973–1974 a Devonian sandstone gas field was discovered beneath salt-bearing deposits. The sedimentary cover is composed of Palaeozoic and Mesozoic strata.

Source rocks have been identified in Devonian, Carbonaceous and Permian strata (e.g. Wushatar; Fig. 30.1) under salt-bearing sediments. Reservoir rocks are associated with sandy deposits, mostly in Mesozoic strata, and also with fractured Palaeozoic carbonate rocks. Peculiar reservoir types have been identified in weathered Archaean and Proterozoic rocks of the basement. Widely occurring here are clayey seals and sealing strata associated with gypsiferous rocks and rock salt.

Traps are represented by structural and non-anticlinal types. Oil and gas accumulations are multilayered and mainly confined to Palaeozoic deposits. Petroleum potential of this basin is estimated as high. At present, 21 gas fields have been discovered. The content of gas finds is typically: methane 70 per cent, nitrogen 30 per cent, sulphur 2.5 per cent and carbon dioxide 0.38 per cent.

Regional geology

Within this basin lie three geological elements; a central uplift and two depressions on both sides of the uplift. The thickness of Palaeozoic carbonate rocks and Mesozoic clastic rocks reaches 8000 m in depressions. Two typical gas fields are Ortaluik and Pulidolosh.

Ortaluik gas field is located in the western depression, and is a faulted anticline with an area of 60 km^2. In 1976 gas was discovered in weathered Proterozoic basement and Lower Permian sandstone with depths in the 1120–2550 m range. The gas reservoir of the Proterozoic basement is a massive fractured metamorphosed rock, with an area of 20 km^2, thickness of 92 m, fracture porosity of 2 per cent and initial daily production rate 21,000 m^3. The Lower Permian gas reservoir, in an anticlinal trap, comprises fractured sandstone with a thickness of 76 m, porosity of 10.2 per cent, depth of 1,120 m, initial pressure of 11.5 MPa, temperature 59°C, initial daily production rate 30,000–50,000 m^3 and methane content of 52.8–83.7 per cent.

Pulidolosh gas field is located in the western depression and is a faulted anticline with an area of 20 km^2. Thickness of Devonian sandstone reservoir is 210 m, depth is 2400 m, porosity 7–18 per cent, permeability 38 mD, initial pressure 25.8 MPa and initial daily production rate 57,000 m^3.

World Atlas of Oil And Gas Basins, First Edition. Li Guoyu.
© 2011 John Wiley & Sons, Ltd. Published 2011 by John Wiley & Sons, Ltd.

Chu-Sarysu Basin

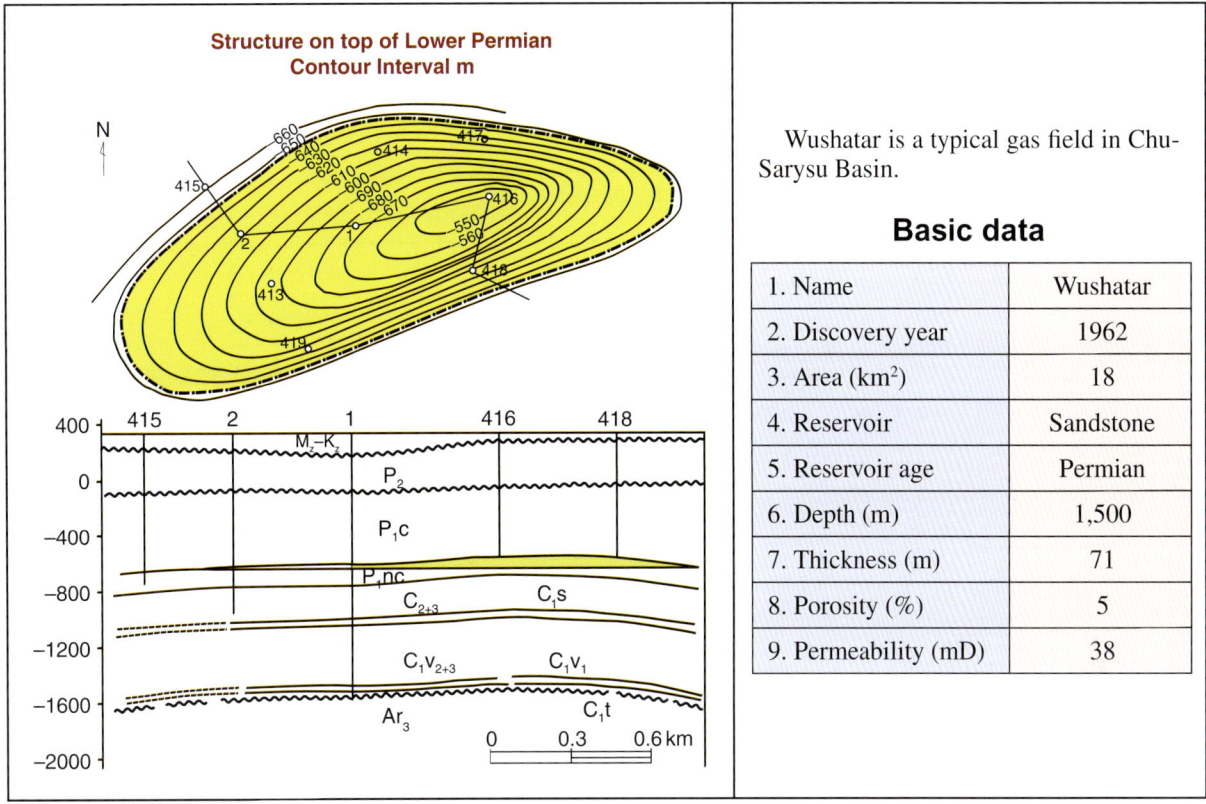

Fig. 30.1 Details of the Wushatar gas field.

Another gas reservoir is in Middle Carboniferous limestone with a depth of 1180 m, porosity of 3.8 per cent, thickness of 70 m, initial pressure of 15.1 MPa, temperature of 59°C and initial daily production rate of 100,000 m³.

TURGAY BASIN

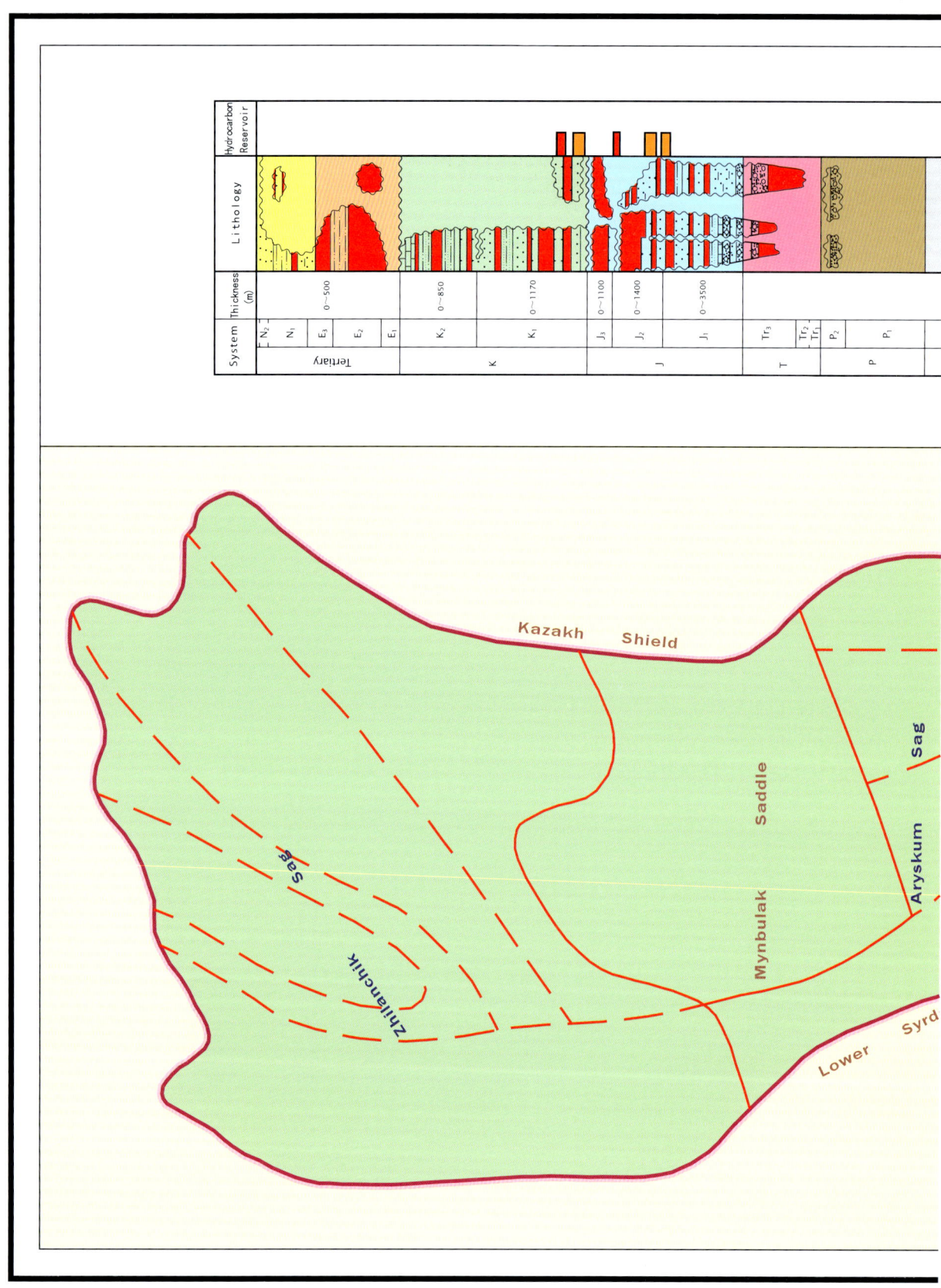

CHAPTER 31

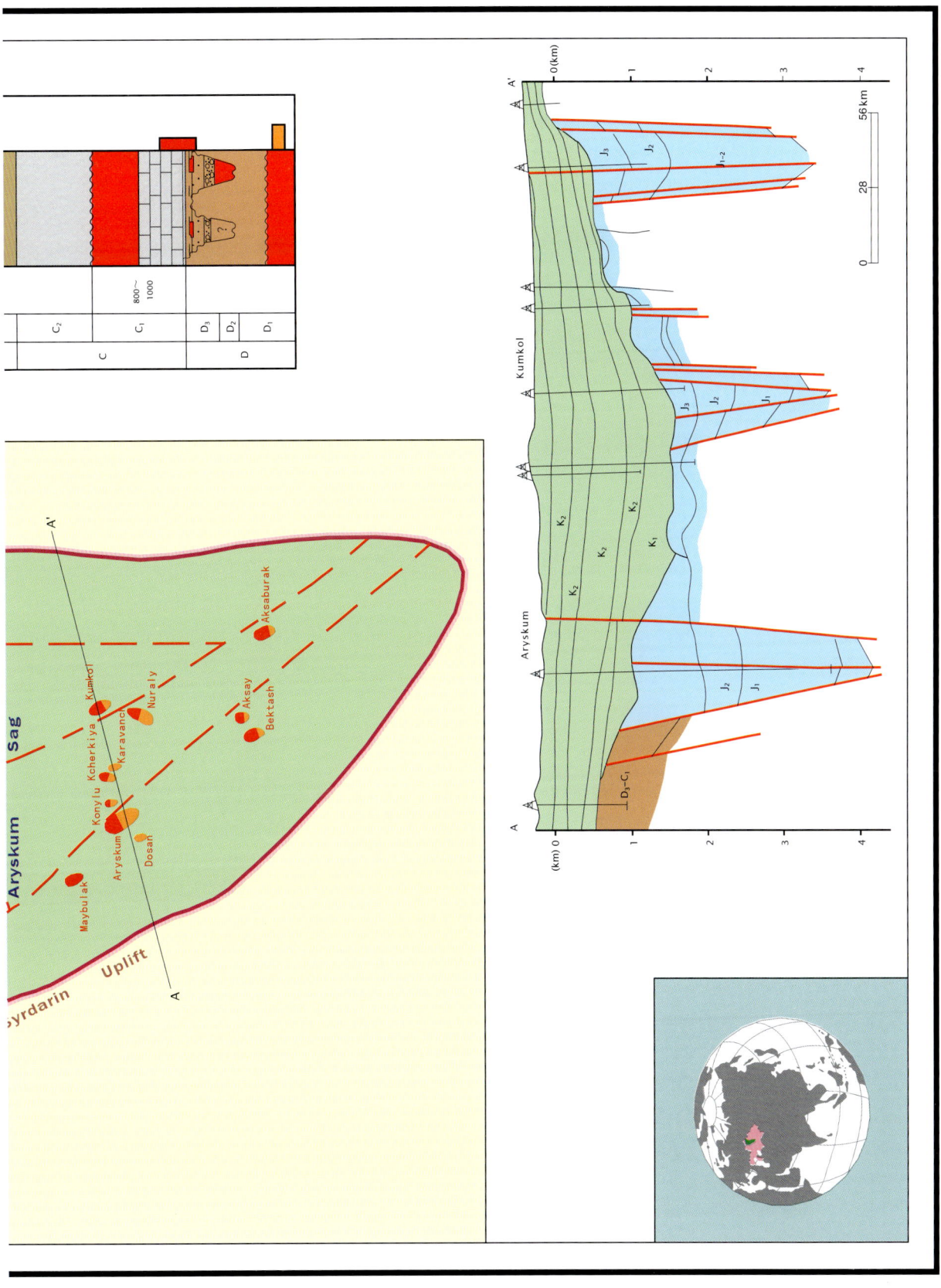

31 Turgay Basin

Geography

The large Turgay Basin lies to the east of the Ural Mountains, with an area of 110,000 km². It forms a link between the West Siberia Basin to the north and the Central Asian terranes to the south. The main Mesozoic depocentre and the primary focus for exploration is the Aryskum Sag in the southeast.

History

Exploration has been undertaken in the basin since the late 1930s. In the 1950s and 1960s drilling activity focused on the northern part of the basin, but only shows were recorded. In the early 1980s drilling activity began in the southern sub-basins and the first discovery was made at Kumkol in 1984 (Fig. 31.1). This is the largest field in the basin. The Kumkol oil field is in an anticlinal trap, with an area of 75 km², average porosity is 19 per cent to 23 per cent for the Jurassic and Cretaceous sandstones, with permeability from 3.5 to 6.5 mD and waxy content 10.8–11.5 per cent. Production began from Kumkol in late 1989 and has risen steadily to a rate of some 5500 t/day in early 1995.

Oil from Kumkol is transported by pipe east to Karakoin, from where the pipeline runs south to the refinery at Shymkent. There are plans to build pipelines from Akshabulak to Kumkol and from western Kazakhstan to Kumkol, and medium to long-term potential exists for export to markets in India and Pakistan. There is little gas-based infrastructure, although a small gas liquefaction plant is planned at Kumkol to supply liquefied natural gas to local markets.

Regional geology

The sedimentary cover comprises Devonian to Tertiary strata. A period of rifting in the Middle to Late Devonian was followed by the development of a Late Devonian to Mid-Carboniferous passive margin sequence, as the area of the basin lay on the margin of the Kazakh continent. The Late Carboniferous–Permian suturing of the Kazakh, Siberian and East European blocks led to significant deformation and erosion. Post-orogenic collapse of the mountain belt led to extension in the mid-Triassic to Middle Jurassic, when synrift volcanics and clastics were deposited in narrow elongated grabens. Thermal subsidence in the late Jurassic and Cretaceous led to a widening of the area of deposition. Phases of wrench movement are recorded in the basin, related to terrane accretion on the southern margin of the Eurasian continent, notably in the Early Paleocene and Neocene.

Source intervals in the basin are all of Jurassic age, although Lower Jurassic mudstones may be overmature for oil generation. The overlying Cretaceous sediments are generally immature. In the north of the basin an Upper Palaeozoic source-rock interval has been identified. Reservoirs are the Middle–Upper Jurassic Kumkol Formation and the Lower Cretaceous Daul Formation clastics. Minor additional reservoirs occur in Middle Jurassic clastics and in weathered and fractured basement. Bitumen and heavy oil reserves have been found in Lower Carboniferous carbonates in the north of the basin. Seals are of Middle Jurassic, Late Jurassic and Early Cretaceous ages.

World Atlas of Oil And Gas Basins, First Edition. Li Guoyu.
© 2011 John Wiley & Sons, Ltd. Published 2011 by John Wiley & Sons, Ltd.

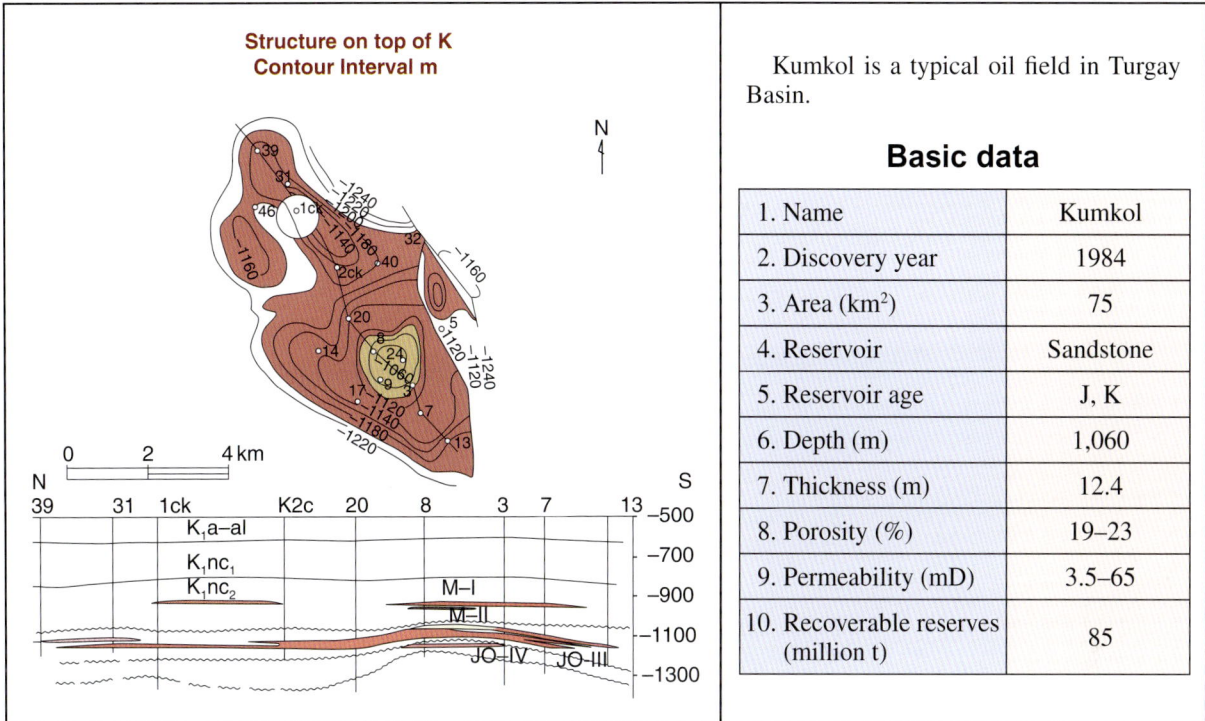

Fig. 31.1 Details of the Kumkol oil field.

Structures in the basin are related to anticlines and faulted blocks resulting fromMiddle Triassic to Middle Jurassic extension, Mid-Jurassic inversion, Late Mesozoic subsidence and Tertiary compression. Four groups of plays have been recognized. The most significant are the Middle–Lower Jurassic Kumkol plays and the Lower Cretaceous Daul plays. Two hydrocarbon systems have been identified: the Jurassic–Cretaceous and Palaeozoic Systems. Oil potential is related to Cretaceous and Lower Jurassic sands and Carboniferous carbonates.

FERGANA BASIN

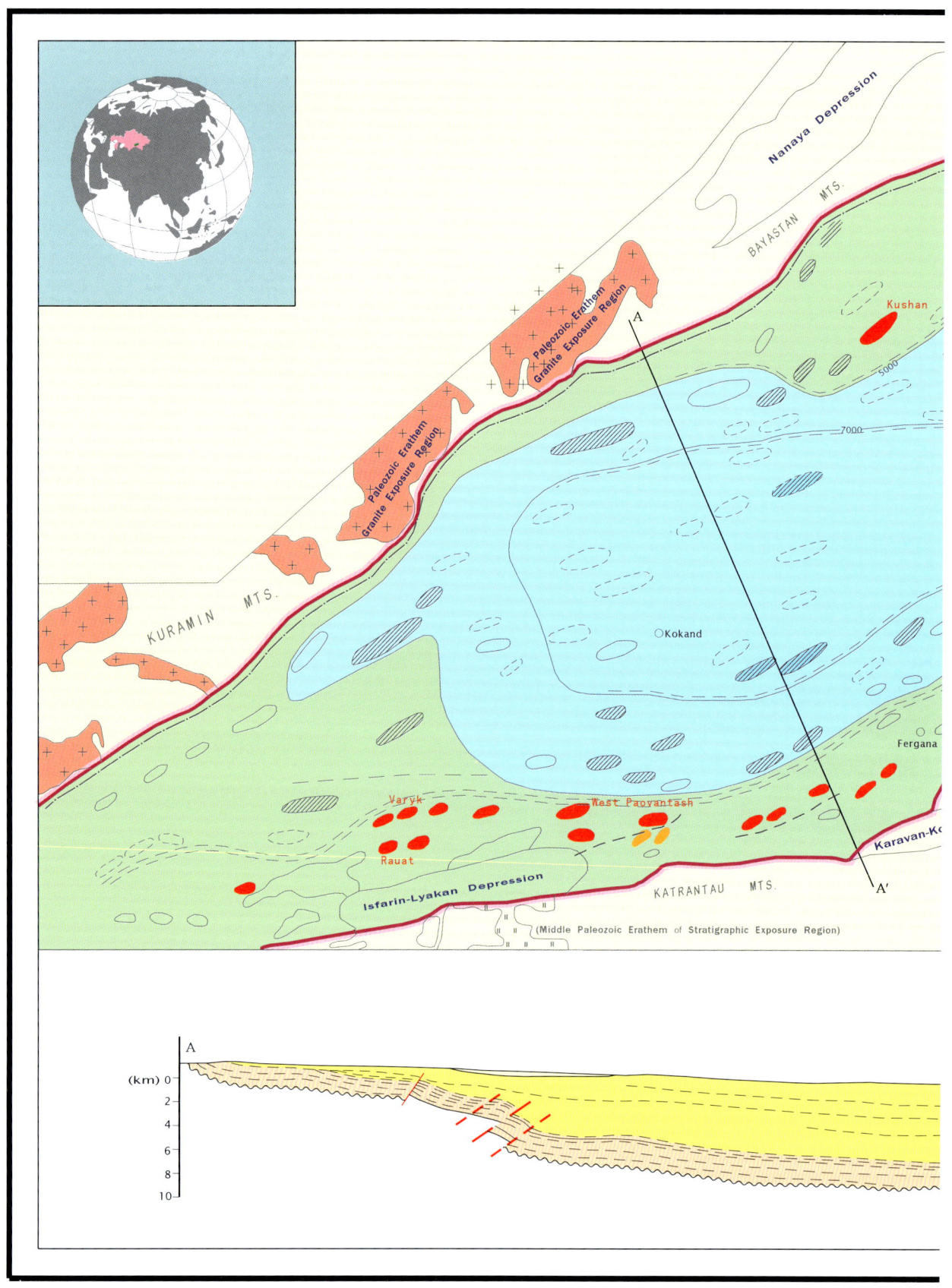

CHAPTER 32

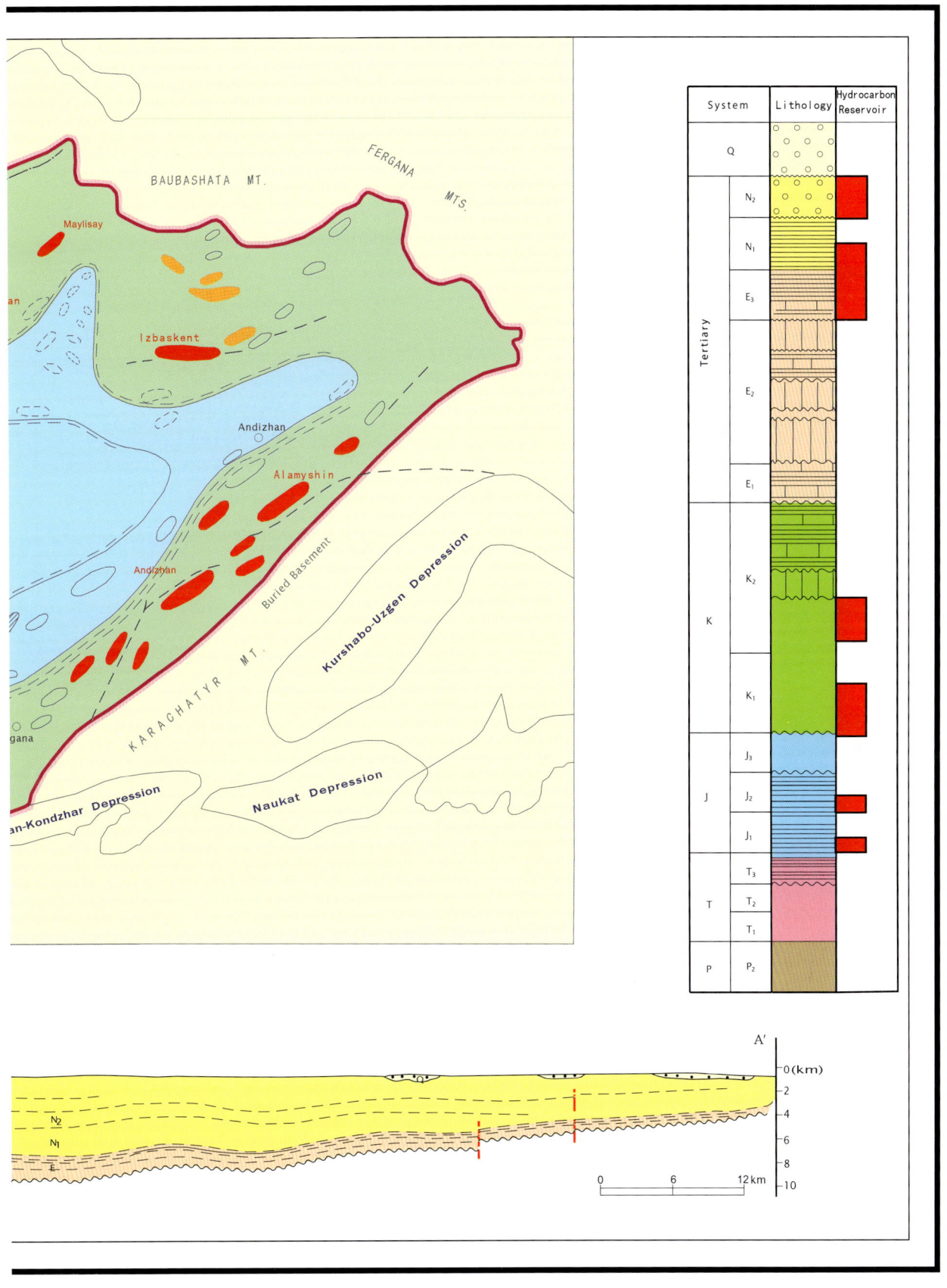

32 Fergana Basin

Geography

The Fergana Basin is one of the large basins of middle Asia and is formed in an intermontane depression. It lies within three countries: Kyrgyzstan, with 47 per cent of the area of the basin; Tajikistan (13 per cent); and Uzbekistan (40 per cent). The centre of the basin lies in Uzbekistan.

History

The basin has a long exploration history. The first wells were drilled before 1900 and the first discovery was Maylisay in Kyrgyzstan in 1901. Later the Andizhan oil field was discovered in 1935, followed by Palvantash oil field in 1943. In 1980 the discovery of oil at Gumkhana (Uzbekistan) was notable in that it was the first find in the Central Graben and further discoveries were made at Makhram (1982, Tajikistan) and Mingbulak (1987, Uzbekistan). Development began in 1909 in the Maylisay field in Kyrgyzstan and in 1936 development began at Andizhan in Uzbekistan. The Maylisu 4-Izbaskent Vostochnyy field was first developed in 1950 and the first wells were planned for the basin's largest field, Mingbulak, in 1993. Oil production in the Fergana Basin began in 1903, reached 2.2 million t in 1964 but declined to 1 million t in 1974. Gas production is thought to have commenced in the late 1950s.

Regional geology

The strata range from Permian to recent in age. The basement of the Fergana Basin is made up of three distinct blocks, each with its own lithological and structural characteristics. The northwestern block consists of block-faulted volcanic rocks. The northeastern block comprises metasediments, folded into antiforms and synforms. The southern basement block consists of a metamorphosed sedimentary complex containing Silurian–Devonian volcanics. The area of Fergana Basin was a separate microcontinent in the Palaeozoic, but after the Middle Carboniferous it was sutured to the Eurasian Plate.

Source rocks are Early–Middle Jurassic and Paleocene strata. Reservoirs in the Fergana Basin are the Paleocene sandstones and carbonates. Additional reservoirs are Palaeozoic, Jurassic, Cretaceous and Neocene strata. Rocks forming seals are found throughout the Fergana Basin succession.

The present distribution, size and style of structural features in the Fergana Basin have resulted from several distinct phases of structural evolution. Structures are thrusts, hanging-wall anticlines, en échelon anticlines, faulted and thrusted anticlines; strike-slip faults; monoclines and anticlines above basement uplifts. The North Soh oil field (Fig. 32.1) is an example of faulted anticline trap discovered in 1957, with an area of $5.5\,km^2$, average porosity 17 per cent, permeability 140 mD, depth 300–1700 m and thickness 16 m. It is a small field with recoverable reserves of 7 million t.

World Atlas of Oil And Gas Basins, First Edition. Li Guoyu.
© 2011 John Wiley & Sons, Ltd. Published 2011 by John Wiley & Sons, Ltd.

Fergana Basin

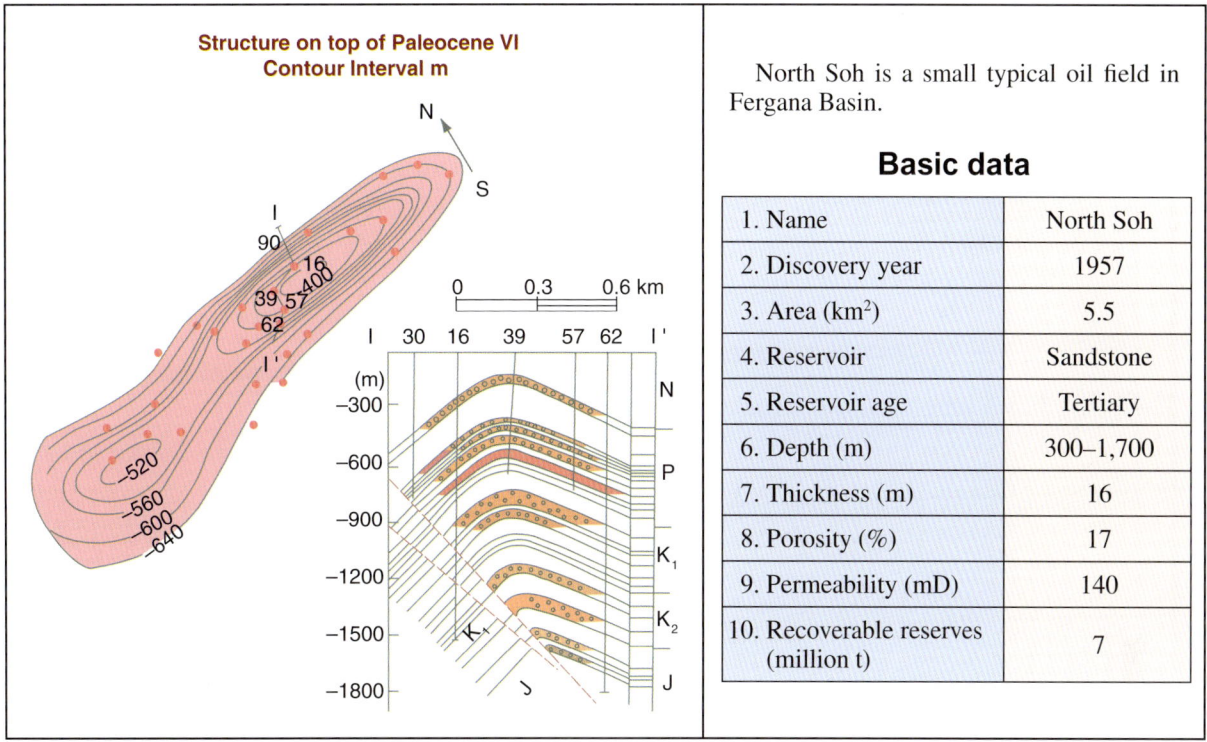

North Soh is a small typical oil field in Fergana Basin.

Basic data

1. Name	North Soh
2. Discovery year	1957
3. Area (km^2)	5.5
4. Reservoir	Sandstone
5. Reservoir age	Tertiary
6. Depth (m)	300–1,700
7. Thickness (m)	16
8. Porosity (%)	17
9. Permeability (mD)	140
10. Recoverable reserves (million t)	7

Fig. 32.1 Details of the North Soh oil field.

AFGHAN-TAJIK BASIN

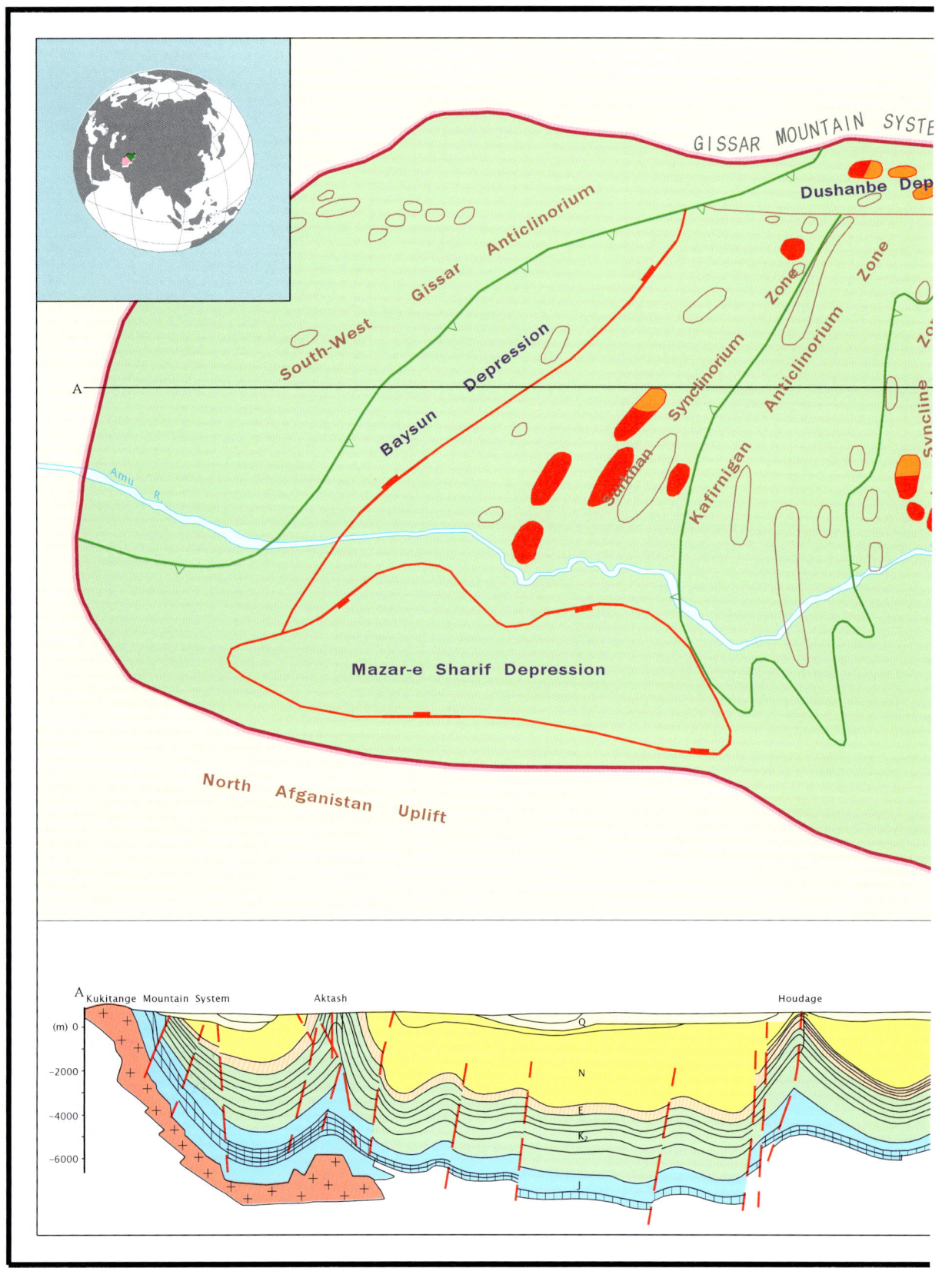

CHAPTER 33

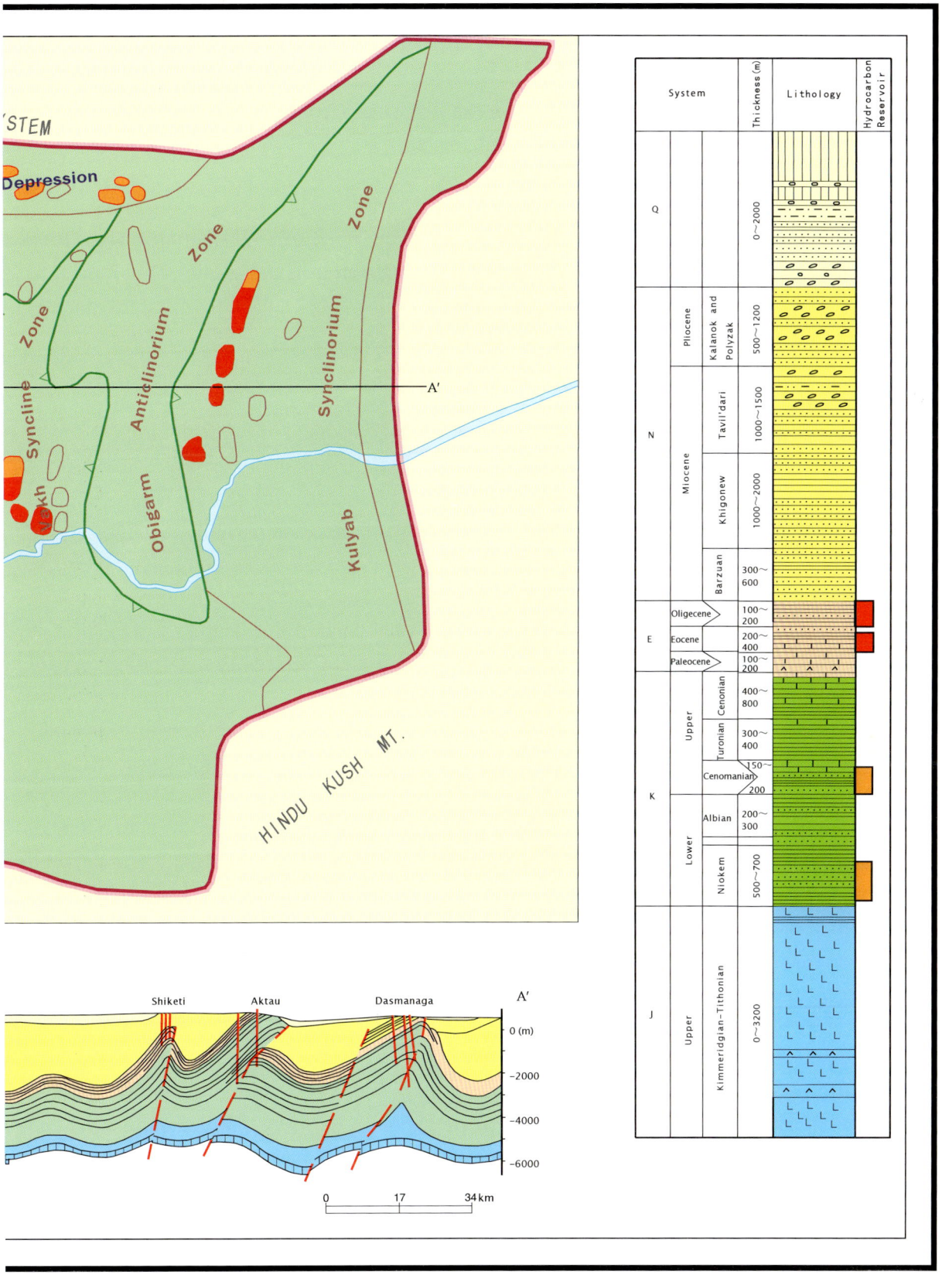

33 Afghan-Tajik Basin

Geography

The intermontane Afghan-Tajik basin lies on the Tianshan Orogenic Belt. It has an area of 100,000 km^2, out of which 31,000 km^2 lies in Afghanistan.

History

Geological surveys were conducted in 1930. Exploration work was concentrated after 1950 in three basins. A total of 24 oil and gas fields were discovered in this basin. There are nine geological elements within this basin, consisting of synclinoriums and anticlinoriums, mainly with a N–S trend. Oil and gas are located in the synclinoriums.

Regional geology

The basin is confined to intermontane depressions of the Tianshan epi-platform orogen. It was formed through the Cenozoic activation of the Palaeozoic Turanian plate. Five basins are confined to intermontane depressions, two of them (Fergana and Afghan-Tadzhik) contain commercial oil and gas accumulations. The basement of the basin comprises folded and partially metamorphosed Palaeozoic rocks at a depth of 4 to 10–12 km. The sedimentary cover consists essentially of Permian, Triassic, Jurassic and Cretaceous terrigenous rocks and Paleocene–Quaternary molasses, totalling 4–8 km in thickness. Some of the basins (Afghan-Tajik) may contain salt-bearing Upper Jurassic sequences. There are multilayered oil reservoir rocks of Jurassic, Cretaceous and Tertiary age, of which the Lower Tertiary dominates. The Jurassic and Cretaceous rocks mainly contain gas while the Tertiary rocks mainly contain oil. The source rock is shale in the strata of these three periods.

Reservoir rocks are represented by Jurassic–Cretaceous sandstones of which the porosity is up to 30 per cent, and Paleocene carbonate–terrigenous deposits. Seventy hydrocarbon fields have been discovered, among them oil fields predominate (54 fields). The original petroleum potential of the basin is estimated as high. A particular reservoir type has been identified in weathered Archaean and Proterozoic rocks of the basement. Clayey seals are Widespread and shale rocks and rock salt seals are common. Traps are represented by structural and non-anticlinal types. Oil and gas accumulations are multilayered and mainly confined to Mesozoic–Cenozoic deposits.

In this basin, four depressions and three structural highs are to be found. The depressions are filled with Meso-Cenozoic deposits. Thicknesses of more than 3 km occur, mostly associated with the Upper Tertiary. Salt domes occur widely and are typically associated with oil traps. There are also occurrences of overthrust structures with distances extending to several thousand metres. Faulted anticlines are a common trap type in this basin (e.g. Fig. 33.1). Fourteen oil and gas fields have been discovered with small reserves. Within this basin more than 182 faulted anticlines (see Fig. 33.2) were formed in the Meso-Cenozoic deposits by strong Tianshan and Pamir orogenic movements. Because of the very complex geological conditions of this basin, oil and gas exploration has encountered many difficulties.

Afghan-Tajik Basin

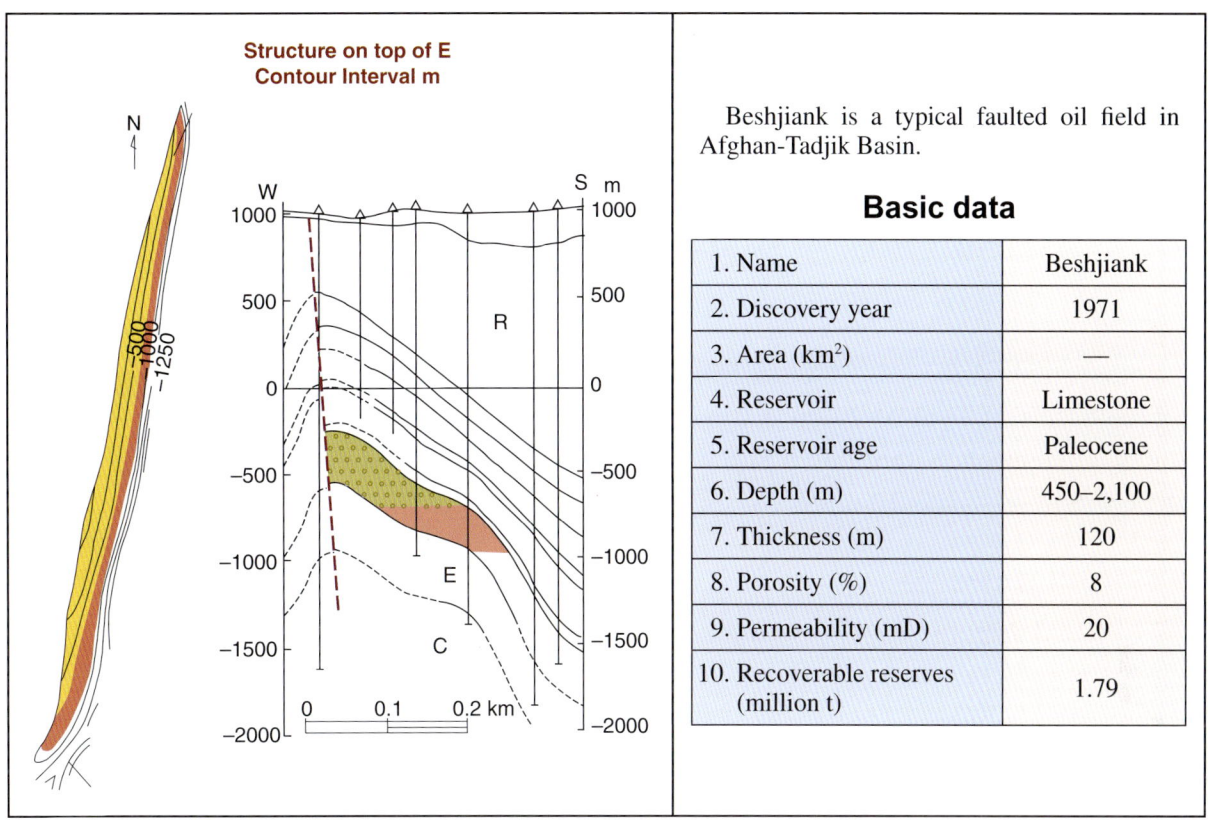

Fig. 33.1 Details of the Beshjiank oil field.

Beshjiank is a typical faulted oil field in Afghan-Tadjik Basin.

Basic data

1. Name	Beshjiank
2. Discovery year	1971
3. Area (km^2)	—
4. Reservoir	Limestone
5. Reservoir age	Paleocene
6. Depth (m)	450–2,100
7. Thickness (m)	120
8. Porosity (%)	8
9. Permeability (mD)	20
10. Recoverable reserves (million t)	1.79

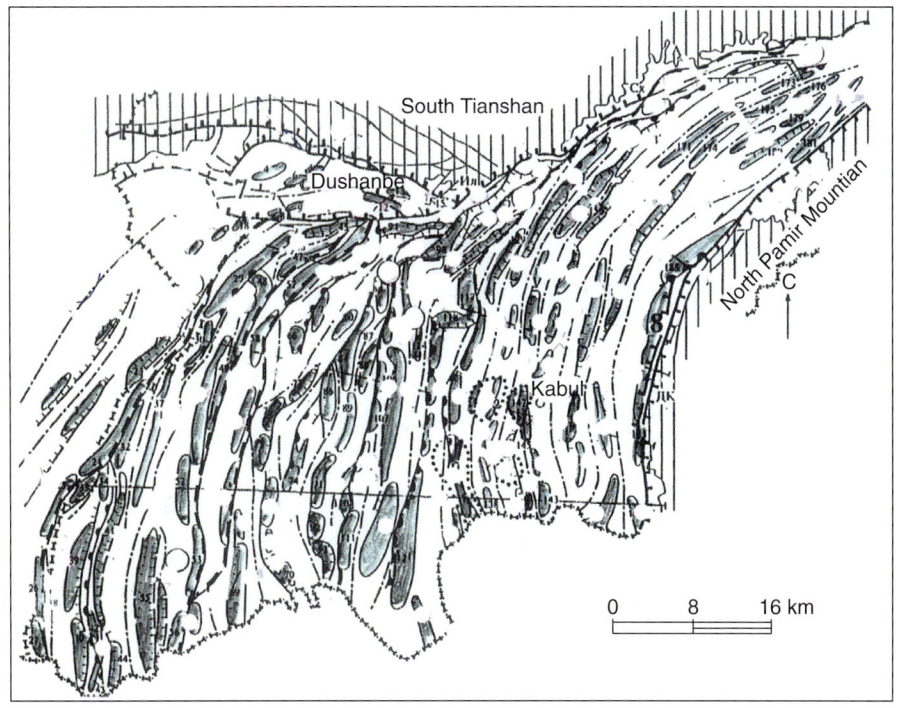

Fig. 33.2 Numerous and closely spaced anticlines in the Afghan-Tajik Basin.

AZERBAIJAN, GEORGIA, ARMENIA AND

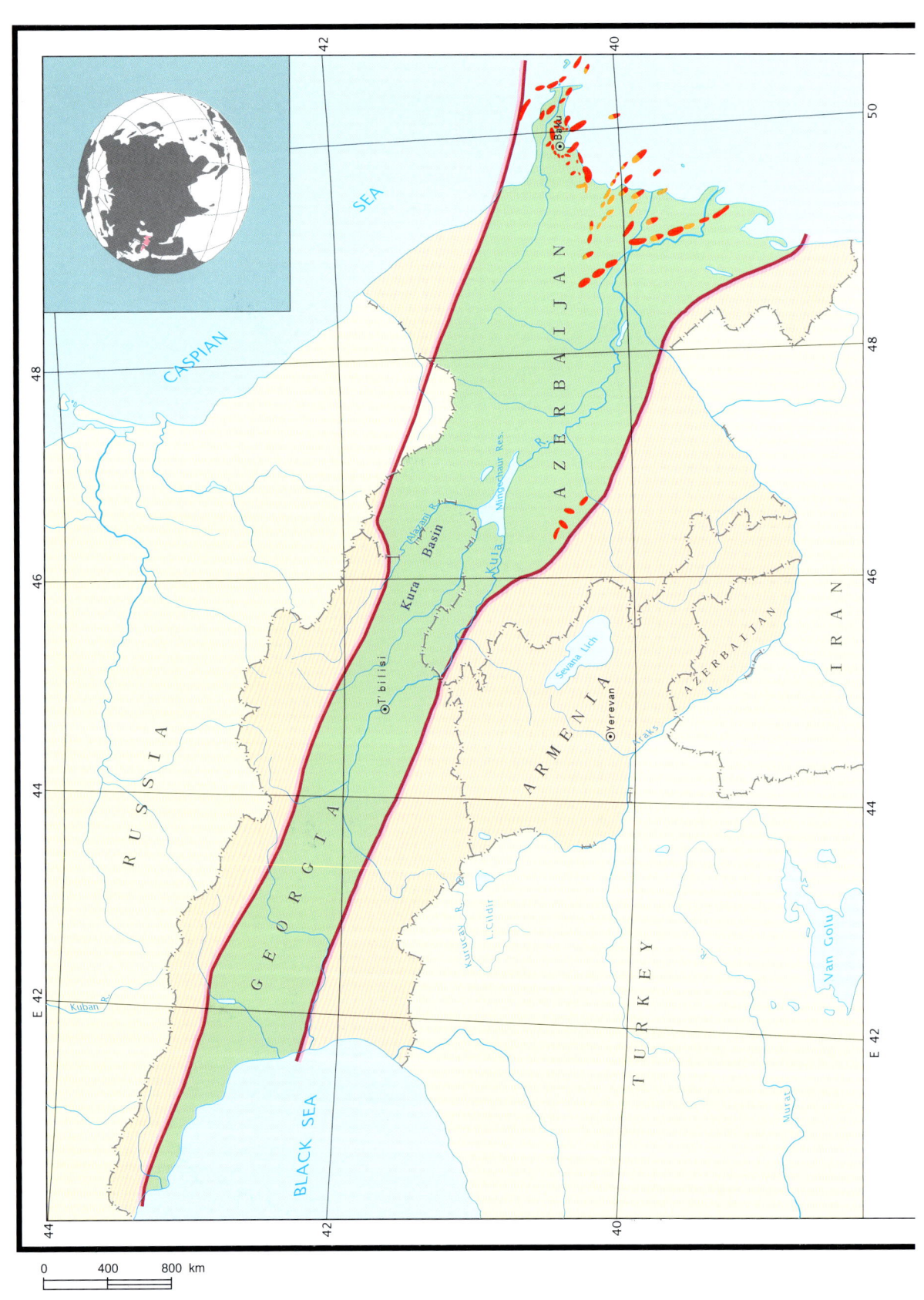

KURA BASIN　　　　　　　　CHAPTER 34

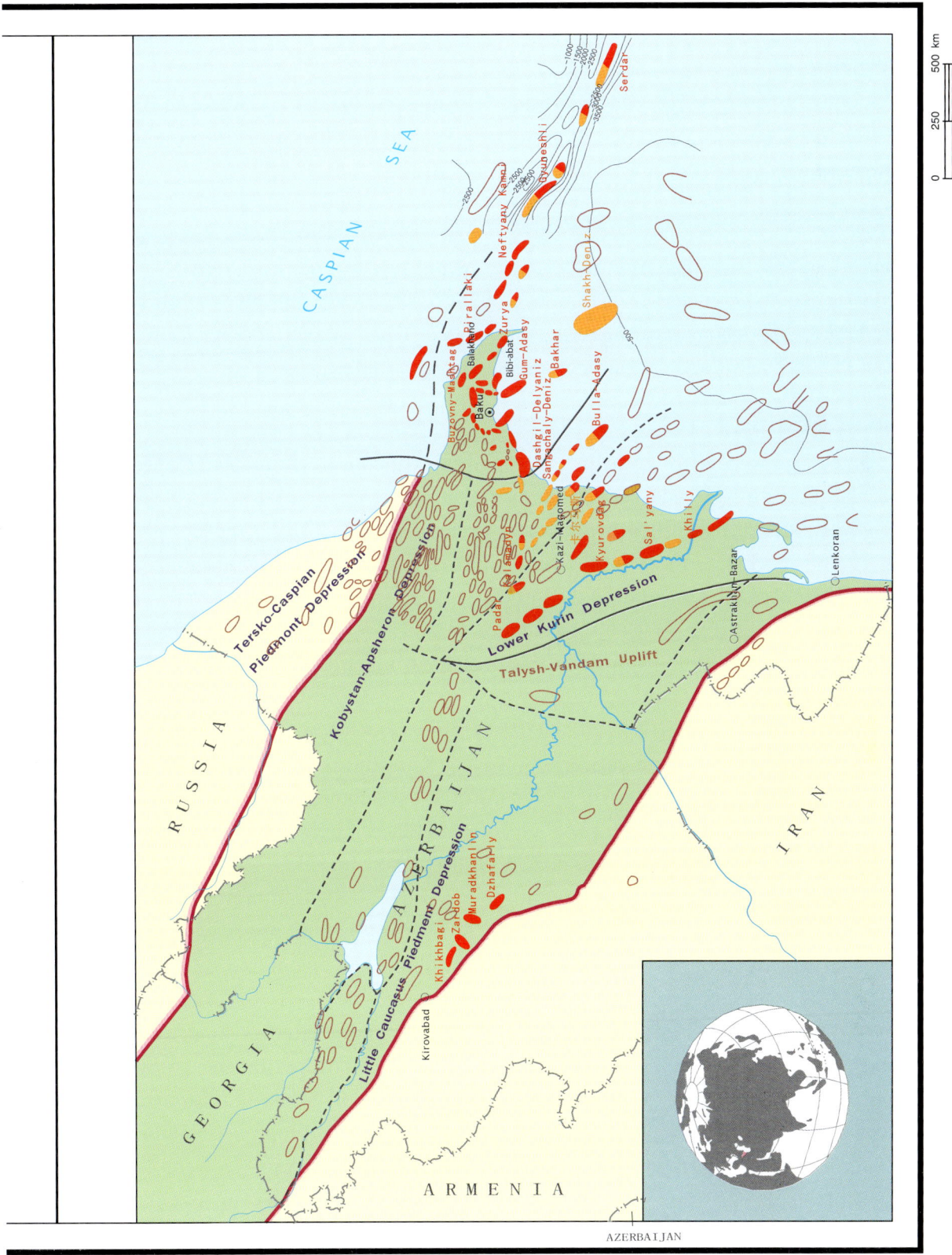

34 Azerbaijan, Georgia, Armenia and Kura Basin

Geography

The countries of Azerbaijan, Georgia and Armenia are located in western Asia, situated principally in the Caucasus Mountain and lying between the Caspian Sea in the east and the Black Sea in the west.

History

Ancient oil and gas seepages were well known, with evidence that local inhabitants used such surface oil for medicinal purposes as long ago as the 11th century, but the first oil field was not discovered until 1858 when oil exploration commenced. This is a well-known oil production area in terms of the early development of the world's petroleum industry.

Regional geology

The oil and gas prospects of Azerbaijan, Georgia and Armenia are dominated by the Kura Basin located to the south of the Caucasus Mountains with an area of 95,000 km^2. This basin is complex and extensively faulted, and filled by Jurassic, Cretaceous and Tertiary rocks with a thickness reaching 20,000 m., Source rocks have been identified in the Jurassic, Cretaceous and Tertiary strata. Reservoir rocks are sandstone and caprock is shale. Traps are represented by anticlinal and mud-volcanic types (e.g. Fig. 34.1). Oil production reached 15.15 million t in 2003. Details of the major oil and gas fields of the Kura Basin are presented in Table 34.1. Petroleum prospects are considered to be good.

Azerbaijan Azerbaijan is a mountainous country with an area of 86,800 km^2 and a population of 8.44 million. The hydrocarbon-bearing Kura Basin is located in this country both on land and offshore. This important oil producing country is famous for being one of the earliest locations of modern oil exploration at the turn of the 20th century. Initial oil exploration began before 1858 and the first oil field Bibi-abat was discovered in 1871. The Bibi-abat field's oil production was 4.01 million in 1880, 10.9 million t in 1901, 10.92 million in 1909, 23.40 million in 1991 and 12.51 million in 1990. By the end of 2007 the proven oil reserves of the Kura Basin were 958.9 million t and oil production was 41.25 million t. The favourable geology of the basin indicates that the petroleum prospects are good.

Georgia Georgia has an area of 68,700 km^2 with a population of 4.52 million. During the period 1868 up to the 1880s many oil seepages were discovered on the ground. The first small oil field was discovered between 1927 and 1930. The oil reservoirs are associated with Tertiary sandstone. During the 1980s oil production reached 3.3 million t per annum. Thereafter, oil production declined sharply. By the end of 2007 the proven oil reserves were 4.79 million t and oil production was 100,000 t only. Future oil prospects of the country are associated with the Black Sea.

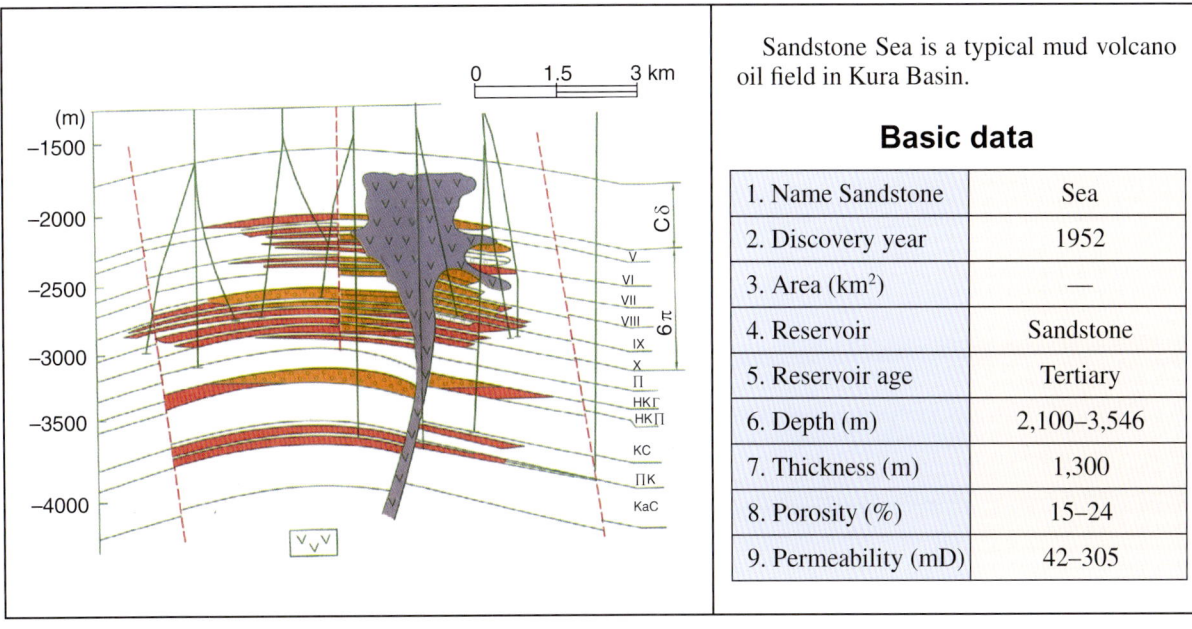

Fig. 34.1 Details of the Sandstone Sea oil field.

Table 34.1 Major oil and gas fields of Kura Basin

Name	Discovery year	Recoverable reserves		Depth (m)	Trap	Reservoir	
		Oil (million t)	Gas (billion m³)			Age	Lithology
Balakhano	1896	336.0	—	1300	Anticline	N	Sandstone
Bibi-abat	1871	280.0	—	1500	Anticline	N	Sandstone
Neftyany Kami	1949	172.0	—	1600	Anticline	N	Sandstone
Sangachaly Deniz	1963	120.0	19	3600	Anticline	N	Sandstone
Peschanyistant	1952	84.0	—	3200	Mud diapir	N	Sandstone
Bakhar	1968	—	103	3900	Anticline	N	Sandstone
Bulla Adasy	1973	—	97	4580	Anticline	N	Sandstone

Armenia   Armenia has an area of 29,800 km² with a population of 3.22 million. It is occupied by a small part of Kura Basin in the northwest. Between 1974 and 1990, four wells were drilled and 3250 km of seismic lines were completed, but no oil and gas fields were discovered. The prospect of oil exploration will be confirmed by future exploratory work.

MIDDLE EAST

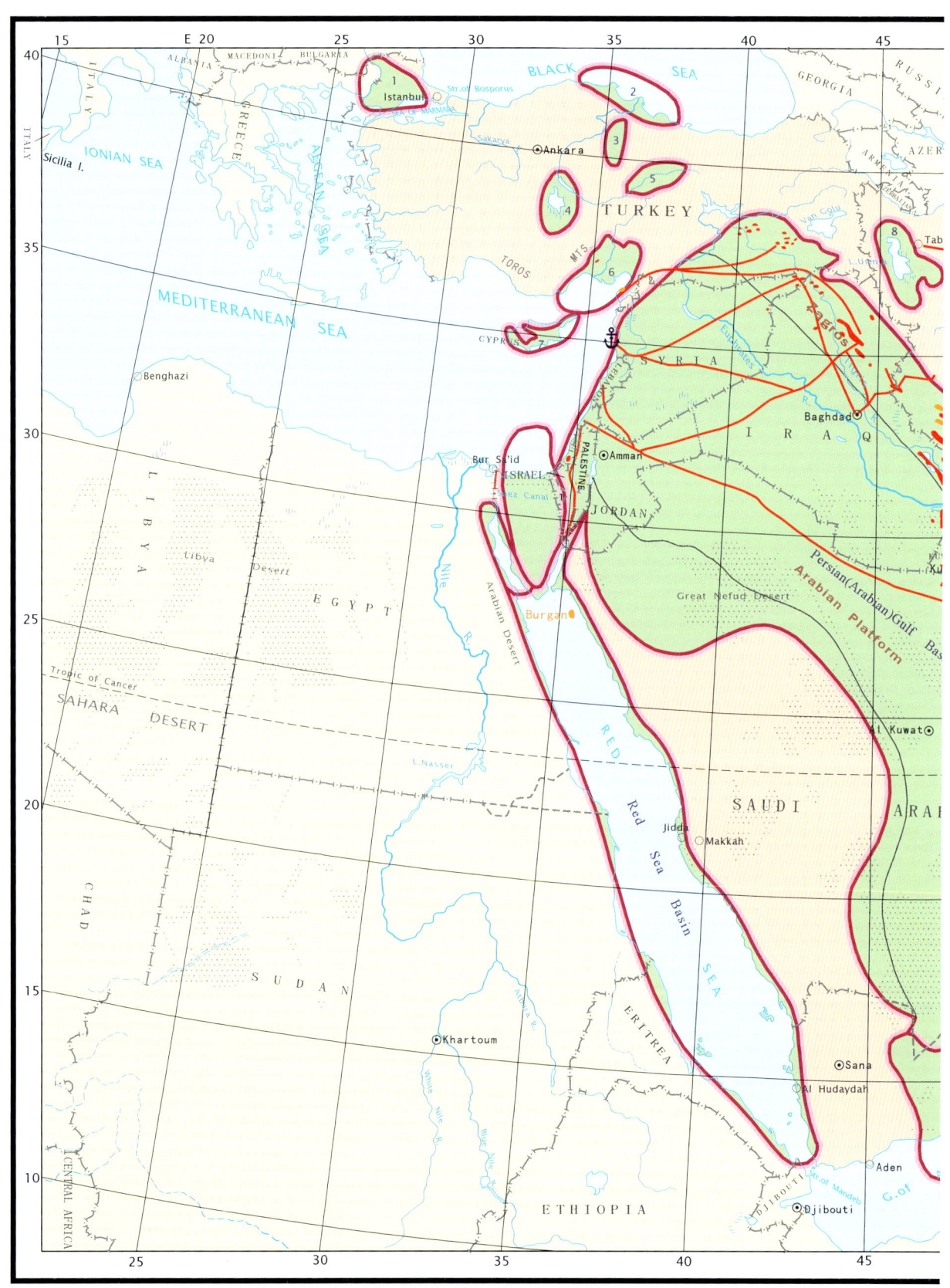

Basins: 1 Thrace 2 Bafra 3 Corum 4 Tuz Golu 5 Sivas 6 Adana 7 Cyprus 8 Tabriz

CHAPTER 35

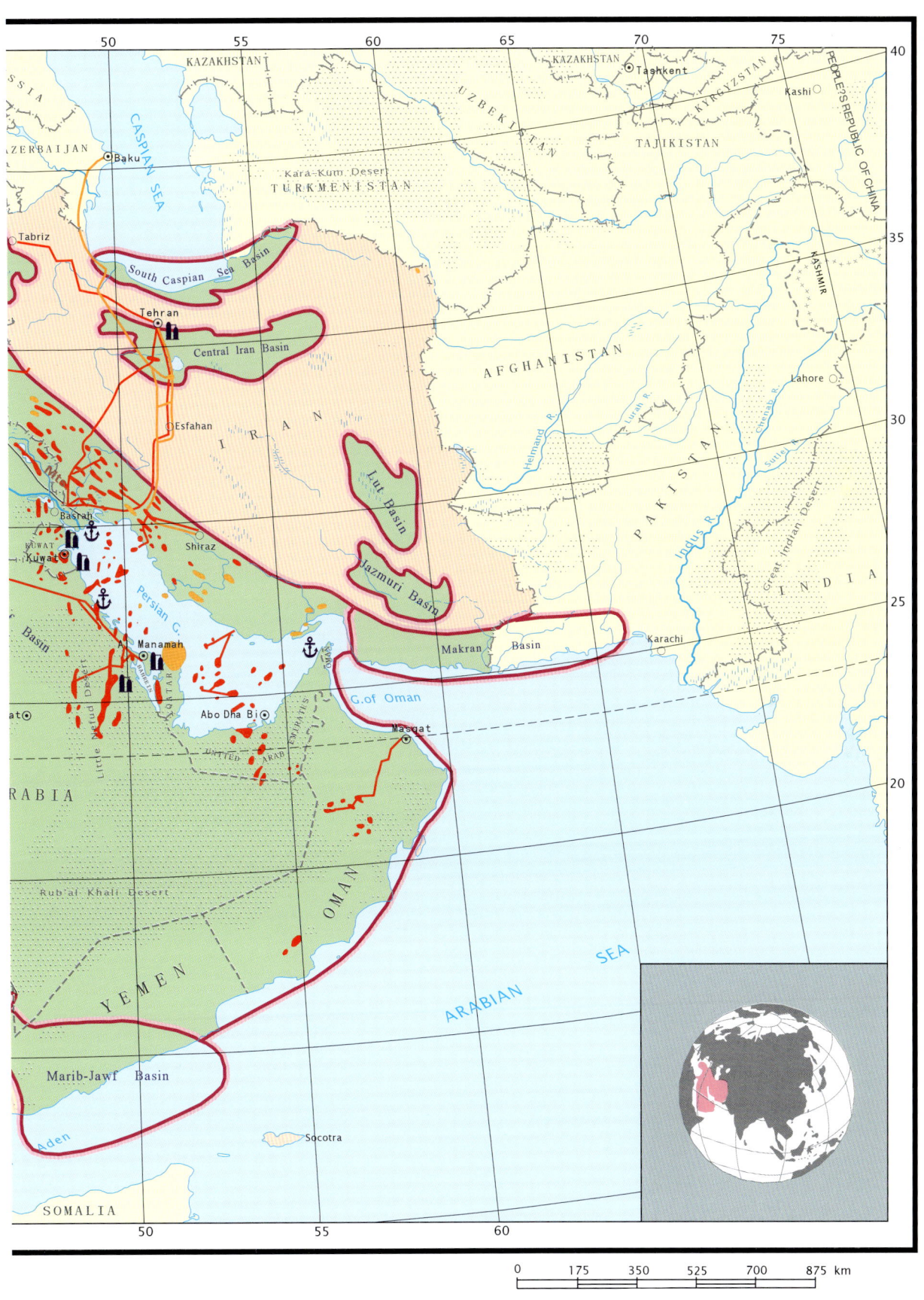

35 Middle East

Geography

The Middle East region of western Asia is bordered by the Arabian Sea, the Caspian Sea, the Black Sea, the Mediterranean Sea, the Red Sea and the Gulf of Aden, having a total area of 6.3 million km^2 and a population of 274 million. There are huge reserves in the petroleum basins of the Middle East region (Table 35.1), and output from the region has been a major contributor to overall global economic development (Table 35.2).

History

The Middle East is the richest petroleum province in the world (Table 35.3) and is unique among oil provinces in terms of the magnitude of its fields. An indication of the size of typical fields is given by the fact that, of the producing or proven fields in the Middle East province, 25 have super giant status (in excess of 5 billion barrels recoverable), while 69 are qualified as giants (500 million to 5 billion barrels). At least 14 of the fields in the region have recoverable reserves in excess of 10 billion barrels, with Ghawar, the largest, having conservatively estimated recoverable reserves of 83 billion barrels. To date, almost 300 oil accumulations have been discovered in the region. Production has been established from approximately 250 reservoirs, of which 200 are Middle Jurassic to Upper Cretaceous in age. Thirty-two productive reservoirs occur in the Oligocene to Miocene.

Table 35.1 Petroleum basins in the Middle East

Country	Basins	Area (thousand m^2)	Major sedimentary rocks		Major reservoir rocks	
			Era	Thickness (m)	Era	Lithology
Saudi Arabia	Persian Gulf	3280	Palaeozoic, Mesozoic, and Cenozoic	11,000	Jurassic, Cretaceous, and Tertiary	Carbonate and sandstone
Iran	Red Sea	570				
	Central Iran	100	Cretaceous and Cenozoic	8000	Tertiary	Carbonate
Iraq	Zagros	450	Palaeozoic–Cenozoic	9000	Cretaceous and Tertiary	Limestone and sandstone
Yemen	Marib-Jawf	240	Palaeozoic–Cenozoic	5000	Jurassic, Tertiary offshore	Carbonate

World Atlas of Oil And Gas Basins, First Edition. Li Guoyu.
© 2011 John Wiley & Sons, Ltd. Published 2011 by John Wiley & Sons, Ltd.

Table 35.2   Oil and gas production, reserves and crude exports in Middle East from 1980 to 2008

Year	Remaining recoverable oil reserves (billion t)	Oil production (million t)	Remaining recoverable gas reserves (trillion m^3)	Gas production (billion m^3)	Crude export (million t)
1980	49	909	21	—	—
1990	90	825	37	119	747
1991	90	810	37	117	691
1992	90	867	43	101	772
1993	90	914	44	112	822
1994	90	925	45	110	825
1995	89	938	45	131	832
1996	92	950	45	132	585
1997	92	985	48	136	909
1998	91	1049	49	140	935
1999	92	1016	49	131	917
2000	93	1073	52	134	947
2001	93	893	55	171	954
2002	93	976	56	174	903
2003	99	1035	71	189	947
2004	99	1109	71	230	981
2005	102	1215	72	286	991
2006	102	1223	22	282	991
2007	102	1201	72	284	991
2008	102	1155	73	337	990

Table 35.3   Statistics of oil and gas in Middle East at the end of 2008

Country	Proven reserves		Production	
	Oil (million t)	Gas (billion m^3)	Oil (million t)	Gas (billion m^3)
Total Middle East	102,197	73,387	1155	337.6
Abu Dhabi	12,630	5620	122	—
Bahrain	17	92	8.5	9.3
Dubai	547	113	5.5	—
Iran	18,650	26,850	195	99.4
Iraq	15,753	3169	118	6.8
Israel	0.26	30	—	—
Jordan	0.13	6	—	—
Kuwait	13,904	572	116	14.1
Neutral Zone	684	28	28.5	—
Oman	753	849	35.7	19.8
Qatar	2083	25,257	42	61.4
Rasal Khaimah	13	33	0.5	—
Saudi Arabia	36,198	7304	445	72.7
Sharjah	205	302	2.7	—
Syria	342	240	19.2	5.9
Yemen	410	478	15.2	—

IRAN AND IRAQ

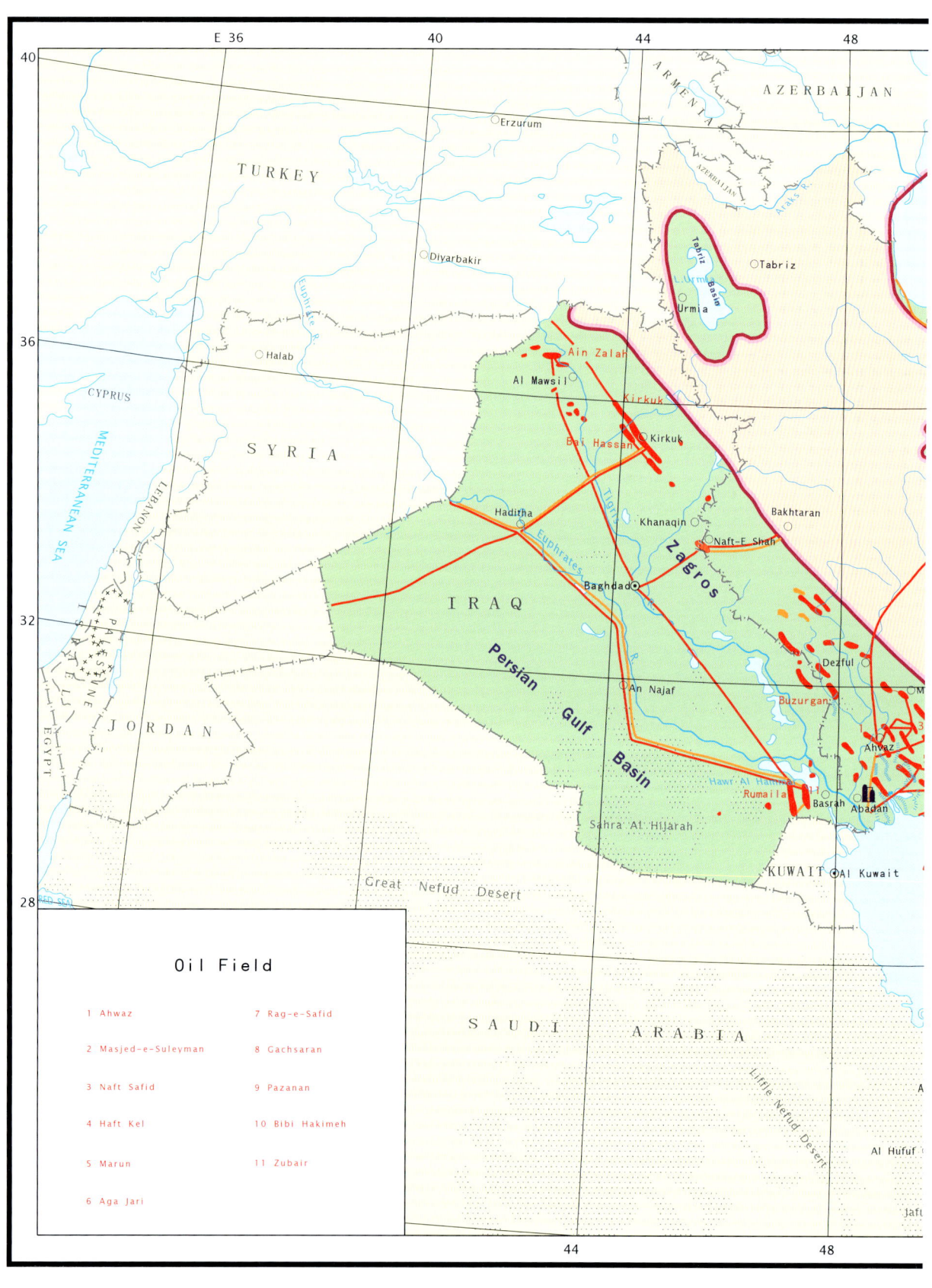

CHAPTER 36

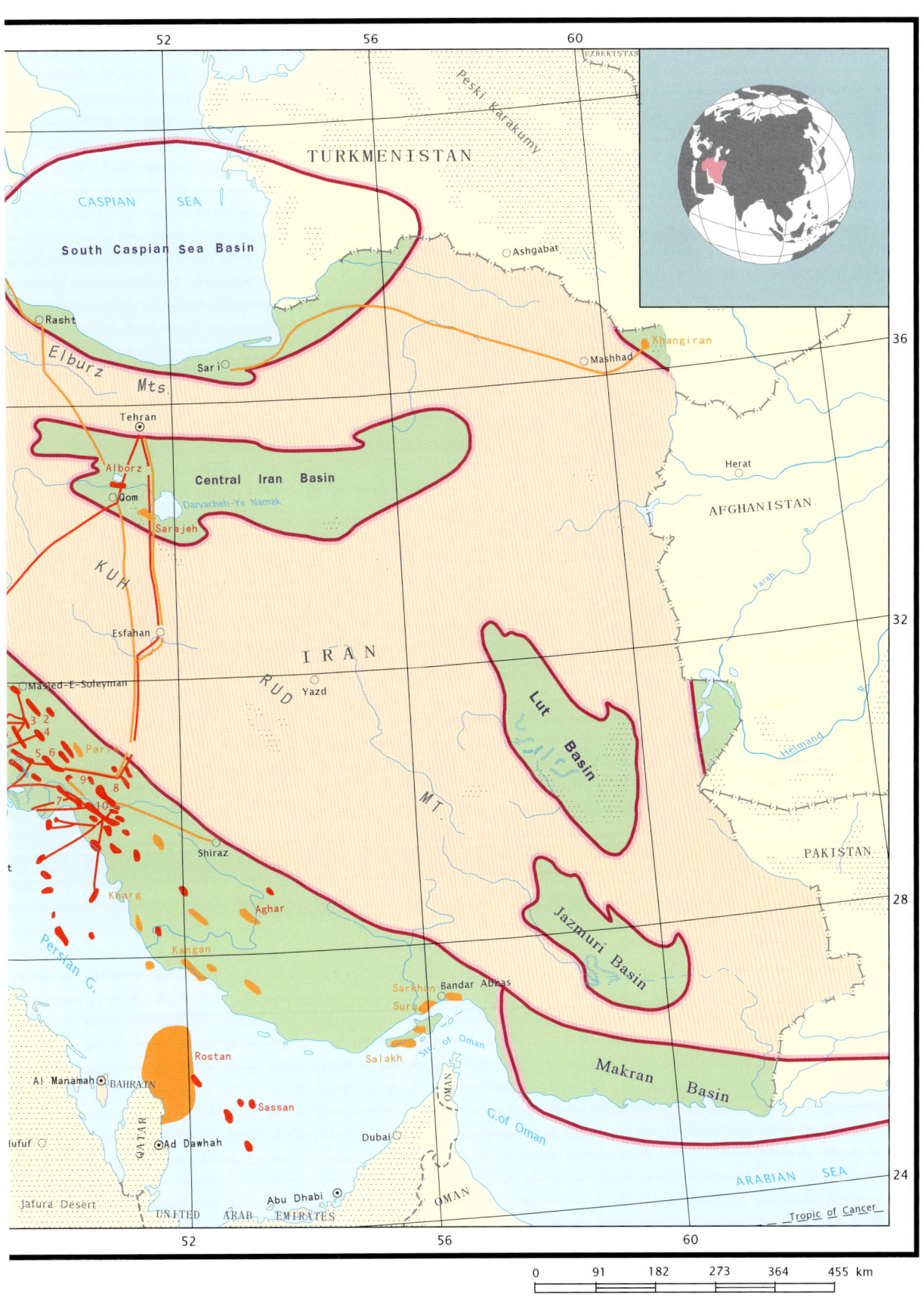

36 Iran and Iraq

Geography

The two countries are located in western Asia, bordering the Persian Gulf to the south. Both countries are mainly associated with the Zagros Orogenic Belt.

History

Despite being the richest oil region in the world, the first large oil field, the Majed-Sulaiman, was discovered in Iran only in 1908. This was almost half a century after important pioneering discoveries in areas such as the Caspian Sea and Romania. This discovery in Iran triggered off the development of the petroleum industry in the Middle East. Iran and Iraq were the largest oil producing countries in the world after the end of the Second World War. Several truly large oil and gas fields have been discovered in these two countries.

Regional geology

The following is summarized from Shannon and Naylor (1989). Iran and Iraq are bisected by the NW–SE trending Zagros Mountains comprised mostly of Mesozoic strata. The countries occupy the Zargos–Tauras basinal area on the southern margin of the Alpine and Himalayan fold belts, which have undergone and are undergoing great tectonic stresses. Iran can be divided into seven tectonic elements, four of which extend into Iraq: the southern portion of the Caspian Sea; the Erbula fold belt; the middle intermontane basin; the Zagros fold belt, the Zagros foreland belt; the Arabia platform; the Lut Massif.

Throughout much of the Middle East region a substantial phase of uplift and erosion at the end of the Turonian removed a considerable amount of cover rocks, especially in the Zagros area and over regional palaeohighs. This phase marks the onset of the Alpine orogeny and marks the reactivation of pre-existing structural elements and the formation of an open marine foredeep parallel to the Zagros Fold Belt. An orogenic front, marked by nappes, developed along the Zagros zone with the nappes emplaced from the northeast.

The Zagros orogenic front is accompanied by a typical Alpine tectonic-related sedimentological assemblage. The platform itself, largely stable and uniform until Campanian times, now became significantly more complex. Both shallow- and deep-water carbonates were deposited and clastic-dominant and mixed environments developed in a variety of the intrashelf basins. By Late Maastrichtian time, the main orogenic pulse had waned and a more uniform depositional pattern had resumed. The area now showed two major depositional trends. These were the northwest–southeast basin parallel to the Zagros chain, with deep-marine conditions predominating and the area to the southwest where a carbonate–evaporate platform emerged with local carbonate shoals trending north–south parallel to the reactivated Mesozoic lineaments. The Qatar–South Fars Arch was still a prominent positive element and was the site for a shallow carbonate–evaporite platform with a central salt pan.

The southern region, centred on the oceanic crustal Oman region, remained stable following the orogeny but the northern region of continental crust began to rise isostatically producing a clastic

World Atlas of Oil And Gas Basins, First Edition. Li Guoyu.
© 2011 John Wiley & Sons, Ltd. Published 2011 by John Wiley & Sons, Ltd.

apron. Around the fringes of the Oman Mountains, uplifted during Upper Cretaceous–Paleocene times, Paleocene and Eocene strata are predominantly of shallow-water carbonate facies with reefal development fringing the mountains. These strata thicken southwards across the site of the previous South Oman High and onto the South Oman Shelf. A deeper water facies, perhaps possessing some source potential, could be anticipated in a southwards direction. This fairly stable pattern persisted through much of the Paleocene, with the gradual narrowing of the open marine basin due to the eastward progradation of the western platform and the southward progradation of the clastics shed from the rising mountains in the northwest of the region. In the Oligocene a eustatic sea-level fall led to the removal of underlying Eocene beds over substantial areas.

By Oligocene to Early Miocene times deposition occurred only on a narrow northwest–southeast evaporitic shelf with a series of deeper basin centres. A major deltaic province, the Ahwaz Delta, prograded eastwards from the Arabian mainland and these deposits, together with the age-equivalent carbonate shelf sequences to the east, the Asmari Limestone, provide significant Tertiary reservoirs in Iran. The limestones are generally sealed by evaporites, which inter-digitate with the clastics and carbonates and which developed as a sabkha deposit in this region but probably as thicker marine evaporites in the southern deeper basin.

The Miocene sequence is succeeded by an upwards coarsening of clastic succession in uppermost Miocene and Pliocene times which heralded the commencement of the Tertiary orogeny. Molasse sandstones and conglomerates are coincident with folding, thrusting and nappe emplacement induced by crustal spreading in the Gulf of Aden and the Red Sea. The Zagros area, the site of the main orogeny, is now a dextral strike-slip zone.

Iran The majority of the area of Iran consists of a huge central plateau dominated by three mountain ranges. The very long Zagros range runs along the western border with the modern state of Iraq. Iran has an area of 1.63 million km^2 and a population of 70 million. Iran retains its importance as an oil and gas province with 9 per cent, or 90 billion barrels, of the world's proven oil reserves. Iranian crude is produced mainly from five onshore fields in the Khuzestan region near the Iraqi border and the Persian Gulf: Ahwaz-Bangestan, Marun, Gachsaran, Agha Jari and Bibi Hakimeh. The large Kirkuk oil field is located in Iran. Discovered in 1927 in an anticlinal trap it had recoverable reserves of 5.2 billion t in a fractured Tertiary reservoir. The oil field is renowned for its very high productive rate per well, which reached several hundred tons. Iran's economic progress has been hugely dependent on oil. At the beginning of 2009 estimated reserves were 57.5 billion barrels, only Saudi Arabia and Kuwait in the Organization of Petroleum Exporting Countries (OPEC) have more. In addition to its oil Iran also has reserves of gas which, at 485 trillion cubic feet, are second only to those of the USSR. Iran has affirmed its key strategic position as an emerging gas exporting nation.

Iraq Iraq is located mainly within the Zagros Mountains, having an area of 441,839 km^2 and a population 29.5 million. Before the first Gulf War, Iraq was, after Saudi Arabia, the second largest oil producer in the OPEC. In August 1990, before production dropped because of the war, output was averaging 3.4 million barrels per day. Income from oil has been used to develop the economy, the potential of which is much better than that of most other Arab oil-based economies because Iraq is relatively populous, has ample agricultural land, a good river system, and several non-oil mineral resources. Apart from oil, agriculture is the mainstay of the economy, employing 40 per cent of the labour force.

SAUDI ARABIA, KUWAIT, BAHRAIN, QATAR, UNITED

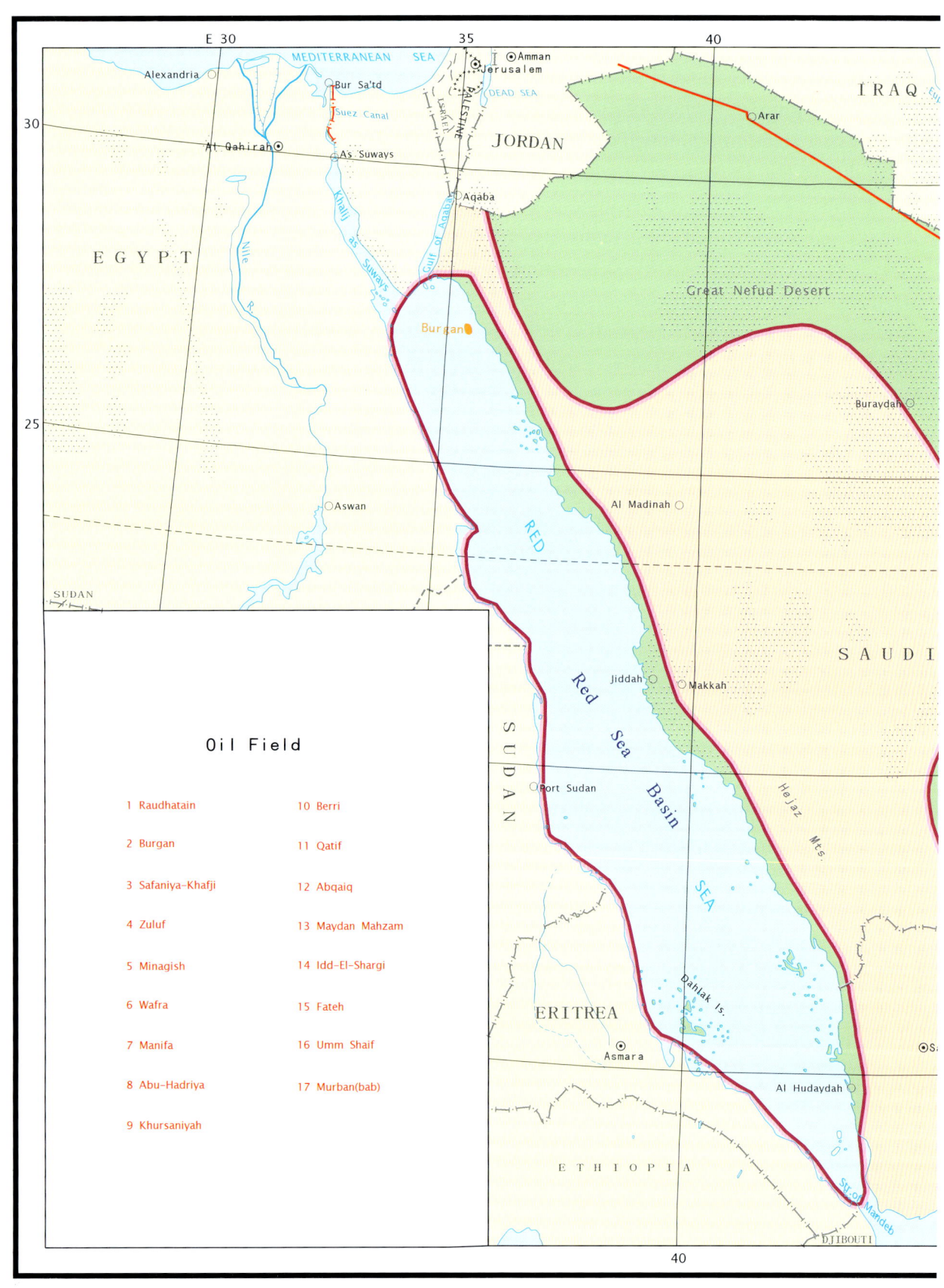

Oil Field

1 Raudhatain
2 Burgan
3 Safaniya-Khafji
4 Zuluf
5 Minagish
6 Wafra
7 Manifa
8 Abu-Hadriya
9 Khursaniyah
10 Berri
11 Qatif
12 Abqaiq
13 Maydan Mahzam
14 Idd-El-Shargi
15 Fateh
16 Umm Shaif
17 Murban(bab)

ARAB EMIRATES, OMAN AND YEMEN   CHAPTER 37

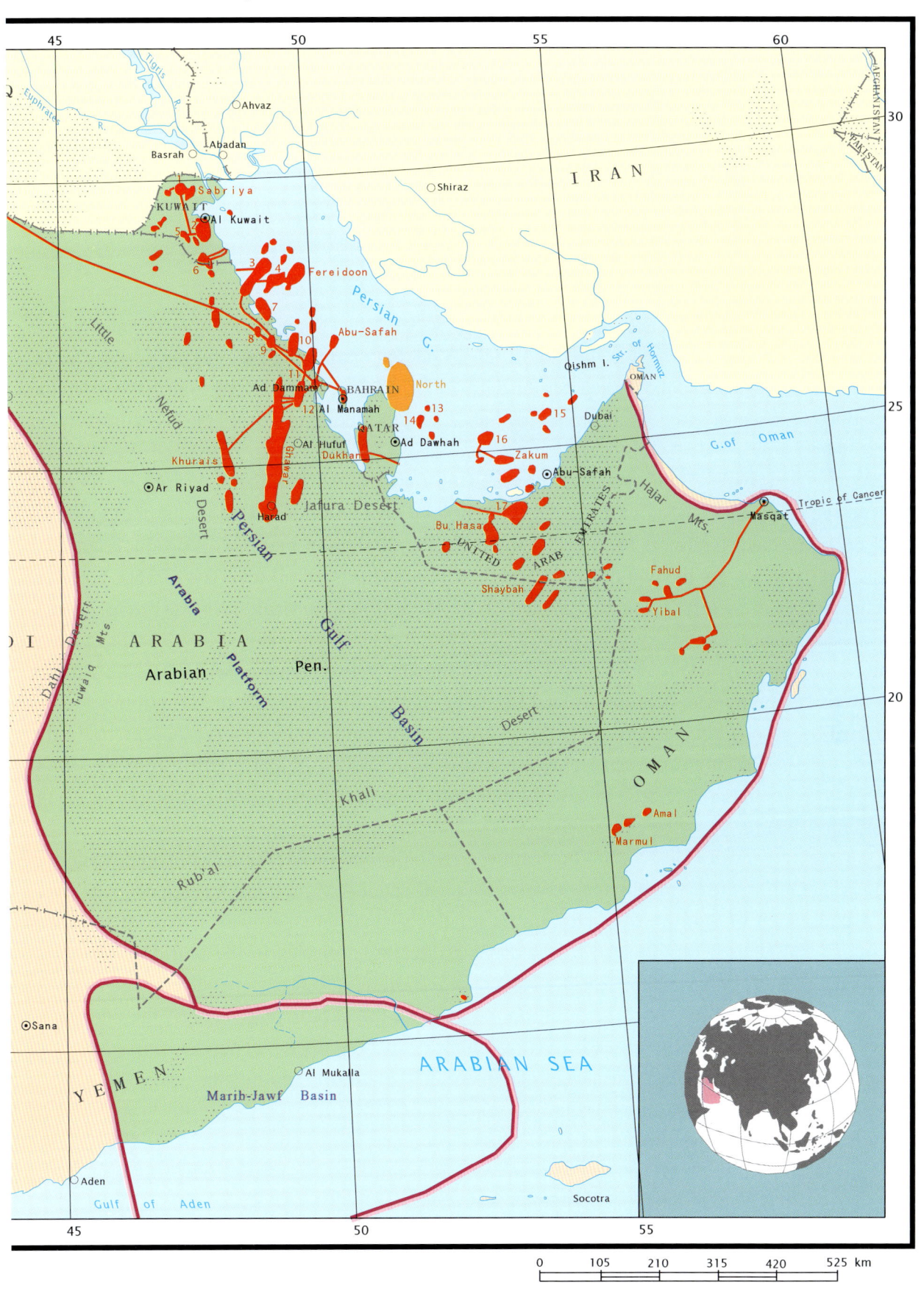

37 Saudi Arabia, Kuwait, Bahrain, Qatar, United Arab Emirates, Oman and Yemen

Geography

These seven countries are located in western Asia, bordering the Persian Gulf to the east, Arabian Sea to the south, Red Sea to the west and Jordan to the north and they occupy by the most productive petroleum area of the Middle East, alongside Iran and Iraq.

History

Oil exploration in these seven countries started in 1924 in Oman. Numerous super giant oil and gas fields are concentrated in the largest oil-producing countries of the region such as Saudi Arabia, Kuwait, Qatar and the United Arab Emirates. Unique geological conditions have led to the formation of the world's richest oil-producing region (see the geological description accompanying maps 39 and 40).

Saudi Arabia Saudi Arabia is located on the Arabian Peninsula, facing the Persian Gulf to the east and Red Sea to the west, with an area of 2.25 million km^2 and a population of 22.5 million. Saudi Arabia is the world's richest oil area, the top ranking oil producer and the largest oil exporter. By the end of 2007, Saudi Arabia had proven oil reserves of 36.1 billion t, ranking first in the world. It had oil production of 445 million t, proven gas reserves of 7.1 trillion m^3 and gas production of 55.3 billion m^3. Annual oil exports totalled 393.8 million t. Oil exploration began in 1933. The first oil field, Dammam, was discovered in 1940. The world's largest oil field, Ghawar, was discovered in 1948. About 50 per cent of the kingdom's area is occupied by desert. The western part comprises Precambrian metamorphosed basement. In the eastern part, sedimentary cover reaches thicknesses of 5000 m of Permian to Tertiary strata. Petroleum potential is very high in new regions with deep potential hydrocarbon-bearing strata, including offshore.

Kuwait Kuwait is located in the northwestern corner of the Arabian Peninsula, bordering the Persian Gulf to the east and lying between Iraq and Saudi Arabia. It has an area of 17,817 km^2 and a population of 3.46 million. Despite its relatively small area, Kuwait is one of the most important and strategically located oil- and gas-producing countries in the world. Exploration commenced in the 1930s and the world's second largest oil field, Burgan, was discovered in 1938, with recoverable oil reserves of 11.5 billion t. However, the explosive growth of the petroleum industry of the country really started in 1961 when the country attained independence from the UK. Kuwait is located on the eastern margin of the Arabian platform. Sedimentary cover comprises Mesozoic and Cenozoic strata with thicknesses reaching 7500 m. Most of the oil was produced from Burgan oil field. By the end of 2007 proven oil reserves of Kuwait were 13.9 billion t, oil production 108 million t, proven gas reserves 572 billion m^3 and gas production 10.5 billion m^3. The petroleum potential remains very high.

World Atlas of Oil And Gas Basins, First Edition. Li Guoyu.
© 2011 John Wiley & Sons, Ltd. Published 2011 by John Wiley & Sons, Ltd.

Saudi Arabia, Kuwait, Bahrain, Qatar, United Arab Emirates, Oman and Yemen

Bahrain Bahrain is a small island in the Persian Gulf. It is connected to the Kingdom of Saudi Arabia in the east by the King Fahd Causeway, which was built in 1986. It has an area of 711.9 km^2 and a population of 75.4 thousand. Oil exploration started early in 1890, and the first oil field, Awali, was discovered in 1930. This oil field has an area of 410 km^2, with depths ranging from 650 and 1,000 m. It has proven oil reserves of 17 million t in multilayered reservoirs of Permian, Jurassic and Cretaceous strata. By the end of 2007 the proven oil reserves were 17 million t, oil production was 1.6 million t and proven gas reserves were 92 billion m^3. Peak annual oil production was 3.83 million t in 1970.

Qatar Qatar is a small peninsula in the Persian Gulf, having an area of 11,437 km^2 and a population of 800,000. It is famous for having the world's largest gas field, the North Field. By the end of 2007 proven oil reserves were 2 billion t, oil production was 40 million t. Proven gas reserves stand at 25.6 trillion m^3 representing about 15 per cent of the world total, with gas production of 51.6 billion m^3. The economy of Qatar depends overwhelmingly on gas and oil production. Oil and gas potential are classified as very high.

United Arab Emirates The seven Emirates of United Arab Emirates (UAE) are located on the eastern flank of the Arabian Peninsula, having a total area of 83,600 km^2 and a population of 4.7 million. Oil exploration activity in the region started in the early 1950s. The oil-producing area of this country is distributed almost equally on land and offshore, lying on the eastern part of the Arabian Plate and covered by Upper Carboniferous to Jurassic, Cretaceous, Tertiary and Quaternary sedimentary rocks, with thicknesses reaching 5000 m. By the end of 2007 proven oil reserves were 12.6 billion t with oil production of 118.3 million t. Proven gas reserves totalled 5.8 trillion m^3 and gas production was 44.6 billion m^3.

Oman Oman is located in the southeastern part of the Arabian Peninsula, bordering Saudi Arabia and the UAE to the north and Yemen to the west. It has an area of 309,500 km^2. Oil exploration began in 1924 and the first oil field, Yibal, was discovered in 1962.

Yemen Yemen is located in the southern portion of the Arabian Peninsula, having an area of 555,000 km^2 and a population of 21.39 million. Oil exploration started in 1930. The first oil field was discovered in 1984. By the end of 2007 proven oil reserves were 410 million t, oil production was 18.5 million t and proven gas reserves stood at 478.5 billion m^3. Oil and gas accumulations are associated with two rock complexes: subsalt Upper Jurassic terrigenous–marine carbonate strata and above-salt Upper Jurassic–Lower Cretaceous terrigenous–marine carbonate strata.

PERSIAN GULF OIL AND GAS REGION

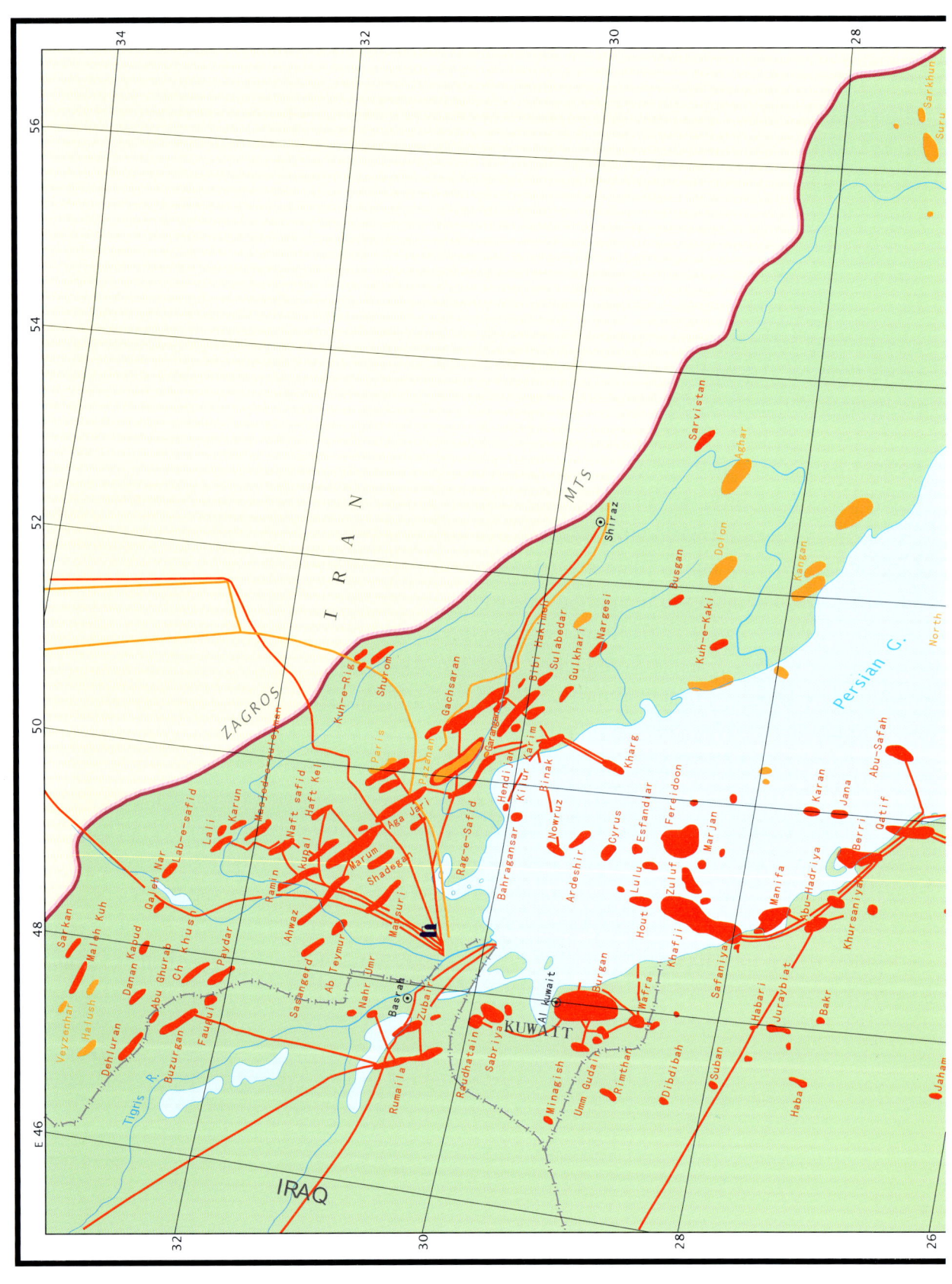

CHAPTER 38

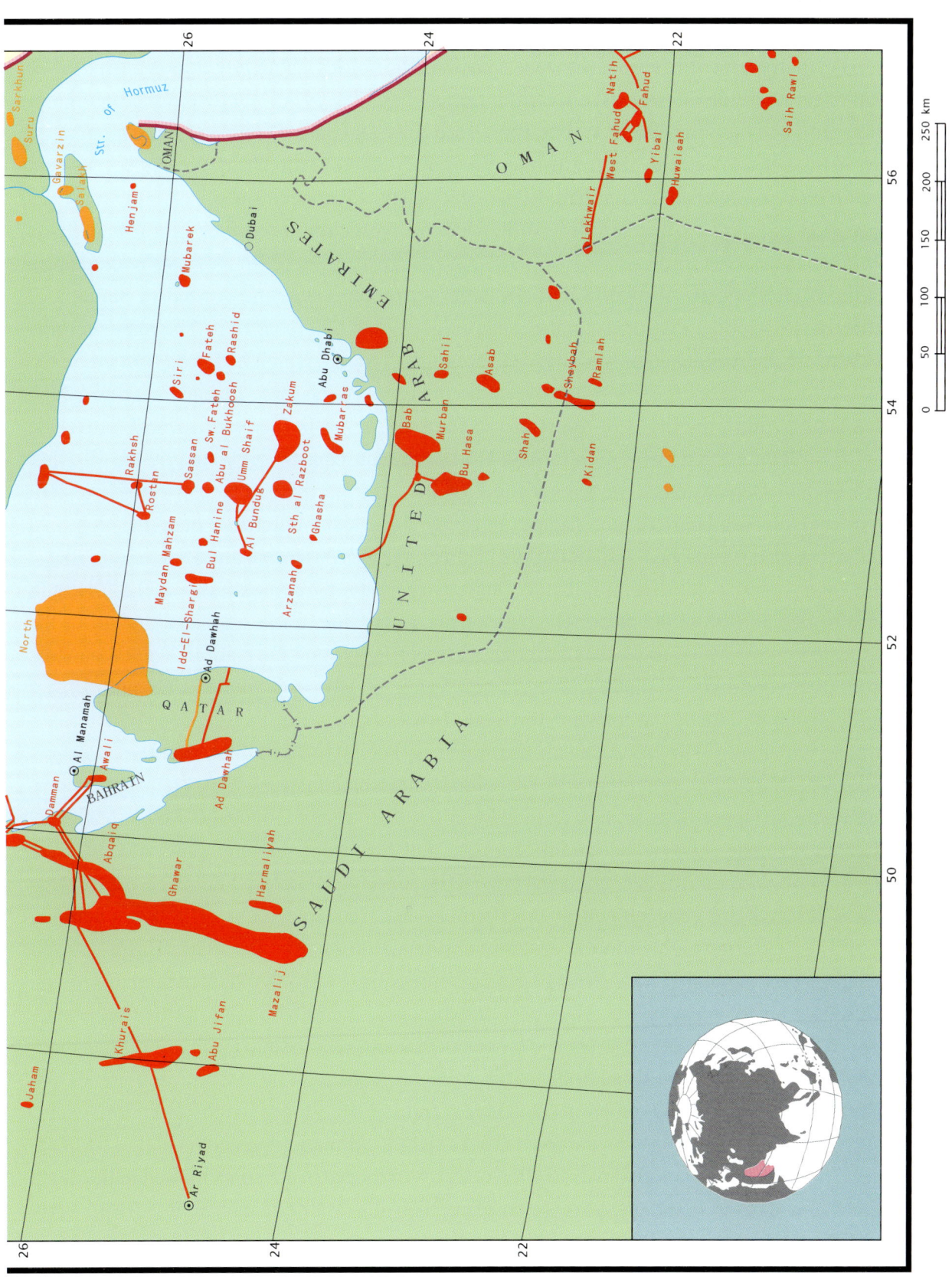

38 Persian Gulf Oil and Gas Region

The Persian Gulf petroleum basin is the richest basin in the world in both production and reserves of oil, and also one of the largest in the world in terms of gas reserves. The region has an area of 3.5 million km^2, and includes Saudi Arabia, Iran, Iraq, Kuwait, Qatar, Oman and the United Arab Emirates. A total of 535 oil and gas fields have been discovered in the basin. Nearly 80 per cent of recoverable oil reserves and more than 30 per cent of gas are concentrated in 40 oil and 6 gas fields classified as unique. Concentration of the bulk of hydrocarbon reserves in a few fields is typical of the basin (Table 38.1).

The basin of the Persian (Arabian) Gulf is a large asymmetric syncline located in the area of conjunction of the Arabian Platform and the Anatolian–Iranian Orogenic Belt. The main tectonic elements here are the Arabian Plate and the Mesopotamian Trough, which had been formed on the platform and folded basement. The conventional boundary between them is considered to be the Euphrates Fault. Within the limits of the platform flank of the basin there are several transverse NE-trending structures such as the Central and North Arabian Uplifts, separated by the El-Khar and Rub-El-Khali Depressions. The folded flank of the basin is divided into the northwestern pre-Taurus and the southeastern pre-Zagros areas by the Mosul outcrop where the depth to basement is less than 4 km. The basement of the basin comprises Archaean and Proterozoic rocks. The sedimentary cover comprises Vendian to Quaternary strata, with thicknesses of 12–14 km in the most subsided portion of the basin. The base of the succession of the southeastern half of the basin is formed by the Vendian–Early Cambrian salt-bearing Ormuz Formation. The Palaeozoic succession is predominantly sandy–clayey. The Permian, Mesozoic, Paleocene and Lower Miocene deposits are mostly carbonates. Terrigenous rocks prevail in the Neocene–Quaternary strata, among which a Middle Miocene salt-bearing sequence up to 1 km-thick occurs.

Oil and gas accumulations are found within a wide stratigraphic interval. Major petroleum complexes are Permian, Upper Jurassic, Lower and Upper Cretaceous and Oligocene–Lower Miocene. Carboniferous, Triassic, Lower and Middle Jurassic and Paleocene–Eocene strata are of lesser importance. Mesozoic deposits account for about 75 per cent of the explored oil reserves, whereas most of gas reserves are concentrated in the Permian and Cenozoic rocks. The overwhelming majority of the fieldsare confined to the eastern subsidence of the Arabian Plate (Basra–Kuwait Depression, Gaza Structural Terrace, Rub-El-Khali Depression) and to the Mesopotamian Trough. In the Mesopotamian Trough hydrocarbon accumulations are confined mainly to the Oligocene–Lower Eocene (Asmari Formation) strata and Upper Cretaceous (Bangestan Formation) limestones at depths of 0.2–3.5 km. On the Arabian Plate, oil-prone horizons are Lower Cretaceous sands and sandstones (Zubair and Burgan Formations) and Upper Jurassic limestones (Arab Formation) occurring at depths of 1.3–3.2 km. Permian carbonate rocks (Khuff Formation) at depths of 3–4.5 km are gas-prone. The bulk of the explored hydrocarbon reserves of the basin are concentrated in the depth interval of 1–3 km, with oil and gas reserves reaching their maximum quantities at 2–3 km depth. Maximum gas reserves are confined to the 3–5 km depth interval (Vysotsky, 1995).

World Atlas of Oil And Gas Basins, First Edition. Li Guoyu.
© 2011 John Wiley & Sons, Ltd. Published 2011 by John Wiley & Sons, Ltd.

Table 38.1 Super giant oil and gas fields in the Middle East

Country	Name	Discovery year	Recoverable reserves Oil (million t)	Recoverable reserves Gas (billion m³)	Basin	Depth	Trap
Saudi Arabia	Ghawar	1948	26,400	924	Persian	2200	Anticline
	Safaniya	1951	5051	330	Persian	1600	Anticline
	Manifa	1957	2380	134	Persian	2300	Anticline
	Abqaiq (Buquiq)	1941	1792	56	Persian	1900	Anticline
	Berri	1964	1680	103	Persian	2200	Anticline
	Zuluf	1965	1484	–	Persian	1700	Anticline
	Khurais	1957	1190	57	Persian	1500	Anticline
	Abu Sa'fah	1963	1050	59	Persian	1900	Anticline
Iran	Marun	1963	1330	1124	Zargos	2200	Anticline
	Ahwaz	1958	1422	336	Zargos	1900	Anticline
	Agha Jari	1938	1260	503	Zargos	1900	Anticline
	Gach Saran	1928	1190	–	Zargos	1200	Anticline
Iraq	Kirkuk	1927	5200	–	Zargos	340–1200	Anticline
	Rumaila	1953	1960	–	Persian	3200	Anticline
	North Rumaila	1961	1120	–	Persian	3200	Anticline
Kuwait	Burgan	1938	10,500	2530	Persian	1400	Anticline
	Raudhatain	1955	120	369	Persian	2400	Anticline
United Arab Emirates	Murban Bab	1954	490	127	Persian	2500	Anticline
	Mubarraz	1969	280	–	Persian	2700	Anticline
	Ummal Dalkh	1973	70	176	Persian	2700	Anticline
	Bu Tini	1980	140	–	Persian	3300	Anticline
Qatar	North	1976	–	24,000	Persian	2700	Anticline
Oman	Fahud	1964	126	–	Persian	600–1853	Anticline

The fields are of structural type, most of them are multilayered. In the Mesopotamian Trough the accumulations are confined to large high-amplitude anticlinal folds stretched along the Zagros Fold System. On the Arabian Plate, the accumulations are mainly associated with local structures that complicate swell-like uplifts of submeridional trend. South of Basra most of the local structures to which oil and gas accumulations are confined are associated with salt diapirism manifestations.

Source rocks are developed almost throughout the entire Silurian to Eocene succession. The main source rocks are those of Early Cretaceous and Late Jurassic age deposited in intrashelf marine basins. These are marine carbonate and terrigenous–marine carbonate rocks with increased organic matter content of essentially sapropelic type. The greatest source potential is associated with the marls and clayey limestones of the Albian Kazhdumi Formation, developed mainly in the central part of the folded flank, as well as clayey limestones of the Callovian–Oxfordian Khanifa and Tuwain Formations occurring in the Central Arabian Uplift and in the north of the

Asian Oil and Gas Basins

Rub-El-Khali Depression. In individual portions of the basin, Late Permian (Khuff Formation), Late Triassic (Kurachine Formation), Middle Jurassic (Sargelu Formation), Late Cretaceous (Gurpi Formation) and Paleocene-Eocene (Pubdekh Formation) rocks are also considered to be source rocks. An additional source of hydrocarbon supply in the southern portion of the basin is provided by Ordovician and Silurian bituminous clayey–siliceous shales. In the extreme northern areas of the basin, source rocks are represented by Devonian bituminous shales.

The petroleum basin of the Persian Gulf is an example of unique hydrocarbon concentration caused by the following factors:

1. prolonged and uninterrupted sagging, which led to the deposition of considerable amounts of predominantly marine sedimentary rocks;
2. regional lateral and vertical distribution of source rock series composed of carbonate and terrigenous–marine carbonate deposits with increased content of sapropelic organic matter;
3. wide development of reservoir rock strata possessing perfect reservoir and filtration properties – their aggregate volume in the basin exceeds 2.5 million m^3;
4. the presence in the succession of regional seals represented mainly by evaporites, and less frequently by clays, clayey limestones, marls with good sealing properties;
5. the availability of numerous large anticlinal traps that had been formed prior to the onset of the main phase of hydrocarbon migration;
6. favourable spatial distribution of the centres of oil/gas generation and zones of oil/gas accumulation which preconditioned short distance and short duration of migration – vertical migration has accounted for the formation of the overwhelming majority of unique oil and gas accumulations.

All these factors have manifested themselves in full in areas located in the vicinity of the Persian Gulf. Within these portions of the basin the bulk of original in-place oil and gas resources are confined. Here, all the giant fields are located. There are six supergiant oil and gas fields in the six large oil producing countries of the Middle East, i.e. the Ghawar oil field of Saudi Arabia (Fig. 38.1), the Burgan oil field of Kuwait (Fig. 38.2), the Kirkuk oil field of Iraq (Fig. 38.3), the Gach Saran oil field of Iran (Fig. 38.4), the Zakum oil field of UAE (Fig. 38.5) and the North Field gas field of Qatar (Fig. 38.6).

Persian Gulf Oil and Gas Region

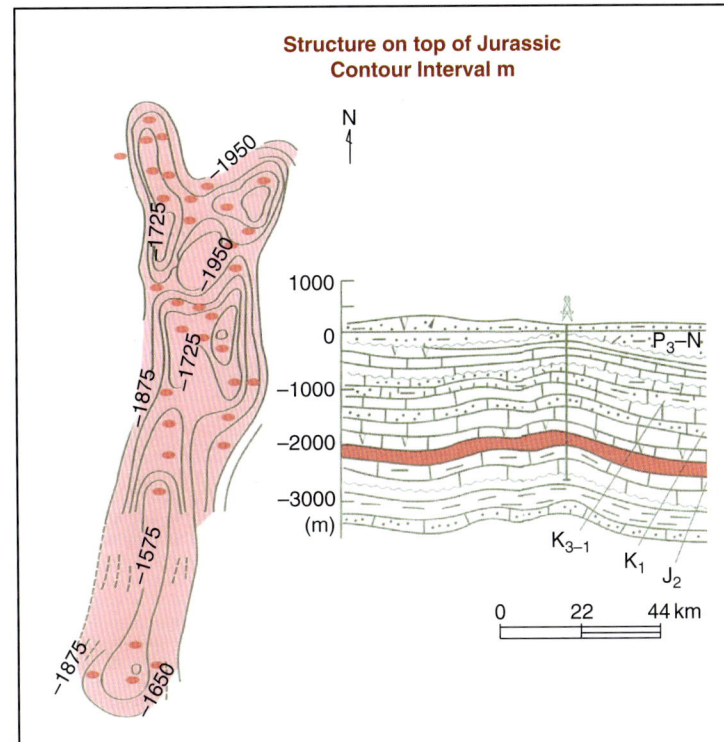

Ghawar is the world's largest supergiant oil field, located in Saudi Arabia.

Basic data

1. Name	Ghawar
2. Discovery year	1948
3. Area (km²)	2,200
4. Reservoir	Carbonate
5. Reservoir age	Jurassic
6. Depth (m)	1,700–2,000
7. Thickness (m)	83
8. Porosity (%)	30
9. Permeability (mD)	4,000
10. Recoverable reserves (billion t)	11.4
11. Peak oil production (million t)	250 (2000)

Fig. 38.1 Details of the Ghawar oil field.

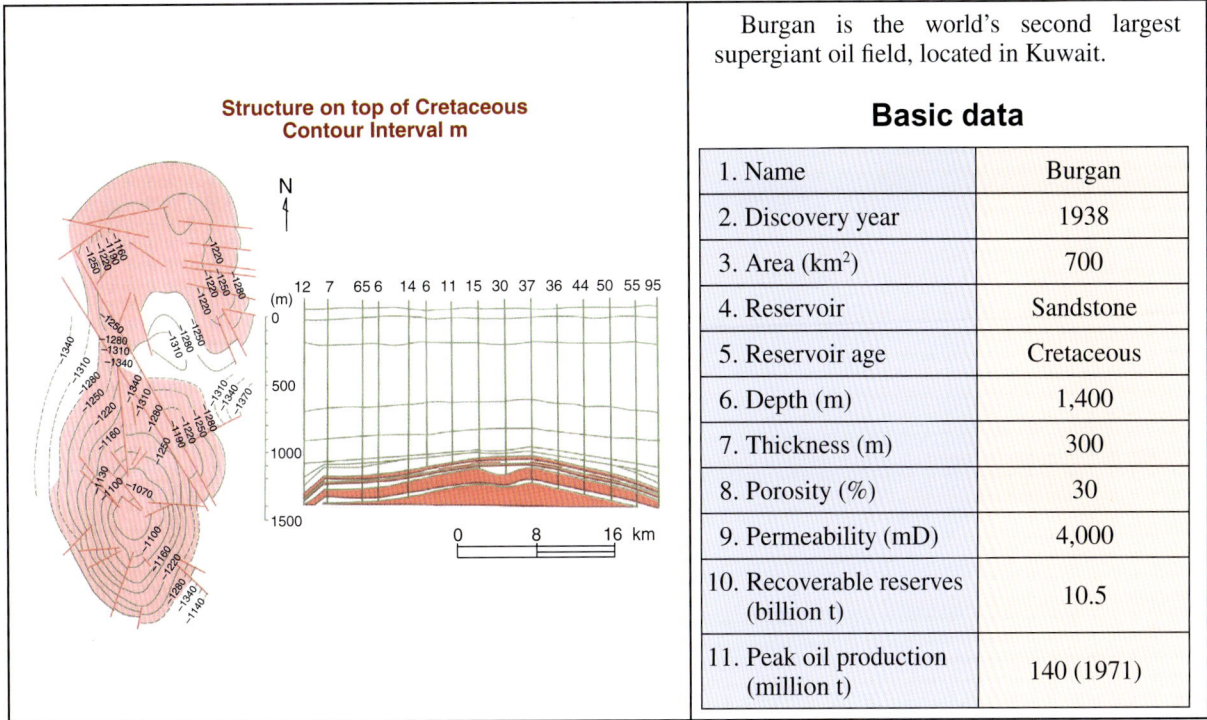

Burgan is the world's second largest supergiant oil field, located in Kuwait.

Basic data

1. Name	Burgan
2. Discovery year	1938
3. Area (km²)	700
4. Reservoir	Sandstone
5. Reservoir age	Cretaceous
6. Depth (m)	1,400
7. Thickness (m)	300
8. Porosity (%)	30
9. Permeability (mD)	4,000
10. Recoverable reserves (billion t)	10.5
11. Peak oil production (million t)	140 (1971)

Fig. 38.2 Details of the Burgan oil field.

Asian Oil and Gas Basins

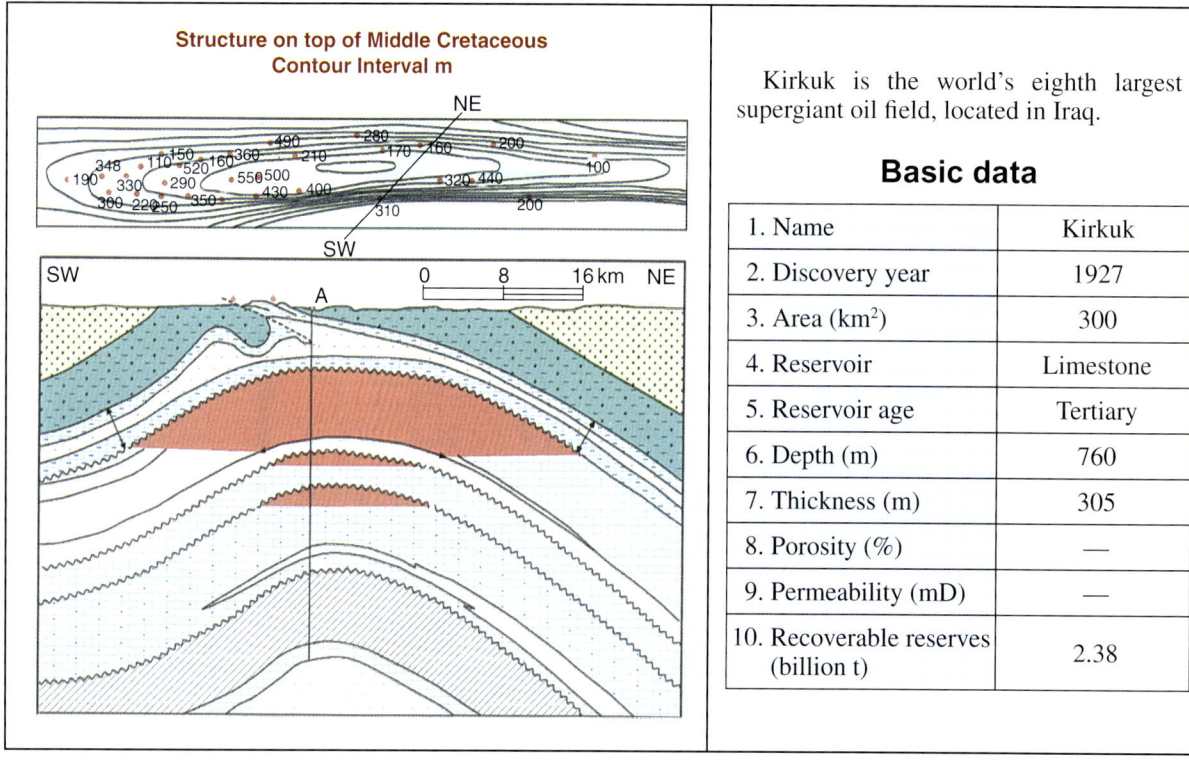

Fig. 38.3 Details of the Kirkuk oil field.

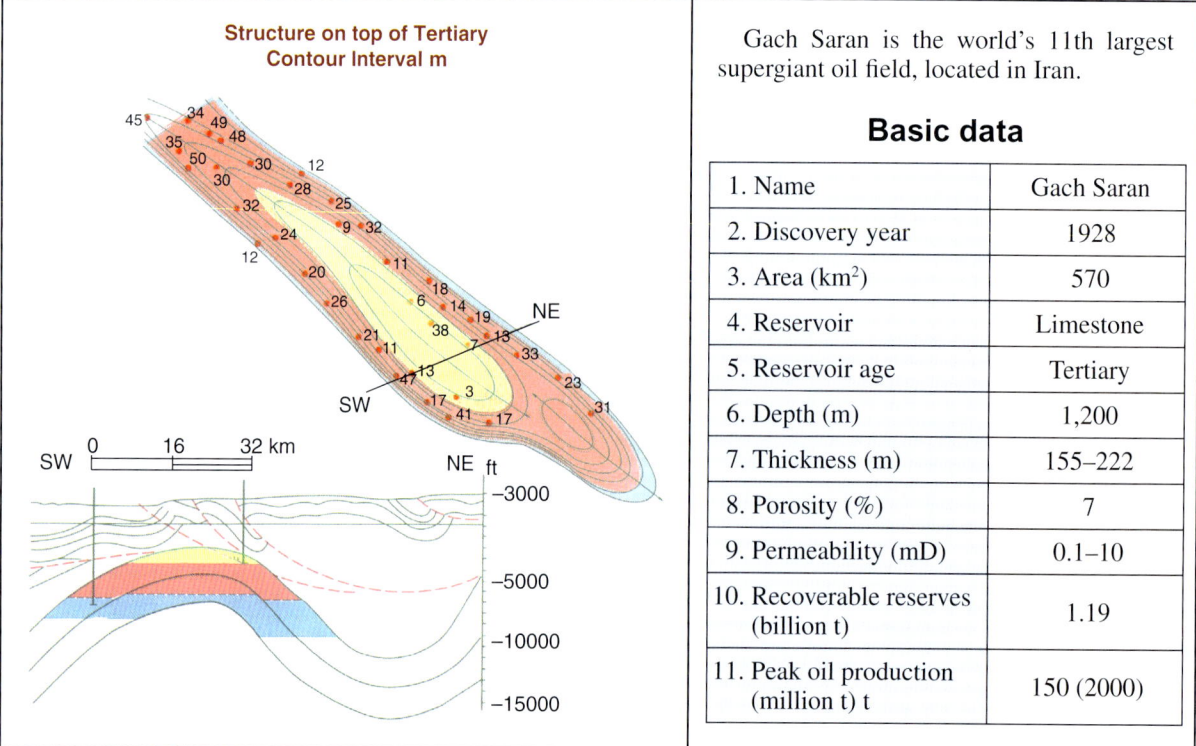

Fig. 38.4 Details of the Gach Saran oil field.

Persian Gulf Oil and Gas Region

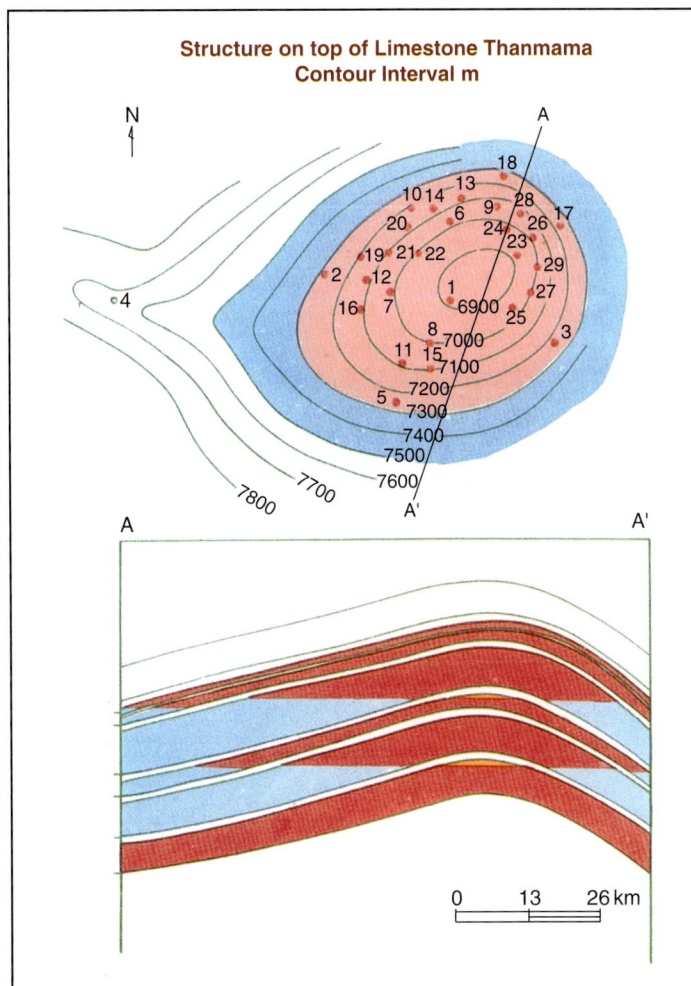

Zakum is the world's fourth largest supergiant oil field, located in UAE.

Basic data

1. Name	Zakum
2. Discovery year	1964
3. Area (km^2)	500
4. Reservoir	Carbonate
5. Reservoir age	K, J
6. Depth (m)	2,700
7. Thickness (m)	143
8. Porosity (%)	14–29
9. Permeability (mD)	1–60
10. Recoverable reserves (billion t)	2.57
11. Peak oil production (million t)	54.7 (2000)
12. Water depth (m)	5–24

Fig. 38.5 Details of the Zakum oil field.

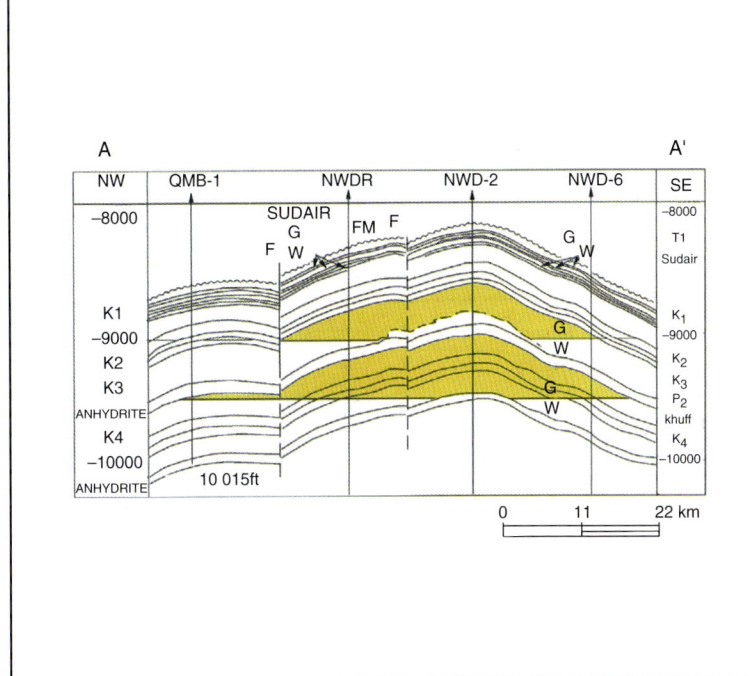

North Field is the world's largest supergiant gas field, located in Qatar.

Basic data

1. Name	North
2. Discovery year	1971
3. Area (km^2)	
4. Reservoir	Limestone
5. Reservoir age	Permian
6. Depth (m)	2,900
7. Thickness (m)	200
8. Porosity (%)	9.5
9. Permeability (mD)	300
10. Recoverable reserves (trillion m3)	22

Fig. 38.6 Details of the North Field gas field.

GEOLOGICAL MAP OF THE PERSIAN GULF OIL

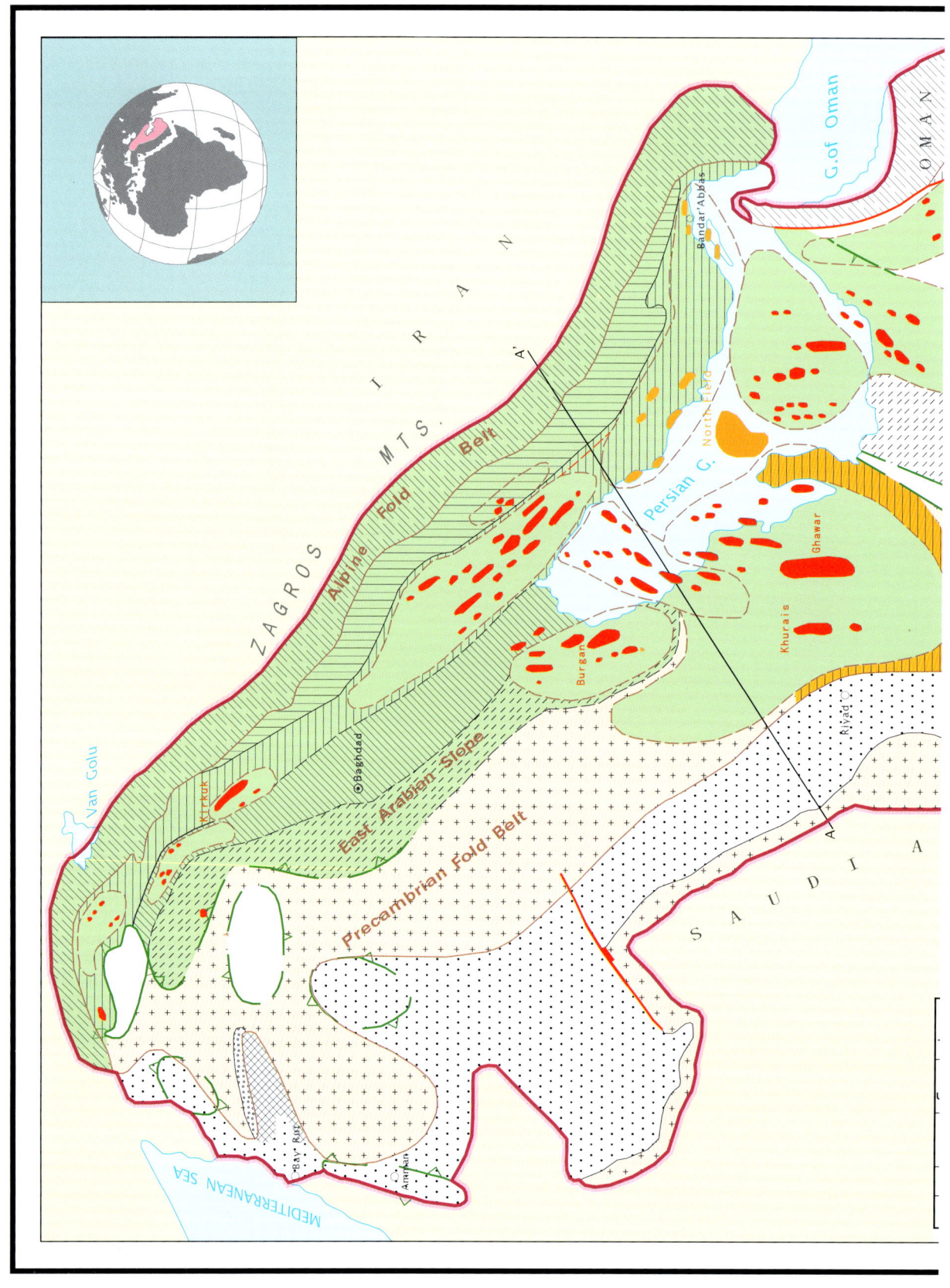

AND GAS REGION CHAPTER 39

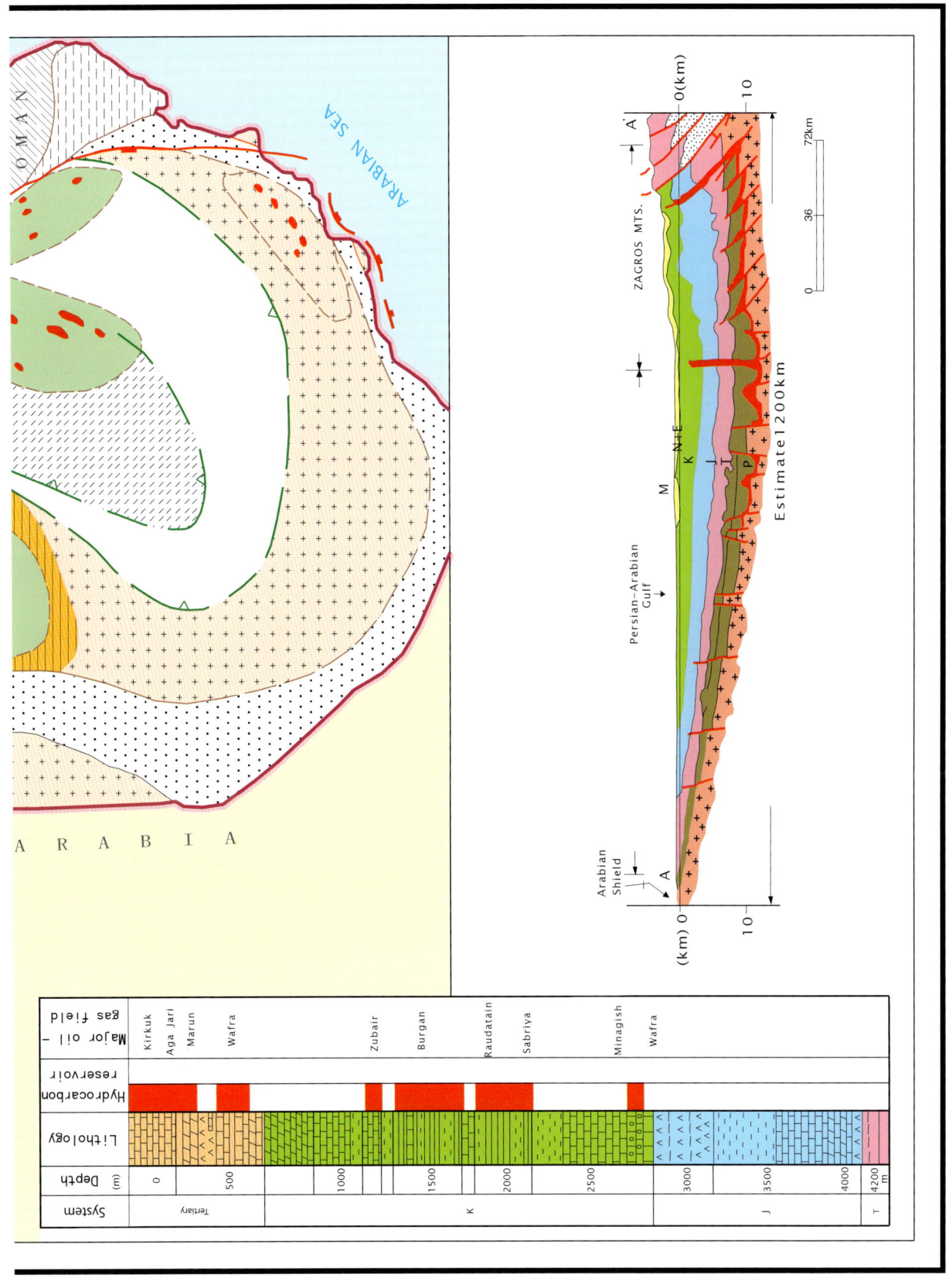

39 Geological Map of the Persian Gulf Oil and Gas Region

Tectonic setting

The Persian Gulf Basin can be considered as a single geological entity in a regional tectonic setting that can be simplified to consisting of three tectonic elements. The first tectonic region comprises the Precambrian massifs of the Arabian Shield. The second tectonic area is the Stable Shelf area, lying to the north and northeast of the Arabian Shield. The third region is the Zagros–Taurus basinal area, which extends northwestwards from the Straits of Hormuz to southeast Turkey.

The Arabian Platform (craton) has an ancient Precambrian basement. The platform was formed in Miocene times when it was separated from the African Platform by the Red Sea and the Gulf of Aden rifts. Within the platform the Arabian–Nubian Shield and the Arabian Plate can be distinguished. The northeastern part of the latter represents an area of peri-cratonic subsidence, which was developing during the entire Phanerozoic Era. Large graben-like depressions occurring on the basement's surface and in the sedimentary cover are filled with predominantly Mesozoic sedimentary rocks. Late Mesozoic rifting accounts for the formation of these grabens.

The Anatolian–Iranian Orogenic Belt is a segment of the Mediterranean Fold Belt formed during the Alpine orogeny. The Anatolian–Iranian Orogenic Belt consists of the southern and northern branches of fold belts. The branches are separated by the basin of the Aegean Sea. Along the southern branch, within the limits of Zagros and East Taurus, there is the Mesopotamian Foredeep that occupies the area lying between the Tigris and Euphrates Rivers, and a considerable portion of the Persian (Arabian) Gulf. The foredeep came into existence at the beginning of the Neocene within the East Arabian peri-cratonic subsidence.

The structure of the orogenic belt is complicated by numerous intermontane depressions formed within the fold belt. The depressions are filled predominantly with Cenozoic and Mesozoic deposits, sometimes also with Palaeozoic rocks. Their total thickness may reach up to 11 km. Nappe basalts are developed in the Cenozoic part of the succession.

The thalassocraton of the Indian Ocean was formed in Late Cretaceous times. In its northern portion a depression corresponding to the Gulf of Oman has been identified. The depression is characterized by suboceanic and oceanic crust. In the Gulf of Oman, pelagic, essentially Cenozoic terrigenous rocks were deposited with probably Cretaceous volcano-sedimentary rock beneath. Total thickness is 4–8 km.

Stratigraphic framework

The Late Carboniferous to Miocene sequence of the Middle East oil province is the world's richest hydrocarbon habitat. The depositional history was dominated by carbonate sedimentation on a very stable, broad platform, bounded to the east by the open Tethys Ocean. The carbonates are as a rule replaced westwards by clastics, which had their source in the uplifted highlands of the Arabian–African continent.

World Atlas of Oil And Gas Basins, First Edition. Li Guoyu.
© 2011 John Wiley & Sons, Ltd. Published 2011 by John Wiley & Sons, Ltd.

Geological Map of the Persian Gulf Oil and Gas Region

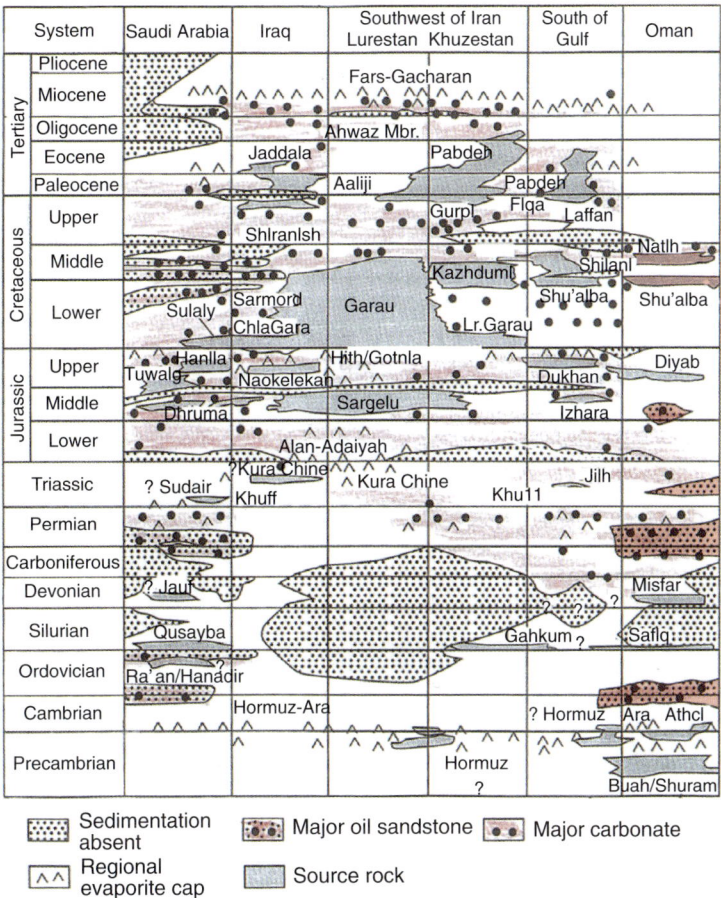

Fig. 39.1 Regional combination of source-rock reservoirs and caprocks of the Persian Gulf (M. W. Hughes Clarke, in Magon and Dow, 1994).

Figure 39.1 shows the generalized stratigraphy of the Middle East region, noting the step-wise evolution of the post-Hercynian sedimentation pattern. From Late Carboniferous to Early Jurassic times a very shallow carbonate platform was in existence, often with central evaporitic depressions and with clastic incursions from the west. From the Middle Jurassic to the Turonian, the platform became more differentiated, with intrashelf basins breaking up the shallow carbonate platform. Clastics were still derived from the west, reaching their maximum development in the early to middle Albian (Magon and Dow, 1994).

A major change in the depositional system occurred in the late Turonian to early Senonian, related to the Alpine Orogeny affecting the Tethys realm. A shale-filled open marine foredeep was formed, with flysch-type sediments derived from the rising orogen in the east. After the late Campanian to early Maastrichtian tectonic activity that resulted in large-scale overthrusting and ophiolite emplacement, stable conditions returned. There was now had a stable carbonate platform in the west and a shaley successor basin in the east, which received its clastic supply from the isostatically rising orogenic belt on its eastern flank.

The depositional basin was considerably narrowed during the early to middle Oligocence, probably in connection with a worldwide fall in sea levels, to rise again in the early Miocene and Pliocene. Late Oligocene to middle Miocene carbonates and evaporites are related to the late Alpine orogenic phase that created the present-day Zagros and Palmyra Fold Belts.

SYRIA, LEBANON, JORDAN, CYPRUS, PALESTINE

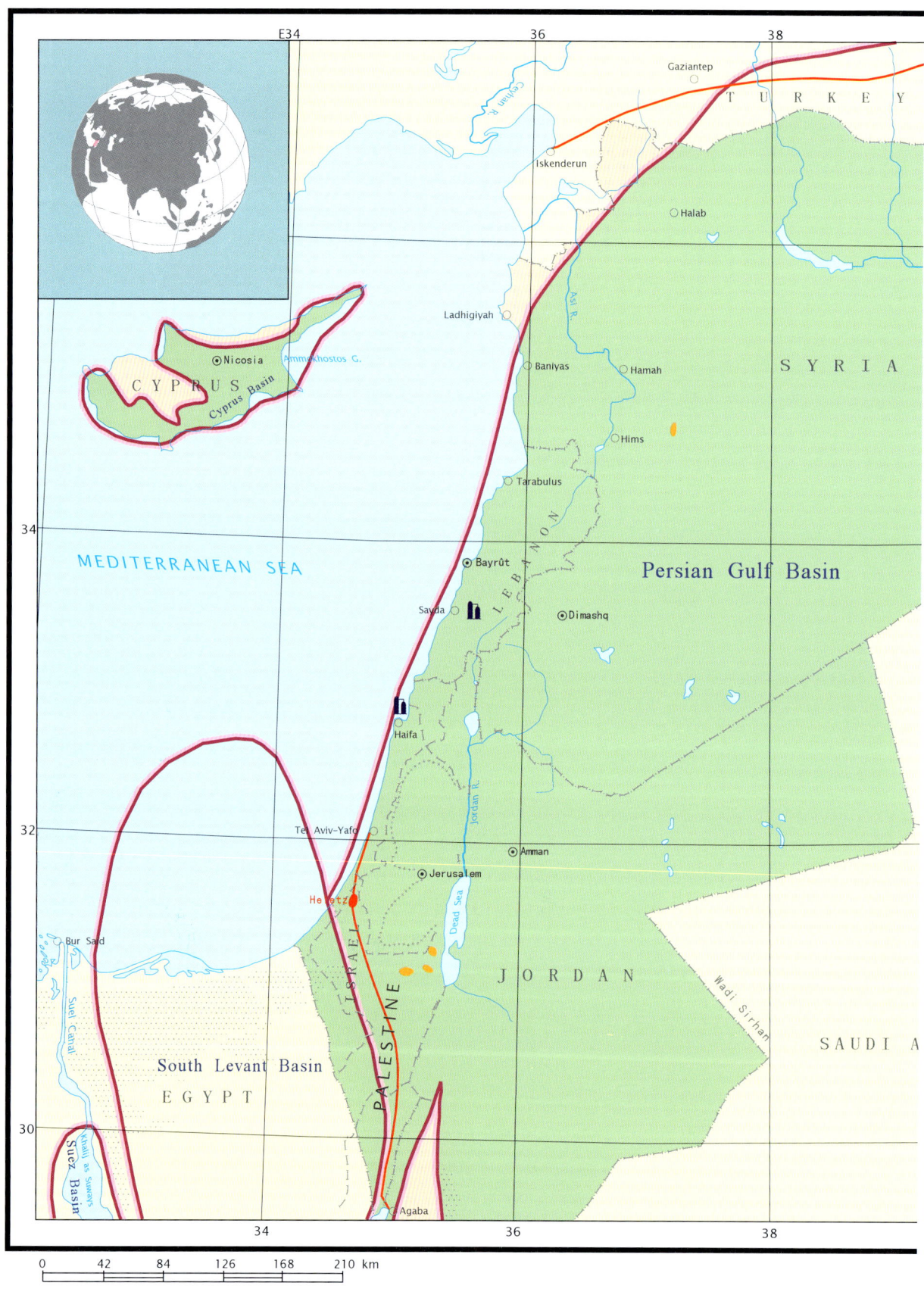

AND ISRAEL CHAPTER 40

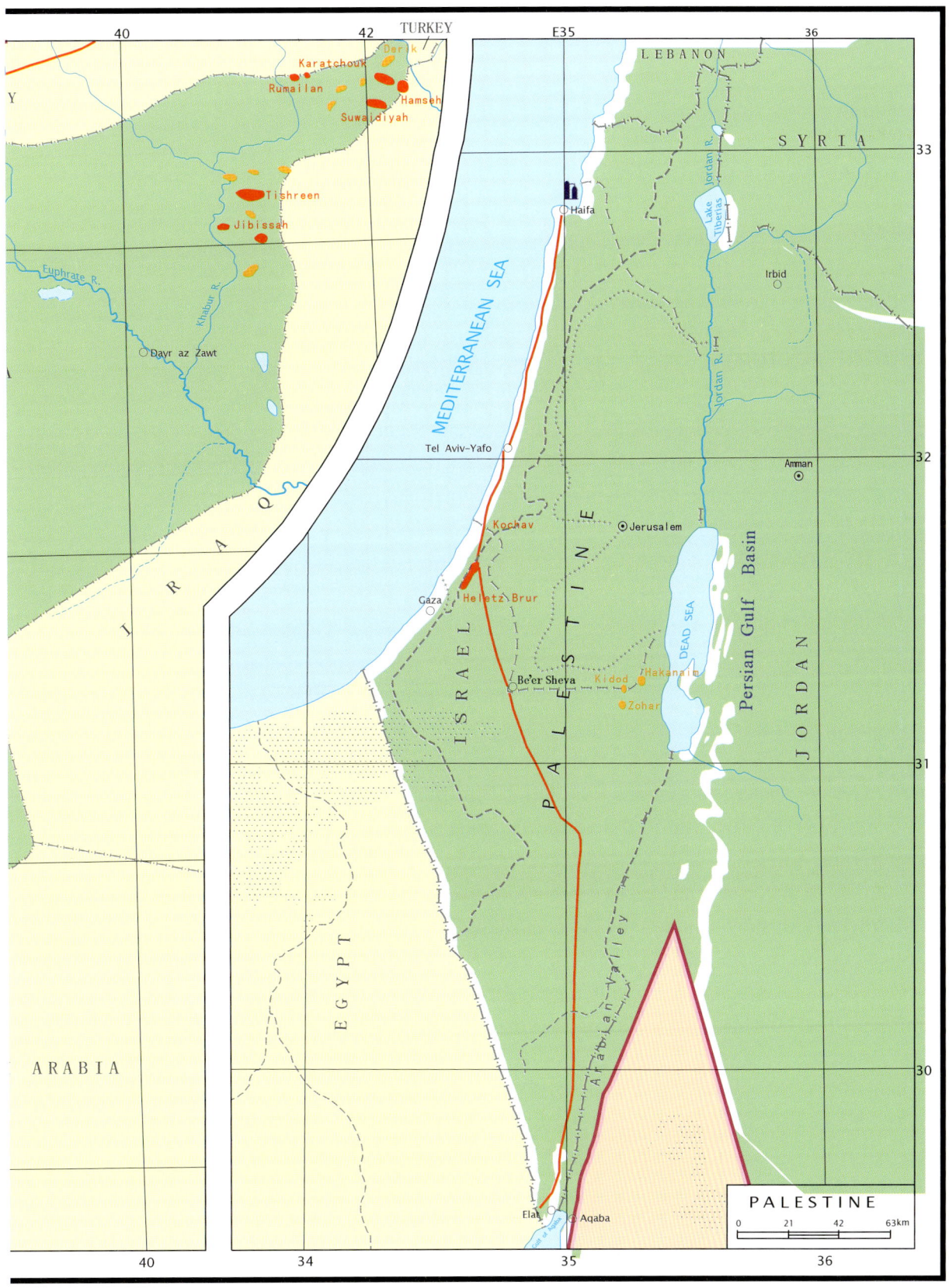

40 Syria, Lebanon, Jordan, Cyprus, Palestine and Israel

Geography

These countries are located in western Asia, facing the Mediterranean Sea to the west.

History

In this area oil exploration started in the early 1930s. Only Syria has commercial oil production due to convenient petroleum geology conditions.

Regional geology

The Arabian Plate is joined along the Neocene collision zone that had been formed due to subduction of blocks of continental plate. On the western and southwestern edges of the Arabian Plate, elements of passive continental margin were formed owing to young divergent movements that resulted in the separation of the Arabian Plate from the African Plate, at first at the continental stage of, then during oceanic rifting.

The Arabian Platform (craton) has an ancient Precambrian basement. The platform was formed in Miocene times when it was separated from the African Platform by the rift system of the Red Sea and the Gulf of Aden. Within the platform the Arabian–Nubian Shield and the Arabian Plate can be distinguished. The northeastern part of the latter represents an area of peri-cratonic subsidence, which was developing during the entire Phanerozoic Era. Large graben-like depressions occurring on the basement surface and in the sedimentary cover are filled with predominantly Mesozoic sedimentary rocks. Late Mesozoic rifting accounts for the formation of these grabens. The rift system terminates the platform in the south, southwest and west. It is traceable via the Gulf of Aden, the Red Sea, the Aqaba Gulf, the Dead Sea and further north to the Turkish border. The Israel–Lebanon–Syrian coastal area lying west of the rift zone together with the Levantian Depression of the eastern portion of the Mediterranean represents an element of the East Mediterranean subsidence of the African Platform.

Syria

Syria has an area $185,180\,km^2$ and a population of 19.5 million. This country is located on the northern part of the Arabian platform, covering the foreland fold belt of the Zagros Mountains, filled by Mesozoic–Cenozoic sediments. Reservoir units range in age from Triassic to Upper Cretaceous, consisting of reef, dolomite, porosity and fractured carbonate with thicknesses more than 400 m, overlaid by shale seals with thicknesses reaching 130 m. Oil exploration started in 1930. The largest oil field, Suwaidyah, was discovered in 1959, with an area of $72\,km^2$. The thickness of the Cretaceous, Jurassic and Triassic oil reservoirs was 260 m, and the recoverable oil reserves were 137 million t. About 90 per cent of oil production of this country is from this oil field. Oil production began in 1968 and reached a peak of around 30 million t annually in 1995. Most of the light oil requiring pre-refining blending is imported from Iraq by pipeline. At the end

World Atlas of Oil And Gas Basins, First Edition. Li Guoyu.
© 2011 John Wiley & Sons, Ltd. Published 2011 by John Wiley & Sons, Ltd.

of 2007 recoverable oil reserves were 342 million t, oil production 19.3 million t, gas reserves were 240 billion m^3 and gas production 5.7 billion m^3.

Lebanon Lebanon is located on the western coastline of the Middle East, bordering the Mediterranean Sea to the west, lying on a narrow coastal plain. It has an area of 10,452 km^2 and a population of 4 million. Oil exploration began in 1947. The Dead Sea fault runs throughout the length of the country from south to north, developing a large graben filled by Mesozoic–Cenozoic sediments with a total thickness in excess of 5000 m. Its neighbouring countries are important oil-producing areas, but Lebanon is a non-oil-producing country. Geologists estimate the oil prospect of this country as low.

Jordan Jordan is located in the northwest part of Arabian Peninsula, bordering Saudi Arabia to the south. It has an area of 89,340 km^2, is about 80 per cent covered by desert and has a population of 5.91 million. Exploration can be dated back to 1930. The first oil field, Hamza, was discovered in 1984 with recoverable oil reserves of 41,000 t. The maximum recoverable oil reserves were 1.64 million t in 1987. During 1987–1995 annual oil production dropped from 25,000 t to 5,000 t. At present oil production has been discontinued.

Cyprus Cypress is an island in the Mediterranean Sea, having an area of 9251 km^2 and a population of 758,000. The whole island is part of the Alpine Orogenic Belt, consisting of Palaeozoic–Cenozoic strata. Carbonate rocks are lightly metamorphosed. Two wells were drilled, with depths of 3297 and 2587 m, in the Cretaceous rocks. Structures are anticlines and overthrusts. At present this country is non-oil-producing.

Palestine Palestine is an ancient country, adjacent to Israel and facing Mediterranean Sea to the west. It is located on the Arabian platform. Sedimentary cover consists of Palaeozoic and Mesozoic rocks. The country is non-oil-producing.

Israel Israel is a highly industrialized and agricultural country, facing the Mediterranean Sea to the west. It lies at the junction of Asia, Africa and Europe, having an area of 15,200 km^2. Seismic work began in 1934. Oil exploration started in 1947. The first oil field, Aeletz-Brur, was discovered in 1955. Some small complex structures were discovered on the northern part of the northwestern Arabian platform. By the end of 2007 proven oil reserves were 200,000 t and gas reserves were 30.4 billion m^3. Oil production was a few thousand tons only.

AFRICAN OIL AND GAS BASINS

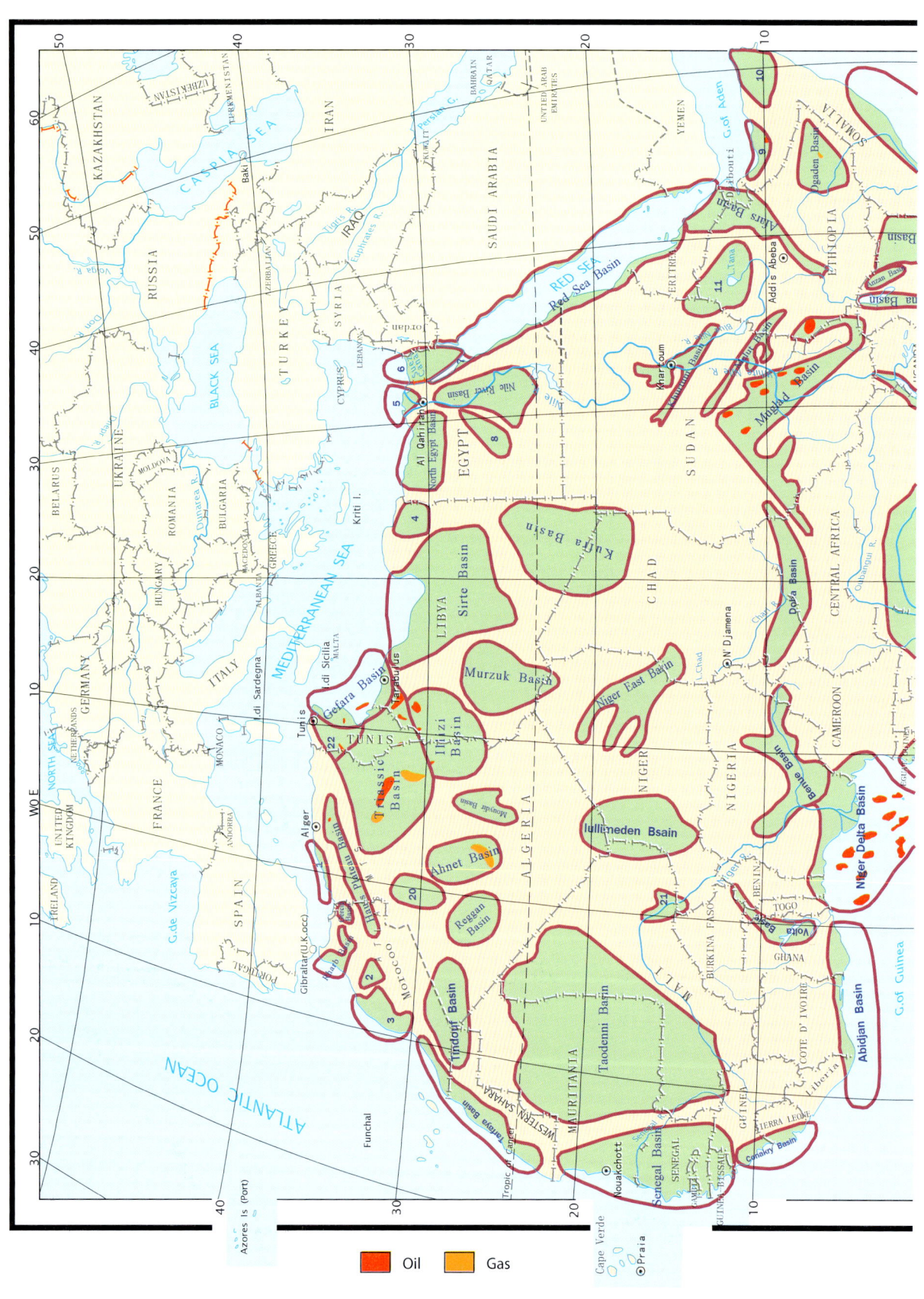

PART III

Part III
African Oil and Gas Basins

Geography

Africa comprises 61 countries. The African region comprises the continent of the same name and adjacent deep-water areas of the Atlantic and Indian Oceans, the Mediterranean and Red Seas, Madagascar and a number of smaller islands. The region together with the islands occupies an area of 30.3 million km^2 and a population estimated at 900 million (2005).

History

The first oil field was discovered in Egypt in 1909, which marks the advent of the development of the oil industry of Africa. The continent of Africa has very substantial oil and gas prospects, but development of the oil industry has been relatively slow, particularly when compared with other regions such as the Middle East. Oil production was 1.4 million t in 1960, 100 million t in 1965 and 270 million t in 1970. Production in 2008 climbed to 462 million t. Over the past decade, countries such as Sudan, Angola and Algeria (Table III.1) have become major oil and gas producing areas, establishing themselves, alongside other producers such as Nigeria, Cameroon, Gabon and Egypt, into significant players in the global oil and gas market. At present, South Africa and East Africa are notable for being the only regions where there is little or no production, although there have been considerable exploration activities in these regions in recent years. Further exploration will depend on the trend of international oil prices.

Regional geology

The The African continent represents the core of ancient Gondwanaland and is a stable craton comprising amalgamated blocks of Precambrian basement. The sedimentary basins of Africa are mainly of two types, sag basins and rifts. The continent comprises flat, albeit relatively elevated, plains of Precambrian basement, composed of igneous and metamorphic rocks of great and diverse ages. Contenent-wide, these basement rocks are locally overlain by a cover of Cambrian to recent shallow marine and continental sediments.. During the Triassic Period there was continent-wide volcanic activity accompanying the tectonic events that resulted in Africa becoming a separate continent. Jurassic, Cretaceous and Tertiary times are characterized by the establishment of rifting on the eastern side of the continent.

The Karoo Basin is clearly syn-depositional in origin, but many of the sag basins, notably those of North Africa, are clearly post-depositional in origin. The uniform stratigraphy of basins such as those of Algeria shows that their sediments were laid down on the gently sloping Saharan platform. Similarly the Murzuk and Kufra Basins of southern Libya, when they existed throughout the Palaeozoic and much of the Mesozoic eras, were northerly sloping embayments open to the

World Atlas of Oil And Gas Basins, First Edition. Li Guoyu.
© 2011 John Wiley & Sons, Ltd. Published 2011 by John Wiley & Sons, Ltd.

Table III.1 Proven reserves and production of oil and gas in Africa (2008)

Country	Proven reserves		Production	
	Oil (million t)	Gas (billion m³)	Oil (million t)	Gas (billion m³)
Total Africa	16,036	13,990	463	184
Algeria	1671	4502	687	93
Angola	1237	269	93	1.6
Benin	1	1	—	—
Cameroon	27	135	4.2	—
Chad	205	—	7.2	—
Democratic Republic of Congo	24	0.9	1.2	—
Congo	219	90	12	—
Egypt	506	1656	34	45.5
Equatorial Guinea	150	36	16	—
Ethiopia	0.05	24	—	—
Gabon	273	28	11.7	0.1
Chana	2	22	0.3	—
Cote D'ivoire (Ivory Coast)	13	28	1.5	—
Libya	5680	1539	86	—
Maurtania	13	28	1.3	—
Mozambique	—	127	—	—
Namibia	—	62	—	—
Nigeria	4961	521	97	27
South Africa	2	—	0.7	—
Sudan	684	84	24.5	—
Tunisia	58	65	4.2	2.2

Tethys Ocean to the north. They did not gain their present closed basinal shape until the middle of the Cretaceous Period.

The genesis of circular sag basins such as those in Africa has long attracted attention and study. One of the most popular models proposes that thermal doming over a mantle 'hot spot' led to the erosion of uplifted crustal rock, followed by cooling and crustal collapse, initially into a rift, followed by gentle subsidence to form a sag basin. More recently it has been suggested that crustal sags may result from 'cold spots' due to mantle cooling, resulting in downwelling and a gentle sagging of the crust. Palaeocurrent data clearly show that the basins post-date their sediment fill, and imply that the northerly slope of the palaeo-Tethys was locally disrupted by crustal sag basins in the mid-Cretaceous.

The Cretaceous Period was a very important time in the history of Africa. The break-up of the Gondwanaland, and the concomitant opening up of the Atlantic Ocean defined the present boundaries of the African continent. As the Cretaceous rifts extended across Gondwanaland, some rifts failed, becoming infilled with thick sequences of sediments that often included organic-rich muds laid down in restricted freshwater–marine environments. Failed rifts are characterized by high heat flow, due to crustal thinning. These failed rift basins are thus often important petroleum provinces. Epicratonic rifts, such as the Sirte embayment of Libya have largely carbonate reservoirs, while the intracratonic rifts, such as those of the Sudan and East Africa, are characterized by terrigenous sediments.

Finally, the present Atlantic and Indian coasts of Africa are defined by rifts that failed and which were extended by sea-floor spreading in oceans. The Atlantic coastal basins are significant petroleum provinces because their sediments prograded out over organic rich muds that were deposited in the narrow anoxic Cretaceous seaway of the proto-Atlantic Ocean. Where the failed rift system of the Benue Trough cross-cuts the Atlantic coast, a vast influx of terrigenous detritus generated the major petroleum province of the Niger delta. The analogous rift basins on the Indian Ocean coast seem to lack such rich source beds. The discovery in more recent years of more new oil and gas fields on the continent serves to confirm the high prospectivity of the continent as a whole, such that many large companies are focusing their exploration work here (Selley, 1997).

EGYPT

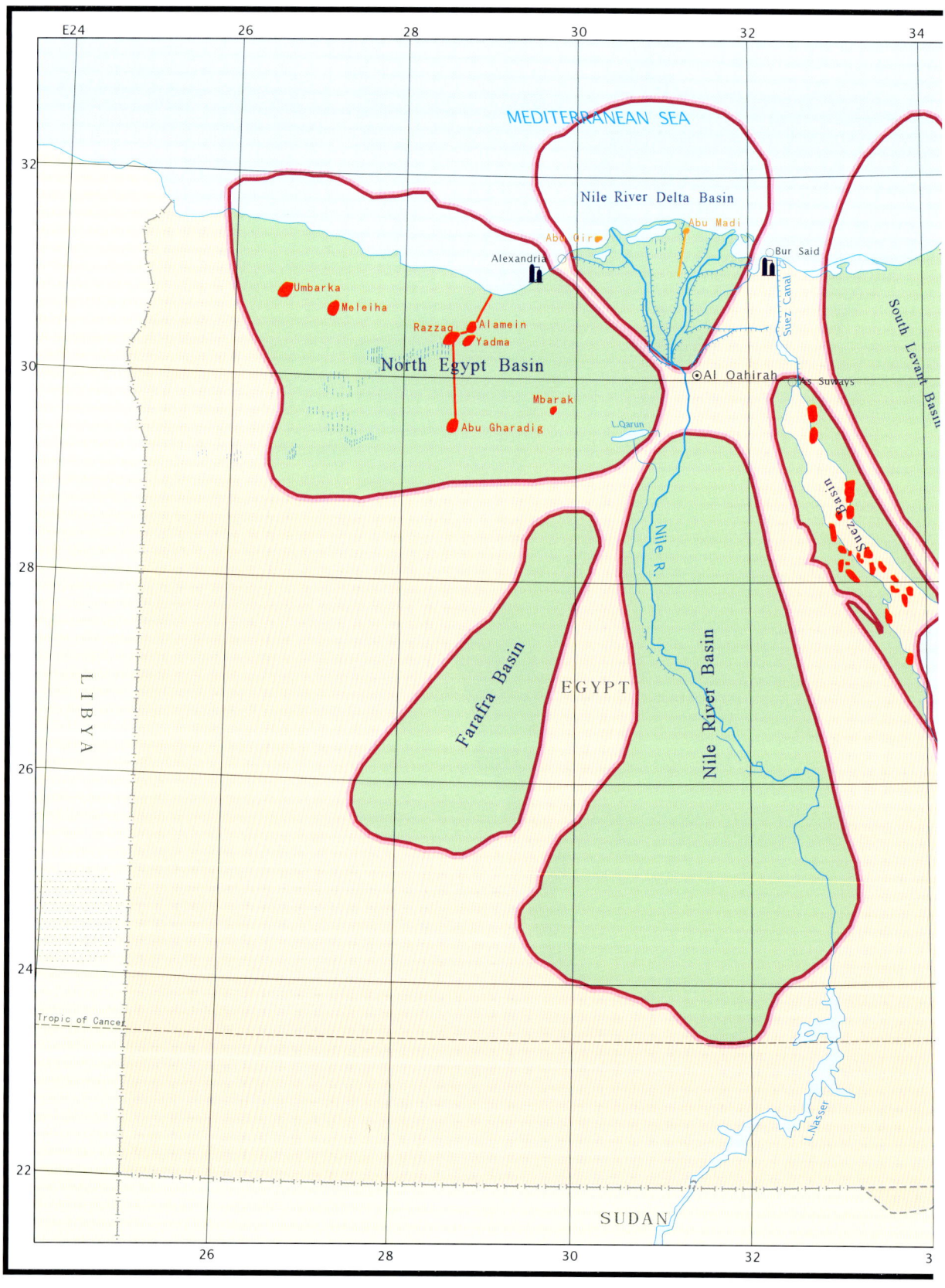

CHAPTER 41

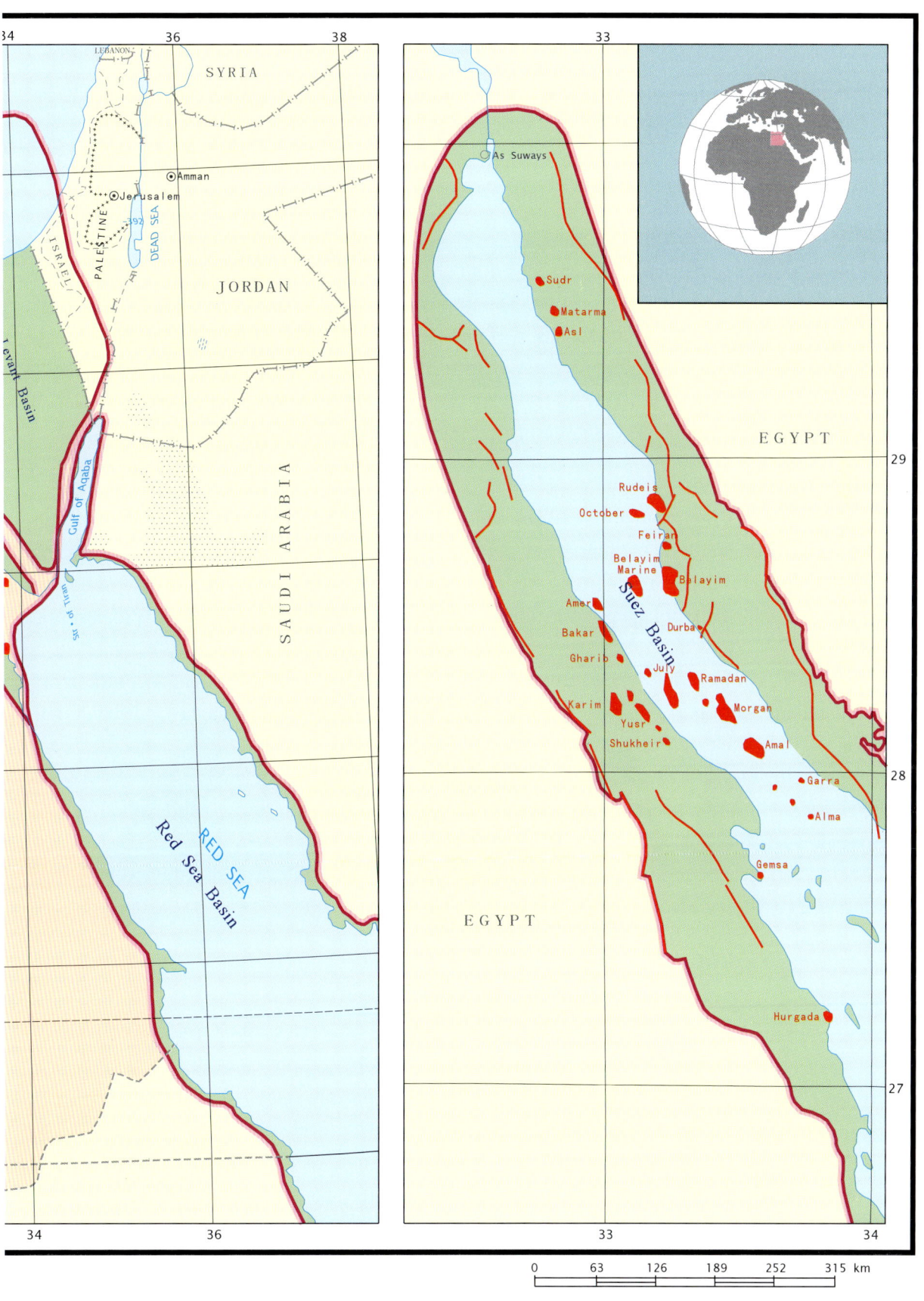

41 Egypt

Geography

Egypt is located in North Africa, with its northern coastline bordering the Mediterranean Sea. The area is 1,000,000 km² and the estimated population in 2005 was 73 million. About 97 per cent of the area of the country is made up of the Sahara Desert.

History

Egypt is one of four ancient civilization centres of the world. In Roman times, there is evidence that oil was used by the inhabitants of the region. The first commercial oil field was discovered in 1886. The first truly large oil field, the El Morgan, was discovered in 1965. Oil exploration activities have been carried out mainly in the Western Desert, the Suez Basin and the Nile River Basin. Oil production showed a steady growth between 1963 and 1995, but the trend since then has been a gradual decline. In 1963, 56 million t were produced, 45.9 million t in 1995 and then declining to 34 million t in 2008.

Regional geology

Basement exposures represent about 10 per cent of the total area of Egypt. The rest of the country is covered by Phanerozoic sediments which increase progressively in thickness northwards following the regional dip of the African Plate towards the Mediterranean Sea. Development of irregular thickness distributions was controlled by basement structures and the post-Precambrian tectonic history.

Tectonically Egypt has experienced five main evolutionary stages, representing a complex history of crustal subduction of arc–trench systems leading to accretion onto the east African continental plate. These stages are as follows:

1. eugeosynclinal flysch sedimentation stage, with island arc volcanism and crustal subduction – these sediments and associated ophiolites were regionally metamorphosed and folded by later orogenic events;
2. orogenic and syn-orogenic intrusion stage, of granite and plutonic granodiorite bodies – these were later partly metamorphosed into gneisses;
3. post-orogenic geosynclinal volcanic stage;
4. post-orogenic foreland basins stage;
5. late orogenic and post-orogenic plutonic intrusions stage.

Within Egypt the basement complex is exposed north and west of the Red Sea and the Gulf of Suez, in the areas of Sinai and the Eastern Desert. In the high mountain areas the Nubian craton was formed of orogenic granites and developed a crust 30 to 40 km thick. In contrast, the crust at the margin of the Red Sea Trough is 20 km thick and consists of alkaline metasomatic granites belonging to the latest plutonic intrusions. The crustal thickness of the east Saharan craton is of similar magnitude, from the Jabal Uwernat area to the Mediterranean coast.

The Mesozoic evolution and development of sedimentary basins of North Africa and Egypt followed the break-up of Pangea since Triassic times. Crustal movements associated with different

World Atlas of Oil And Gas Basins, First Edition. Li Guoyu.
© 2011 John Wiley & Sons, Ltd. Published 2011 by John Wiley & Sons, Ltd.

Table 41.1 Major sedimentary basins of Egypt

Basin	Area (thousand km²)	Major sedimentary rock		Reservoir	
		Age	Thickness (m)	Age	Lithology
Farafra	43			J, K, R	Sandstone
Nile River	170	Pz–Kz	3500	E_3, N_1	Sandstone
Nile River Delta	55	R	3800	R	Sandstone
North Egypt	132	R	13,500	Mz	Sandstone
South Levant	74	Mz, R	5500	Pz, K, R	Sandstone
Suez	30	Pz–Kz			Sandstone

phases of rifting and opening of the Atlantic Ocean and the Tethys, and eustatic sea-level changes, have influenced tectonics and sedimentation in the Egyptian basins. The Triassic is represented by a single transgressive depositional cycle restricted in its distribution to the northern uplift of southern Egypt. The following Jurassic sequence is also transgressive in nature as it is clastic-dominated at the base and is carbonate-dominated at the top. Evidence of the Jurassic transgression is not commonly found in the cratonic areas because of the residual effect of the Hercynian uplift of central and southern Egypt. Late Jurassic marine clastic sediments, however, were discovered in the subsurface in the Dakhla Basin, restricted to the Mesaha Trough. The Late Jurassic–Early Cretaceous boundary in Egypt is associated with alkaline volcanic activity.

The six sedimentary basins of Egypt are either intracratonic or pericratonic basins, or rifts basins. that were developed following crustal attenuation and fracturing of uplifted areas and subsequent faulting and subsidence. Symmetrical or asymmetrical basins are infilled by terrigenous clastics followed by shallow-marine clastics, as well as carbonates and, in some cases, evaporites.

The Gulf of Suez Basin is a tensional tectonic rift that forms the northern extension of the Red Sea Graben. The basin is 60 to 80 km wide and consists of two major tilted blocks found on each side of the rift. The rift is considered to be a post-Eocene feature that became a fully developed basin during the Miocene with sediment thicknesses of more than 3700 m. The Senonian and Eocene carbonates are considered to be the main hydrocarbon source rocks in the Gulf of Suez Basin. Oil was found in the Miocene, Cenomanian, Albian and Carboniferous sandstone reservoirs (Table 41.1). Porous and fractured limestones and reefs of the Miocene, Eocene and Late Cretaceous are also good reservoirs. The estimated oil reserves of the Gulf of Suez amount to 520 million t. The October oil field in the northern part of the Suez Basin, discovered in 1977 in a faulted anticlinal trap, has accumulated oil production of 411 million t.

The Nile River Basin has an area of 170,000 km² with thicknesses of up to 3500 m of Palaeozoic and Mesozoic sediments. The North Egypt Basin has an area of 132,000 km² with up to 13,500 m thickness of Tertiary sediments. At present only a few small oil fields have been discovered in the North Egypt Basin. In the Nile River Basin 53 exploration wells have been finished without commercial result and so deeper exploration is needed in the future. Source rocks have been identified in marine and continental sediments. reservoir rocks are associated with sandstones, mostly in Tertiary strata. Seals are associated with mudrocks, and traps are mainly multiple and represented by structural and non-anticlinal types.

LIBYA

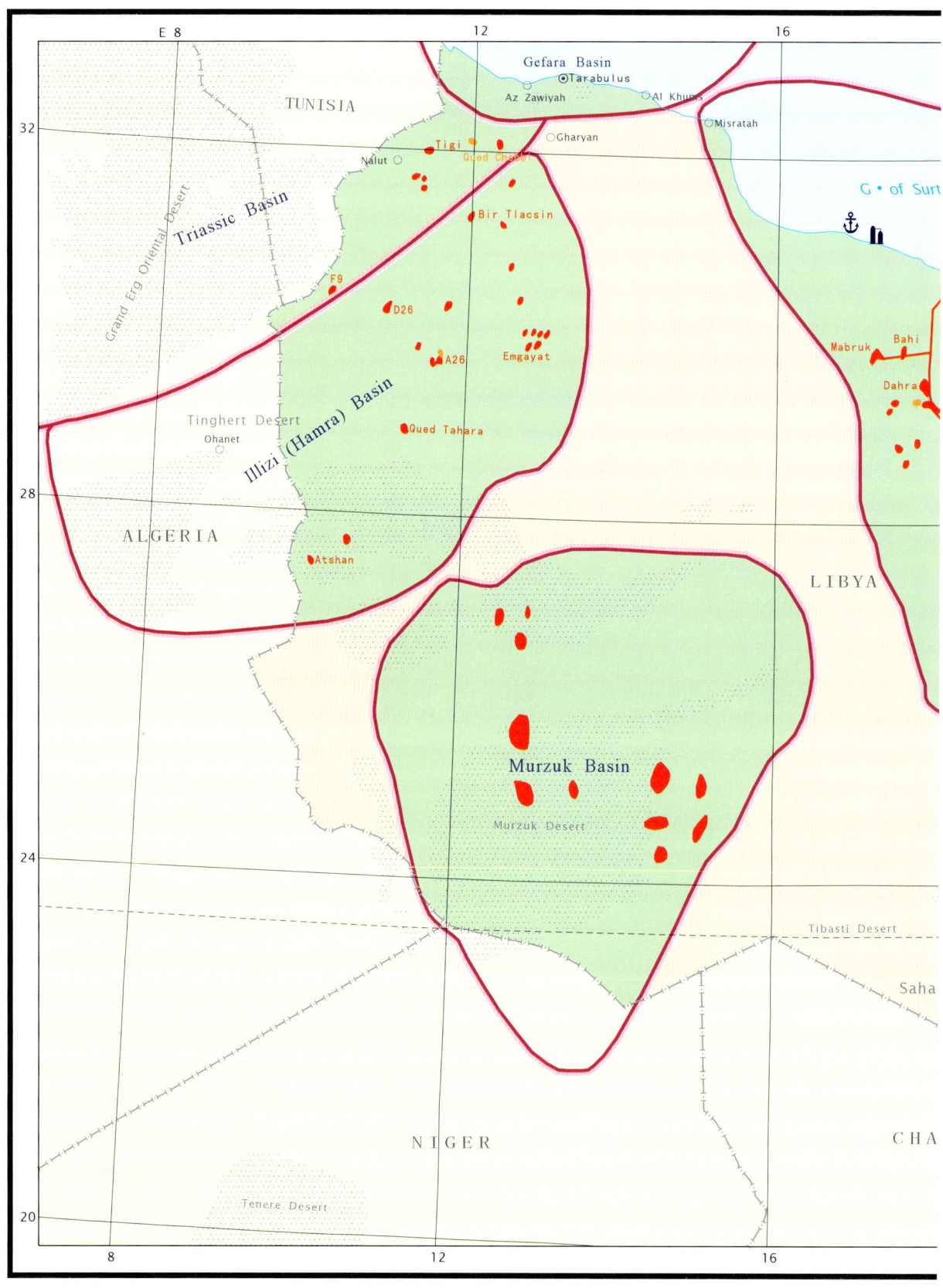

CHAPTER 42

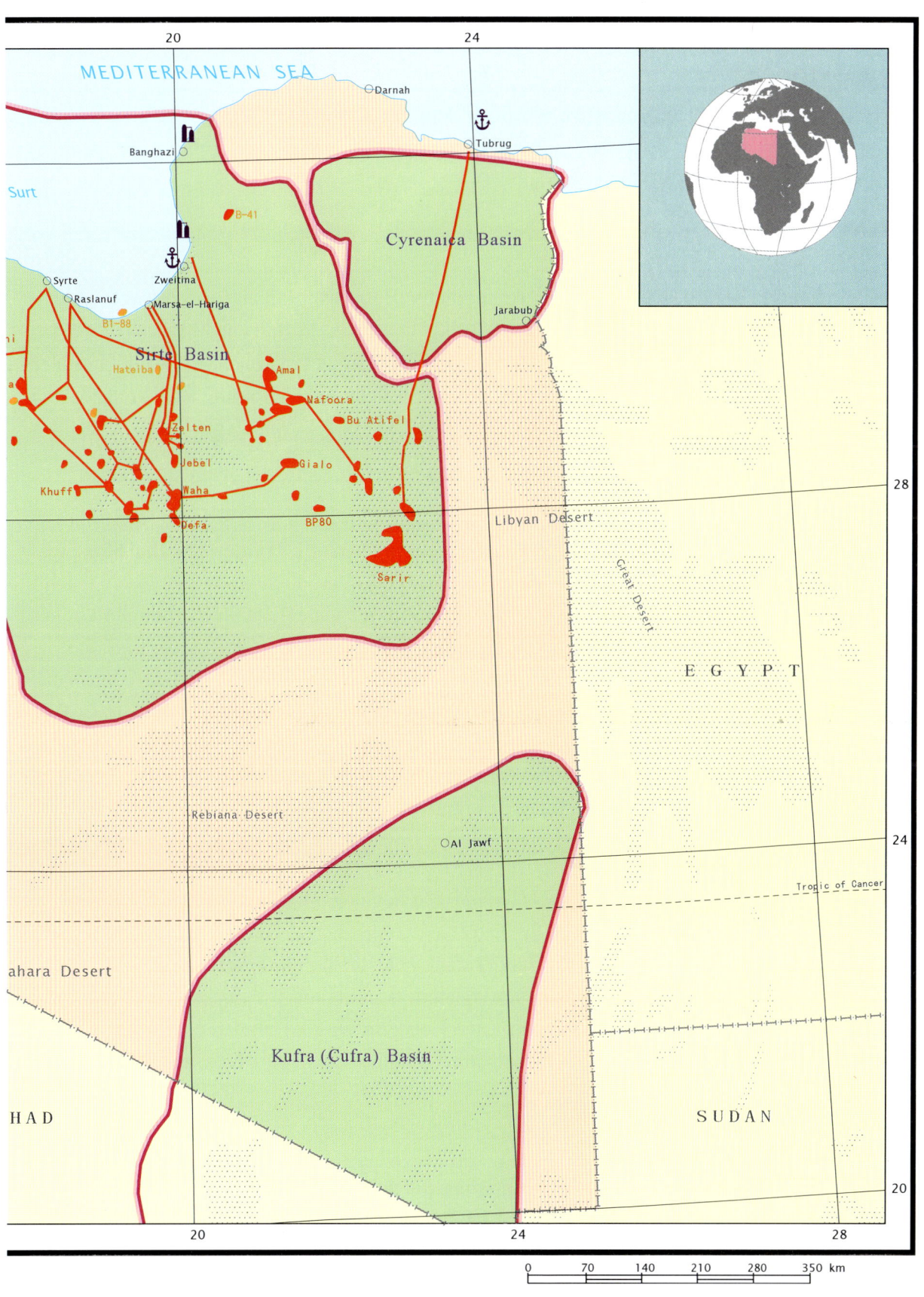

42 Libya

Geography

Libya is one of the large countries of North Africa that border the Mediterranean Sea to the north. It is located between Egypt and Sudan and occupies an area of 1.75 million km^2 with an estimated population of 5.9 million (2005).

History

Before the 1950s, very little oil exploration was completed. Since then, however, the pace of oil and gas exploration activities has increased significantly. Between 1955 and 1959, 10 oil and gas fields were discovered, including Amal, Nasser, Intisar D, Intrsor A, Baha and Mabruk. This was quickly followed by production drilling and crude oil production started in 1961. Libya has now become one of the most important oil producing countries in the world, holding an estimated 4 per cent of the world's oil reserves.

Reserves and production

In 1960 proven oil reserves stood at 300 million t. At the start of oil production in 1961, annual output was a mere 870,000 t. In 1965, oil reserves were estimated at 1.7 billion t, with gas reserves of 0.21 trillion m^3. Within 10 years of the commencement of oil production, output was to reach a peak of 159 million t in 1970. After that year, production started to decline, with production of just 70 million t in the year 2000. As recently as 2008 annual production remained at 86 million t.

The comprehensive blacklisting of the Libyan Government by the Western powers contributed to the minimal level of oil exploration in the country for several years, and this serves to explain the modest history of production compared to the country's vast production potential. Following the recent resolution of most of these international disputes and embargoes, it is expected that Libyan exploration and production will increase significantly.

Regional geology

This description of the regional geology is based on Selley (1997). Libya is located on the foreland belt of the African Shield and can be divided into two distinctive parts. The north, where the Sirte and Cyrenaica Basins are located, is underlain by unstable continental shelf, whereas the south, where the Murzak and Cufra Basins are located, is underlain by stable craton. The sedimentary cover comprises Palaeozoic, Mesozoic and Cenozoic strata. Detailed examination of the stratigraphy of Arabia and North Africa reveals great similarities, in particular in the Palaeozoic strata. This uniformity of stratigraphy is common to all the basins of northwest Africa and characterizes structural evolution of the main sedimentary basins, although there is occasional local variation in the stratigraphy.

The basement includes several sedimentary sequences that have undergone regional metamorphism in general proportion to their age. The identification of the Precambrian–Cambrian boundary thus presents something of a problem. Geologists have long remarked on the similarity of the

World Atlas of Oil And Gas Basins, First Edition. Li Guoyu.
© 2011 John Wiley & Sons, Ltd. Published 2011 by John Wiley & Sons, Ltd.

Table 42.1 Major sedimentary basins of Libya

Basin	Area (thousand km²)	Major sedimentary rock		Reservoir		Remarks
		Age	Thickness (m)	Age	Lithology	
Kufra (Cufra)	210	Pz	2600	Pz	Sandstone, limestone	Extends into Chad
Cyrenaica	65	S, D, C, Mz–Kz	40,000	R	Sandstone	
Illizi (Hamra)	250	Pz	6000	O, S	Sandstone	Extends into Algeria, Tunisia
Murzuk	259	Pz, K	3000	Pz	Sandstone	Extends into Niger
Sirte	519	K, R	5000	K, R	Sandstone, reef, basement rock	

Lower Palaeozoic stratigraphy in Arabia and the Sahara. There is a remarkably uniform sequence of facies that may be traced from basin to basin, all the way from the Atlantic Ocean to the Arabian Gulf. Two main facies were deposited during much of the Mesozoic Era. A thick section of marine Mesozoic sediments crops out in the complex folds of the Atlas Mountains. These can be traced southwards in the subsurface in the Algerian sedimentary basins. Over much of the Sahara, however, the Palaeozoic Strata are unconformably overlain by non-marine sandstones, variously known as 'Continental Post-Tassilian' and 'Continental Intercalaire' in Algeria, 'Continental Mesozoic' in Libya, and 'Nubian' in Sudan, Egypt and Arabia. The dateable marine Mesozoic strata of the northwest will be described first, followed by an account of the non-marine Mesozoic sediments.

In the Atlas Mountains and in the subsurface of the Algerian basins, the Mesozoic sequence begins with a series of evaporites. The Stephanian evaporites of the Bechar basin are an example. Evaporites are notoriously difficult to date palaeontologically, so it is quite possible that evaporite deposition commenced in the Atlas area in the Permian Period or earlier. The evaporites include both anhydrite and halite, and are interbedded with dolomites, red shales and occasional sandstones. The evaporites provide a regional barrier to petroleum migration across the Algerian basins, acting as the seal to the Hassi Messaoud oil field. Basal sands (dated as Triassic) between the Hercynian unconformity and the evaporites serve as petroleum reservoirs in the Hassier R'Mel and other Algerian fields.

The petroleum geology presents specific features of note. The sedimentary cover consists of Palaeozoic and Mesozoic rocks. Source rocks have been identified in the marine sediments. Reservoir rocks are associated with both sandstone and carbonates (Table 42.1). Seal rocks are shales. The traps are represented by structural and unconformity types. On the whole, both the oil and gas prospects of this region have been estimated as being very high (Shannon and Naylor, 1989).

SIRTE BASIN

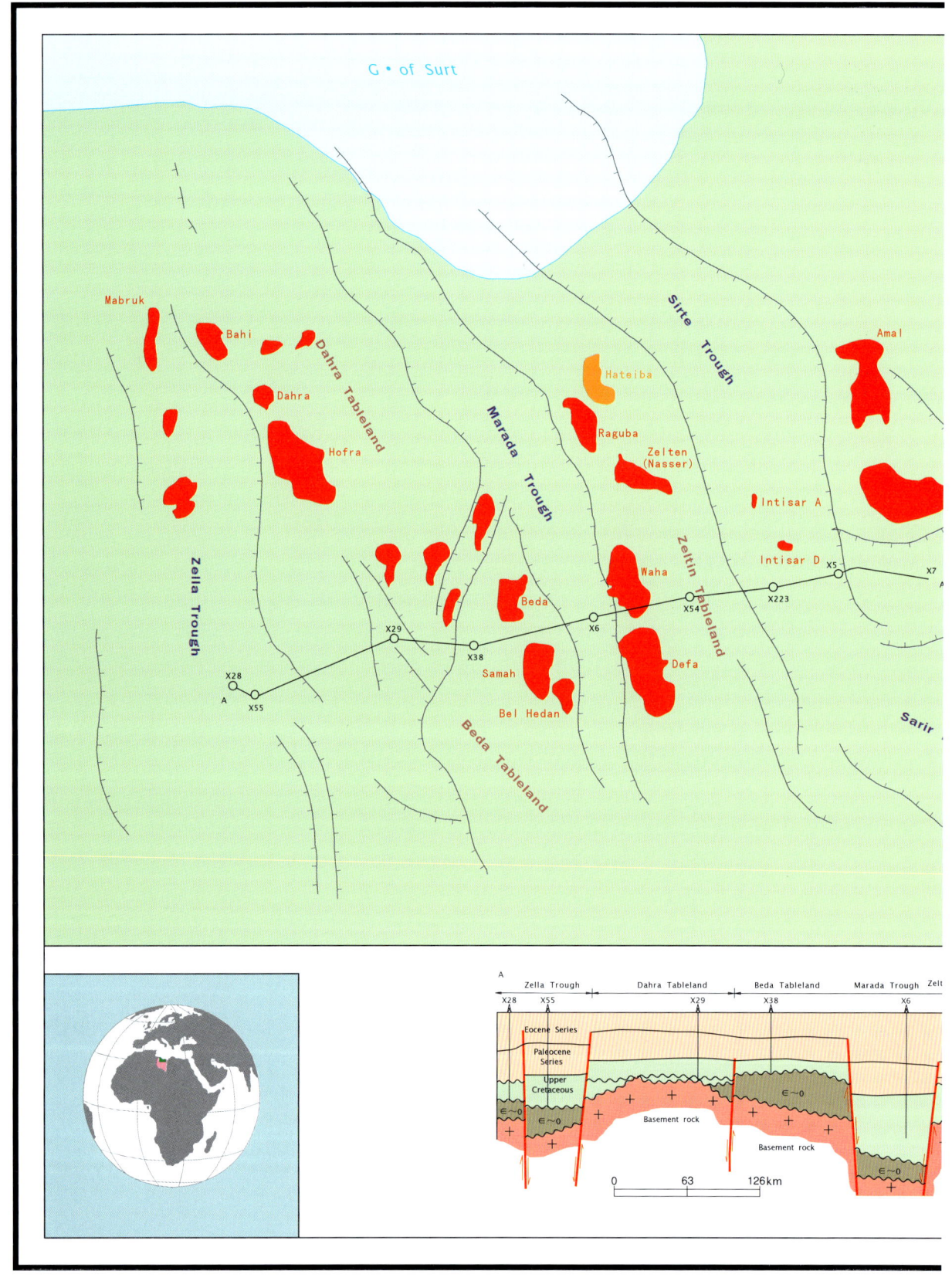

CHAPTER 43

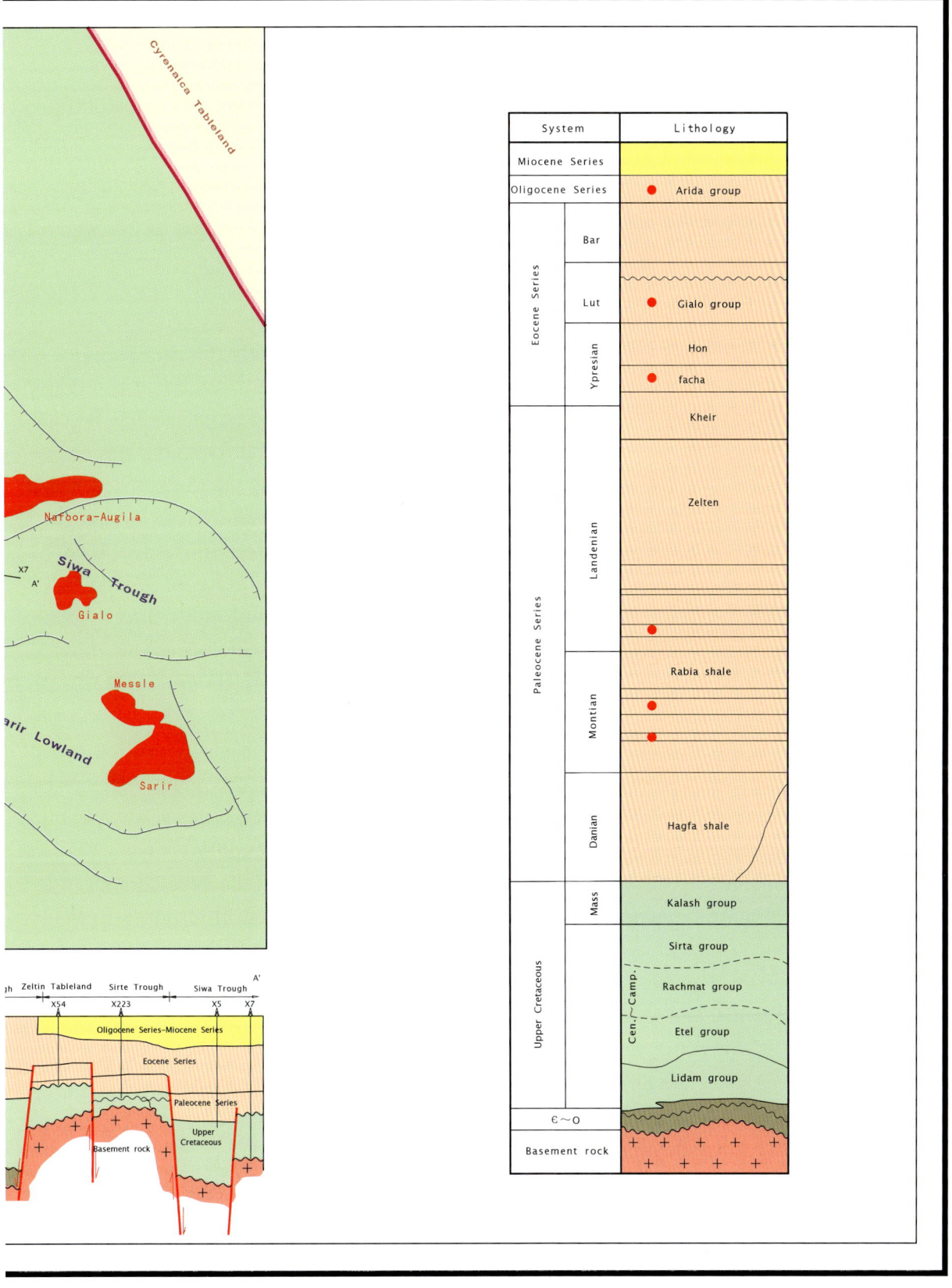

43 Sirte Basin

Geography

The Sirte Basin is located in northern Libya. It has an area of 519,000 km², including 404,000 km² on land and 115,000 km² on the offshore shelf.

History

The first oil field was discovered in 1958. During the period 1958–1970 a total of 20 large oil fields were discovered in this basin. It is estimated that about 96 per cent of total oil production of Libya emanate from this basin.

Regional geology

This description of the regional geology is based on Selley (1997). The Sirte Basin is a major sedimentary basin and extends southwards from the Gulf of Surt in the Mediterranean Sea. It occupies a collapsed north–south trending positive feature, the Tibesti–Sirte Arch. This arch was a positive feature throughout much of Palaeozoic and Mesozoic times, separating the Murzuk and Kufra embayments. With the break-up of the African continent in the Cretaceous Period, the northern part of the Tibesti–Sirte Arch collapsed to form the Sirte Basin. Thus the floor of the Sirte Basin is a major unconformity, above which is a thick sequence of Late Cretaceous to recent sediments.

The floor of the Sirte Basin is a major regional unconformity. This unconformity directly overlies Precambrian basement in the basin centre and progressively younger formations moving westwards and eastwards. The oldest rocks encountered above the Sirte Basin unconformity are of Cenomanian age.

The floor of Sirte Basin is extensively faulted. These faults exerted a strong control on initial sedimentation. Immediately above the Sirte unconformity Early Cretaceous sands thin out by non-deposition or truncation on to the crests of the fault blocks. Late Cretaceous sands and thin Paleocene shales in the troughs pass up into reefal carbonates on the crests of the horsts.

The progressive opening up of the Atlantic Ocean was associated with the development of an extensive rift system across much of Africa. The African rifts developed in response to radial lithospheric divergence emanating from Abong M'Band in central Cameroon, forming a positive feature that separated the Murzuk and Kufra embayments from Cambrian times up until the mid-Cretaceous collapse. Rifts, and failed rift basins, are normally accompanied by volcanic activity, and in this regard the Sirte Basin was no exception.

The petroleum geology of the Sirte Basin is characterized by the following features:
1. sedimentary cover is composed of Palaeozoic and Ceno-Mesozoic rocks;
2. source rocks have been identified mainly in Cretaceous shales;
3. reservoir rocks are associated with sandstone and carbonate (Table 43.1 and Fig. 43.1) rocks from basement to Oligocene strata;
4. seal rocks are associated with regional shale and evaporite rocks;
5. traps are represented by faulted anticlines.

World Atlas of Oil And Gas Basins, First Edition. Li Guoyu.
© 2011 John Wiley & Sons, Ltd. Published 2011 by John Wiley & Sons, Ltd.

Table 43.1 Major oil and gas fields of Sirte Basin, Libya

Name	Discovery year	Recoverable reserves		Depth (m)	Trap	Reservoir	
		Oil (million t)	Gas (billion m³)			Age	Lithology
Sarir C	1961	910.0	—	2600	Anticline	K	Sandstone
Amal	1959	595.0	—	3000	Anticline	K	Sandstone
Gialo	1961	490.0	—	600	Anticline	E	Carbonate rock
Nasser	1959	308.0	3.8	1800	Anticline	K	Carbonate rock
Defa	1960	252.0	—	1600	Anticline–reef	E	Carbonate rock
Augila	1966	252.0	—	2400	Stratum	K	Sandstone
Intisar D	1967	210.0	—	2800	Reef	E	Carbonate rock
Messla	1971	210.0	—	2600	Stratum	K	Sandstone
Waha	1960	196.0	—	2000	Anticline	E	Carbonate rock
Sarir L	1966	160.0	—	2700	Anticline	K	Sandstone
Intisar A	1967	168.0	—	2800	Reef	E	Carbonate rock
Bu Attifel	1968	168.0	—	4300	Anticline	K	Sandstone
Raguba	1961	140.0	—	1600	Anticline	K	Carbonate rock
Bahi	1958	84.0	—	1800	Anticline	E	Carbonate rock
Dahra Hofra	1959	70.0	—	1000	Reef	E	Carbonate rock
Mabruk	1959	70.0	—	1200	Anticline	E	Sandstone
Samah	1962	70.0	—	2700	Anticline	K	Carbonate rock
Hateiba	1963	—	28	2600	Anticline	K	Sandstone

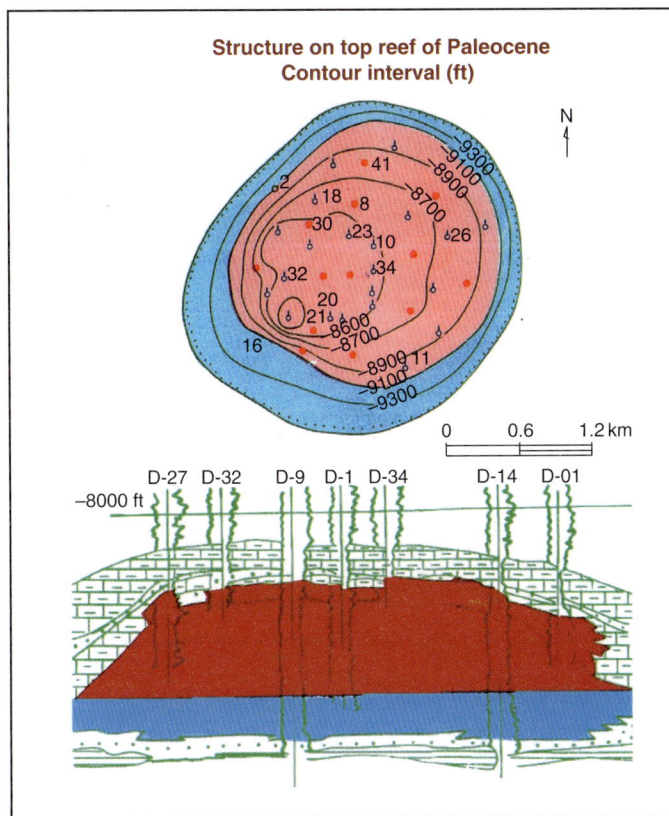

This is a famous big oil field of Libya, with initial oil production rate of 9,996 t per day from well D1-103.

Basic data

1. Name		Intisar D
2. Discovery year		1967
3. Area (km²)		25
4. Reservoir		Carbonate
5. Reservoir age		Tertiary
6. Depth (m)		2,726
7. Thickness (m)		229
8. Porosity (%)		22
9. Permeability (mD)		87
10. Recoverable reserves (million t)		210

Fig. 43.1 Details of the Intisar D oil field.

ALGERIA, MOROCCO AND TUNISIA

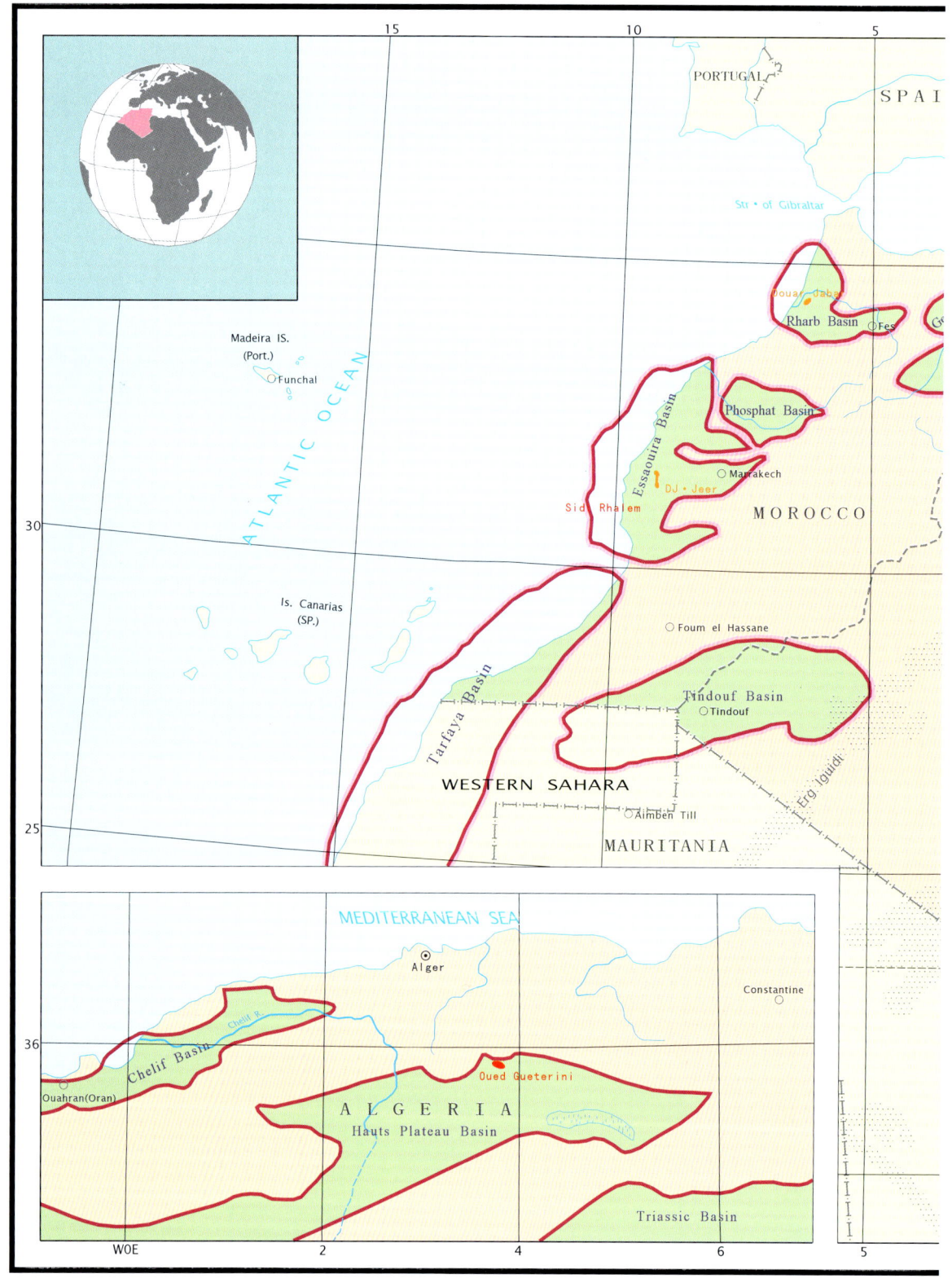

Oil-gas field: 1 Tin Fouye 2 Hassi Mazoula 3 La Reculee 4 Ohanet 5 Tiguentourine 6 El Adeb Larache 7 Ouan Taredert 8 In Amenas 9 Gassi Touil

CHAPTER 44

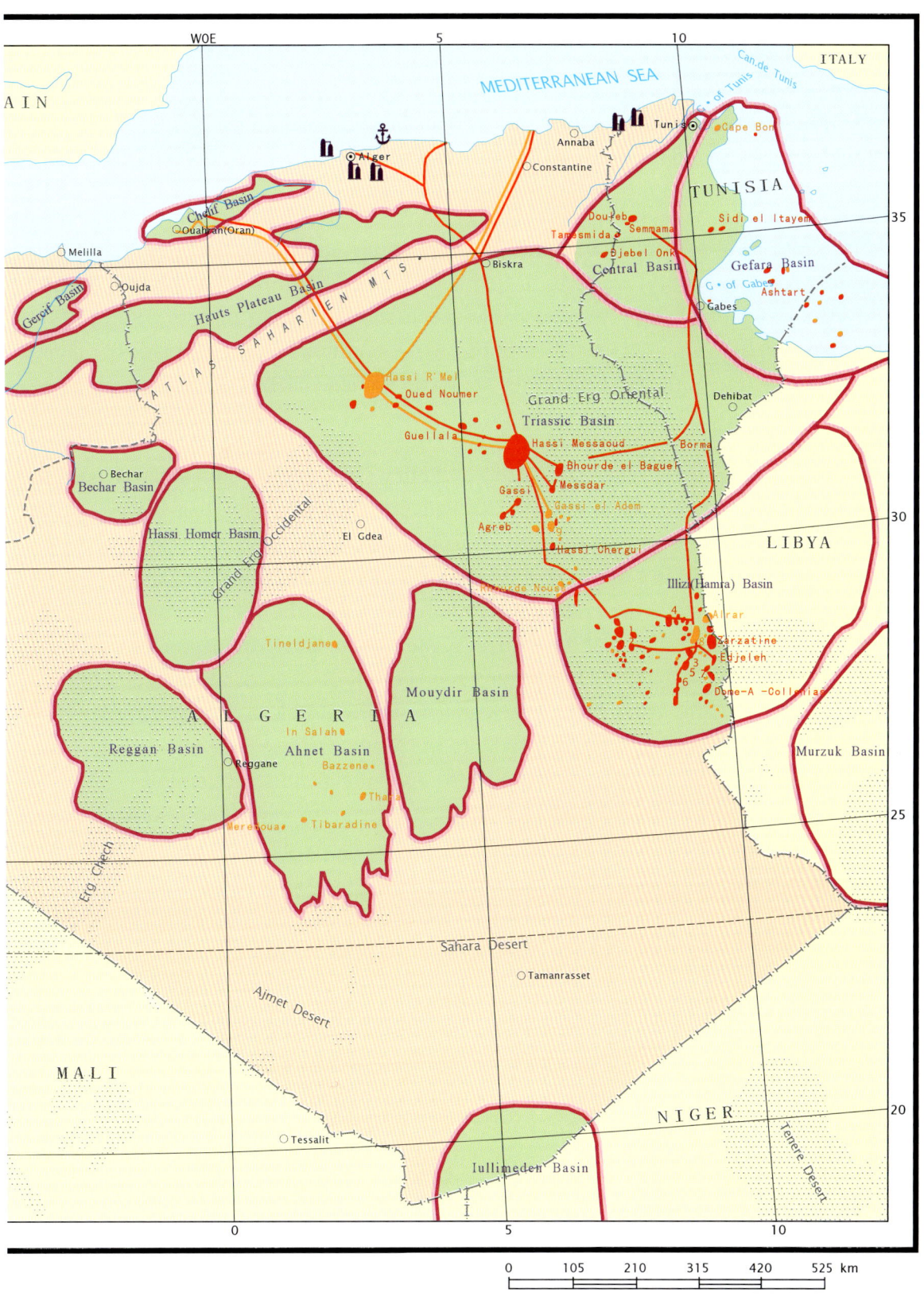

44 Algeria, Morocco and Tunisia

Geography

The contiguous countries of Algeria, Morocco and Tunisia are located in the North Africa, bordering the Mediterranean Sea to the north, and east in the case of Tunisia.

History

There is a long history of oil seepages in the Atlas Mountains, indicating the presence of subsurface accumulations. The first oil well was drilled in Algeria in 1892, in Morocco in 1890 and in Tunisia in 1909. At present, only Algeria has a successful petroleum industry. Since 1968 Algerian oil production has increased from 42 million t to 67.7 million t in 2007. Details of oil and gas reserves and production of the three countries are given in Table 44.1.

Regional geology

This description of the regional geology is based on Selley (1997). The sedimentary basins of northwest Africa share a remarkably uniform Early Palaeozoic stratigraphy, that became more varied regionally until the extensive Late Carboniferous marine regression. Northwest Africa consists of the Saharan Shield. This is a vast area of Precambrian continental crust that has undergone little structural deformation since the end of the Proterozoic Era. The Phanerozoic sediments were deposited on the southern shores of the Tethys Ocean. In the Atlas Mountains the sedimentary succession attains its maximum thickness, with rocks of all geological periods represented. It is believed, however, that the Atlas Fold Belt accreted onto the African Plate during late lateral crustal movement as the Tethys Ocean closed into the Mediterranean Sea, because the Atlas Mountains are young additions, and thus not really part of the mainland Africa craton.

It is not easy to clearly define the sedimentary basins of the northwest Sahara. Some are easily recognized as quadripetally closed basins, but many are northerly plunging embayments that originally opened out into the former Tethys Ocean, to become subsequently closed. The main

Table 44.1 Key data for Algeria, Morocco and Tunisia in 2007

Statistic	Country		
	Algeria	Morocco	Tunisia
Area (k^2)	2,381,741	459,000	164,150
Population (thousand)	33,800	30,340	10,020
Oil proven reserves (million t)	1671	0.1	58
Annual oil production (million t)	67	—	4.2
Gas proven reserves (billion m^3)	4502	1.5	65
Annual gas production (billion m^3)	93	—	2.2
Annual oil consumption (million t)	12	—	—
Annual oil exports (million t)	55	—	—

World Atlas of Oil And Gas Basins, First Edition. Li Guoyu.
© 2011 John Wiley & Sons, Ltd. Published 2011 by John Wiley & Sons, Ltd.

Table 44.2 Major oil and gas basins

Country	Basin	Area (thousand km²)	Major sedimentary rock Age	Major sedimentary rock Thickness (m)	Reservoir Age	Reservoir Lithology	Remarks
Algeria	Ahnet	150	EC	5000	T, O, D	Sandstone	
	Hassi Homer	71	Pz	5000	D	Sandstone	
	Hauts plateau	99	Pz, Kz		E_2	Limestone	Extends into Morocco
	Mouydir	110	Pz–Kz	7500	O–C, T	Sandstone, Limestone	
	Reggan	90	Pz	5000	Pz	Sandstone	
	Tindouf	88	Pz	9000	Pz	Sandstone, Limestone	Extends into Morocco,
	Triassic	440	E–R	6000–7000	O–D	Sandstone	Extends into Tunisia
Morocco	Essaouira	710	Mz, Kz	5000	J	Carbonate rock	
Tunisia	Central	55	Mz		K	Limestone	Extends into Algeria
	Gefara (Jeffara)	159	Mz, R		K, R	Sandstone, Carbonate	Extends into Liberia

central Saharan sedimentary basins from west to east are the Tindouf, Reggan, Ahnet, Mouydir, Murzuk and Kufra. The structure and tectonic evolution of these basins will now be defined. When the Ahnet and Mouydir basins are traced northwards it becomes more difficult to resolve basin architecture beneath the ever-thickening sequence of monoclinally dipping Mesozoic strata. This region is best described under the broad regional term of 'the Triassic Salt Basin'. The northern limit of the Triassic Salt Basin is defined by a major tectonic feature at the foothills of the Atlas Mountains.

Algeria Algeria has an area of 2,381,741 km², more than 80 per cent of which is desert. It is the second largest country in Africa in terms of size. It has a long Mediterranean coastline. There are three large sedimentary basins in the country with a combined area of more than 700,000 km², namely, Ahnet, Mouydir and Triassic (Table 44.2). These basins are infilled with Palaeozoic and Mesozoic rocks. Oil exploration started in 1892. In excess of 207 oil and gas fields have been discovered, with most of the oil produced to date coming from 12 giant fields. By the end of 2007 annual oil production reached 67.7 million t and gas production was 93.6 billion m³.

Morocco Morocco has an area of 459,000 km² with a population of 30.34 million. Oil exploration and exploitation has been developed only sporadically and by the end of 2007 proven oil reserves were 110,000 t and gas reserves 1.5 billion m³ only (Table 44.2). It is classified as a non-oil-producing country.

Tunisia Tunisia has an area of 164,150 km² and a population of 10.02 million. The first oil well was drilled in 1909. By the end of 2007 proven oil reserves reached 58 million t with annual oil production of 4.2 million t (Table 44.2). The oil and gas prospects have been estimated as high.

MAURITANIA, WESTERN SAHARA, SENEGAL,

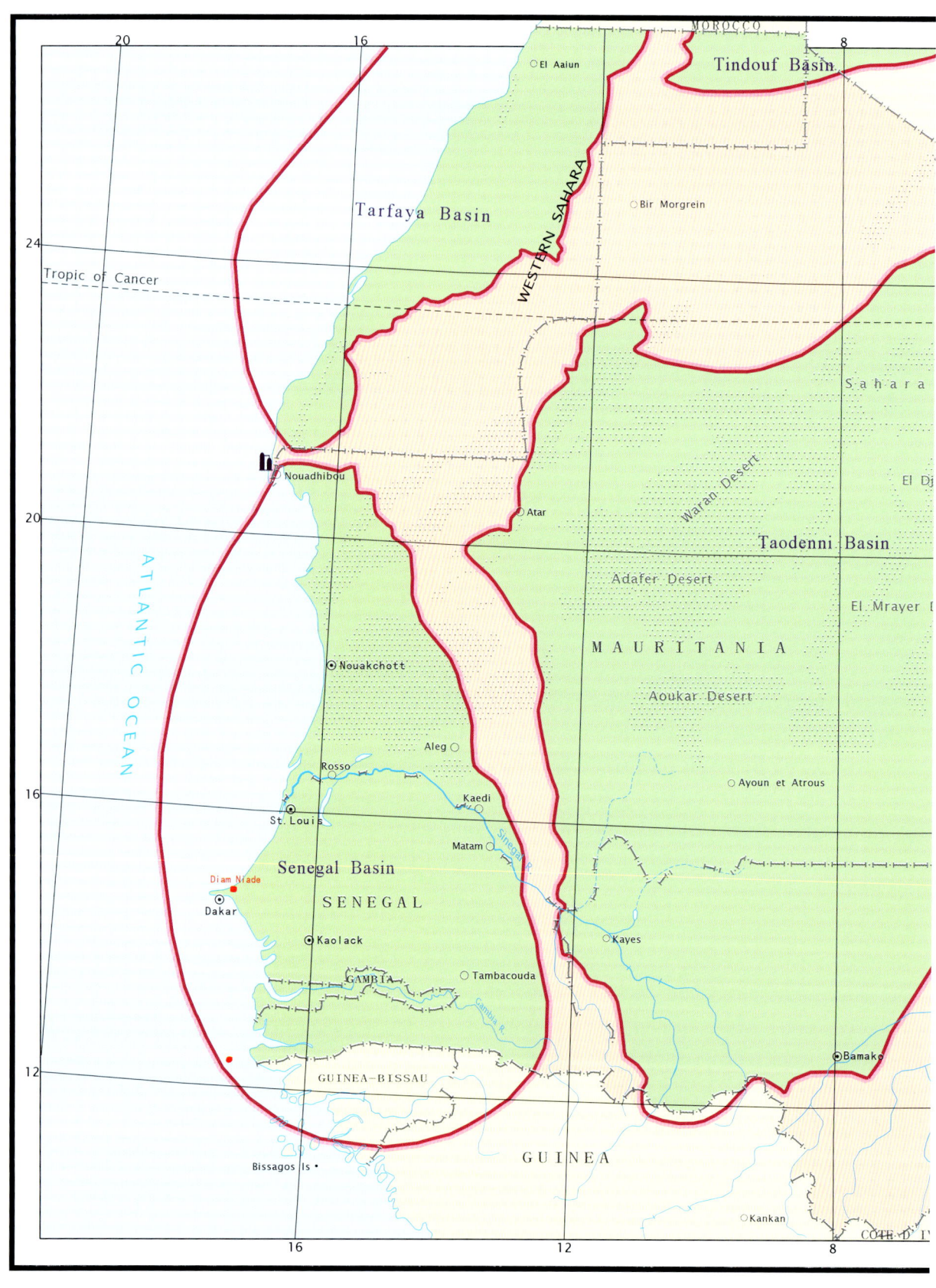

198

GAMBIA, MALI AND BURKINA FASO   CHAPTER 45

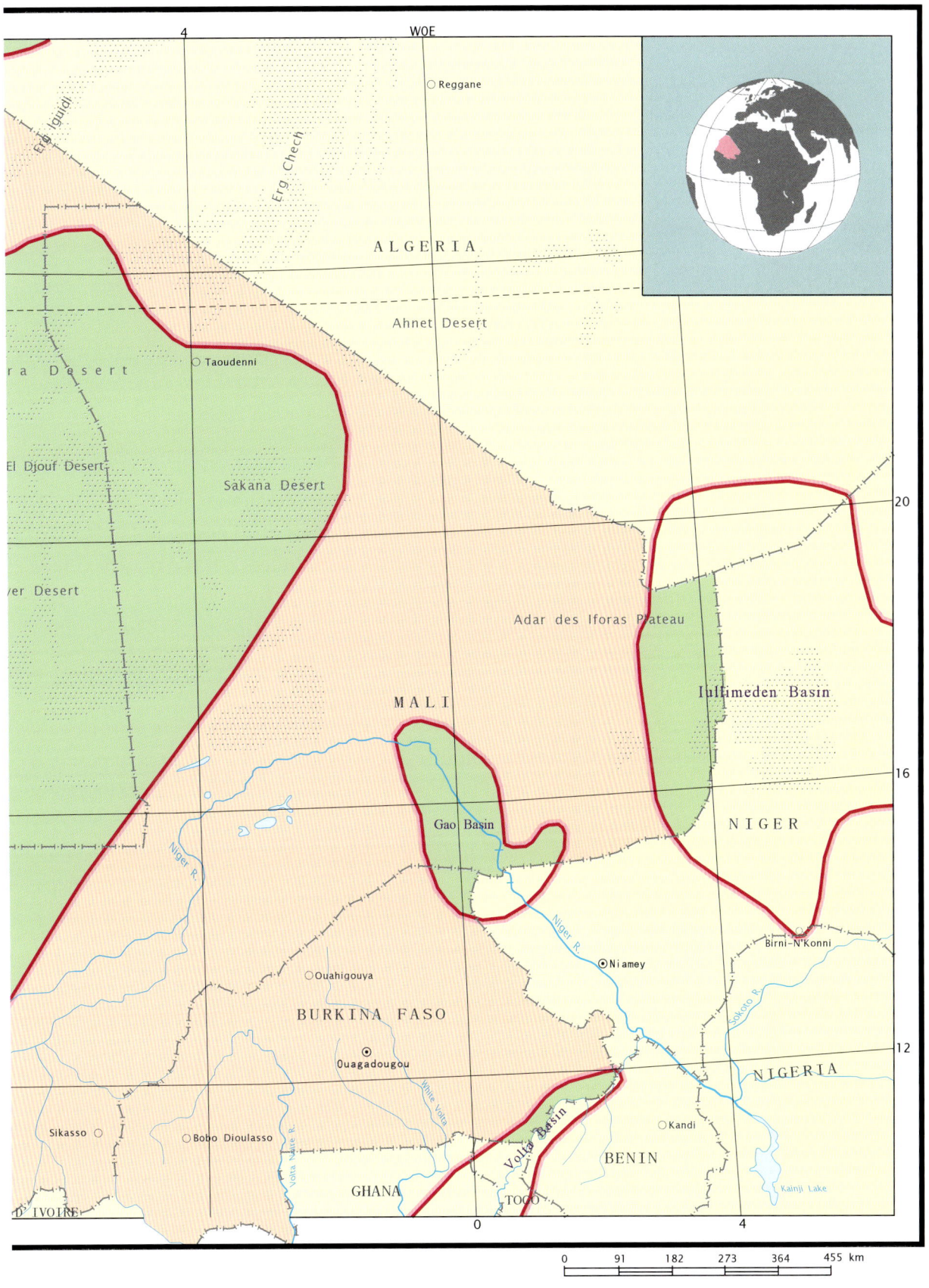

45 Mauritania, Western Sahara, Senegal, Gambia, Mali and Burkina Faso

Geography

Mauritania, Western Sahara, Senegal, Gambia, Mali and Burkina Faso are located in northwest Africa.

History

In spite of the proximity of many of these countries to oil-producing areas to the east and north, oil exploration started very late. Oil exploration in Senegal started in the 1950s. Because these countries, with the exception of Senegal, are relatively economically backward, oil exploration has been very spasmodic, with very little commercial quantities discovered. Only Mauritania has a meaningful oil industry with annual crude production of 1.3 million t in 2008.

Regional geology

Northwest Africa consists mainly of an eastern craton with Precambrian basement, and a western continental margin. The Iullimeden Basin has been variously described as a cratonic basin, a graben and as a half-graben–synclinal structure. The origin of the Iullimeden Basin may be linked to crustal extension with formation of the depression resulting mainly from loading by the sediment and water that collected in the Benue Trough and Bilma Graben structures. The resultant surface depression exceeds the width of the initial graben by more than fivefold.

The Precambrian basement rocks of the Iullimeden Basin are well known as a result of aerial magnetic surveys that have been carried out in the past by French companies. These rocks are divided into three categories. In the Liptako region of western Niger the basement is exposed at the surface forming a vast peneplain lying about 200–300 m above sea level. The lithostratigraphy of the Palaeozoic strata that crop out in the Tim Mersoi basin on northern/northeastern margin of the Iullimeden Basin is described by Greigert (cited in Selley, 1997). The Lower Palaeozoic, Cambro-Ordovician sediments are essentially marginal marine. They are overlain by the Schistes à graptolites which are the lateral equivalent of the Silurian Tanesouft Shales of North Africa. The overlying Devonian sediments are also predominantly sandstones with non-marine deposits characteristic of the Middle and Upper Devonian. Uplift and folding occurred during the Permian. Permo-Triassic, Jurassic, Cretaceous, Tertiary and Quaternary sediments infill the different basins, but correlation of the different strata has nor been established between them. Details of the major sedimentary basins in the six countries are provided in Table 45.1.

Most of the area of these six countries is covered by the Sahara Desert. As mentioned, only Mauritania has a meaningful oil industry. The other five countries are classified as non-oil-producing, but there may be minor oil and gas prospects for these countries.

Mauritania Mauritania is located in northwest Africa facing the Atlantic Ocean to the west. It occupies an area of 1,030,000 km^2 with a population of 3 million. Oil exploration began in

World Atlas of Oil And Gas Basins, First Edition. Li Guoyu.
© 2011 John Wiley & Sons, Ltd. Published 2011 by John Wiley & Sons, Ltd.

Table 45.1 Major sedimentary basins

Country	Basin	Area (thousand km²)	Major sedimentary rock		Reservoir		Remarks
			Age	Thickness (m)	Age	Lithology	
Mouritania	Taodenni	1230	Pz	3500	Z	Limestone	Extends into Mali, Guinea
Senegal	Senegal	563	Pz–Kz	8000	K	Sandstone	Extends into Mauritania,
Western Sahara	Tarfaya	264	Mz, Kz	2000–6000	Mz, Kz	Sandstone, Limestone	Extends into Morocco
Mali	Gao	49	Mz, R	1600			Extends into Niger
Nigeria	Iullimeden		Pz, Mz, Kz			Sandstone	Extends into Niger

the Taodenni Basin in 1957. Recently, the main oil exploration activity is being carried out in the Senegal Basin. Since 2001 three new oil fields have been discovered, namely, Banda, Chinguoti and Tiof, providing further impetus for the development of the petroleum industry in this country.

Western Sahara Western Sahara is also located in northwestern Africa. The area is 266,000 km² with a population of 270,000. In the southeastern part of the country, part of the Precambrian basement is exposed at the surface. The sedimentary basin is located on the continental shelf.

Senegal Senegal is located in northwestern Africa with its entire coastline on the Atlantic Ocean. It has an area of 196,722 km² and a population of 11.9 million. Over 100 wells were drilled in the country during the 1950s, both onshore and offshore. However, the results were not encouraging. After a lull, there has been renewed interest in the country with some active exploration activities are taking place offshore. There have been mixed results so far.

Gambia Gambia is located in West Africa. A relatively small country, it has an area of 10,380 km² and a population of 1.5 million. Prior to 1991 a total of three wells were drilled in the country, two of them being onshore. All wells were dry and subsequently abandoned.

Mali Mali is located in West Africa. It is a landlocked country with an area of 1,241,238 km² and a population of 13.5 million. In the second half of the 1960s a total of four wells were drilled in the Iullimeden Basin, all of which turned out to be dry. In the Taodenni Basin, one well was completed to a depth of close to 2500 m in the Cambrian strata. There has been some renewed exploration interest in this country with appreciable seismic surveys being carried out in the past 5 years.

Burkina Faso Burkina Faso (formerly Upper Volta) is a landlocked Sahel country that is almost entirely covered by the Sahara Desert. The terrain is mostly flat with undulating plains and isolated hills in the west and southeast. The area is 274,200 km² and estimated population of 13.2 million. This country is located on the West African Shield. Given the absence of any significant sedimentary rocks, the country is not considered a viable candidate for oil and gas exploration.

GUINEA, GUINEA BISSAU AND SIERRA LEONE

CHAPTER 46

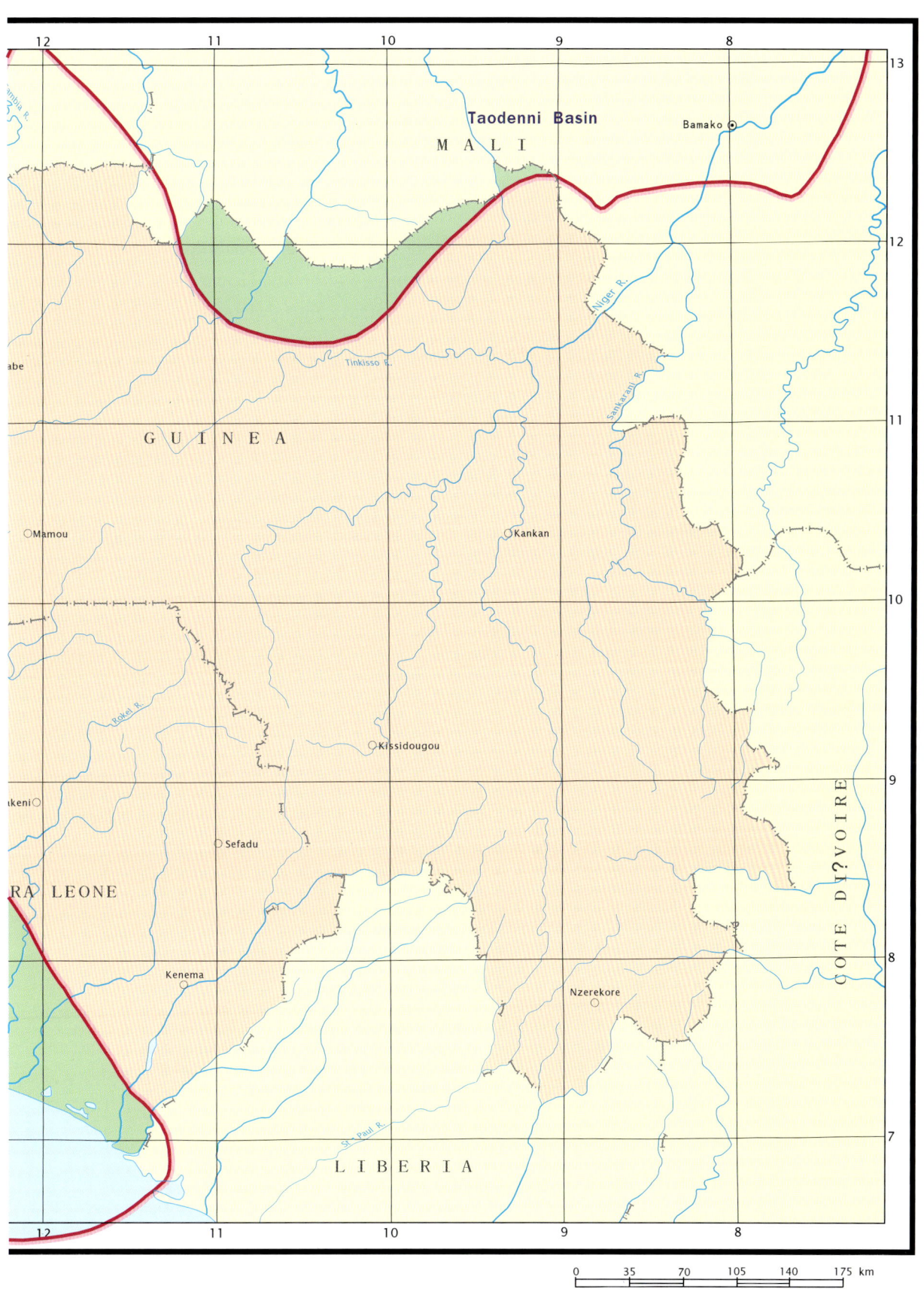

46 Guinea, Guinea Bissau and Sierra Leone

Geography

Guinea, Guinea Bissau and Sierra Leone are located on the West African coast facing the Atlantic Ocean to the west.

History

The complex geological conditions and previous unsuccessful oil and gas exploration has limited the development of the petroleum industry of these three countries. However, the region has substantial occurrences of sedimentary rocks which are showing future prospects for oil and gas exploration.

Regional geology

Western Africa is characterized by sedimentary basins distributed between basement massifs. Figure 46.1 shows the major sedimentary basins and massifs in northwestern Africa. From a petroleum geology perspective, the presence of ancient rivers cutting through these massifs would account for the infill of these sedimentary basins and thus they would have interesting hydrocarbon potential. The Eburneen and Nigerian Massifs are aligned parallel to the Guinea–Nubian lineament zone, which cuts through the amalgamated Benin–TogoBasin to Benue Trough and marks the southern tip of the Air Massif in neighbouring Niger. Faults with a NW–SE trend mark the western limit of the Benin–Togo Basin. The fault bounded Gao-Ansongo Trough in Mali is regarded as a NW–SE trending graben feature. Like the Benue Trough, it is a sediment-filled fault-bounded structure. It tilts toward the southwest and the Cretaceous infill attains a thickness of almost 4000 m.

Guinea   Guinea is located in northwest Africa facing the Atlantic Ocean to the west. It has an area of 245,857 km^2 and a population of 9.4 million. A small portion of the Senegal Basin lies in this country. The sedimentary cover consists of Mesozoic rocks. Seismic exploration has been carried out on the continental shelf. One well was drilled in the Cretaceous strata but no subsequent exploration activity was carried out due to the absence of oil and gas occurrences. The country is classified as being non-oil-producing.

Guinea Bissau   Guinea Bissau is located in West Africa and is one the smallest countries on the continent of Africa. Its coastline lies along the Atlantic Ocean. The area is 36,125 km^2 with a population of 1.59 million. Seismic surveys have been carried out both onshore and offshore. Four onshore wells and 10 offshore wells were drilled prior to 1991. All 14 wells were subsequently abandoned having shown no appreciable hydrocarbon deposits. Guinea Bissau is partly located in the Senegal Basin. The country has a continental shelf (water depth between 1200 and 3000 m) with an area of approximately 40,000 km^2. This is a rift basin filled by Meso-Cenozoic sediments. In the offshore basin more than 20 salt diapir structures have been identified which may have formed traps. In the past 5 years exploration blocks have been allotted to some

World Atlas of Oil And Gas Basins, First Edition. Li Guoyu.
© 2011 John Wiley & Sons, Ltd. Published 2011 by John Wiley & Sons, Ltd.

Guinea, Guinea Bissau and Sierra Leone

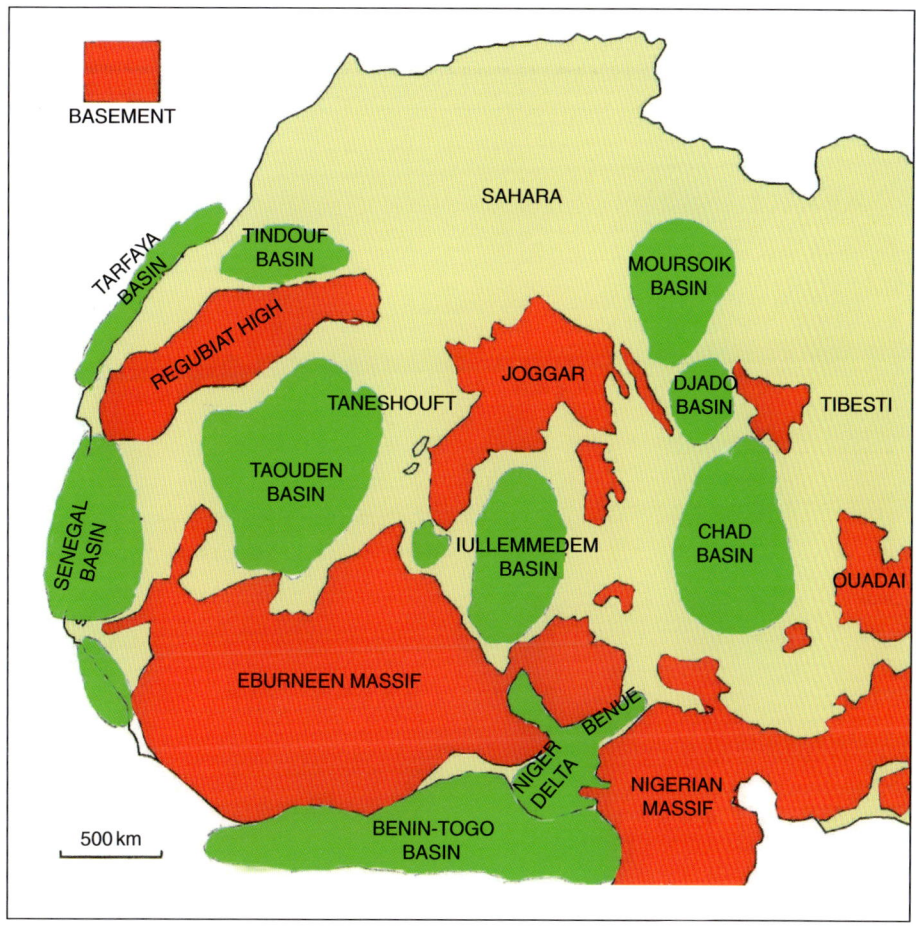

Fig. 46.1 Major sedimentary basins and massifs of West Africa (Selley, 1997; after Dikouma, 1990).

foreign companies, particularly in the offshore regions bordering Senegal and there are grounds for optimism about prospectivity. Currently, the country is non-oil-producing.

Sierra Leone Sierra Leone is a country located on the Antlantic coast of West African coast. The country borders Guinea to the north, with an area of 71,326 km^2 and a population of 3.7 million. This country is located on the northwestern Africa craton. In most areas the old Archaean and Protozoic rocks are exposed on the surface. The continental shelf is very narrow with an area of a mere 27,000 km^2. However, the sedimentary cover on the shelf has a thickness of up to 6000 m, suggesting that the geological conditions exist for oil exploration. Some oil exploration blocks have been granted to regional and multinational oil companies. However, at the present time the country is non-oil-producing with some prospects for future hydrocarbon production.

LIBERIA, COTE D'IVOIRE, GHANA, TOGO AND

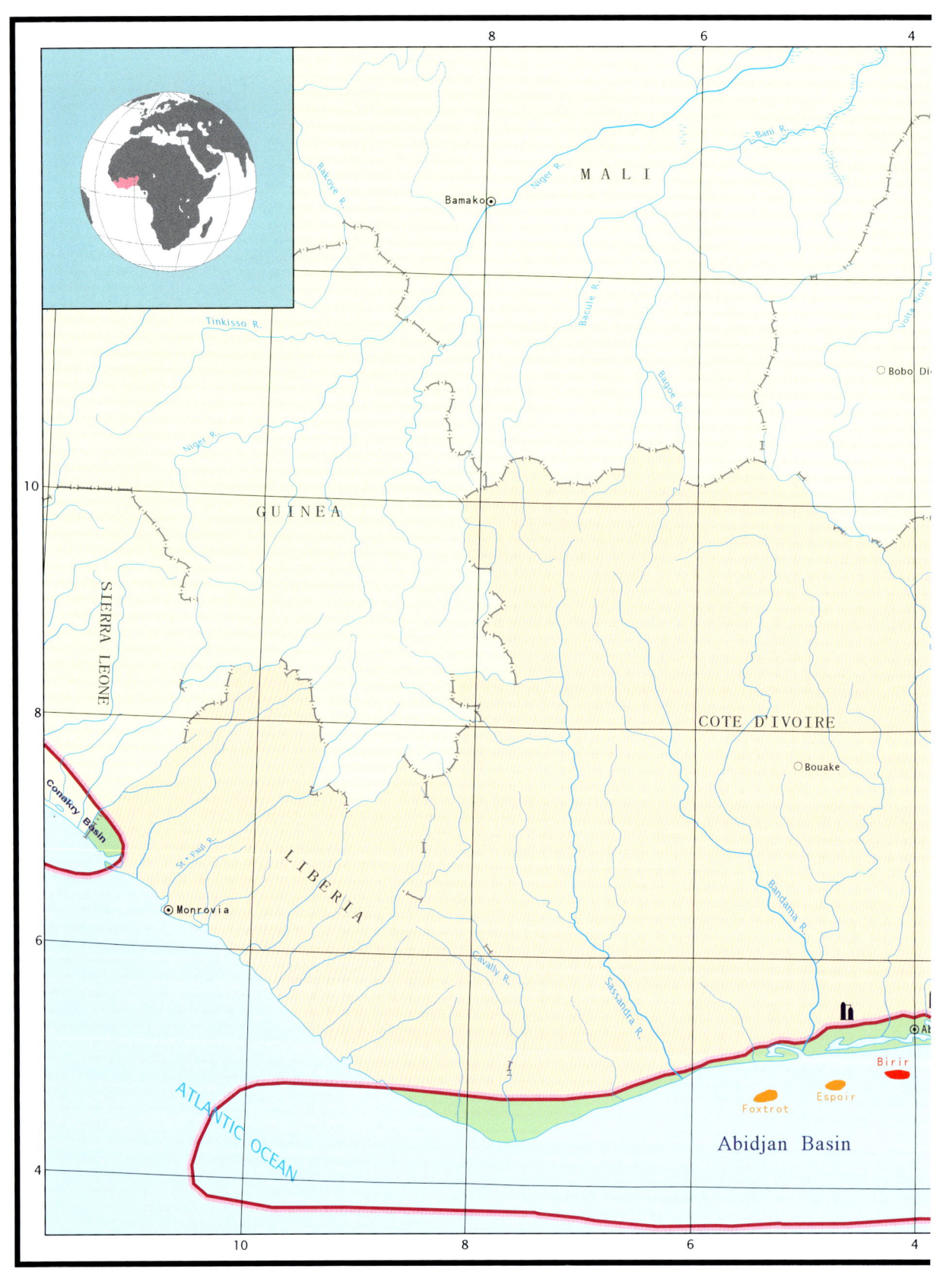

BENIN CHAPTER 47

47 Liberia, Cote D'ivoire, Ghana, Togo and Benin

Geography

Liberia, Cote D'ivoire, Ghana, Togo and Benin are located on the southern coast of West Africa.

History

In these countries, with the exception of Ghana, oil exploration began only in the second half of the 20th century. However, exploration in Ghana is known to have started as far back as 1896.

Regional geology

The Volta, Abidjan and Niger Delta Basins are located in southern West Africa. They have a uniform regional geological background. Benin, Togo and Cote D'ivoire are associated with the Niger Delta. The Niger Delta is situated in a crucial location with regard to the opening of the South Atlantic, lying in a pivotal position within the Gulf of Guinea between the passive divergent margin basins of the equatorial coast to the south and the basins of the coast westwards, which experienced relative lateral motion and wrench modification during the separation of Africa and South America.

The onshore basins of neighbouring Nigeria lie along the rifted valleys of the Niger and Benue Rivers and are mentioned here only insofar as they impinge on the history of the delta region. The Abidjan Basin is associated with Cote D'ivoire, Liberia and Ghana. Source rocks with a total area of $120,000\,km^2$ have been indentified in Upper Cretaceous sediments. Reservoir rocks are associated mostly with Cretaceous sandstone (Table 47.1). Cap rock is shale. Traps are represented by faulted structures. Birir oil field was discovered in 1974, being formed by a faulted anticline. Reservoirs are Cretaceous sandstone with a porosity in the 18–20 per cent range and a permeability of between 200 and 500 mD. The Volta Basin is located in Ghana with a total area of $140,000\,km^2$. This basin is covered by Palaeozoic sediments including Cambrian, Ordovician, Silurian and Devonian strata. Surface oil seepages first indicated the presence of subsurface hydrocarbons. The first exploration well was finished in 1977 at a depth of 1167 m in Cambrian sandstones but without oil and gas shows (Selley, 1997).

Liberia Liberia is a country located on the western end of the south coast of West Africa, with an area of $111,370\,km^2$ and population of 3.48 million. Four wells were drilled in the country

Table 47.1 Major sedimentary basins

Country	Basin	Area (thousand km²)	Major sedimentary rock		Reservoir		Remarks
			Age	Thickness (m)	Age	Lithology	
Cote D'ivoire	Abidjan	210	J, K, R	2800	K	Sandstone	Extends into Liberia
Ghana	Volta	140	Z, Pz				Extends into Togo, Benin

World Atlas of Oil And Gas Basins, First Edition. Li Guoyu.
© 2011 John Wiley & Sons, Ltd. Published 2011 by John Wiley & Sons, Ltd.

before 1981 up to a maximum depth of 3172 m. During the 1980s seismic surveys were completed which mapped interesting subsurface structures. Sedimentary cover consists of the Jurassic–Cretaceous and Tertiary strata. Subsequently, a further four wells were drilled but were found not to be hydrocarbon bearing. Presently, Liberia is classified as being non-oil-producing.

Cote D'ivoire

Cote D'ivoire (also referred to as Ivory Coast) is a country located on the southern coast of West Africa. Its coastline runs along the Gulf of Guinea. It has an area of 322,463 km^2 and a population of 18.4 million. The petroleum industry in the country is known to have commenced in 1953. The Abidjan oil field was discovered in 1974. The Abidjan Basin is the main basin in the country, measuring about 21,000 km^2. The basin is infilled by Devonian to Tertiary sediments. The reservoirs consist of Cretaceous sandstone with average porosity of 25 per cent. The seal is interbedded shale. Traps are associated with graben and faults. By the end of 2008 proven oil reserves were 13 million t with a modest annual oil production of 1.5 million t.

Ghana

Ghana is located on the southern coast of West Africa, lying along the Gulf of Guinea to the south, with Cote D'ivoire on its western border and Togo to the east. It has an area of 238,537 km^2 and an estimated population of 22 million. Initial oil exploration began in 1896 mainly in the southwest region bordering the Niger Delta basin to the south. The first oil field was discovered in 1970 and put into production in 1978. Geologically Ghana is divided into two parts. The inland portion consists mainly of the West African craton formed of Palaeozoic sediment. Offshore is a marginal sea basin, infilled with Ceno-Mesozoic rocks. By the end of 2008 proven oil reserves were 2 million t with annual oil production of 300,000 t. The past two years have witnessed a dramatic change in the petroleum industry development of the country. The extremely successful development of the offshore Jubilee field means that Ghana is set to join the ranks of significant African producers, ranking behind only Nigeria in West Africa. The Ceno-Mesozoic marginal sea basin extends to depths between 1000 m and 1700 m with commercial oil deposits discovered at subsurface levels of 1000 m to 3000 m. Estimates vary as to the ultimate production potential of the country. Initial production levels are estimated to be about 120,000 barrels/day, with the most optimistic peak forecasts estimated at 500,000 barrels/day within the decade. Whatever the outcome, Ghana will definitely become a major regional and continental oil player.

Togo

Togo is also located on the southern coast of West Africa, located between Ghana to the west and Benin to the east. It has an area of about 36,785 km^2 and a population of 5.2 million. Oil exploration began in 1970. Two wells were finished in southern offshore waters up to a depth of 3828 m. Oil shows were found in Cretaceous and Tertiary sandstone and limestone. Up to the present time there have been no discoveries of commercial quantities of oil in the country and it is still classified as a non-oil-producing nation.

Benin

Benin is a narrow coastal country located in West Africa and lying between Togo to the west and Nigeria to the east. It has an area of 112,622 km^2 and an estimated population of 8.4 million. The southern part of the country lies in the northern portion of the Niger Delta Basin. Early oil exploration started in 1950 with the first oil field discovered in 1968. Despite its proximity to West Africa's giant producer, Nigeria, exploration in Benin has not been significantly successful and present proven oil reserves are about 1.09 million t. The country is still classified as a non-oil-producing nation.

NIGER, NIGERIA, CAMEROON, SAO TOME AND

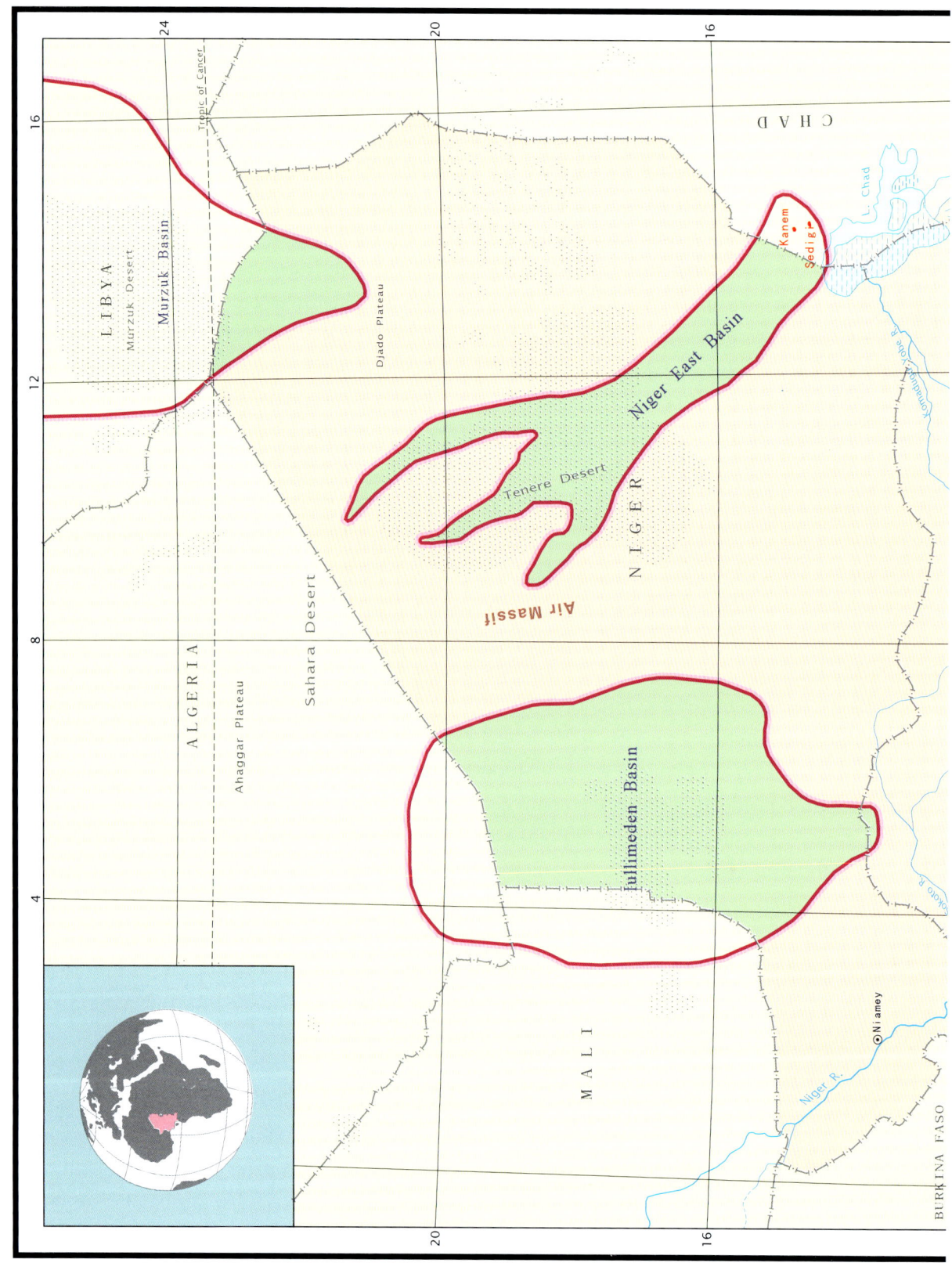

PRINCIPE AND EQUATORIAL GUINEA CHAPTER 48

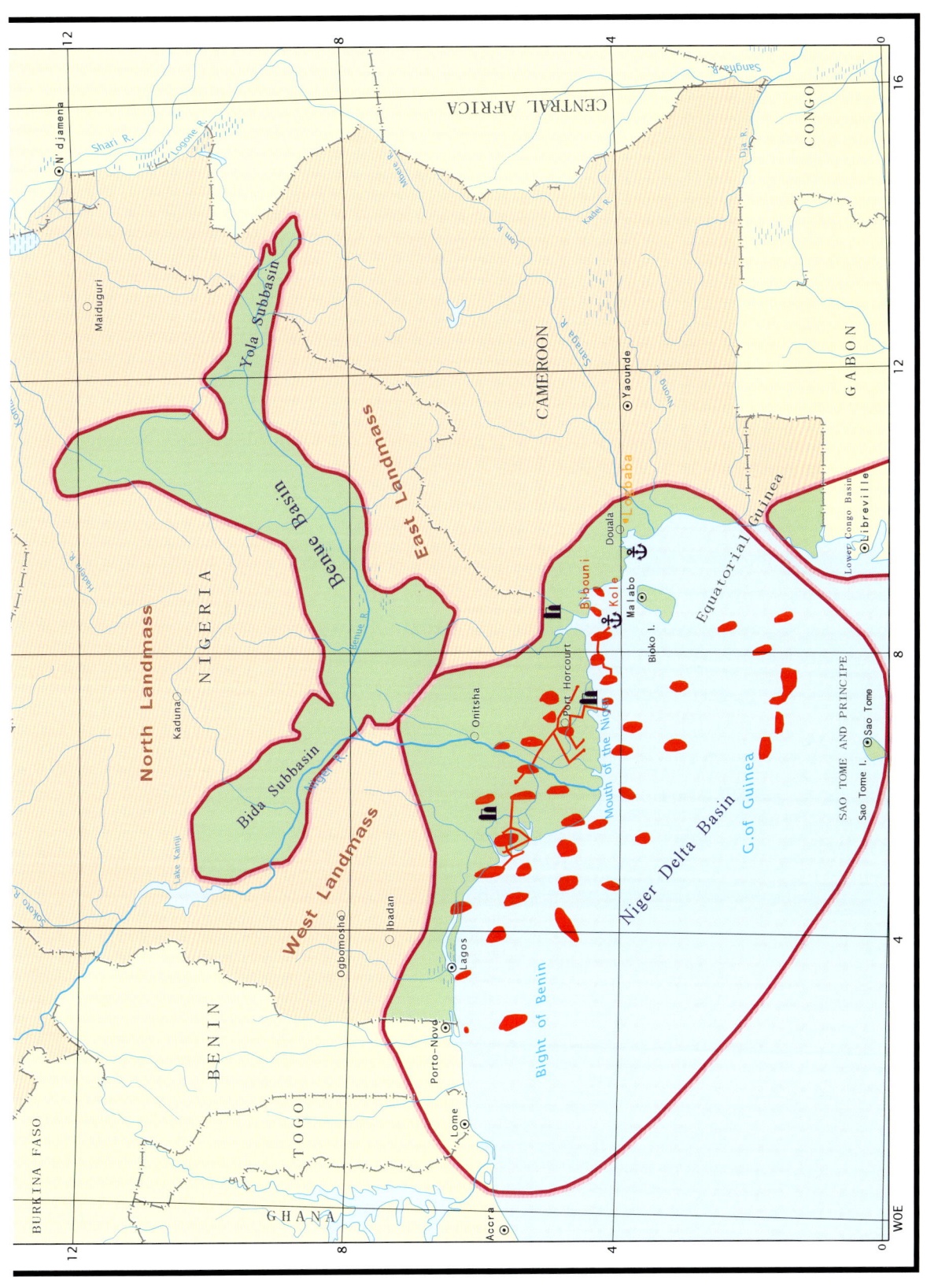

48 Niger, Nigeria, Cameroon, Sao Tome and Principe and Equatorial Guinea

Geography

Niger, Nigeria, Cameroon, Sao Tome and Principe and Equatorial Guinea are located in the east of the region referred to as West Africa. The eastern portions of Cameroon actually abut into the central African region.

History

The petroleum industries of these countries are entirely based on the prolific Niger Delta Basin. Oil exploration began in the 1950s with very rapid commercial finds. Nigeria is so far the largest producer in the region and also the largest producer in Africa ahead of Libya and Angola. It also ranks in the top 10 oil- and gas-producing countries worldwide. Details of the major sedimentary basins are provided in Table 48.1.

Regional geology

From the Niger Delta in Nigeria, heading south to the Cape of Good Hope in the Republic of South Africa, a series of basins exist along the west coast of Africa. The Niger Delta is the largest Tertiary oil-producing area in West Africa. The sediments in these the West African basins are principally supplied by the River Niger, which is 4100 km long and rises in the mountains of Sierra Leone to the west. The largest tributary is the Benue River with which it has its confluence in central Nigeria. The present-day Niger and Benue Valleys are developed along areas of Mesozoic and Cenozoic cover sediments that separate massifs of exposed basement rocks, a number of separate basinal areas are developed along these two principal valleys, including the Benue Basin, the Yola and Bida Sub-basins and, of course, the Niger Delta Basin (Selley, 1997).

Niger Niger is a landlocked country lying to the north of Nigeria and about 62 per cent of its area is covered by the Sahara Desert. The country has an area of 1,267,000 km² and an estimated

Table 48.1 Major sedimentary basins

Country	Basin	Area (thousand km²)	Major sedimentary rock		Reservoir		Remarks
			Age	Thickness (m)	Age	Lithology	
Niger	Iullimeden	300	Pz, Mz	200	Mz	Sandstone	Extends into Mali, Algeria
	Niger East	115	K, R	1200			Extends into Chad
Nigeria	Benue	173	K, R	6000			Extends into Cameroon
	Niger Delta	500	R	4000	R	Sandstone	Extends into Cameroon, Benin, Sao Tome and Principe

World Atlas of Oil And Gas Basins, First Edition. Li Guoyu.
© 2011 John Wiley & Sons, Ltd. Published 2011 by John Wiley & Sons, Ltd.

population of 11.4 million. Although the Iullimeden and the Niger East Basins are widely distributed within this country, no commercial oil and gas fields have been discovered yet. Oil exploration began in the 1950s. Nine wells were completed between 1962 and 1964 with depths varying between 569 m and 1993 m. All these wells were dry.

Nigeria With a land area of 923,768 km^2 and a population estimated at 144 million, Nigeria is the biggest oil producing country in Africa, ranking above Libya, Angola and Algeria. Annual oil production reached 97 million t in 2008. Historically, surface oil seepages were observed out of Cretaceous rocks, but initial exploration activities between 1908 and 1914 proved unsuccessful. Oil exploration in the Benue Basin between 1937 and 1941 was also unsuccessful. Oil was first struck in Oloibiri, to the east of the Niger Delta in 1956. The first offshore oil field, Okan, was discovered in 1964 in the shallow waters of the western portion of the Niger Delta. Thereafter, there were several oil field discoveries which served to underpin the rapid rise of the country to global oil production prominence.

Cameroon Cameroon, lying immediately to the east of Nigeria, has an area of 475,422 km^2 and a population of 16.4 million. Most of the discovered oil fields are offshore in the Rio del Rey and Douala/Kribi-Compo Basins. There have also been significant finds onshore including the Logone-Birni Basin to the north of the country. Cameroon started oil exploration in 1947. The first oil field was discovered in 1954. There were subsequent important oil field discoveries in the country. However, the country's petroleum development has been inefficient, with the result that several of the producing oil fields are rapidly reaching maturity, without the corresponding discovery of replacement resources. Production levels have declined in recent years. By the end of 2007 proven oil reserves stood at 27.39 million t. Oil production was 4.2 million t and proven gas reserves stood at about 110 billion m^3.

Sao Tome and Principe The country has an area of 1001 km^2 and a population of a mere 157,000. The country consists of more than ten small islands lying in the Atlantic Ocean some 250 km to the west of Gabon and Equatorial Guinea. Most of the islands are volcanic in origin. Bituminous sand as well as oil seepages were known to occur out of the Cretaceous rocks. One-hundred shallow wells were drilled in the exposed Cretaceous rocks at various times with generally mixed results. It has now been confirmed that the petroleum potential of this country derives from several deep-water fields. Exploration activities have proceeded at widely differing speeds by the operators of these blocks, but China's Sinopec commenced exploratory drilling in 2009. Although the country is currently classified as non-oil-producing, it has immense prospects.

Equatorial Guinea Equatorial Guinea has an area of 28,051 km^2 and a population of 540,000. It consists of the coastal enclave of Rio Muni and the principal island of Bioko, along with four smaller islands. Onland there are large tracts of exposed metamorphosed basement, with wide distributions of Cretaceous volcanic rocks. The offshore area comprises rift basins infilled by Cretaceous marine carbonates and sandstone. Oil exploration commenced in the early 1960s, but it was not until 1991 that commercial production started from Mobil's Alba oil field. By the end of 2008 proven oil reserves stood at 150 million t, with annual oil production of 16 million t.

NIGER DELTA BASIN

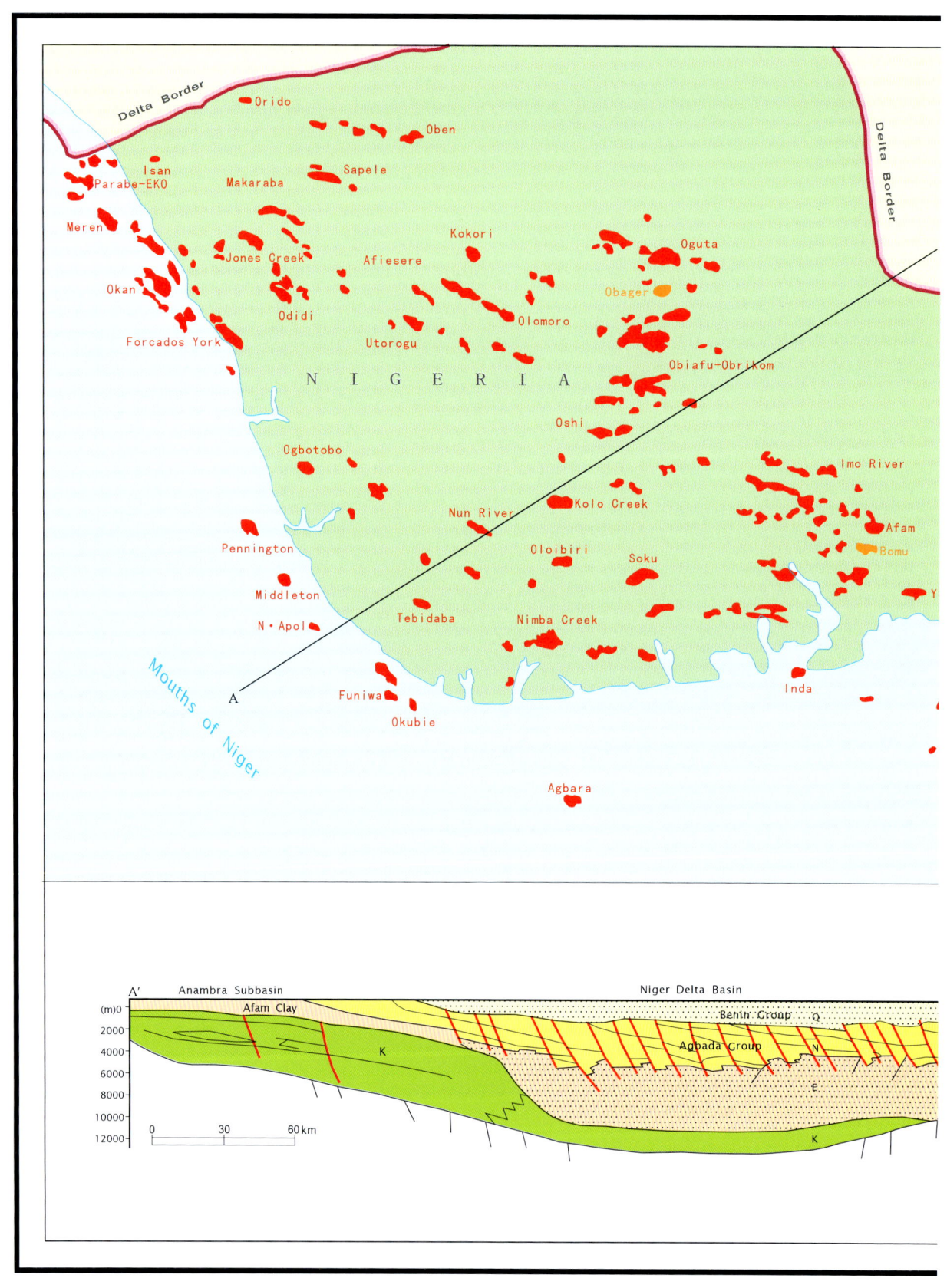

CHAPTER 49

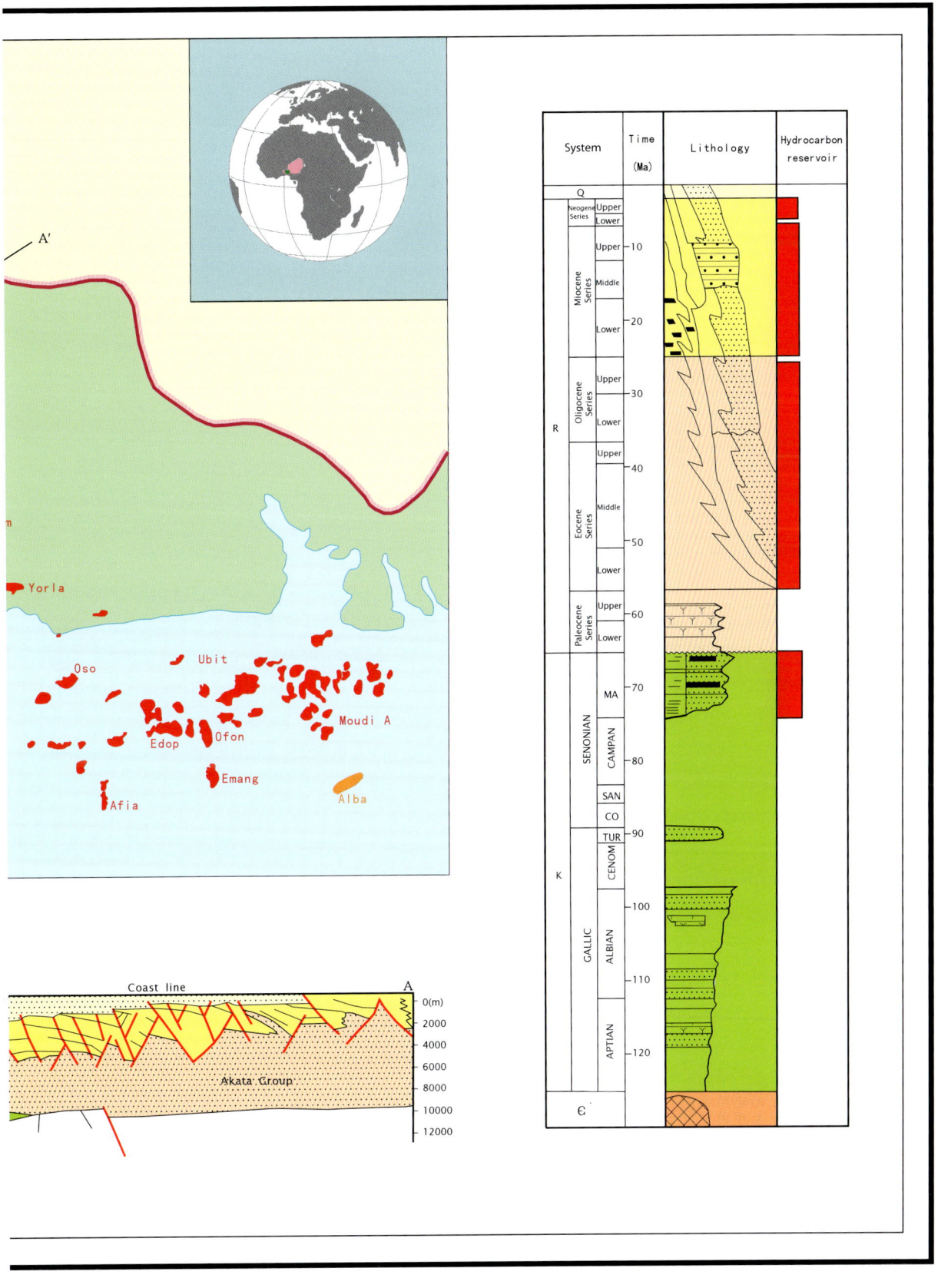

49 Niger Delta Basin

Geography

The Niger Delta Basin occupies the Gulf of Guinea continental margin in equatorial West Africa. The Niger Delta, with a total area of about 75,000 km^2 and a clastic sediment fill up to 12,000 m thick, is one of the largest in Africa. The tectonic setting and geological evolution of the Niger Delta Basin transcend and pre-date the post-Eocene regressive clastic wedge that is conventionally ascribed to the delta. The megastructure is the southern Benue Trough. Apart from the fact that Cretaceous strata are exposed along the northern margins of the Niger Delta, and have been intercepted on its eastern and western flanks, the probable occurrence of 5–6 km of Jurassic and Lower Cretaceous strata beneath the delta, as well as the location of the Niger Delta on the oceanward extension of the Benue Trough, have regional tectonic and stratigraphic implications for the delta.

History

The history of oil exploration in Niger Delta Basin extends to more than 50 years. Since 1952 over 592 oil and gas fields have been discovered with some 240 of these put into commercial production.

Regional geology

The Niger Delta Basin occupies the coastal land and oceanward part of a much larger and older tectonic feature, the Benue Trough, which is a NE–SW folded rift basin that runs diagonally across Nigeria. The Niger Delta is the youngest sub-basin in the Benue Trough, a region whose stratigraphic and palaeogeographical evolution has been controlled by southward shifting deltaic depocentres together with westward post-deformational displacement of depocentres, and northward directed marine transgressions. The incidence of Precambrian tectonics in the structural evolution of the Niger Delta was largely limited to movements along the Equatorial Atlantic Ocean fracture zones that extend beneath the delta and determined the initial locus into which the proto-Niger built its delta. As the delta advanced onto oceanic crust, repeated subsidence of the oceanic basement created more space for the thick sedimentary pile of the prograding Cenozoic Niger Delta.

The sedimentary cover comprises mainly Cretaceous and Oligocene–Miocene deltaic deposits. Source rocks are Tertiary Akata Formation shales with a geothermal gradient of 1.3–1.8°C/100 m. reservoir rocks are associated with Tertiary Agbaba Formation sandstones. The reservoir is multilayered with thicknesses of 15–45 m, porosity of 14.4 per cent and permeability of 576 mD. Seals are represented by the Ataka Formation shales as well as faults, and traps are mostly rollover structures,

Growth faults triggered by penecontemporaneous deformation of deltaic sediments are the common structure in the Niger Delta. They are generated by rapid sedimentation and gravitational instability during the accumulation of the paralic Agbada Group deposits and the continental Benin Group sands over the mobile, undercompacted Akata Group pro-delta shales during growth faulting and related extension. These also account for the diapiric structures on the continental slope of the Niger Delta in front of the advancing depocentre of paralic sediments. Growth faults in the Niger Delta are restricted to the paralic Agbada Group. They comprise the major structure-building

World Atlas of Oil And Gas Basins, First Edition. Li Guoyu.
© 2011 John Wiley & Sons, Ltd. Published 2011 by John Wiley & Sons, Ltd.

Niger Delta Basin

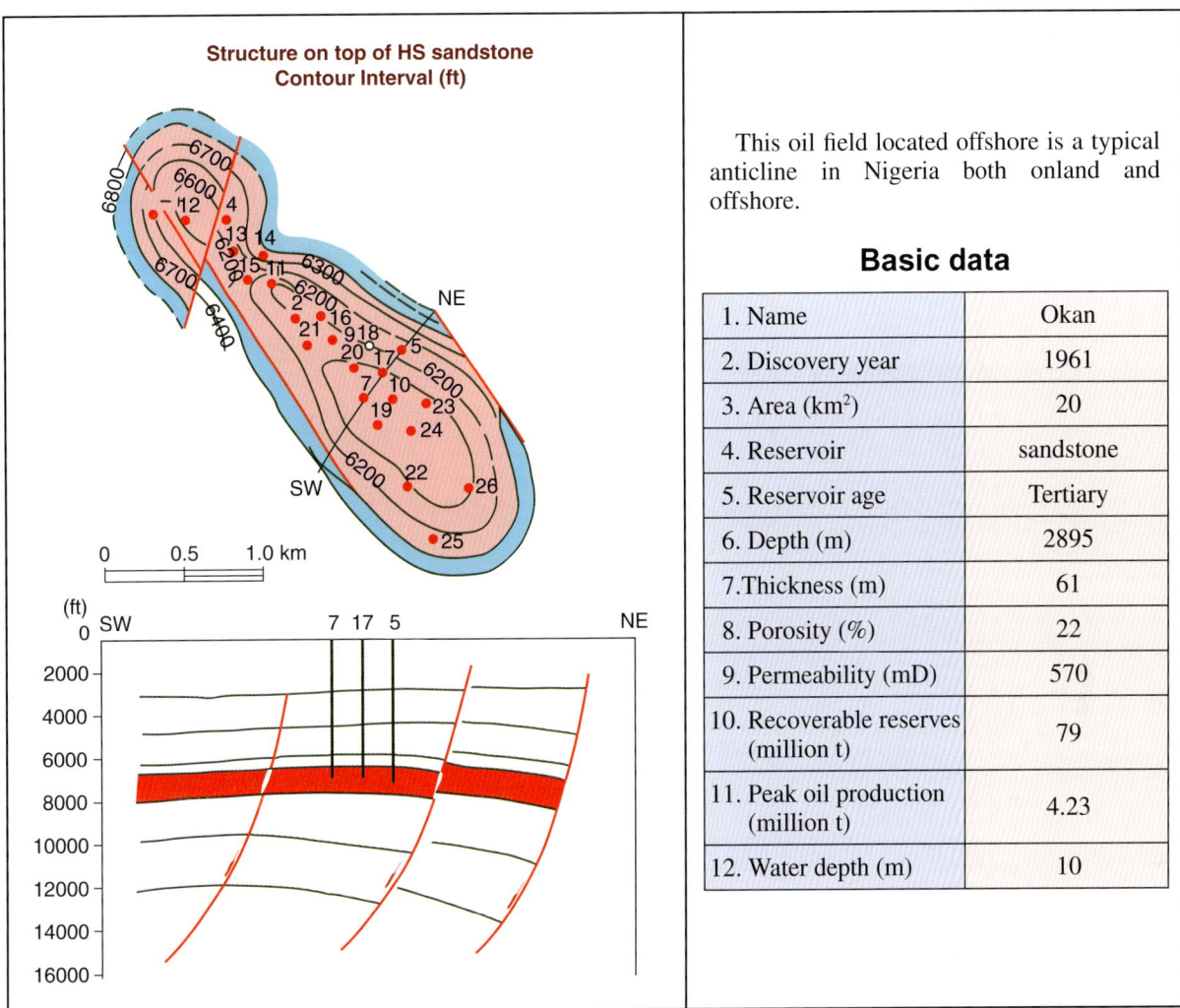

Fig. 49.1 Details of the Okan oil field.

faults, some of which bound the depocentre belts. There also exist steep, parallel crestal faults, which cut the rollover structures as well as antithetic faults. Associated with the structure-building faults are the rollover anticlinal structures. Gowth faults and rollover structures are the dominant hydrocarbon traps in the Niger Delta (Halbouty, 1991).

The Niger Delta Basin holds enormous petroleum reserves, with the delta ranking as seventh in global production. A few giant oil and condensate fields, with reserves exceeding 70 million t occur in the Niger Delta, examples of which are the Edop, Jones Creek, Nembe, Meren, Delta South and Okan (Fig. 49.1) fields. Oil and gas reserves in the Niger Delta Basin are concentrated in sandstone reservoirs throughout the paralic Agbada Group. Because of sand–shale alternations and the repetitive nature of the traps, most oil fields in the Niger Delta have multiple, stacked reservoirs, with oil-column heights ranging from 15 to 50 m. Mature Eocene and Miocene shales of the Akata and Agbada Groups constitute the major source rocks. Niger Delta crudes originated mostly from land plant material, hence they are high in resins and waxes, with significant contributions of structureless organic matter from marine sources. The low sulphur contents of Nigerian crudes (below 0.4 per cent) also confirm land plant source material.

CONGO, DEMOCRATIC REPUBLIC OF CONGO AND

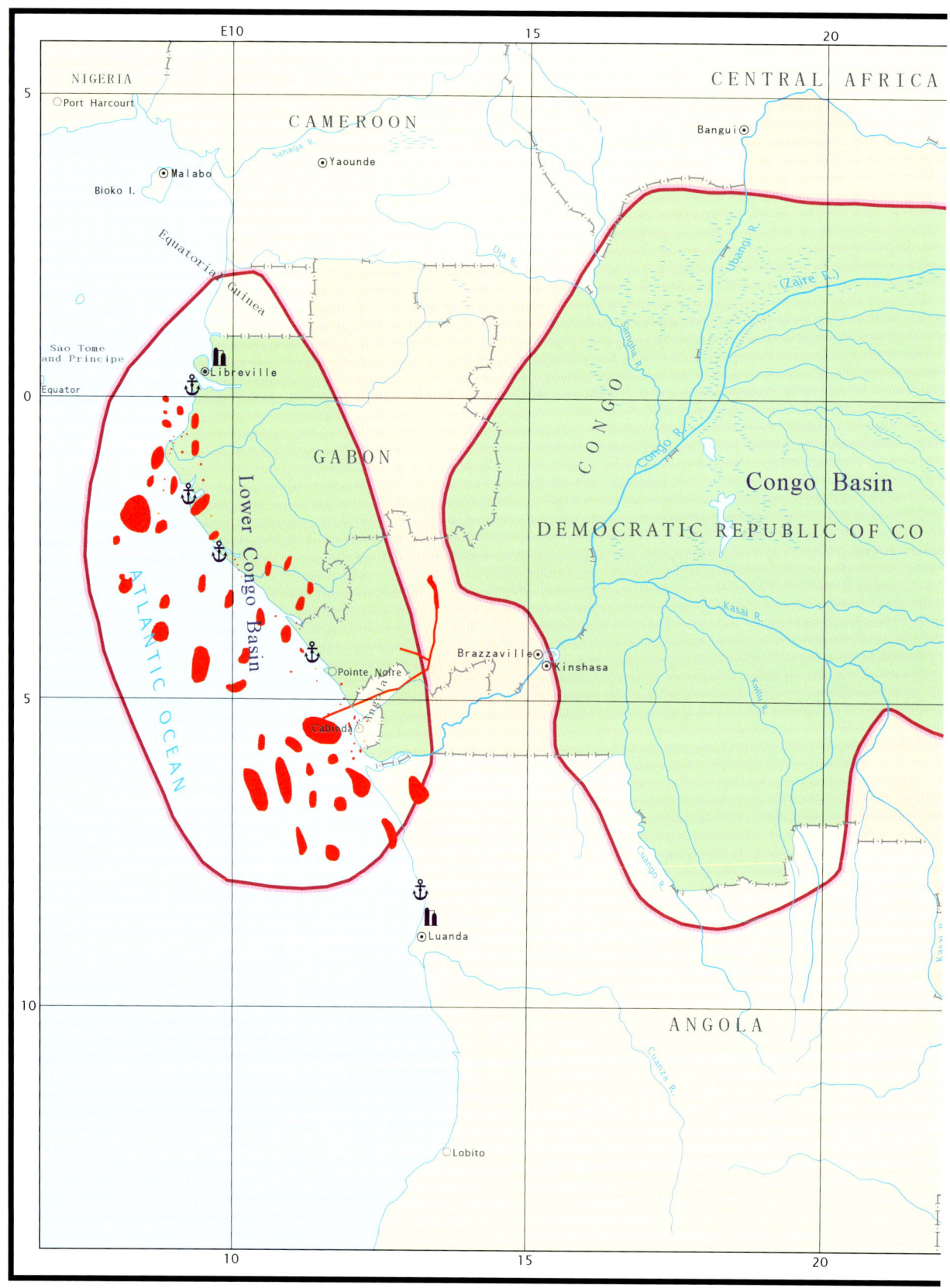

Basin: 1 Mobutu-sese-seco 2 Tanganyika

GABON
CHAPTER 50

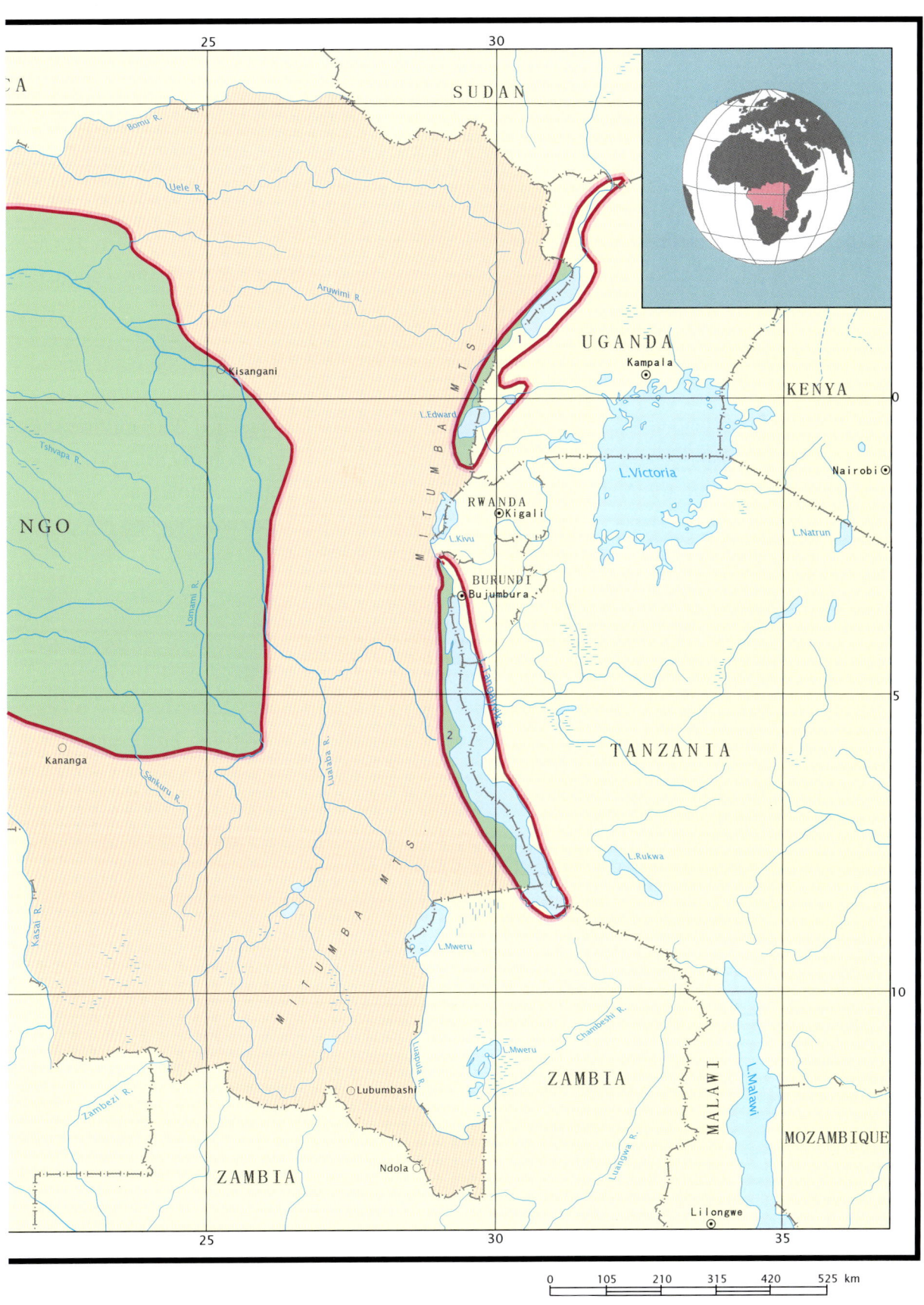

219

50 Congo, Democratic Republic of Congo and Gabon

Geography

Congo, Democratic Republic of Congo and Gabon are located in the western portion of central Africa and all face the Atlantic Ocean to the west. Congo is sometimes referred to Congo Brazzaville, while the Democratic Republic of Congo (formerly Zaire) is commonly referred to as the DRC. In some reference books Gabon is classified with West African oil-producing nations.

History

Early oil exploration started in 1928 in Gabon. Later in the 1950s oil exploration commenced in the Congo and also in the DRC. Many large oil and gas fields were discovered both onshore and offshore, mainly in Gabon and Congo. Annual oil production has reached 23.5 million t in both territories in the past 3 years. Currently, there is very little oil production in DRC, primarily because exploration has been curtailed by political instability and violence. However, with the existence of the huge Congo Basin, with an area of 1.4 million km², there are encouraging prospects in the future. The major oil and gas basins are listed in Table 50.1.

Regional geology

This description of the regional geology is based on Selley (1997). In the west of this area a large metamorphic massif is exposed on the surface, mainly in Gabon. To the west, lying offshore, is the Lower Congo Basin with area of 590,000 km², which is infilled by Ceno-Mesozoic rocks. To the east there is the main onshore Congo Basin with an area of 1.4 million km² comprising Palaeozoic sediments.

The Lower Congo Basin has its southern flank where the River Congo discharges into the Atlantic Ocean. Geographically it extends southwards from Equatorial Guinea, through Gabon, across Congo, Cabinda (Angola) and the DRC. Sediments crop out along the coast continuously in the Lower Congo Basin. The sedimentary succession in the Lower Congo Basin comprises Palaeozoic, Mesozoic and Cenozoic strata.

The northwest–southeast trending Lambarene Horst separates the Lower Congo Basin in the west from the Congo Basin in the east. The Lower Congo Basin reaches its maximum thickness and

Table 50.1 Major oil and gas basins

Country	Basin	Area (thousand km2)	Major sedimentary rock		Reservoir		Remarks
			Age	Thickness (m)	Age	Lithology	
Congo, DRC	Congo	1400	Pz	8000	Pz	Sandstone	Extends into Angola, Gabon
Gabon	Lower Congo	590	J, K, R	3100	K, R	Sandstone	Extends into DRC, Angola, Equatorial Guinea

World Atlas of Oil And Gas Basins, First Edition. Li Guoyu.
© 2011 John Wiley & Sons, Ltd. Published 2011 by John Wiley & Sons, Ltd.

extent in Gabon, before narrowing and thinning south into Congo. The Lower Congo Basin is more important economically than its sister Congo Basin. The lower Congo Basin contains up to 18 km of sedimentary fill. The tectonic and sedimentary history of the Lower Congo Basin can be grouped into pre-salt, salt and post-salt phases, which are well-known very important geological events locally.

The Lower Congo Basin is a major petroleum province with production from both beneath and above the salt. The pre-salt strata are described above. Although the post-salt sequence contains Cretaceous black shales, gas chromatography reveals that the post-salt hydrocarbons are of non-marine origin. Furthermore they show identical chromatograms to those found in pre-salt reservoirs that were clearly sourced from the lacustrine shales of the Cocobeach Group. Migration from the Cocobeach source rocks has been facilitated by the laterally extensive Gamba Sands. Petroleum has escaped across the salt horizon, either where the seal is breached by local salt solution, or where it is breached by listric faults. Both these mechanisms have allowed pre-salt petroleum to migrate up into post-salt shallow- and deep-water sand reservoirs. The main traps are provided by the Aptian salt domes. Halokinesis has taken the form of both diapirs and salt walls. The salt walls occur on the uplifted, landward side of listric faults that extend out towards the Atlantic Ocean.

Congo Congo has an area of 342,000 km^2 and a population of 3.86 million. By year end 2007 proven oil reserves of 219 million t were estimated with annual oil production of 12 million t. Proven gas reserves were estimated at 90 billion m^3. All major oil and gas fields are located offshore.

Democratic Republic of Congo (Previously Known as Zaire) The DRC has an area of 2,344,883 km^2 with an estimated population of 57.5 million. By year end 2007 the proven oil reserves were 24.6 million t, with annual oil production of about 1,000,000 t. Proven gas reserves were 90 billion m^3. Although the country straddles one of the largest sedimentary basins, the results of oil exploration have only been modest.

Gabon The country has an area of 367,667 km^2. The population is 1.36 million. By the end of 2008 proven oil reserves were 273 million t with annual oil production of 11.7 million t. Proven gas reserves totalled 28 billion m^3. The large sedimentary basins, Congo and Lower Congo, present a very high prospect for oil and gas exploration in the future. As mentioned earlier, political instability and a permanent state of civil strife and insurrection in most of the DRC has made that country an unattractive target for exploration activity. This is likely to improve with stable political conditions.

ZAMBIA, ANGOLA AND MALAWI

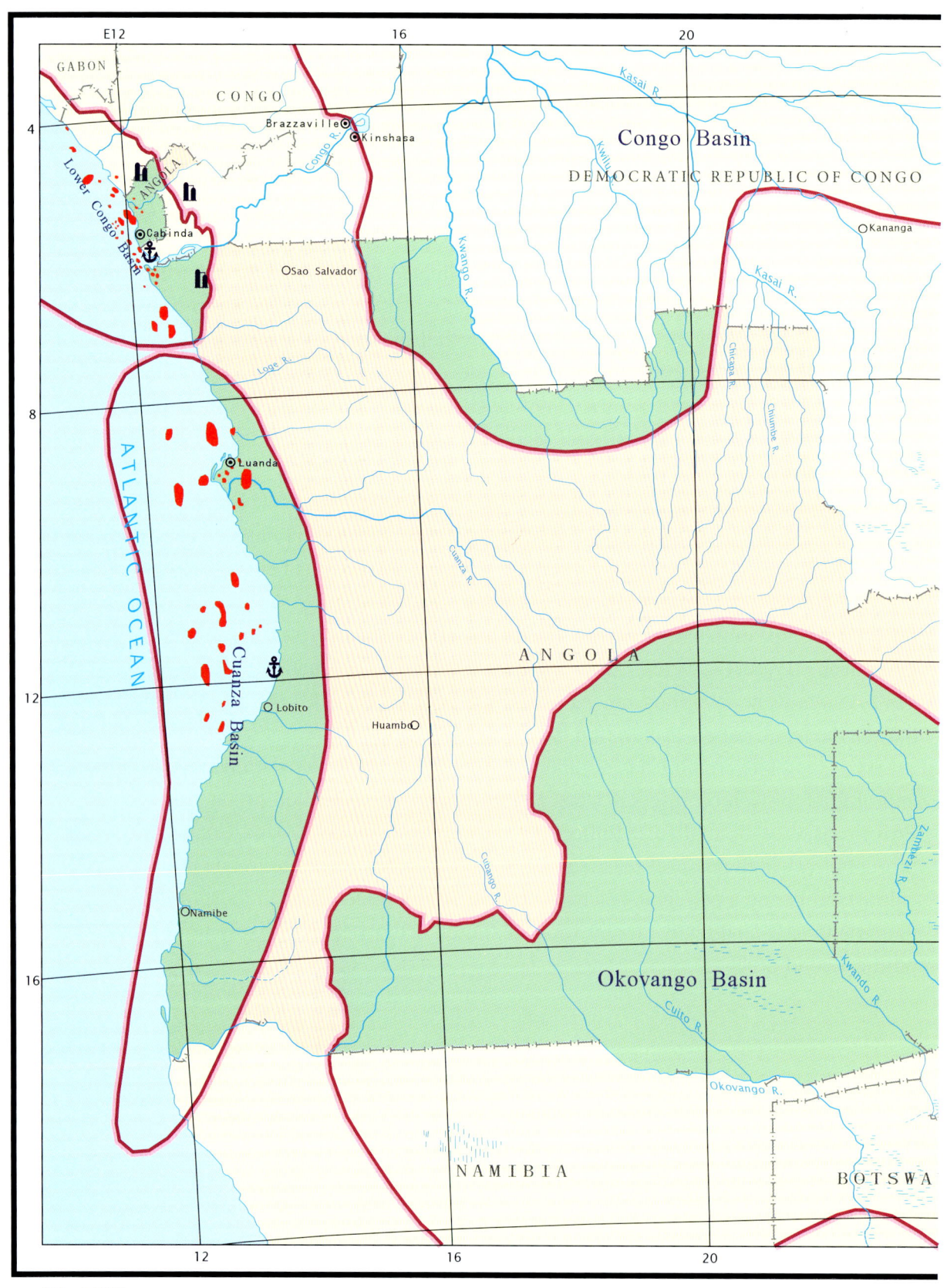

Basin: 1 Malawi

CHAPTER 51

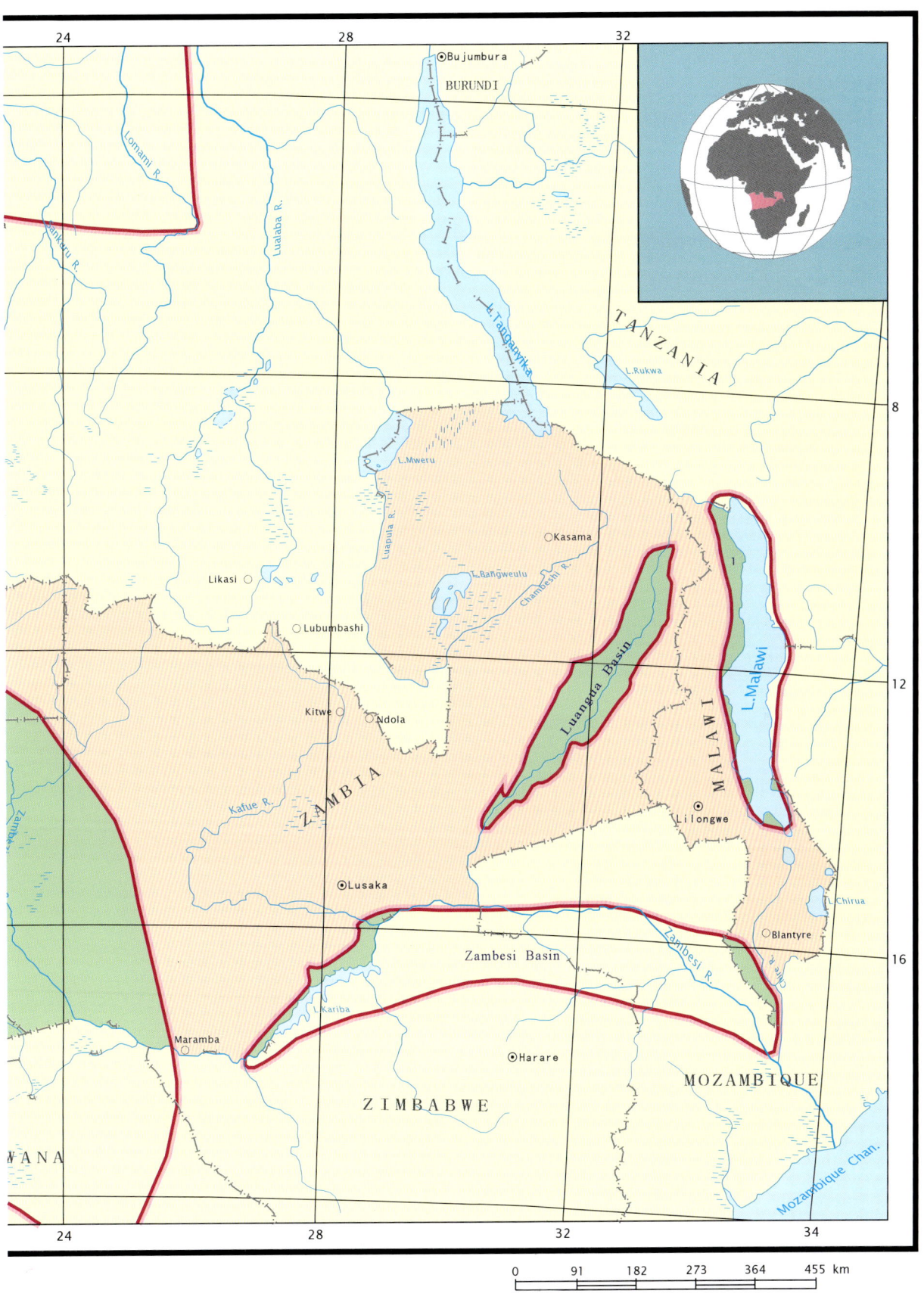

51 Zambia, Angola and Malawi

Geography

Zambia, Angola and Malawi are located in southern Africa. Whereas Angola has a long Atlantic Ocean coastline, both Zambia and Malawi are in the interior and are both landlocked.

History

Among these three countries only Angola is oil producing. The oil exploration history is relatively short, beginning only in the 1920s. Details of the major sedimentary basins are listed in Table 51.1.

Regional geology

This region consists of many sedimentary basins, some of which are very important. The East African rift system and Precambrian crystalline basement are among the most significant geological features. The most prospective area is the Cuanza Basin, although there are thought to be many exciting prospects in the Okovango Basin as well. The Cuanza Basin starts as a narrow coastal strip of sediment only a few kilometres wide, before it widens out where the Cuanza River enters the Atlantic Ocean. The Cuanza (or Kwanza) basin is also referred to as the Angola Basin.

The sedimentary and structural history of the Cuanza Basin is very similar to that of the basins to the north. The sediments of the Cuanza Basin can be divided into the pre-salt, salt, and post-salt phases described earlier. The pre-salt sequence consists of interbedded clastics and volcanics of the Lower Cuvo Formation. The clastics include continental red sandstones and shales with occasional thin coals. The volcanics consist of basaltic lavas and tuffs. Angola became the earliest oil-producing country in southwest Africa after initiation of oil exploration in the Cuanza Basin prior to 1920. Offshore exploration started in 1969 and to date 61 oil fields and 6 gas fields have been discovered in Cuanza Basin (Selley, 1997).

Table 51.1 Major sedimentary basins

Country	Basin	Area (thousand km²)	Major sedimentary rock		Reservoir		Remarks
			Age	Thickness (m)	Age	Lithology	
Angola	Cuanza	300	K_1, R	2700	K, R	Carbonate Rock	Extends into Namibia
	Okovango	920	Pz, K		P, T		Extends into Zambia, Botswana, Namibia
Zambia	Luangua	40	P, T	4000		Clastic rock	
Malawi	Malawi	48	K–R				Extends into Tanzania, Mozambique

World Atlas of Oil And Gas Basins, First Edition. Li Guoyu.
© 2011 John Wiley & Sons, Ltd. Published 2011 by John Wiley & Sons, Ltd.

Zambia, Angola and Malawi

The Okovango Basin is located in the Congo Craton. It extends northwards into southern Angola and may well continue into western Zambia. The eastern boundary in Namibia may be a buried basement updoming trending in a northeast direction from Grootfontein to Mashsri. This updoming is breached about half way along its length by a southeast-trending valley about 350 m deep. Although having formed initially as a late Proterozoic sedimentary basin, it was also a centre of deposition in the Permo-Jurassic and again during the Late Cretaceous to Tertiary. The present form is still basin shaped. The hilly to mountainous rim on three sides of the basin reaches elevations of 1500–1700 m in the north in Angola and 1400–2100m in the west and south. From the lowest point in the basin, the Etosha Pan, having an average elevation of 1084 m, the surface of the basin gradually rises to 1150 m above the Grootfontein–Mashari updoming in the southwest, to 1250 m at the base of the rim in the west and to 1200 m at the base of the rim in the south.

Zambia Zambia has an area 752,614 km^2 with an estimated population of 12.34 million. About 50 per cent of this country is covered by forest. Two main sedimentary basins are located within Zambia. Part of the Okovango Basin is located in western Zambia. The Luangua Basin is located in eastern Zambia. The great East African rift runs through the northeast of Zambia. Excluding these two basins, most of Zambia consists of Precambrian crystalline basement. The Luangua Basin is infilled by Permian, Triassic, Cretaceous and Tertiary strata. The sedimentary cover of Okovango Basin consists of Upper Permian, Triassic and continental Cretaceous strata. All exploration carried out to date has shown little potential commercial prospectivity.

Angola Angola has an area of 1,246,700 km^2 with a population of 15.9 million. Angola is one of the major oil-producing countries of Africa. Oil production reached 93 million t in 2008. Following up sighting of oil seepages in the Cretaceous rocks of the Cuanza Basin, onshore oil exploration commenced in the 1920s. Twenty shallow wells were drilled after which further exploration activity was suspended. Through its Portuguese subsidiary, Petrofina recommenced exploration which resulted in discovery of commercial quantities in 1955 in the Benfica oil field, with oil being produced out of Cretaceous carbonate rocks. This is considered to be the first commercial oil field in this part of Africa. Angola has two sedimentary basins, namely, Okovango and Cuanza (sometimes spelt as Kwanza). Only the northern half of the Okovango Basin is located in Angola. The southern half is located in Namibia and Botswana. Production in the Cuanza Basin catalysed the development of the Angolan petroleum industry, which was delayed due to the lengthy civil war in that country. The Cuanza Basin is infilled with Jurassic, Cretaceous and Tertiary sediments. Oil is produced mainly from Cretaceous and Tertiary strata. Oil production reached 10 million t in 1984, 45 million t in 2004, rising rapidly to 93 million in 2008.

Malawi Malawi has an area of 118,486 km^2 and a population of 12.34 million. This country is mainly covered by a Precambrian crystalline basement with thin Cretaceous sediments at its edges. The geology is very complex, and very few geological surveys exist. At the present time Malawi is classified as a non-oil-producing country.

SUDAN, CENTRAL AFRICA AND CHAD

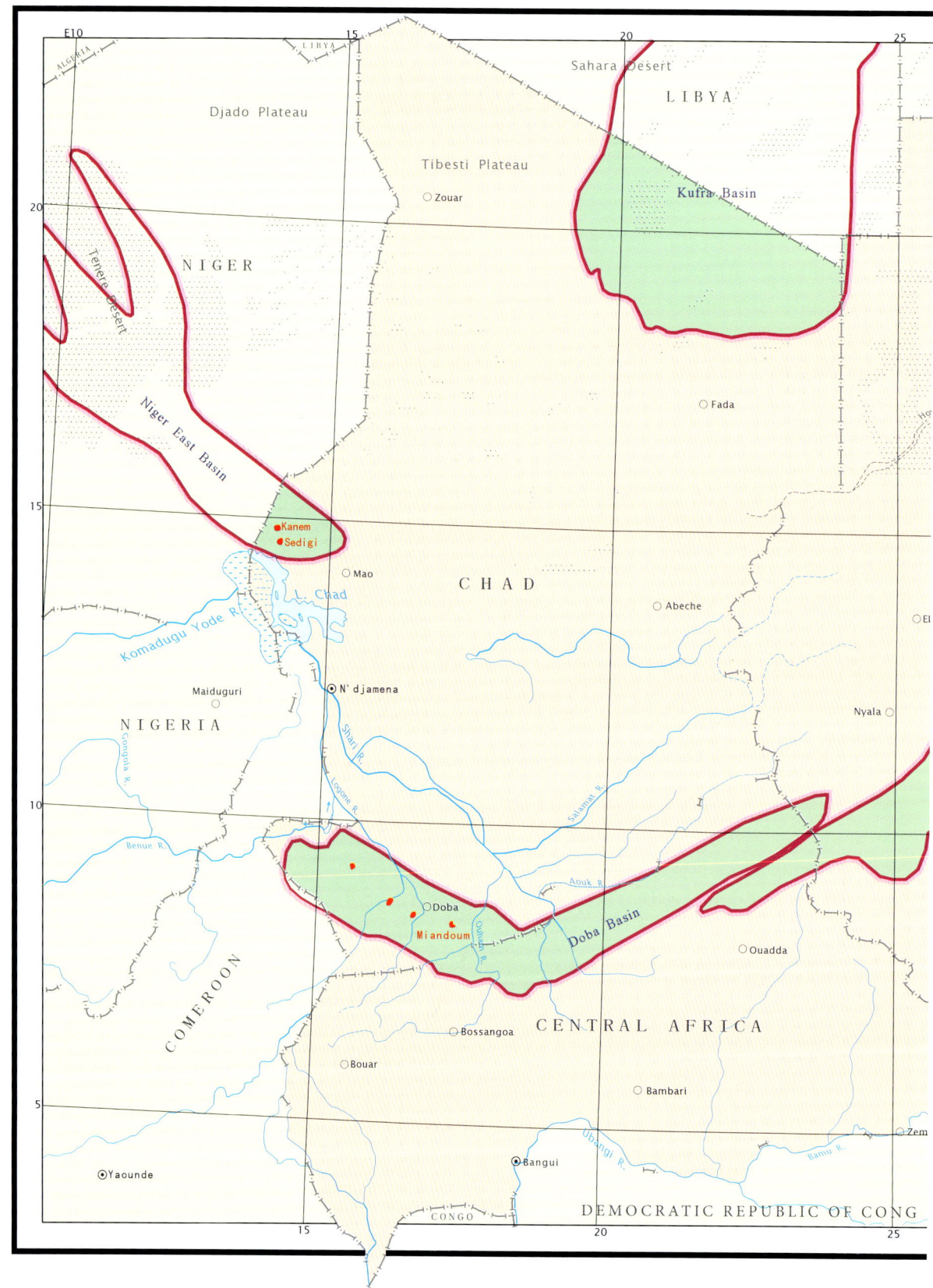

CHAPTER 52

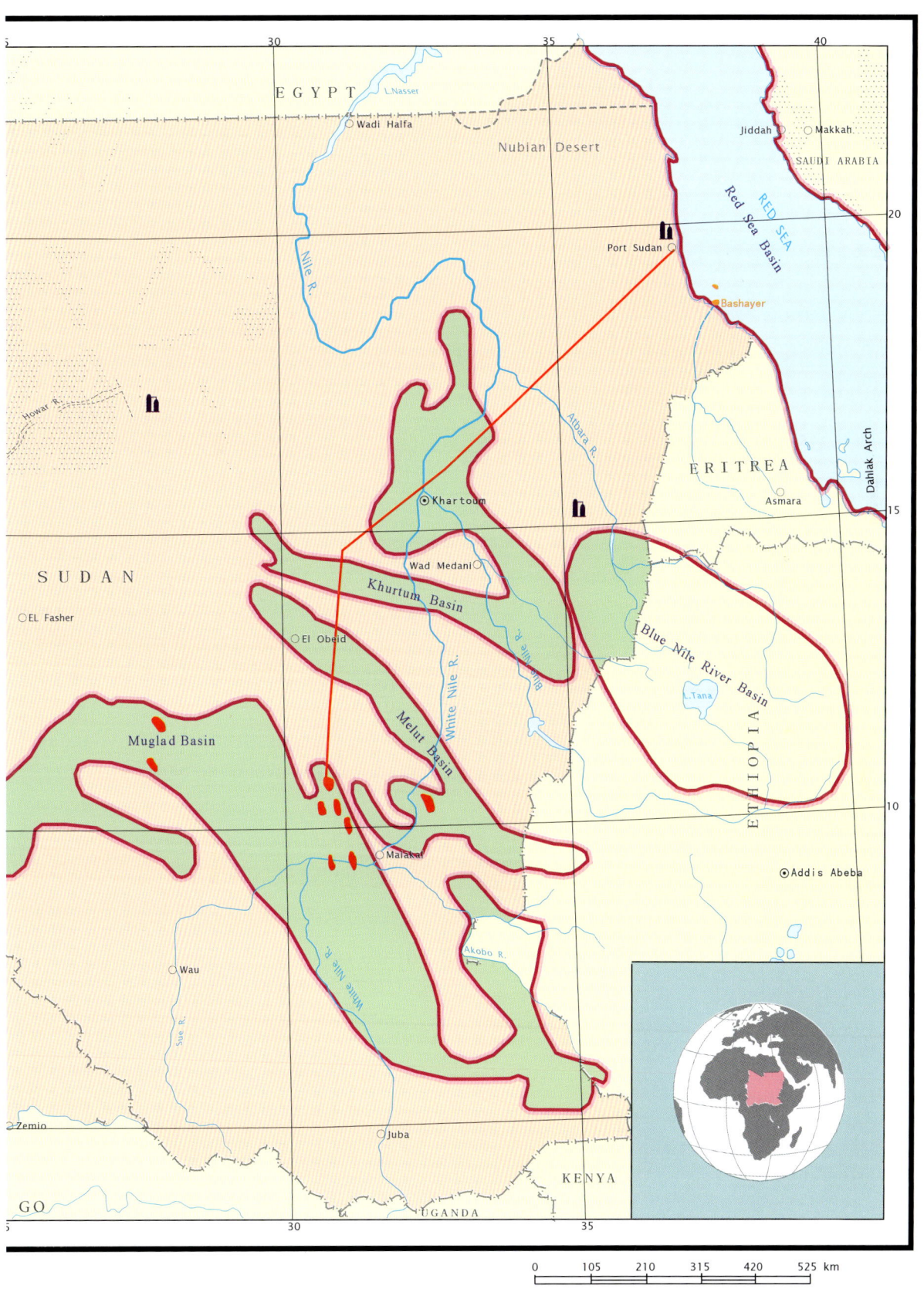

52 Sudan, Central Africa and Chad

Geography

Sudan is located in North Africa. Central Africa and Chad are geographically located in central Africa. Chad shares a border with Nigeria in the Lake Chad region, and also with Cameroon to the south.

History

In this area, both Sudan and Chad are oil-producing countries. Oil exploration activity began in the 1950s and the 1960s respectively. In the past decade in particular, the development of the petroleum industry of both countries has been very rapid. Details of the major sedimentary basins are provided in Table 52.1.

Regional geology

This description of the regional geology is based on Selley (1997). The three most pronounced geological features are the East African rift, infilling with Cretaceous and Tertiary sediments and the existence of a widespread Precambrian crystalline basement, which is exposed at the surface. The rift system in these three countries is associated with several oil- and gas-producing basins, such as Muglad. Cretaceous strata up to 6000 m in thickness infill the rift system, in which source rocks have been identified, as well as sandstone reservoirs. The Muglad Basin, the Doba Basin and half of the Kufra Basins are located within these countries. The petroleum geology of these areas is represented by the Sudanese rift system and the Kufra Basin.

The Sudanese rift system Bahr el-Arab rift comprises two major structures, the Baggara Graben and the Sudd Graben. The Baggara Graben covers the area between the Nuba Mountains in the east and Central Africa in the west. In the north it is defined by the faulted Mesozoic deposits south of the Darfur Dome. North of this line the Mesozoic deposits crop out over a distance of more than 200 km. South of the faulted zone, Mesozoic sediments are found below the surface

Table 52.1 Major sedimentary basins

Country	Basin	Area (thousand km²)	Major Sedimentary rock		Reservoir		Remarks
			Age	Thickness (m)	Age	Lithology	
Sudan	Khurtum	100	Mz				
	Melut	50	R				Extends into Ethiopia
	Muglad	230	K1,R	13,000	K,R	Sandstone	Extends into Central Africa Ethiopia
Chad	Doba	100	K	4000–6000	K	Sandstone	Extends into Central Africa, Sudan

World Atlas of Oil And Gas Basins, First Edition. Li Guoyu.
© 2011 John Wiley & Sons, Ltd. Published 2011 by John Wiley & Sons, Ltd.

and are known from borehole records, and comprise remnants of sandstone and laterite cropping out along Bahr el-Arab.

The Sudan Graben is bounded in the north by the Nuba Mountains and the Akoke ridge in the northeast. This rifting phase was accompanied by minor volcanism which occurred 82 million years ago, dated from a dolerite sill in the northwest Muglad Basin. The end of this phase is marked by the deposition of an increasingly sand-rich sequence that ended with the deposition of thick Paleocene sandstone of the the Amal Formation. The final rifting phase began in the late Eocene–Oligocene and is characterized by thick lacustrine and floodplain claystone and siltstones. Late Eocene basalt flows are recorded in the Melut block near Ethiopia.

The rifting was followed by an intracratonic sag phase during the middle Miocene, with very gentle subsidence accompanied by little or no faulting. Limited outcrops of volcanic rock in the area southeast of Muglad dated at 5 million years and 2.7 million years indicate that minor volcanism occurred locally. This is associated with the extensive volcanism of Jebel Marra, Tagabo and Meidob along the Zalingei Fold Belt.

The Kufra Basin is the last and most easterly of the Saharan basins. It occupies a large part of southeast Libya, though it extends northeastwards into Egypt, and southeast into Sudan and southwest into Chad. It is separated from the Murzuk Basin to the west by the Tibesti–Sirte Arch, and from the Sirte Basin to the north by the Calanscio Arch. Its eastern limb forms the western edge of the Arabian–Nubian Shield to the east. The Precambrian Ennedi Craton defines its southern rim.

Sudan Sudan has an area of 2,905,871 km^2 with an estimated population of 35.39 million. About 52 per cent of the area of Sudan is covered by Precambrian crystalline basement and Tertiary volcanic rocks. East of the rifting basins are petroleum provinces. Heglig and Unity oil fields are the highest producing among all of the fields in the region. The reservoir rocks are associated with Cretaceous sandstone. By the end of 2008 the proven oil reserves were 484.9 million t with annual oil production of 23 million t.

Central Africa Central Africa is a landlocked country with an area of 622,984 km^2 and a population of 3.9 million. The whole area of this country mainly consists of Precambrian crystalline basement. About 12 per cent of the area is covered by dense forest. The eastern part of the Doba Basin is located in Central Africa. Just as in Sudan, the Doba Basin is of a rift type. It is infilled by Cretaceous rocks with thicknesses reaching 6000 m. Very little oil exploration has been carried out in this country with only one dry well on record. Central Africa is classified as non-oil-producing, although the Doba Basin is considered to have decent prospects for the future.

Chad Chad has an area 1,284,000 km^2 and a population of 9.8 million. Three sedimentary basins are to be found in Chad, although only a relatively small proportion of each basin lies within its borders. Significantly, half the Doba Basin lies within Chad and this represents the main oil-producing area of the country. The Lake Chad Basin, bordering Nigeria, is another area of considerable exploration activity. By the end of 2008 the proven oil reserves were 205 million t, with annual oil production of 7.5 million t.

MUGLAD BASIN

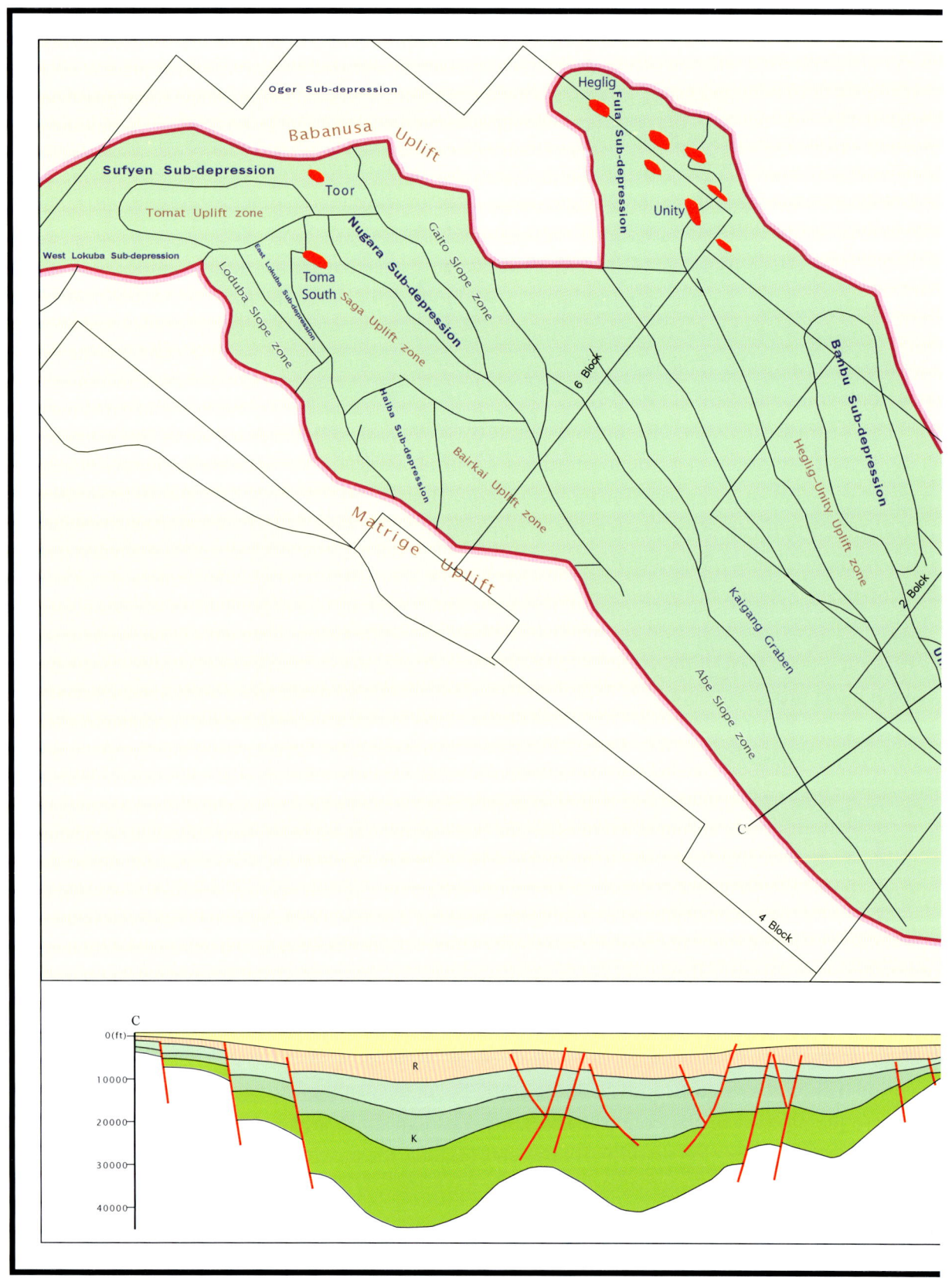

CHAPTER 53

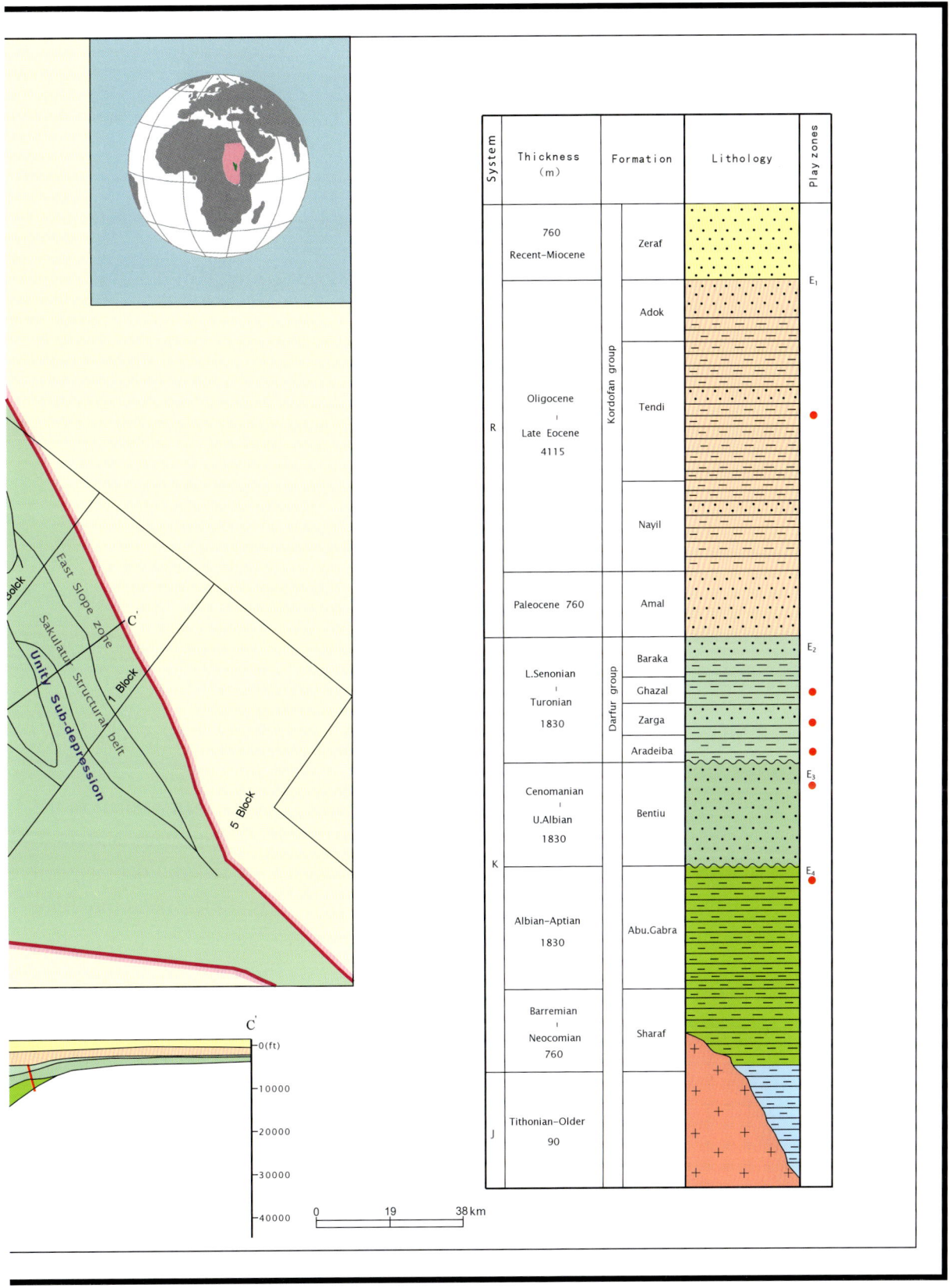

53 Muglad Basin

Geography

The Muglad Basin is located in southern Sudan with a total area of 220,000 km². It is the largest basin in Sudan. The whole basin is infilled with Cretaceous and Tertiary sediments locally reaching a thickness of 13,000 m.

History

Oil exploration in this basin started in 1973 resulting in the discovery of two oil fields, namely, Toma South and Toor. This basin is now the main oil-producing area of the Sudan. The major oil and gas fields are listed in Table 53.1 (Halbouty, 1990).

Regional geology

This basin is located in the East African rift system. The concept of plate tectonics has revolutionized the interpretation of the geological history in Sudan. Previously, the geological map of Sudan was dominated by four main units, namely:

1 the Precambrian–Cambrian basement complex rocks which covered 50 per cent of the map area;
2 Palaeozoic continental deposits scattered in the northwestern corner;
3 Mesozoic Nubian Sandstone Formation cropping out in the northern part of Sudan and covering about 30 per cent of the total area;
4 Tertiary Umm Ruwaba deposits covering the central and southern parts.

The Cretaceous and Tertiary sedimentary cover up to 13,000 m thick was thought to have filled shallow synclinal basins.

A major exploration programme started in the western and southern regions of Sudan. In the 12 years that followed significant amounts of geological and geophysical data were acquired. These included extensive aeromagnetic and gravity surveys, 58,000 km of seismic data as well as 86 exploratory wells being drilled. These intensive and detailed investigations shed more light on the history and development of the rift basins of Sudan. The productive and prospective

Table 53.1 Major oil fields of Muglad Basin

Name	Discovery year	Recoverable reserves		Basin	Depth	Trap	Reservoir	
		Oil (million t)	Gas (billion m³)				Age	Lithology
Heglig	1970s	—	—	Muglad	—	Anticline	C	Sandstone
Unity	1980s	—	—	Muglad	1750–2700	Anticline	C	Sandstone
Abu Gabra	—	—	—	Muglad	—	Anticline	C	Sandstone
Sharaf	—	—	—	Muglad	—	Anticline	C	Sandstone

World Atlas of Oil And Gas Basins, First Edition. Li Guoyu.
© 2011 John Wiley & Sons, Ltd. Published 2011 by John Wiley & Sons, Ltd.

Muglad Basin

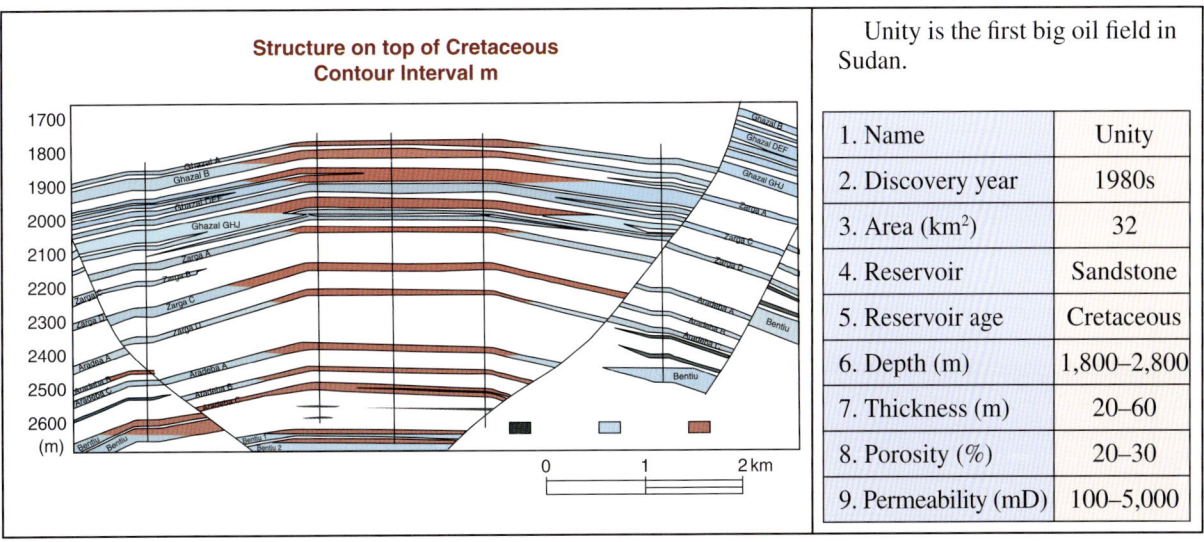

Fig. 53.1 Details of the Unity oil field.

structures resulting from extensional movement and the compressional forces created within resulted in complex structures formed by rotated fault blocks, drape folds and reverse drag folds. These structures produced oil traps.

Petroleum was discovered in the three rift systems explored: the Bahr el-Arad rift, the White Nile rift and the Blue Nile rift. The first oil was recovered from a well in the Muglad Basin, the first significant oil flow occurred in the Abu Gabra Basin and the first important discovery was made in Unity Basin (Fig. 53.1). Oil was discovered in the Cretaceous Abu Gabra Formation sandstone of the Muglad Basin, in the Bentiu Formation of the Babanusa Basin, and in sandstone of the Darfur Group in the Unity Basin. In te Tertiary strata oil was discovered in the sandstone of the Amal Formation. The reservoirs range from quartz arenites and wackestones to arkosic arenites and sandstones. These include sandstones deposited in fluvial channels, lacustrine delta-plain-distributary channels and delta-front facies.

The basins within the Babanusa Trough are defined by extensive faulting systems extending in NW, NE and E–W directions. The grabens and horsts indicate a step-like subsidence of separate blocks. The intensity of the faulting and the subsidence increases southward, where it attains a depth of more than 5 km at the Unity oil field, and an estimated 11 km south of Bantiu oil field.

The Kurkur hills, which form a basement uplift in the northern margin of the Baggara Graben, define the northern margin of Abu Gabra Trough. Its western boundary is represented by a major NW fault system extending southward along the SE sharp bend of Bahr el-Arab. It extends southeast defining the western margin of the Sudd Graben, connecting with the northerly extension of the Aswa line. There are some indications that this fault system is active in the Sewar Trough. Near Sewar, there is a zone of hot ground water, extending along the fault line, as confirmed by boreholes located along this line.

ETHIOPIA, SOMALIA, DJIBOUTI AND ERITREA

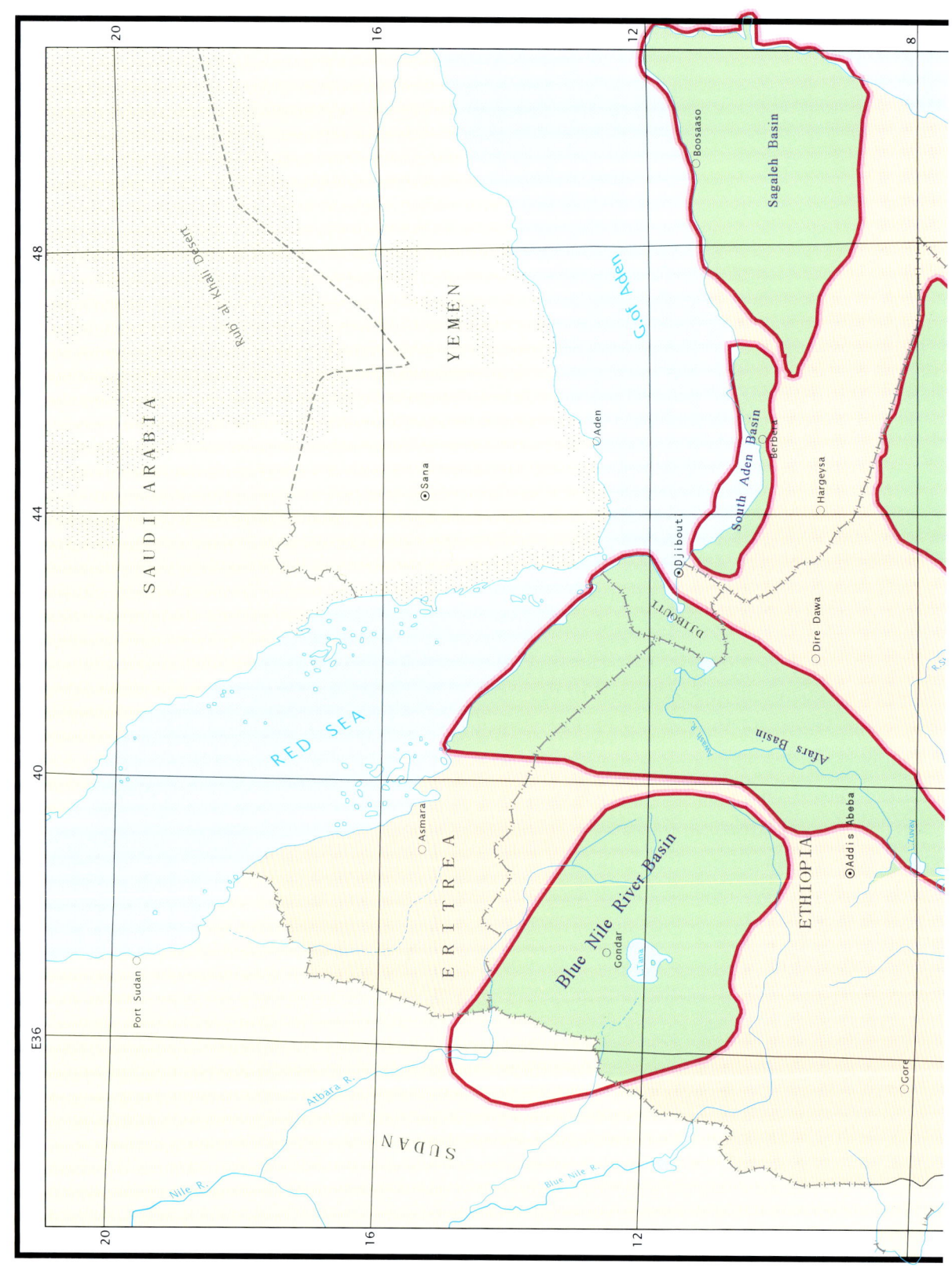

CHAPTER 54

54 Ethiopia, Somalia, Djibouti and Eritrea

Geography

Ethiopia, Somalia, Djibouti and Eritrea are located in East Africa, in the area commonly referred to as the Horn of Africa, bordering the Indian Ocean to the east and the Gulf of Aden and the Red Sea to the north.

History

Initial oil exploration started in Somalia in 1918. In spite of continuous exploration work being carried out over a long period, no commercial deposits of oil and gas have been found. In spite of the presence of many sedimentary basins in the region, none of theses countries presently produce any oil or gas. Given the proximity of the Arabian Pennisula, there are grounds for optimism about future prospectivity.

Regional geology

The basement comprises Precambrian crystalline rocks, and the sedimentary cover comprises Jurassic and Cretaceous strata with a maximum thickness of 10,000 m, including 3000 m of Jurassic carbonates. The first epeirogenic movement to affect the region formed a series of intersecting basins separated by structural highs, and these controlled the pattern of sedimentation throughout. For example, two sedimentary basins, separated by a basement ridge (the Bur-Acaba Uplift of northern Somalia) are present in Somalia. These are the Mesozoic Luug Mandea Basin, which trends NNE–SSW, and the Somali Coastal Basin, trending parallel to the Indian Ocean coast. These contain both Mesozoic and Tertiary sediments. The northern part of the Somali Coastal Basin is overlain by the E–W elongated Somali Embayment structure, which also contains both Mesozoic and Tertiary sediments.

In the Somali Coastal Basin over 2000 m of predominantly greenish shales, occasionally interbedded with limestone bands, and containing Middle Jurassic to Early Cretaceous ammonites were penetrated in the Brava-1 well. This sequence, now called the Brava Formation, is correlatable with the Upper Jurassic shales that crop out in coastal Kenya. The Brava Formation is unconformably overlain by the Gumburo Group (Upper Cretaceous). It contains over 1000 m of light grey shales, sandstones and siltstones, with rare limestone interbeds deposited in an inner middle neritic environment with strong deltaic influence. The Late Palaeocene to Middle Eocene Barren Beds include a maximum thickness of about 3000 m of predominantly shallow marine sandstones partially intercalated with siltstones and shales. Lignitic shales occur in the upper part of this unit, indicating an inner neritic nearshore depositional environment with deltaic influence. The Somali Merca Formation (Miocene to Pliocene) consists of mainly medium- to fine-grained sandstones and white to cream microcrystalline limestones, with thin shale and anhydritic interbeds of littoral to shallow marine environment. These older formations do not crop out because they are covered by recent alluvial and aeolian sediments along the coastal belt. The description of the formations

World Atlas of Oil And Gas Basins, First Edition. Li Guoyu.
© 2011 John Wiley & Sons, Ltd. Published 2011 by John Wiley & Sons, Ltd.

Table 54.1 Major sedimentary basins

Country	Basin	Area (thousand km²)	Major sedimentary rock		Reservoir		Remarks
			Age	Thickness (m)	Age	Lithology	
Ethiopia	Afars	145	Mz Kz	1900			Extends into Jibouti, Eritrea
	Blue Nile	190	R				Extends into Sudan, Eritrea
	Ogaden	130	Mz	2400	T,J,K	Sandstone	
Somalia	East Africa	690	PR	12,000	J,K	Sandstone	Extends into Kenya, Tanzania, Mozambique
	Sagaleh	140	Mz	3000			
	South Aden	26	Mz Kz				

is therefore based on subsurface data. The major sedimentary basins of the four countries are listed in Table 54.1 (Selley, 1997).

Ethiopia Ethiopia has an area of 1,103,600 km² with a population of 77.4 million. Three main sedimentary basins exist in this country, namely the Afars, the Blue Nile and the Ogaden Basins all of which are infilled with Meso-Cenozoic rocks. The geography is represented by plateaus of predominantly volcanic lava. Before the Second World War, two wildcat wells were drilled in the Ogaden Basin. The first well was abandoned in 1950 at a depth of 3100 m while the second well was abandoned in 1954 at a depth of 2750 m. In subsequent periods, 34 wells were to be drilled in this basin. The Cabup condensate field was discovered in 1974. Proven gas reserves are placed at 24.9 billion m³. At present though, Ethiopia is a non-oil-producing country.

Somalia Somalia has an area of 637,657 km² with a population of 8.86 million. The East Africa, Sagaleh and South Aden Basins are to be found within the country. These consist of Meso-Cenozoic rocks. Oil exploration in the country started in 1918 and has been largely unsuccessful. Over 50 wells are known to have been drilled with some oil and gas shows but no commercial prospectivity. The country is classified as non-oil-producing.

Djibouti Djibouti (really an enclave) has an area of 23,200 km² and a population estimated at 712,000. A small portion of the South Aden Basin is to be found in Djibouti. While a few geological surveys have been undertaken, no wildcat or exploratory drilling has been carried out. It is still a non-oil-producing country.

Eritrea Eritrea has an area of 124,320 km² and a population of 4.56 million. The country faces the Red Sea to the northeast. Any prospectivity would perhaps be associated with beds to be found in the Red Sea, but presently it is a non-oil-producing country. The existence of sedimentary basins shows potential for prospection in the future.

KENYA AND UGANDA

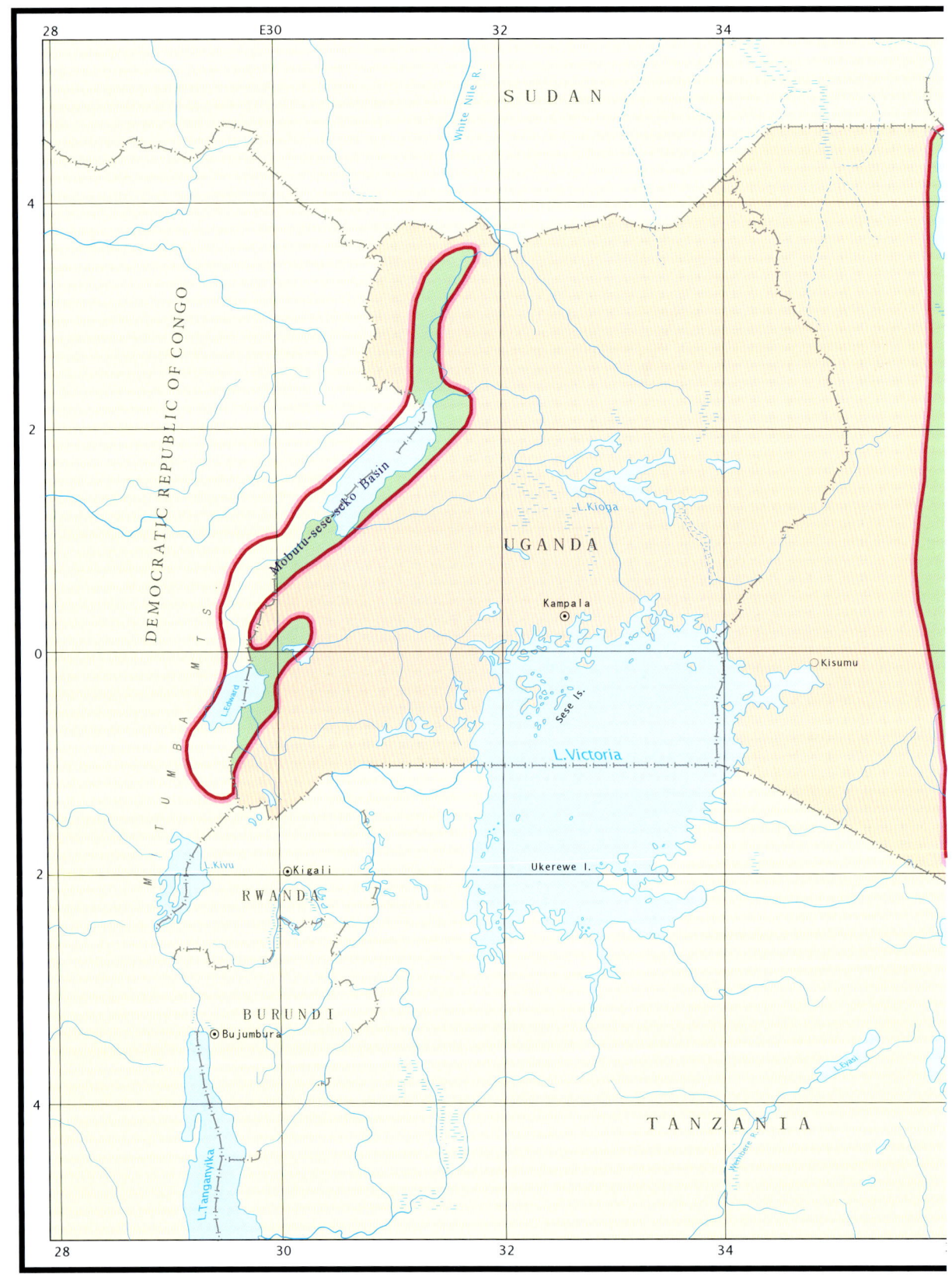

CHAPTER 55

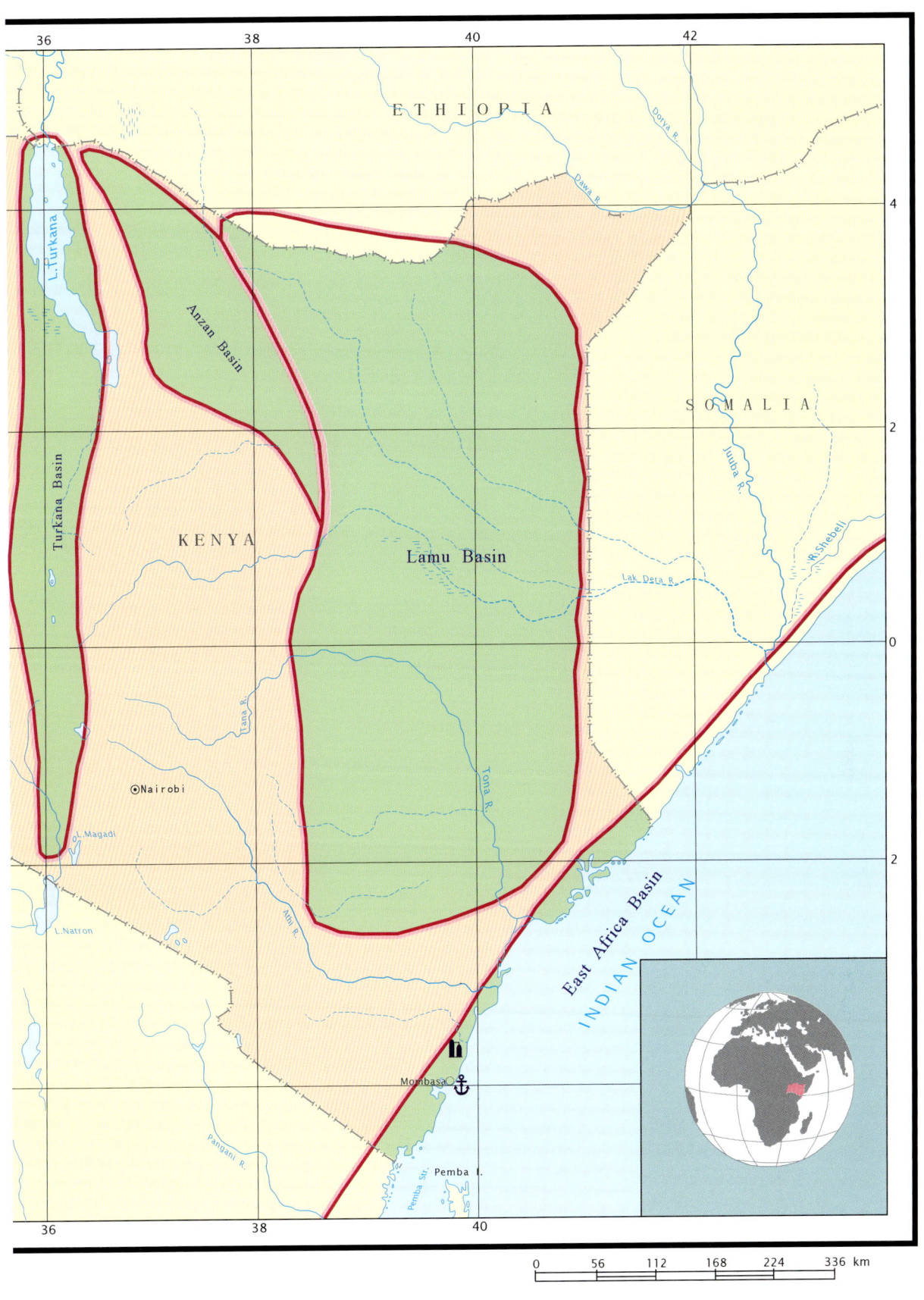

55 Kenya and Uganda

Geography

Kenya and Uganda are located in East Africa, bordering the Indian Ocean to the east and sharing borders with the Democratic Republic of Congo to the west.

History

Oil exploration commenced in 1913. During the long periods of exploration that followed, only limited success was encountered and no meaningful oil and gas fields have been discovered. However, the situation has changed significantly in Uganda where two foreign companies have established the presence of commercially viable reservoirs in the Albertine Basin located in the Mobutu-sese-seko Basin in the uppermost arm of the Great Rift Valley. The long-term prospects of Uganda are considered to be very good. Presently, neither country is oil-producing.

Regional geology

Little exploration has been undertaken in Kenya and Uganda because of the complex geology. The basement comprises Precambrian crystalline rocks and the sedimentary cover comprises Cretaceous and tertiary strata. In the coastal East Africa Basin several Mesozoic structures have been mapped. The Somali Coastal Basin is the northern arm, and the Anza Graben to the northwest failed to open. This is when Madagascar is considered to have migrated southwards.

The oldest and most extensive outcropping sedimentary rocks, locally called the Duruma Series or Duruma Sandstones, are of Karroo age, faulted to the west against the basement, while to the east they either disappear under the cover of post-Karroo rocks, or are faulted against them. Salt diapirs detected by geophysical data offshore are thought to be of this age and equivalent to the Mandawa salt basin of southern Tanzania. To the north, Karroo rocks thin beneath the cover of Neogene non-marine deposits, underlying the Cretaceous and Tertiary sediments in the southern part of Anza Graben. Karroo rocks are overlain by limestones of Middle Jurassic age which marks the beginning of the post-Karroo marine phase. Although partly good source rocks are at least developed locally in the region and reservoir potential is recognized, no commercial oil has been reported so far, despite drilling in most major structures. Reservoir potential is poor in these rocks, but traps are abundant, although sealing is a problem. Petroleum may have migrated into early traps, however, and then remigrated into younger traps during later movements.

Maturation within the post-Karroo sequence is believed to have been slow. Because of a low geothermal gradient in Kenya, Lower Cretaceous source rocks have been found to be overmature, while Mid-Cretaceous source rocks are just at the oil window. Tertiary shales are immature in most wells except in Tanzania, where Tertiary development is extraordinary. Here a burial curve reconstruction has proven that the maturity of source beds within the Pugu–Musanga structural axis is higher than one would expect for normal subsidence to their present depths. This is suggestive of Mid-Miocene inversion which is believed to have resulted in the removal of several thousands of metres of strata. Tertiary beds are suspected to have generated oil that migrated into

World Atlas of Oil And Gas Basins, First Edition. Li Guoyu.
© 2011 John Wiley & Sons, Ltd. Published 2011 by John Wiley & Sons, Ltd.

Table 55.1 Major sedimentary basins

Country	Basin	Area (thousand km²)	Major sedimentary rock		Remarks
			Age	Thickness (m)	
Kenya	Lamu	230	J,R	6000	Extends into Ethiopia, Somali
	Turkana	50	Kz		
	Anzan	60	Kz	6000	
Uganda	Mobutu-sese-seco	32	R	1200	Extends into Congo

traps within the vicinity, but Miocene inversion might have caused significant redistribution and eventual loss of hydrocarbons by erosion or leakage. Hence structural traps in this region are considered to be more risky than stratigraphic traps which need detailed geophysical mapping and are expensive to explore. The major sedimentary basins are listed in Table 55.1.

Kenya Kenya has an area of 582,646 km² and a population of 34.26 million. It is a plateau, mainly consisting of lava and basement. The east branch of the Great African Rift runs through the west and centre of this country. Sedimentary basins are of the rifting type. Based on the experience of the discovery of oil and gas fields in the rifting basins of Sudan and Egypt to the north, there is some optimism regarding future discoveries in this country. Inital oil exploration began in the 1950s. Some shallow wells were drilled in Lamu Basin. After the discovery of the large Helig and Unity fields in Sudan, seven further exploratory wells were drilled in the Anza Basin. Although these did not lead to the discovery of any major finds, prospects are considered modest for the future.

Uganda Uganda has an area of 241,038 km² and a population of 27.21 million. Lake Victoria, the most significant geographical feature of the country, has an area of 69,400 km² and is the second largest freshwater lake in the world. The Mobutu-sese-seco Basin is located on the western edge of this country. Thirty-two rift basins are distributed over East Africa as a whole, and the Mobutu-sese-seco Basin is one of them. This basin is infilled with Tertiary sediments. Oil exploration work was initiated in 1913. As mentioned above, Uganda's oil and gas prospectivity has improved dramatically in the past few years following the successful exploration and evaluation of drilling programmes being carried out in that country.

TANZANIA, RWANDA AND BURUNDI

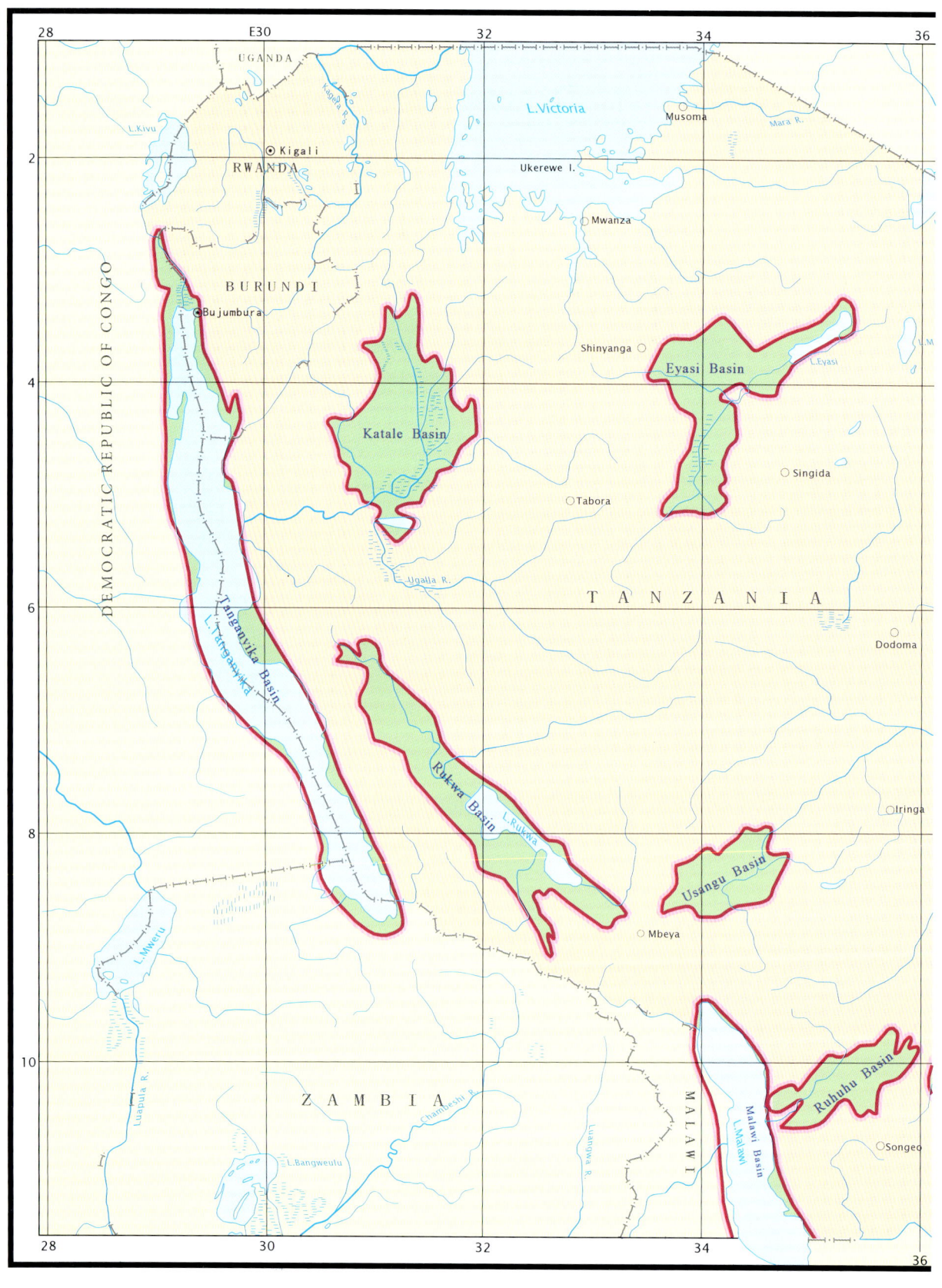

CHAPTER 56

56 Tanzania, Rwanda and Burundi

Geography

Tanzania, Rwanda and Burundi are located in eastern and central Africa.

History

The existence of sedimentary basins in the region is well documented. Oil exploration in the region can best be described as spasmodic. However, there have been some interesting successes in Tanzania which suggest the existence of viable oil and gas systems. Although none of the countries is presently oil-producing, the prospects are good for Tanzania in particular.

Regional geology

Tanzania, Rwanda and Burundi are located in the East African Rift system. The Precambrian crystalline basement is exposed at the surface over extensive areas. The sedimentary cover comprises Jurassic, Cretaceous and Tertiary strata. The sedimentary basins of the East African Rift system vary greatly in structural style and character of their fill. Hence identifying source rocks, reservoir rocks, seals and traps is not straightforward. The main contrast is between the deep and structurally simple basins of the western branch of the rift and the shallower and more complex nature of those in the eastern and Ethiopian sections.

The East African Rift is the key feature, which resulted in massive structural inversions and intensified eastward tilting of the present onshore areas, where they were accompanied by rapid subsidence and deposition. The amount of volcanicity appears to correlate with the complexity of faulting. Areas where the faults are many but small are the foci of volcanic activity, such as the Baringo area. Considerable research has been carried out into the underlying controls on the variations in basin structure and volcanicity in different parts of the rift. The resulting structure looks very similar to the Tanganyika type of basin. In contrast, areas with lower strain rates have a more brittle crust which fails in more places giving many, but small, faults which develop over a wider area.

The main sub-basins distinguished in coastal Tanzania include the Selous–Ruvu–Tanga rift basin, the south of which the Mandawa salt basin forms part, containing a NNW–SSE Karroo fault trend. Both basins are Mesozoic to Tertiary in age and are crossed by a number of N–S structural highs, while E–W is another remarkable fault trend in the basins. The major structural features are also rift-like. Periodic movements along bounding faults led to the cyclic deposition of continental, fluviatile and lacustrine sequences.

At the end of the Early Jurassic, tectonism along bounding faults became less intense and a shallow marine transgression followed during the Middle Jurassic as a result of continued subsidence and tilting. An Early Cretaceous overall regression, associated with a number of tectonically influenced sea-level fluctuations, is notable. A major regional unconformity is recognized at the base of the Aptian and Lower Albian.

Table 56.1 Major sedimentary basins of Tanzania

Basin	Area (thousand km²)	Major sedimentary rock Age	Major sedimentary rock Thickness (m)	Remark
Eyasi	15	R	<1000	
Katale	20	R	<1000	
Ruhuhu	8	K–R	3000	
Rukwa	18	K–R	3000	
Selous	53	K–R	3000	
Tanganyika	40	K–R	3000	Extends into Burundi, Congo, Zambia
Usangu	7	Mz, R		

Tanzania Tanzania has an area of 945,087 km² and a population of 37.38 million. Tanzania is famous for its Kilimanjaro Mountain with an altitude of 5,895 m. Seven sedimentary basins are to be found in the country with individual areas ranging from 7,000 to 53,000 km² (Table 56.1). Also half the East Africa Basin lies within Tanzania, in which the Songo Songo gas field was discovered. Initial exploration commenced in 1952 but in the quarter of a century thereafter exploration has been sporadic. Forty shallow wells and four wildcat wells had been drilled by 1956, along with one exploration well. The Songo Songo discovery offshore has been the major find so far and this gas field was developed in 2008. There have also been some further gas finds onshore in Upper Cretaceous strata. The active Wingayongo oil seep is further evidence of good prospectivity. At present Tanzania is classified as a non-oil-producing country.

Rwanda Rwanda has an area of 26,000 km² and a population of 7.4 million. It is landlocked and characterized by exceptionally hospitable spring-like weather all year round. The only significant hydrocarbon has been the occurrence of methane (22 per cent), dissolved in the waters of Lake Kivu at a depth of about 300 m. Most of the gas (77 per cent) is carbon dioxide. Other than this, there are no reports of any oil or gas exploration in this small country.

Burundi Burundi occupies an area equal to 27,834 km², of which 25,650 km² is land. Population is 7.4 million. Geologically, most of the country is Precambrian basement and no oil and gas exploration has been reported.

MOZAMBIQUE, COMOROS, MADAGASCAR, SEYCHELLES,

MAURITIUS AND REUNION — CHAPTER 57

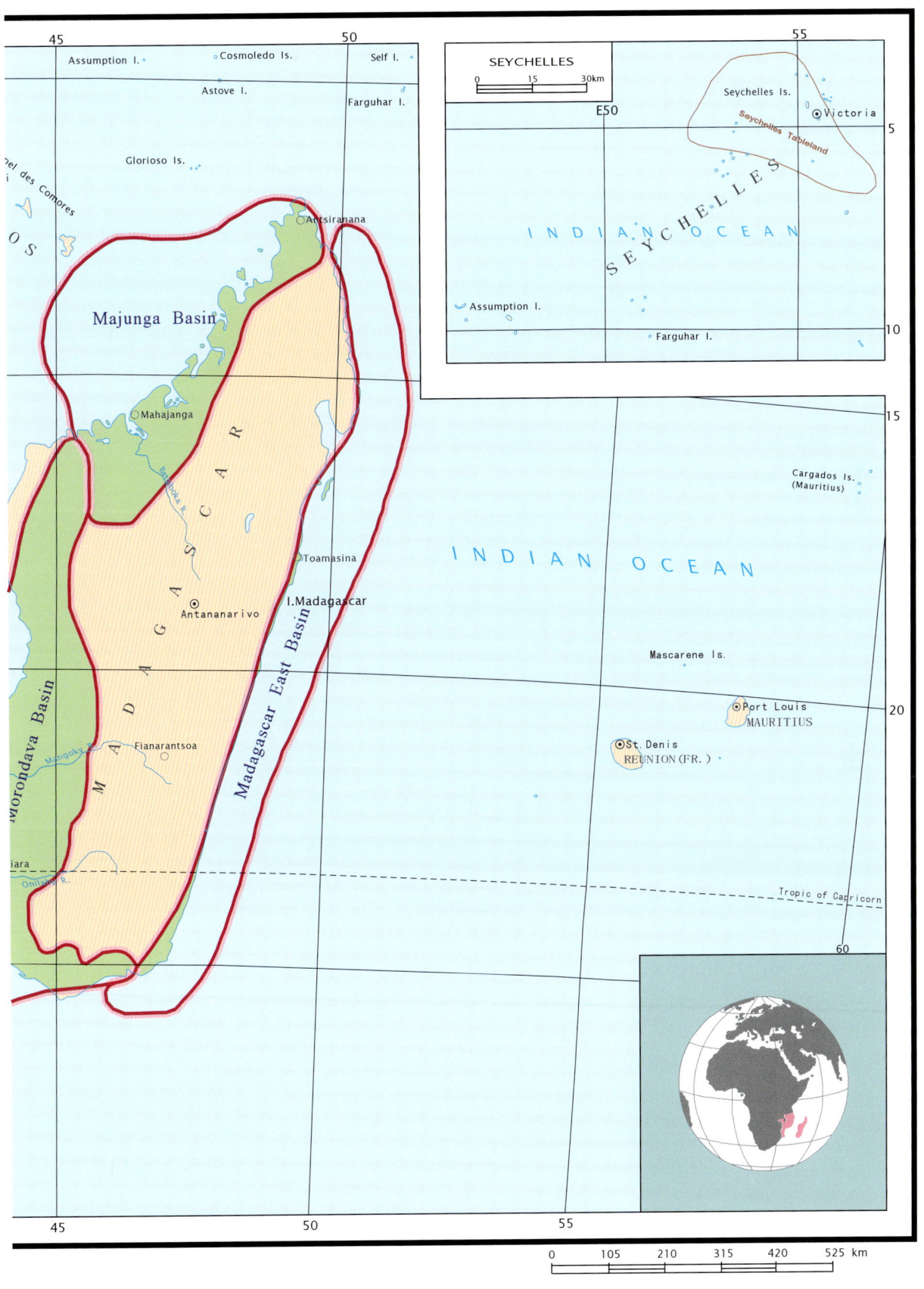

57 Mozambique, Comoros, Madagascar, Seychelles, Mauritius and Reunion

Geography

Mozambique, Comoros, Madagascar, Seychelles, Mauritius and Reunion are located in the eastern part of southern Africa. With the exception of Mozambique, these are islands in the Indian Ocean.

History

Four large basins occur in this region and for this reason this area was considered to have prospection potential a long time ago. Oil exploration started as far back as 1901 in Madagascar. More recently, there has been significant exploration work resulting in interesting gas shows in Mozambique.

Regional geology

The eastern margin of Africa includes Madagascar, Mozambique, Somalia, Kenya and Tanzania. The development of the eastern margin of Africa began with the formation of an intracontinental rift during the early stages of fragmentation of eastern Gondwanaland. Marine influence increased throughout the Jurassic to the establishment of full marine conditions in the region during the Middle Jurassic.

Mozambique Mozambique is located on the east coast of southern Africa, bordered by the Mozambique Channel to the east, and sharing borders with Tanzania to the north and Zambia, Malawi, Zimbabwe and South Africa and Swaziland to the west and south. It has an area of 801,600 km^2 and a population of 2030 million. By the end of 2008 proven reserves of gas were 127.4 billion m^3. At present this country is non-oil-producing but with the presence of the huge Mozambique Basin (1.2 million km^2; Table 57.1), it is considered to be highly prospective. The Mozambique Basin is filled by Ceno-Mesozoic sediments with local thickness up to 10,000 m. Reservoirs are Cretaceous sandstones. Traps are associated with faulted anticlines.

Comoros The volcanic islands of Comoros have an area of 2326 km^2 with a population of 780,000. No oil exploration has taken place up till now and any future prospection is likely to be linked to the ocean basins around the islands.

Madagascar Madagascar is a large island located in the Indian Ocean off the eastern coast of southern Africa. This country has an area of 590,710 km^2 with a population of 18.6 million. The central plateau consists of Precambrian crystalline basement, around which occur three large sedimentary basins, known as Madagascar East, Majunga and Morondava. Oil and gas accumulations are multilayered and mainly confined to Mesozoic deposits. The petroleum potential of these relatively large basins is estimated as high. These three sedimentary basins were earlier recognized as prospective basins, but exploration has proved to be a very long drawn-out process.

World Atlas of Oil And Gas Basins, First Edition. Li Guoyu.
© 2011 John Wiley & Sons, Ltd. Published 2011 by John Wiley & Sons, Ltd.

Mozambique, Comoros, Madagascar, Seychelles, Mauritius and Reunion

Table 57.1 Major sedimentary basins of Mozambique and Madagascar

Country	Basin	Area (thousand km²)	Major sedimentary rock		Reservoir		Remarks
			Age	Thickness (m)	Age	Lithology	
Mozambique	Mozambique	1210	Mz–Kz	10,000	K	Sandstone	Extends into South Africa
Madagascar	Madagascar East	176	KQ				
	Majunga	198	Mz–E_2				
	Morondava	165	Mz–E_2		K, R	Arkose	

Oil exploration is known to have started in 1901 with more than 100 wells drilled near the many seepages of tar sand and heavy oil. No significant oil or gas fields have been discovered.

Seychelles The Seychelles are a collection of islands in the Indian Ocean. They comprise 115 granite and coral islands with a combined area of just 455 km² and a population of 85,000. Oil exploration started in the 1960s. Three wells were finished and abandoned. There is a Northeast Seychelles deepwater ocean basin with an area of 600,000 km². Sedimentary cover comprises Ceno-Mesozoic rocks. Perhaps this basin has prospects for oil exploration in future.

Mauritius Mauritius consists of many volcanic islands located in the Indian Ocean, with a collective area of 2040 km² and a population of 1.25 million. The presence of an arc-shaped ocean sedimentary basin on the south side of the island suggests future oil and gas interest.

Reunion reunion comprises volcanic islands with a combined area of 2512 km² and a population of 700,000. There is no meaningful interest presently in oil and gas prospection.

NAMIBIA, ZIMBABWE, BOTSWANA, SOUTH AFRICA,

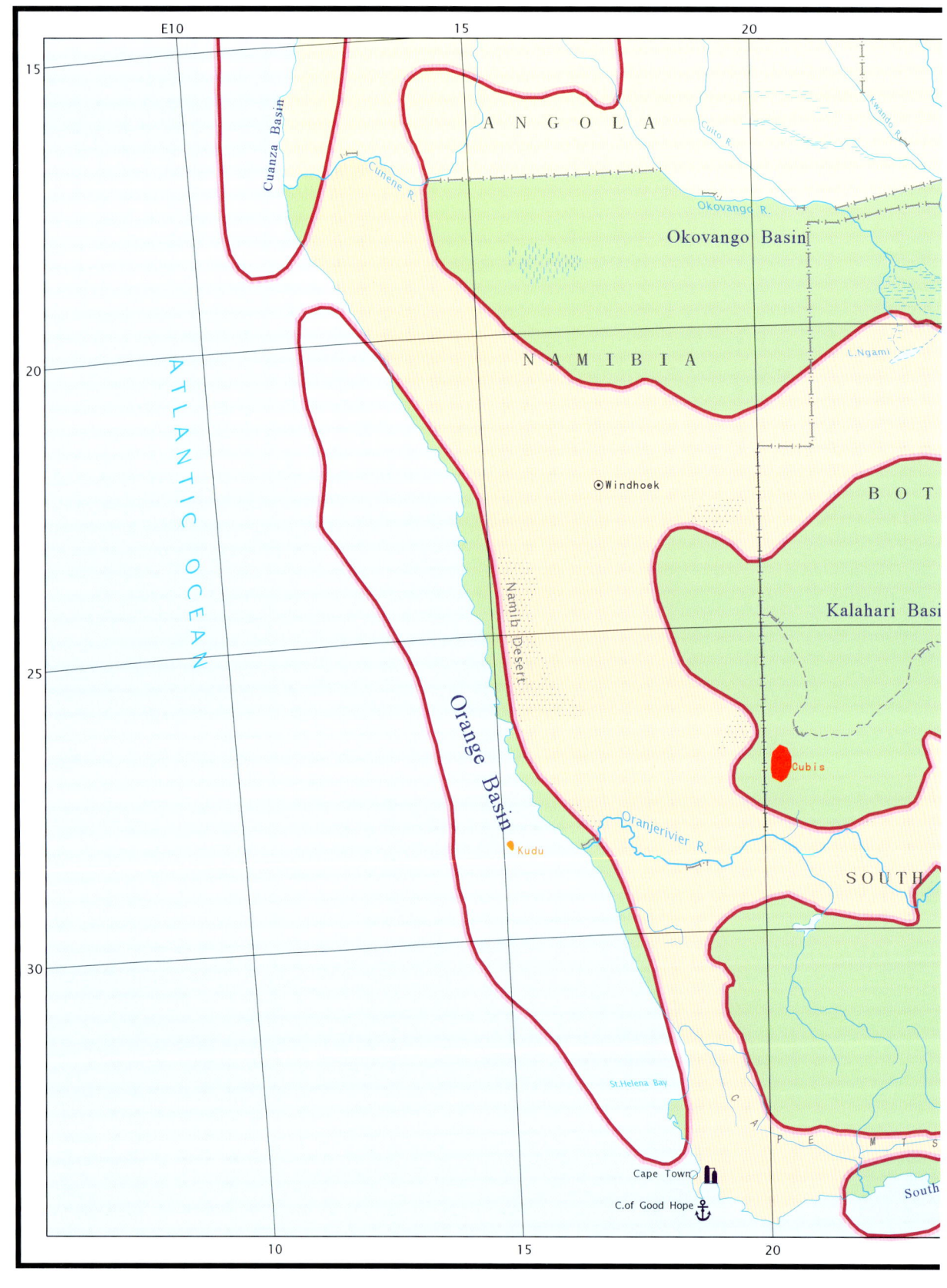

SWAZILAND AND LESOTHO   CHAPTER 58

58 Namibia, Zimbabwe, Botswana, South Africa, Swaziland and Lesotho

Geography

Namibia, Zimbabwe, Botswana, South Africa, Swaziland and Lesotho are all located in southern Africa.

History

This region was earlier recognized as a prospective area. Oil exploration started in the 1940s. Much of the motivation was South Africa's ambitions to become a regional industrial powerhouse, and also because during the apartheid era there were many economic sanctions and trade embargoes applied to that country. In spite of intensive exploration work on some of the large basins in the region, no commercial finds were recorded. We consider this to be abnormal and we believe that the region still has prospects for future successful exploration.

Regional geology

Seven significant sedimentary basins are to be found in this region (Table 58.1). These basins are all filled by Palaeozoic and Mesozoic sediments.

Namibia Namibia has an area of 824,269 km^2 with a population 2.03 million. About 50 per cent of the area is covered by sedimentary basins while the other 50 per cent consists of Precambrian crystalline basement. Significantly, it is just to the south of the major oil producing nation of Angola. Between 1959 and 1989 there was sporadic oil and gas exploration. The large Kudu gas field was originally discovered offshore in the Atlantic waters of the west coast of the country. Investment by some well known international companies has made this a very commercial resource with plans to use the gas to power a major electric power plant. Offshore drilling continues in this country but no integrated petroleum industry has been established.

Zimbabwe Zimbabwe has an area of 390,380 km^2 with a population of 13.1 million. About 80 per cent of the area is covered by Precambrian crystalline and metamorphosed rocks. Parts of some sedimentary basins lie in the country. The Zambesi Basin is filled with Permian, Triassic and Cretaceous sediments. It is a non-oil-producing country.

Botswana Botswana occupies an area of 581,730 km^2 with a population of 1.64 million. About 80 per cent of the Kalahari Basin is located in this country, with an area of 440,000 km^2 and a relatively modest thickness of Cretaceous and Tertiary strata.

South Africa South Africa occupies almost all the southern tip of the African continent, with an area of 1,219,090 km^2 and a population of 47.4 million. Various types of sedimentary basins are located in the country. The Karoo Basin is a Palaeozoic basin with an area of about 630,000 km^2. The Orange Basin is a rifting basin, infilled by Cretaceous sediments. The South Cape basin is rifting type too. Oil exploration started in the 1940s, as mentioned earlier. Significant exploration work was carried out. 261 exploration wells and 63 appraisal wells were completed and out of this

Table 58.1 Major sedimentary basins of southern Africa

Country	Basin	Area (thousand km²)	Major sedimentary rock Age	Major sedimentary rock Thickness (m)	Reservoir Age	Reservoir Lithology	Remarks
Namibia	Orange	360	Mz–R	700	K	Sandstone	Extends into South Africa
Zimbabwe	Sabi	4	C–J				
Zimbabwe	Tuli	20	C–J				Extends into South Africa, Botswana
Zimbabwe	Zambesi	30	K–R	1800			Extends into Zambia, Mozambique
Botswana	Kalahari	440	Pz, K–R	2000	Pz	Sandstone	Extends into Namibia, South Africa
South Africa	Karroo	630	Pz		P	Sandstone	Extends into Lesotho
South Africa	South Cape	50	Mz, Kz	8000	K	Sandstone	

program 20 gas fields and 9 oil fields were discovered. However, no truly large oil or gas fields have been discovered, which is an anomaly given the presence of large basins. However, in spite of the limited discovery of commercial hydrocarbon deposits, the country is still considered to be a highly prospective area.

Swaziland Swaziland has an area of 17,363 km² with a population of 1.13 million. It is located to the east of the Karoo Basin and is partly within the Mozambique Basin. Surface cover consist of Precambrian crystalline basement is known to crop out at the surface. In the east of the country some Palaeozoic cover sediments have been found. The oil prospects are presently unknown.

Lesotho This country has an area of 30,344 km² with a population of 2.2 million and is located in the centre of the fossil-rich Karoo Basin. Most of the area of the country is covered by volcanic lava. One exploration well was completed in 1974 to a depth of 1652 m with no oil or gas shows.

EUROPEAN OIL AND GAS BASINS

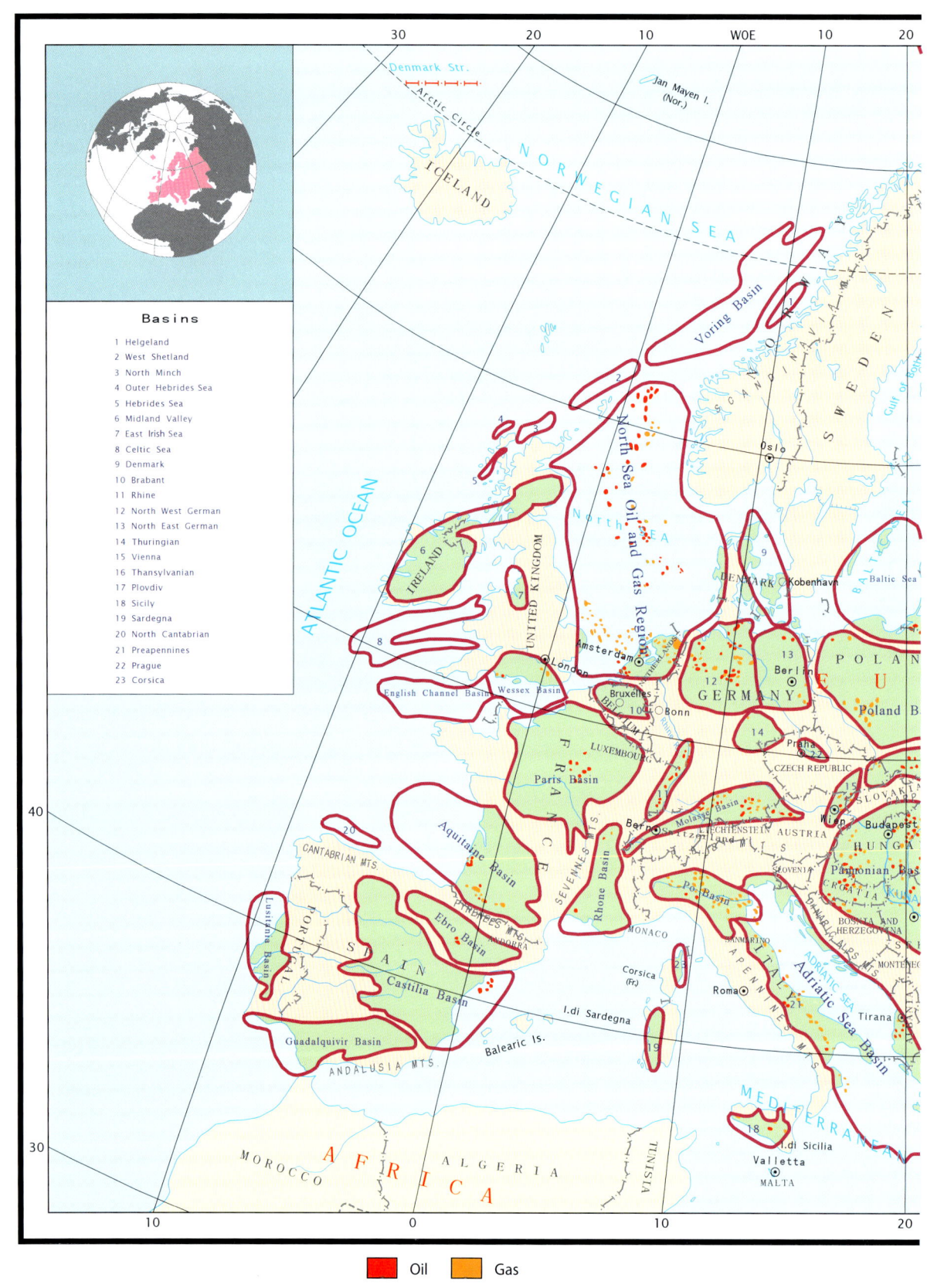

PART IV

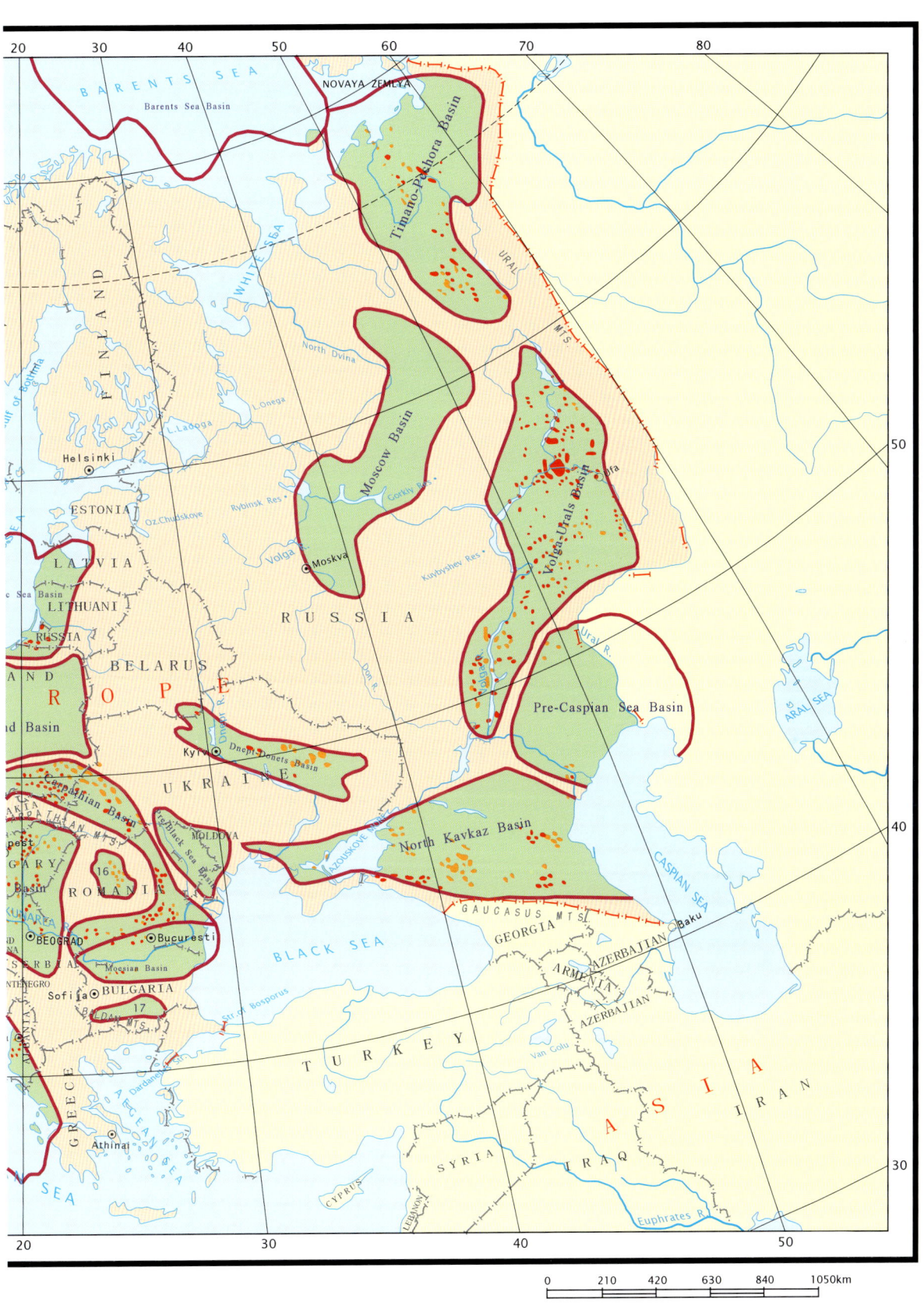

Part IV

European Oil and Gas Basins

Geography

The continental Europe is located on the western part of the Euro-Asian continental land mass, facing mostly the Atlantic Ocean to the west and the Adriatic and Mediterranean Seas in the south. Its northern borders are dominated by the Arctic Ocean and various seas such as the North Sea and the Baltic Sea, with a long history in the development of petroleum industry. The European Region covers the territory from the Atlantic Ocean in the west to the Urals in the east. In the north, it includes some parts of the Arctic Ocean, and in the southwest some parts of the Mediterranean. The southeastern boundary of the region runs through Caucasus. The area of the European Region within the above limits amounts to about 10.5 million km^2 (offshore areas excluded). Europe comprises 45 countries.

History

Several sedimentary basins are located in the continent, ranging in area from $5000\,km^2$ to $575,000\,km^2$ (North Sea Basin). The ancient oil-producing regions of Kavkaz and the Carpathians were in Europe and there are documented records of the famous Kavkaz gas fire in ancient times. In 1857, oil was produced in Romania and this is considered to be the first official record of oil production in the world. The Volga-Urals and North Sea Basins are the two largest oil-producing areas in Europe. Peak oil production of the Volga-Urals Basin reached 226 million t in 1975. Peak oil production of the North Sea Basin reached 286 million t in 2000. By the end of 2008 proven oil reserves were 11 billion t with annual production of 707 million t. Proven gas reserves totalled 53 trillion m^3 with annual gas production of 926 billion m^3 (Table IV.1).

Regional geology

The European Region, together with offshore parts of the adjacent oceans and seas, forms part of the Eurasian Plate. This region is renowned for its long history of geological study, contributing much to the global geological nomenclature. The northern, western and part of the southern (Black Sea) peripheries of the plate are characterized by the development of oceanic crust, but most of the plate is classified as continental crust. The northern and western edges of the continent represent a passive continental margin that is fringed mainly by thalassocratons of the Atlantic and Arctic Oceans. The southern margin of the Eurasian Plate is in contact with the African Plate and also, partially, the Arabian Plate, both of which are being subducted beneath the Eurasian Plate.

The European Region is formed of diverse structures. In the area of continental crust development, the following two main types of tectonic elements have been distinguished: platforms that differ in the age of their basement, and mobile orogenic belts representing different ages of collision of present-day and ancient lithospheric plates. The orogenic belts consist of nappe-thrusted structures, intermontane depressions and foredeeps.

World Atlas of Oil And Gas Basins, First Edition. Li Guoyu.
© 2011 John Wiley & Sons, Ltd. Published 2011 by John Wiley & Sons, Ltd.

Table IV.1 Proven reserves and production of oil and gas in major countries of Europe (2008)

Country	Proven reserves		Production	
	Oil (million t)	Gas (billion m³)	Oil (million t)	Gas (billion m³)
Total Europe	11,035	53,468	707	926.8
Denmark	145	61	14	9.3
Germany	37	175	3	15.4
Italy	55	94	4.9	8.5
Netherlands	13	1415	1.7	85.7
Norway	915	2312	106	99
Turkey	41	8.4	2	0.2
UK	467	342	70	73
Belgium	27	0.2	1.7	–
Romania	82	63	4	6
Russia	8219	47,572	488	643.8
Ukraine	54.1	1104	5	—

The platforms are subdivided into ancient (Precambrian) and young. Amongst the ancient platforms are the East European, Barents–Kara and the hypothetical Eria platforms, as well as epi-Baikalian ones such as Timano–Pechorsky, Mid-European, Moesian and Apulian. Young platforms include the epi-Hercynian West and South European and, partly, the Central Eurasian (Scythian) and epi-Caledonian platforms of the British Isles.

The Mid-European Orogenic Belt was formed during the Hercynian orogeny. Many sedimentary basins have formed within it. This belt stretches from Iberia to Dobrudzha and passes into the Uralo-Okhotsk Orogenic Belt. It serves as a basement of the West European platform. Within the outcropping orogenic belt, intermontane depressions of Upper Silesian type developed and were filled with Upper Carboniferous deposits. A marginal trough composed of Upper Carboniferous rocks can be traced along the outcropping orogenic belt. The Ural Orogenic Belt was also formed during the Hercynian orogeny. The Mediterranean Orogenic Belt stretches from Gibraltar to the Caucasus of the European Region.

Deep-water basins are developed in the abyssal areas of the Atlantic Ocean which may become a prospective region in the future (Vysotsky *et al.*, 1995).

ROMANIA, SERBIA, MONTENEGRO, SLOVENIA, CROATIA, BOSNIA

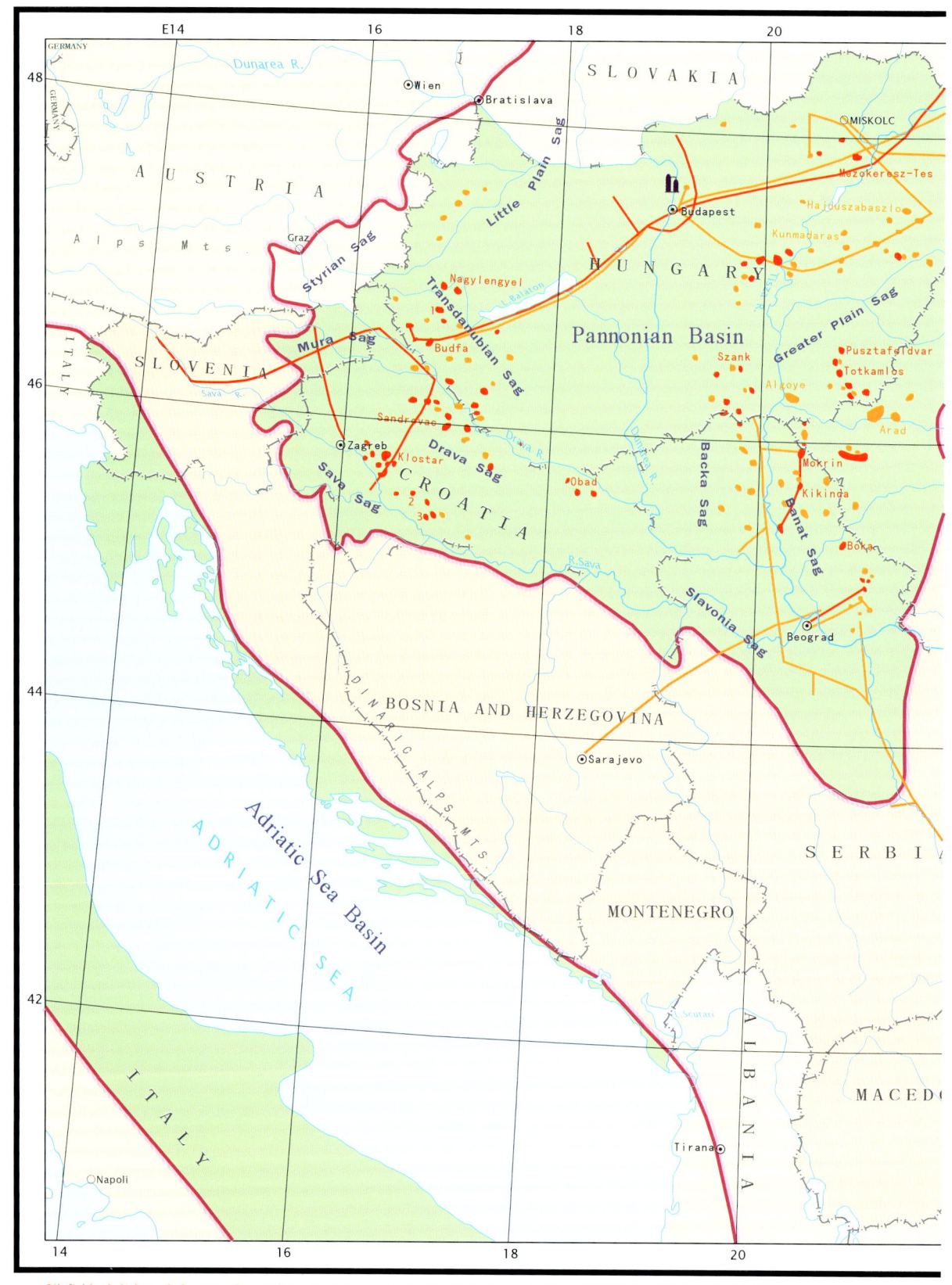

Oil field: 1 Nahot 2 Struzec 3 Lipovljani 4 Urlarti 5 Boldesti 6 Baicoi 7 Aricesti 8 Bustenari 9 Moreni 10 Aninoasa

AND HERZEGOVINA, HUNGARY, BULGARIA AND MACEDONIA   CHAPTER 59

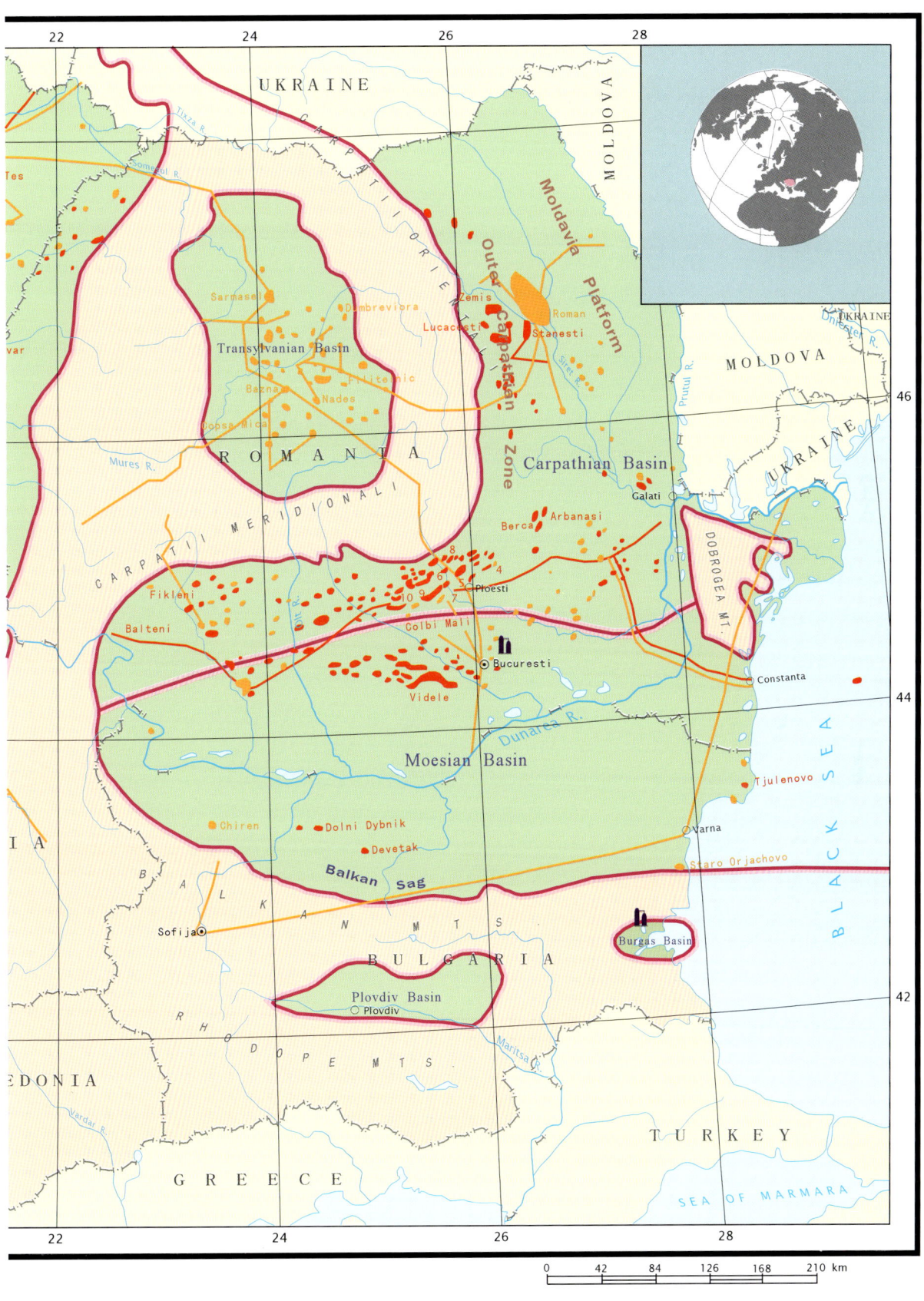

59 Romania, Serbia, Montenegro, Slovenia, Croatia, Bosnia and Herzegovina, Hungary, Bulgaria and Macedonia

Geography

Romania, Serbia, Montenegro, Slovenia, Croatia, Bosnia and Herzegovina, Hungary, Bulgaria and Macedonia are located in southern Europe and all lie to the south of the Alps mountain range, on or around the Apennine peninsula and the Balkan peninsula.

History

Among the nine countries, Romania is the largest oil-producing country with the longest oil-production history, since 1857. Other countries such as Croatia, Hungary, Serbia and Bulgaria are smaller oil- and gas-producing countries. Europe is characterized by complex geological conditions and therefore relatively small production (excluding the Volgo-Urals and the North Sea Basins).

Regional geology

There are seven sedimentary basins distributed over this region. The Carpathian Basin has an area of 770,000 km^2; the Moesian Basin has an area of 120,000 km^2; the Transylvanian Basin has an area of 28,000 km^2; the Adriatic Sea Basin has an area of 180,000 km^2; Panonian Basin has an area of 210,000 km^2; Burgas Basin has an area of 2,000 km^2 and Plovdiv Basin has an area of 5,000 km^2. These basins are infilled by Devonian, Triassic, Cretaceous and Tertiary sandstones and carbonates with thicknesses ranging between 3000 m and 12,000 m. The Moinesti oil field, located in the Carpathian Basin in Romania, is an anticline with Tertiary oil-bearing sandstone lying at a subsurface depth of 1050 m, with a thickness of 102 m, porosity of 15 per cent and permeability of 31 mD, and has maintained production through successful enhanced recovery.

Romania Romania is located in the northern part of the Balkan peninsula with an area of 237,500 km^2 and a population of 21.62 million. The Carpathian mountain arc lies in the centre of this country. Romania is the oldest oil-producing country in the world, a production accomplishment made possible because of the influences of five sedimentary basins, including the Transylvanian, Carpathian and Moesian Basins. In 1900, oil production stood at 219,000 t annually, peaked at 15.16 million t in 1976 and then steadily declined to 4.6 million in 2008.

Serbia Serbia is located in the northwest part of the Balkan peninsula and has an area of 88,300 km^2 with a population of 9.9 million. It is associated with both the Panonian and Adriatic Sea Basins. It has a long history of oil exploration, since the late 1850s, while gas was initially produced in 1918. A total of 13 oil fields and five gas fields have been discovered. By the end of 2008, the proven oil reserves were 10.6 million t with an annual oil production of 740,000 t. At the same time, proven gas reserves stood at 30.5 billion m^3.

World Atlas of Oil And Gas Basins, First Edition. Li Guoyu.
© 2011 John Wiley & Sons, Ltd. Published 2011 by John Wiley & Sons, Ltd.

Romania, Serbia, Montenegro, Slovenia, Croatia, Bosnia and Herzegovina, Hungary, Bulgaria and Macedonia

Montenegro Montenegro is located in the Balkan peninsula, occupying an area of 13,800 km^2, with a population of 620,000. There is no record of any petroleum exploration having taken place there.

Slovenia Slovenia is located on the northern coast of the Adriatic Sea, bordering Italy to the west. It has an area of 56,538 km^2 and a population of 4.4 million. A part of the Alps mountains occupies much of the country. Again, there is no record of petroleum exploration and its potential in oil and gas production is unknown.

Croatia Croatia is located on the northeastern shore of the Adriatic Sea and has an area of 56,538 km^2 with a population of 4.4 million. It is occupied by the southwestern part of the Panonian Basin. Oil exploration work was initiated in the 1920s, with a relatively large oil field, the Benicaanci, discovered in 1969. A total of 35 oil fields and 17 gas fields have been discovered in Croatia. By the end of 2008 proven oil reserves were 10.84 million t with an annual oil production of 740,000 t. At the same time, proven gas reserves were 30 billion m^3.

Bosnia and Herzegovina Bosnia and Herzegovina occupy the northwestern part of the Balkans and lie partially in the Alps mountains. It has an area of 51,129 km^2 and a population of 3.84 million. The petroleum potential is unknown.

Hungary Hungary is a landlocked country in central Europe, bordered by Austria to the west. It has an area of 93,030 km^2, with a population of 100.6 million. The Pannonian Basin occupies most of the area of Hungary, where the covered rocks are Triassic, Cretaceous and Tertiary sediments with thicknesses reaching 5000 m. Oil exploration started in 1942 and the first oil field was discovered in 1963. The large gas field Hajdusnobosnlo is located in the central part of Hungary and has an area 40 km^2, with proven gas reserves of 25 billion m^3. The Pliocene sandstone and oolitic limestone reservoirs are of modest thickness up to 50 m, the porosity is 25 per cent and the permeability 800 mD. During 1937–1991, approximately 110,000 km of seismic surveys were conducted and several exploration and development wells were drilled. By 1991, a total of 126 oil and gas fields had been discovered with oil reserves in place estimated at 264 million t and gas reserves at 314.8 billion m^3. All the oil and gas fields are small to medium sized. By the end of 2008 proven oil reserves stood at 2.7 million t with an annual oil production of 720,000 t. It is estimated that gas reserves are 8 billion m^3.

Bulgaria Bulgaria is located in southern Europe, bordering Romania to the north, with an area of 110,993 km^2 and a population of 7.74 million. The development of the economy of the country is heavily dependent on its consumption of imported oil, most of which is from Russia. The Government keeps control of the rate of economic growth so as to mitigate the effects of high international oil prices. The Provdiv Basin and the southern part of Moesian Basin are located in Bulgaria. Both basins consist of Ordovician to Tertiary sediments. Tertiary sediments in Plovdiv Basin have thicknesses reaching 2500 m. Oil exploration commenced in 1925, while the first gas field, Blizank, was discovered in 1949. By 1991, a total of 25 oil and gas fields had been discovered in the Moesian Basin. The large Dolnidabonik oil field was discovered in 1962, with depths between 3300 m and 3500 m. Reservoirs are Permian carbonate. Proven oil reserves were 9.4 million t at that time. By the end of 2008, the proven oil reserves stood at 2.05 million t with an annual oil production of 50,000 t, and proven gas reserves of 5.6 billion m^3.

Macedonia The Balkan country of Macedonia has an area of 25,713 km^2, with a population of 2.05 million. The terrain is mostly mountainous with an average altitude of 2000–2500 m. The potential of oil and gas reserves is unknown as only little exploration work has been carried out.

CARPATHIAN BASIN

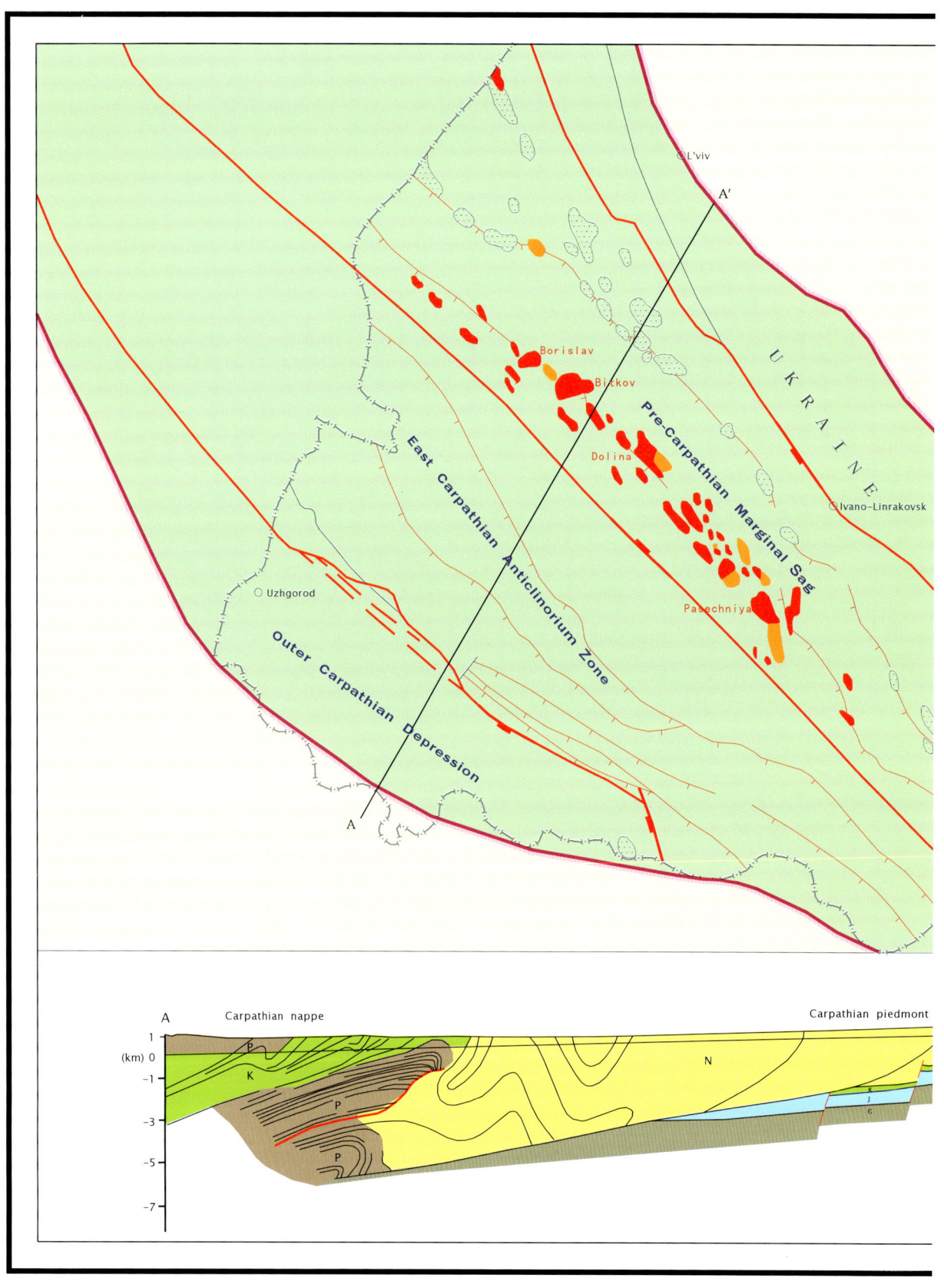

CHAPTER 60

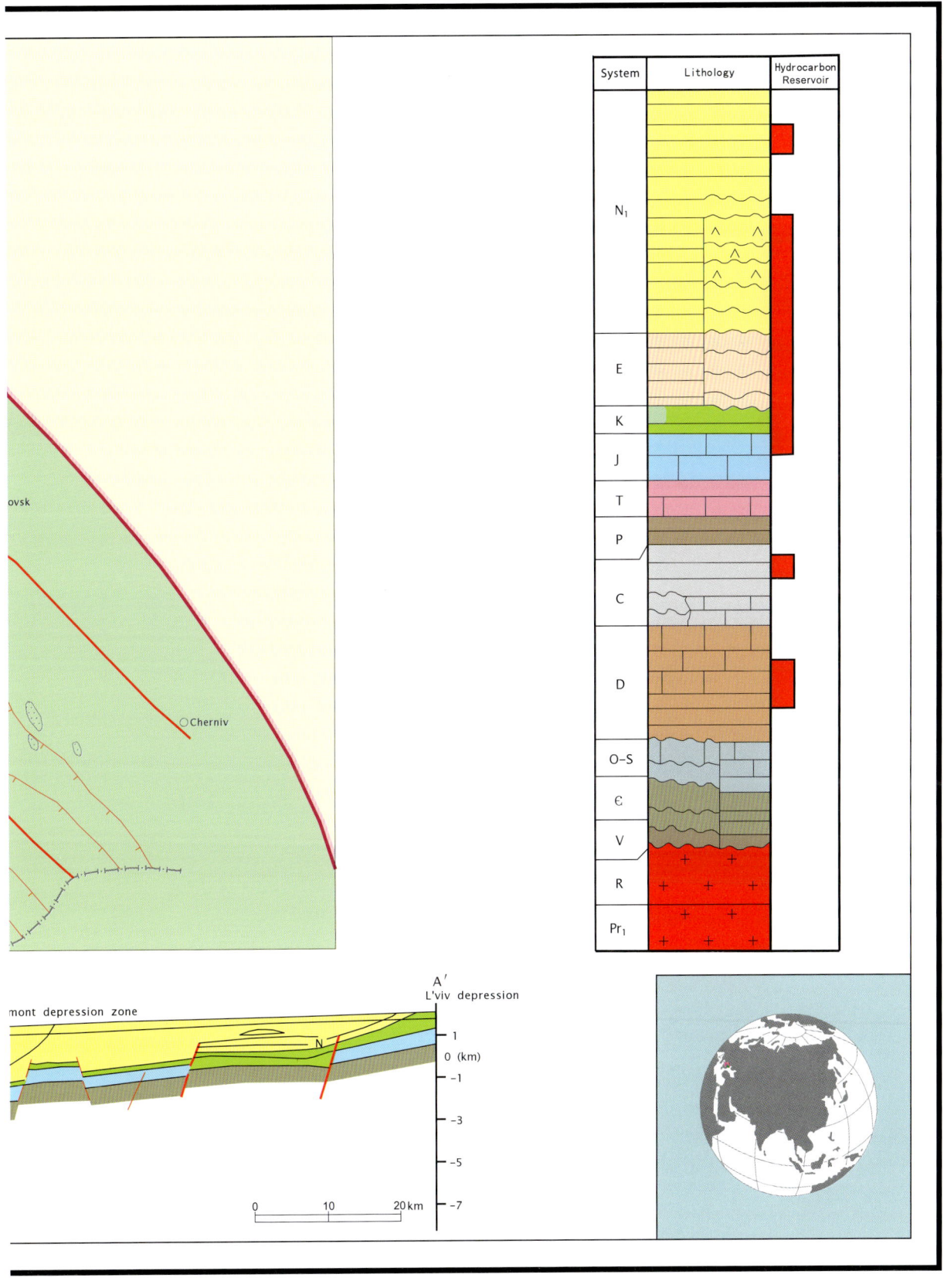

60 Carpathian Basin

Geography

The Carpathian Basin is located in the western part of the Ukraine and has an area of 88,000 km².

History

Oil exploration started in 1860 at the two shallow oil-bearing sites of Borislav and Bitkov (Figs 60.1 and 60.2). The Borislav oil field is an anticlinal trap at a depth of 265–875 m in a sandstone reservoir 18–80 m thick, porosity of 8–10 per cent, permeability of 0.1–180 mD and maximum annual production of 3000 t per well. The Dashav oil field was discovered later and began development in 1924 to reach a maximum annual production of 2.75 million t, which subsequently declined until increasing again to 2.7 million t in 1970.

Regional geology

The Carpathian Basin is an important oil-producing basin in Europe, which is located on the Moesian and Moldavian platforms. The basement of the basin occurs at depths of between 2 km

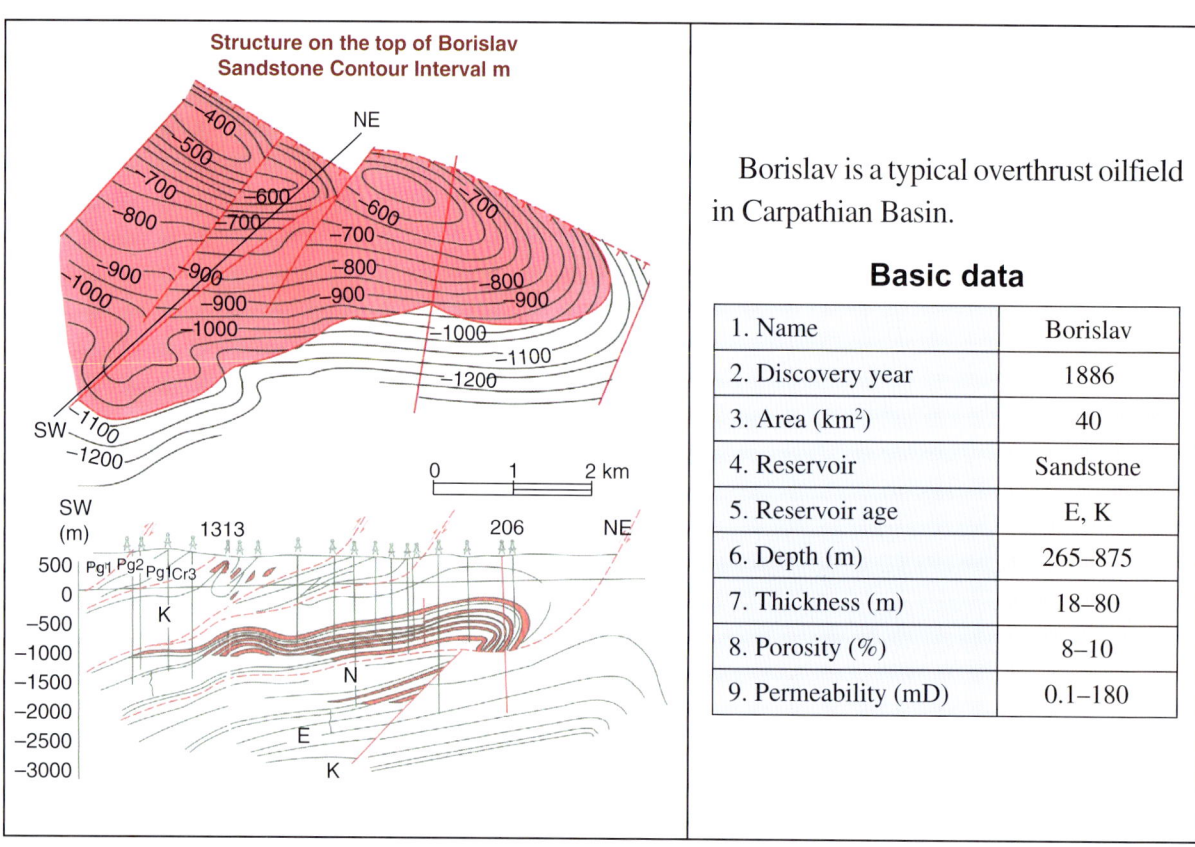

Fig. 60.1 Details of the Borislav oil field.

World Atlas of Oil And Gas Basins, First Edition. Li Guoyu.
© 2011 John Wiley & Sons, Ltd. Published 2011 by John Wiley & Sons, Ltd.

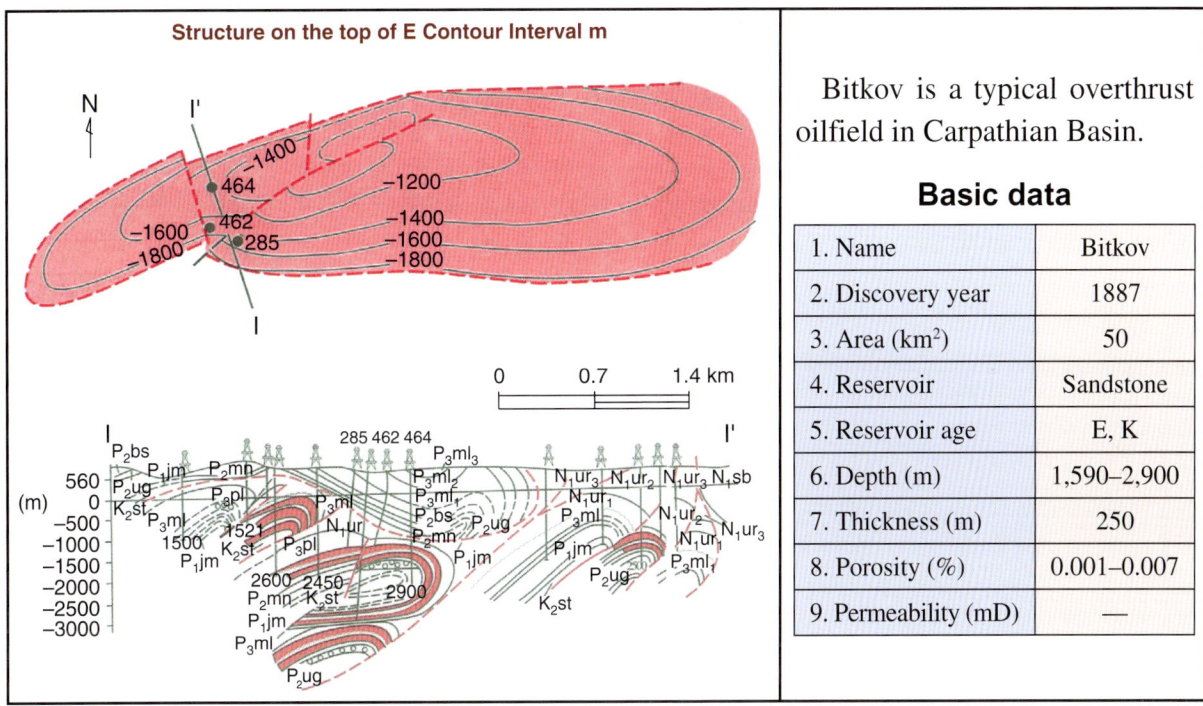

Fig. 60.2 Details of the Bitkov oil field.

and 9 km. Within the limits of the foredeep, the basin's basement is found at depths of 12 km or more. The sedimentary cover of the basin comprises of Palaeozoic, Mesozoic and Cenozoic deposits, including Paleogene and Mesozoic flysch, and a Neogene orogenic complex of molasses and salt-bearing sequences.

Productive layers are Neogene sandstones and Paleogene flysch in the folded flank of the foredeep,. The Pliocene deposits are essentially gas-generating, while those of Miocene and Oligocene produce oil. Gas and oil fields are confined to strongly faulted folds. In the platform portion of the basin, the productive beds are Palaeozoic, Mesozoic and Cenozoic deposits. The latter are mostly gas-generating, whereas the rocks of Mesozoic and Palaeozoic are essentially oil producing.

Oil and gas fields are associated with uplifts of platform type. Geological data indicates that there is a large trend of complex anticlines in the overthrust belt of the inner Carpathian Basin. Giant oil and gas fields are found in faulted anticlinal traps of this trend. There are also numerous non-structural traps. On the platform, oil- and gas-generating rocks are represented by Triassic limestones and Jurassic and Cretaceous clayey rocks. Out of the 236 oil and gas fields discovered in the basin, oil fields prevail (166). Most of the fields are concentrated in the trough and northern portion of the Moesian platform.

Main elements of the basin structures are foredeeps such as the Alpine, Carpathian and Caucasian foredeeps, which usually comprise a folded and a platform side. In some cases the basin may include parts of a nappe-folded area (as, for instance, a portion of the Outer Carpathians in the north of the Pre-Carpathian Basin, or a folded area lying in front of the Pre-Carpathian–Balkanian Basins). Platform areas occupy the most important place in the basin's structure extending in individual cases far beyond the limits of marginal toughs (Meosian and Scythian platforms).

POLAND, CZECH REPUBLIC AND SLOVAKIA

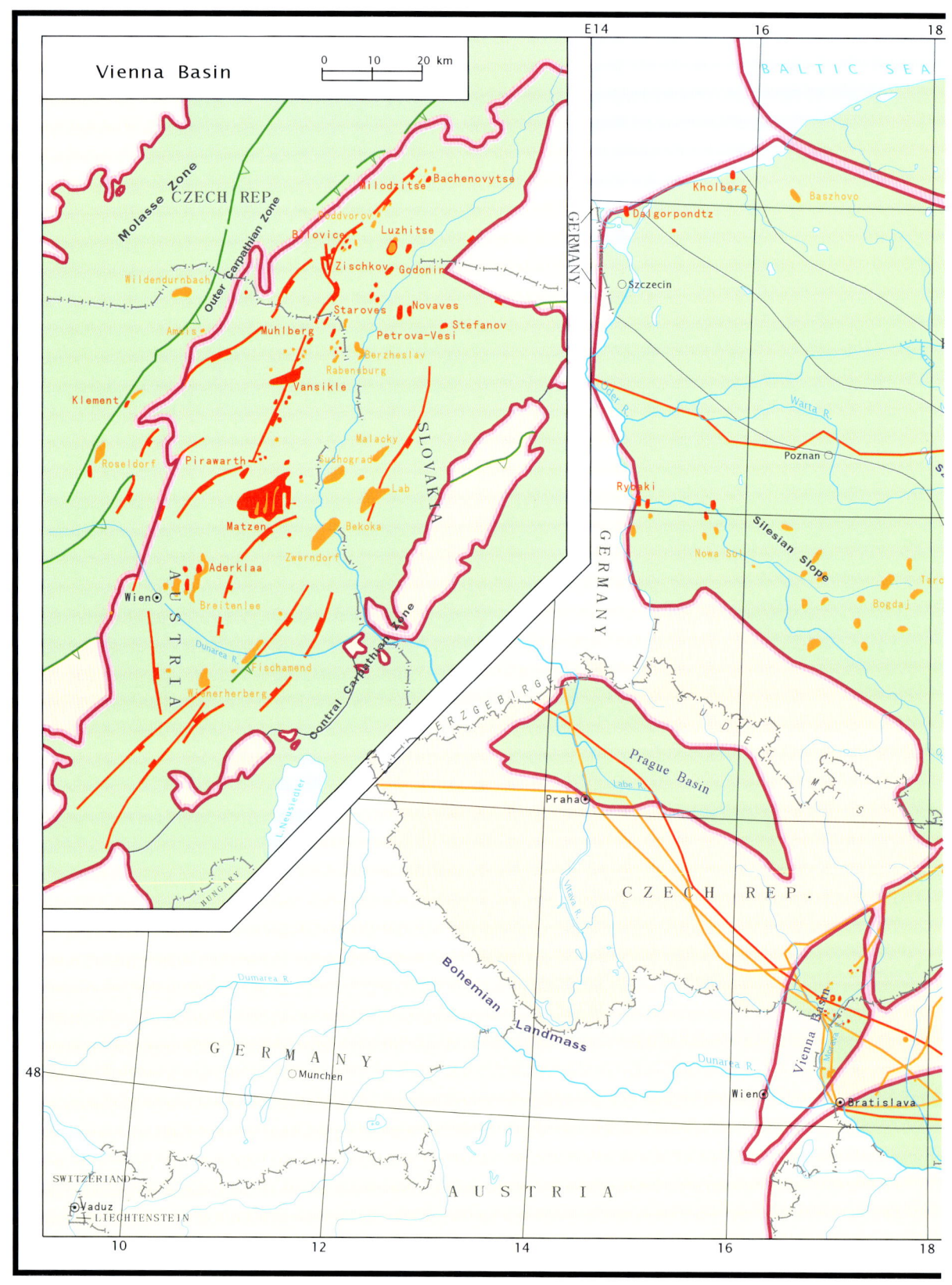

Oil field: 1 Palavovice 2 Grobla 3 Strachocina

CHAPTER 61

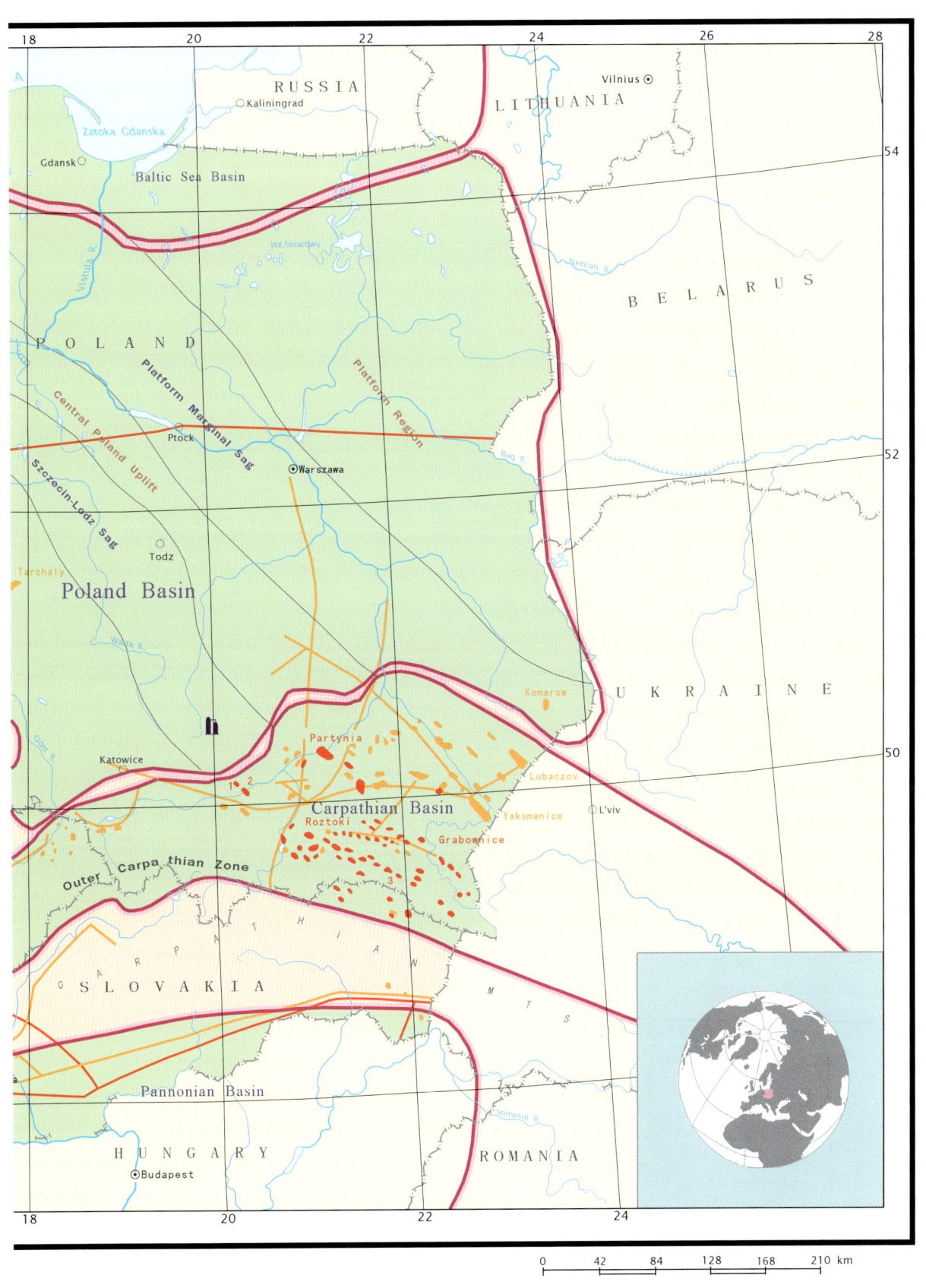

61 Poland, Czech Republic and Slovakia

Geography
Poland, Czech Republic and Slovakia are located in central Europe. Poland borders the Baltic Sea to the north while both the Czech Republic and Slovakia are landlocked.

History
Oil exploration began over a century ago in this region. Six sedimentary basins are found in the three countries: the Baltic, Poland, Carpathian, Pannonian, Vienna and Prague. The vast majority of the oil and gas fields discovered are small to medium sized. The potential of oil reserves and production are limited. The key data of these countries are provided in Table 61.1.

Regional geology
The Belarus platform lies in eastern Poland, consisting of Precambrian basement. Almost all of Poland is occupied by the Middle Europe platform. The Sudeten and Carpathian Mountains lie in the south. Sedimentary comprises Palaeozoic, Mesozoic and Cenozoic strata. The traps are anticline and non-anticline in structure. The three sedimentary basins (Table 61.2) to be found in this region (and their approximate sizes) are the Baltic Sea Basin (40,000 km^2), the Poland Basin (230,000 km^2) and the Prague Basin (14,000 km^2).

Poland Poland is located in central Europe, is bordered by the Baltic Sea to the north and has an area of 312,685 km^2 with a population of 38.16 million. It is the largest oil producer of the three countries. Poland has two sedimentary basins, Carpathian and Poland. The Bobrka oil field was discovered as far back as 1858. Oil was discovered in the Outer Carpathian Zone of the Carpathian Basin in 1854, with a productive well in a Cretaceous sandstone reservoir. Maximum depth of the exploration well reached 7541 m. A further six oil fields and 36 gas fields were discovered in this zone. In the Carpathian Basin itself, six oil fields and 12 gas fields were also discovered. Reservoirs are Triassic, Jurassic and Cretaceous sandstones and carbonates, with porosity of 5–20 per cent and permeability 2–100 mD. A further 21 oil fields and 45 gas fields

Table 61.1 Key data for Poland, Czech Republic, and Slovakia (by the end of 2008)

Index	Country		
	Poland	Czech	Slovakia
Area (km^2)	312,685	78,866	49,035
Population (thousand)	38,160	10,230	5380
Proven oil reserves (million t)	13	2	1
Oil production (million t)	0.8	0.2	—
Proven gas reserves (billion m^3)	164	3.9	14.1

World Atlas of Oil And Gas Basins, First Edition. Li Guoyu.
© 2011 John Wiley & Sons, Ltd. Published 2011 by John Wiley & Sons, Ltd.

Table 61.2 Major sedimentary basins

Country	Basin	Area (thousand km²)	Major sedimentary rock Age	Major sedimentary rock Thickness (m)	Reservoir Age	Reservoir Lithology	Remarks
Poland	Baltic Sea	40	Pz–Kz	1500–4000			Extends into Lithonia, Russia, Latvia, Sweden
	Poland	230	Pz-Mz	10,000–12,000	P, Mz	Carbonate rock, sandstone	
Czech Republic	Prague	14	Mz				

were discovered. By the end of 2008 proven oil reserves were 13.2 million t with an annual oil production of 850,000 t. At the same time, proven gas reserves stood at 164.8 billion m³.

Czech Republic The Czech Republic is situated in central Europe. It is landlocked and borders Germany to the northwest and west, Poland to the northeast, Austria to the south and Slovakia to the east. The Prague Basin lies within this country occupying an area of 14,000 km². Basement consists of Precambrian crystalline rocks. Structurally, most of the basins are anticlines. The sedimentary cover comprises all stages of the Phanerozoic and Proterozoic deposits of marine origin. Mesozoic and Cenozoic strata with thicknesses up to 6,500 m are also present. Source rocks have been identified in the Palaeozoic carbonate. Reservoir rocks are associated mostly with Middle Devonian and Lower Carboniferous sandy deposits. Caprock is shale in non-anticlinal structural traps. By the end of 2008 proven oil reserves were 2.05 million t with an annual oil production of 225,000 t, and proven gas reserves were 3.9 billion m³.

Slovakia Slovakia is located in central Europe with much of its northern border being the Carpathian Mountains. It has a population of 5.38 million. Slovakia is located on only a small part of the Vienna Basin, Outer Carpathian Basin and Panonian Basin.,. Oil exploration started in the Vienna Basin during 1915–1916. From that time until 1948, three small oil fields were discovered with a cumulative production of 550,000 t. A few small oil fields have been discovered in Outer Carpathian Zone of the Carpathian Basin. By the end of 2008 proven oil reserves were 1.23 million t with an annual oil production of only 5000 t, while proven gas reserves were 14.1 billion m³.

GERMANY, LUXEMBOURG, SWITZERLAND AND

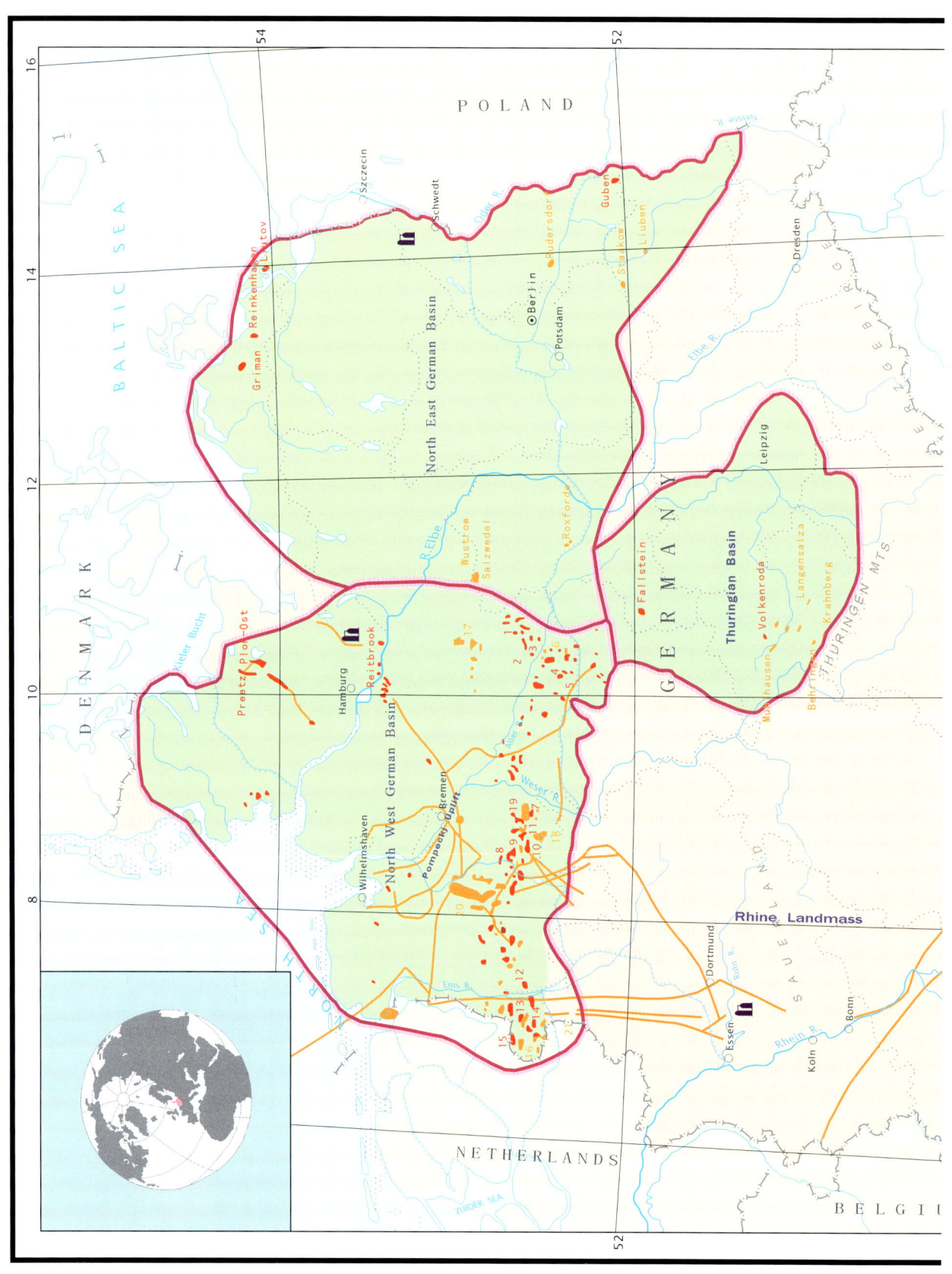

270

LIECHTENSTEIN

CHAPTER 62

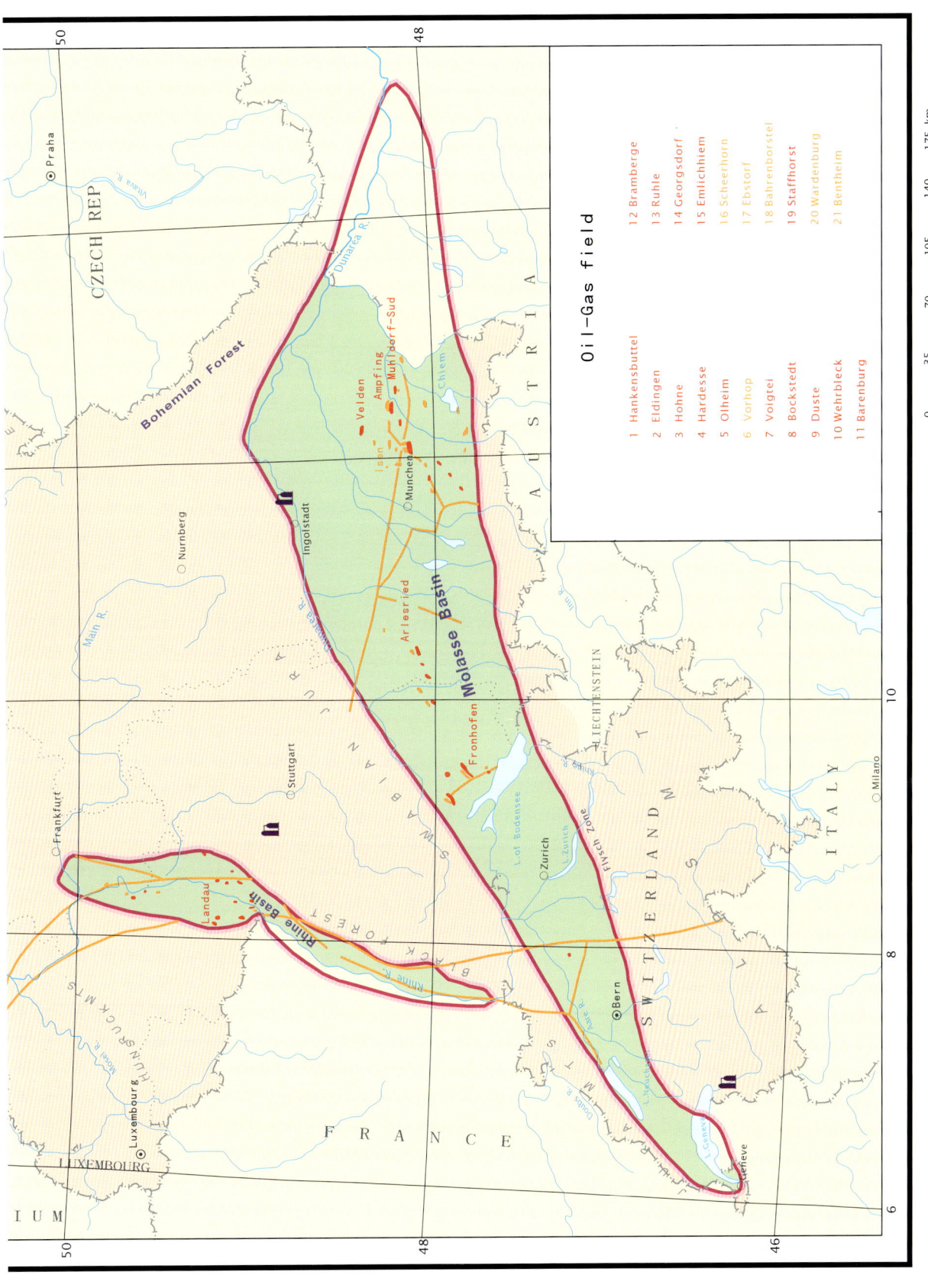

62 Germany, Luxembourg, Switzerland and Liechtenstein

Geography

Germany, Luxembourg, Switzerland and Liechtenstein are located in western and central Europe, with Germany bordering both the Baltic Sea and the North Sea in the north. Much of their southern borders are occupied by the Alps Mountains in the south. Among these four countries, Germany is known to have had oil production in the past.

History

In Germany, oil seepages are known to have been observed as far back as 1545, while in Switzerland, oil sands were identified in 1735. For long periods at various times in the past, Germany and Switzerland devoted resources to oil exploration but only in Germany did a petroleum industry develop. Peak oil production reached at 9 million t in 1967. By the end of 2008, remaining reserves were estimated at a mere 3.3 million t.

Regional geology

Six sedimentary basins are found in this region, Molasse, North East German, North West German, Thuringian, Rhine (Table 62.1) and North Sea (a small portion only). Structurally, the Middle European platform lies in the north while the Russian platform and Baltic Shield are located in the east. Sedimentary cover consists of Palaeozoic, Mesozoic and Cenozoic strata. Reservoirs are Carboniferous and Permian carbonates and sandstones as well as Jurassic, Cretaceous and Tertiary sandstones. Traps are small sized.

Germany Germany is located in the heart of Europe. With a long history of oil exploration; early production in Germany is estimated to have commenced in 1545. Due to the complex

Table 62.1 Major sedimentary basins

Basin	Area (thousand km²)	Major sedimentary rock		Reservoir		Remarks
		Age	Thickness (m)	Age	Lithology	
Molasses	58	Pz–R	6000	J, K, Ez	Sandstone	Extends into Switzerland and Austria
North East German	62	C–R	7000	P	Sandstone	
North West German	56	C–R	C,T,K		Sandstone, Carbonate	Extends into Netherlands
Thuringian	26	C–Mz	2000	P, T	Sandstone, dolomite	
Rhine	10	P–R	3000	T,J,E_3	Sandstone	Extends into France

World Atlas of Oil And Gas Basins, First Edition. Li Guoyu.
© 2011 John Wiley & Sons, Ltd. Published 2011 by John Wiley & Sons, Ltd.

geological conditions and the development of several small salt domes, the country's discovered oil and gas fields are small-sized with limited reserves and production. Peak oil production reached 9 million t in 1967. By the end of 2008, the proven oil reserves were 37.8 million t with annual oil production of 3 million t. Proven gas reserves stood at 175 billion m^3 with annual gas production of 17 billion m^3.

In Germany, the development of oil resources began with oil seepages, and so-called 'tar pools' in the Hanover region were exploited initially. Geophysical prospecting later showed these to be marginal to salt domes. Reservoirs were developed in Jurassic and Lower Cretaceous rocks, in the Hanover region and elsewhere. Subsequently, similar deposits were found at Reitbrook near Hamburg. This confirmed the proposition that oil could also occur in economic quantities over non-piercing salt domes, in this case emplaced in Upper Cretaceous and Tertiary strata. The later stages of exploration have relied on seismic reflection survey, and more than 80 oil fields have been found. In 1930, a major incursion of oil into potash mines in Thuringia, derived from the underlying Zechstein dolomite, drew attention to Permian strata as an oil source, and oil production was established from that level, as well as from Cretaceous strata overlying Liassic source rock shales at Heide, in Holstein. Subsequently, oil discoveries in the Zechstein have been few and far between. In the Rhine Graben, asphalt and oil seepages provided the initial incentive for locating the Pechelbronn oil field complex in France. In northern Germany the mining of oil-impregnated Wealden Sands also took place at this time. The discovery of the Groningen gas field in the Rotliegendes sands from the German frontier has acted as a major incentive for deep gas exploration in Permian strata, and numerous salt structures and fault traps have been drilled. Some 90 gas fields have been found.

Luxembourg Landlocked Luxembourg is located in western Europe and is separated from the North Sea by Belgium. It shares borders with France and Germany. It has an area of 2586 km^2, with a population of 460,000. The essential geological feature is Palaeozoic basement overlain by Mesozoic rocks with thicknesses up to 750 m. According to data obtained from exploratory drilling, the Permian strata comprise sandstones and dolomites, the Triassic strata comprise sandstones and the Jurassic strata comprise shales, limestones and evaporites. Although many faults have developed and have been identified, no oil and gas discoveries have been made.

Switzerland Mountainous Switzerland is located in central and southern Europe and has an area of 41,284 km^2 with a population of 7.5 million. About 60 per cent of the area is covered by southern part of the Alps. The Molasse Basin has an area of 58,000 km^2, of which about 15,000 km^2 is located in Switzerland. This basin was infilled by Jurassic, Cretaceous and Tertiary sediments. Oil exploration in this country was initiated in 1735 by digging for surface Cretaceous tar sands. A total of 21 wells have been drilled in modern times but with no oil or gas discoveries.

Liechtenstein Liechtenstein is a landlocked country located in the Alps, in central Europe, and occupies an area of a mere 160 km^2. The only significant geology is that carbonate sediments exist there but the country itself does not appear to have any petroleum potential.

AUSTRIA, ITALY, ALBANIA, GREECE, SAN MARINO

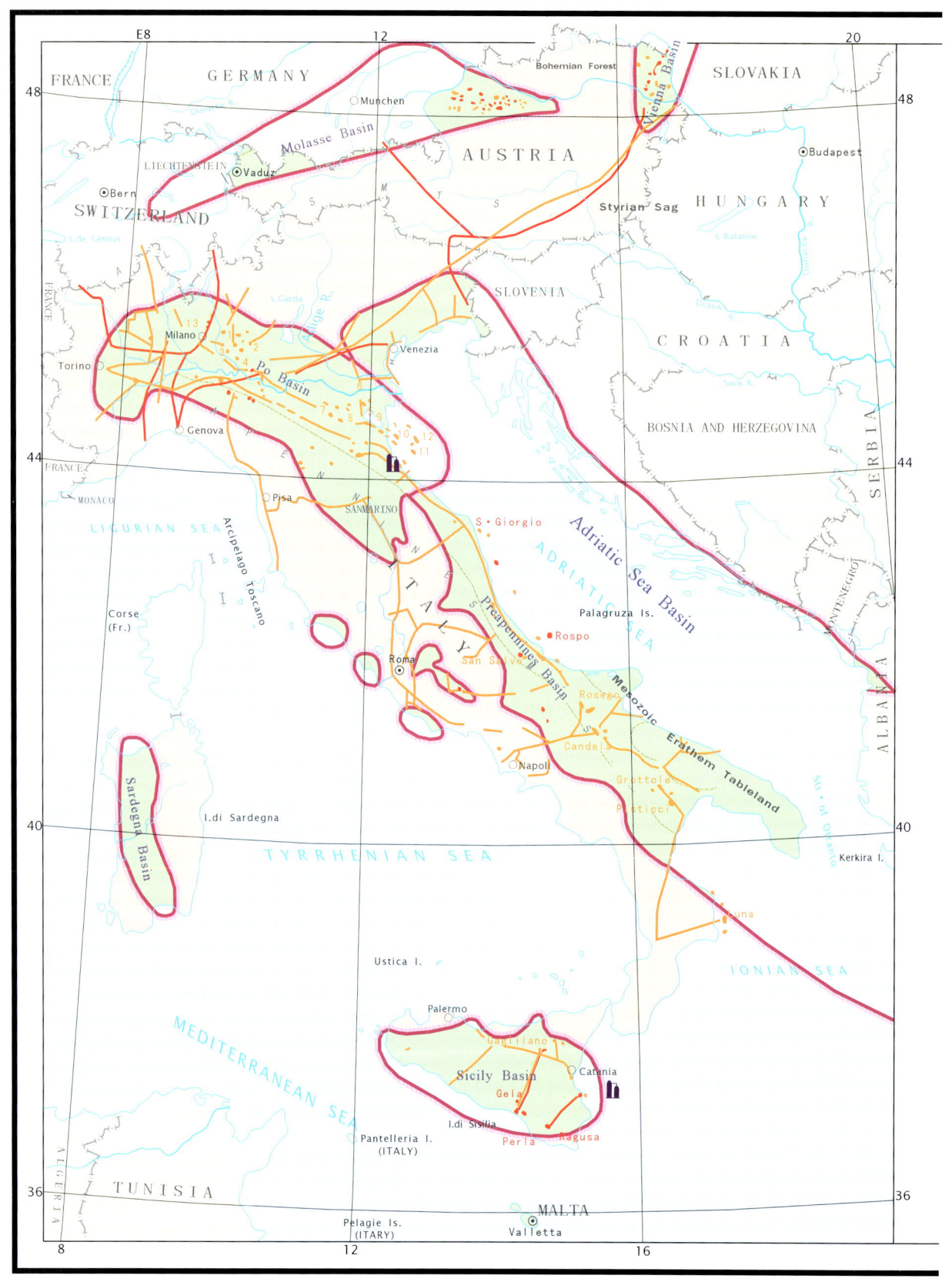

Gas field: 1 Brugherio 2 Ripalta 3 Caviaga 4 Cortemaggiore 5 Alfonsine 6 Sabbioncello 7 Minerbio 8 Salva 9 Ravenna 10 Ravenna marine

AND MALTA   CHAPTER 63

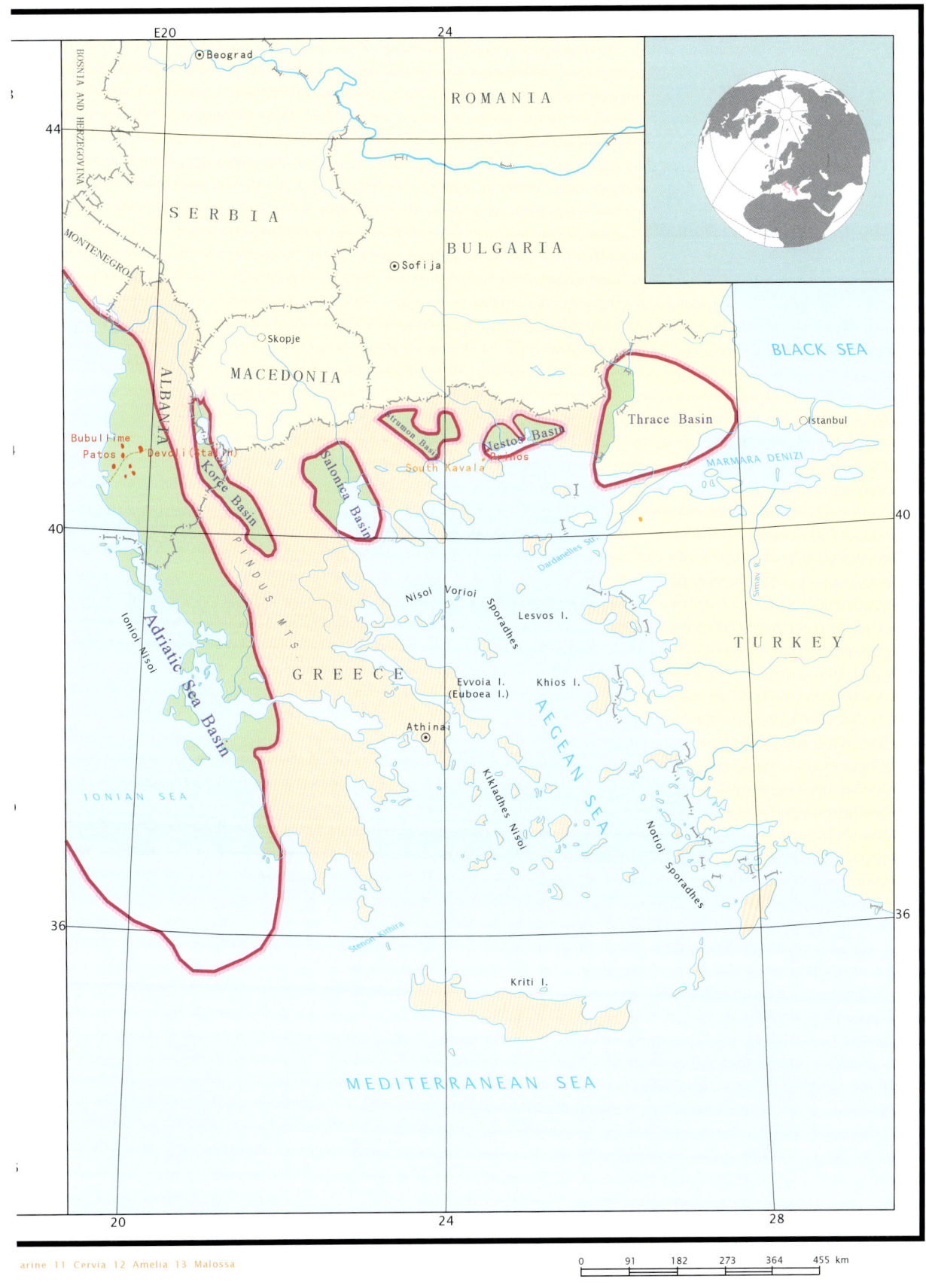

63 Austria, Italy, Albania, Greece, San Marino and Malta

Geography

Austria, Italy, Albania, Greece, San Marino and Malta are located in south and central Europe, and are variously bordered by the Aegean Sea, the Adriatic Sea and the Mediterranean Sea. The Alps and the Apennines mountains lie in this area.

History

Of these countries, Italy is known to have been an oil producing country in ancient times. During Roman times, oil seepages were observed in the Apennine Mountains. Austria, Albania, Italy and Greece are all currently oil- and/or gas-producing countries.

Regional geology

Southern and central Europe are formed by diverse structures. Two main types of tectonic elements are identifiable, i.e. basement platform and orogenic belts. In the young Alpine Orogenic Belt, the basins are infilled with Mesozoic and Cenozoic sediments. Several, mostly small, sedimentary basins are found in this region (Table 63.1).

Austria Austria is located in central Europe and has an area of 83,871 km^2, with a population of 8.2 million. Parts of both the Molasse and Vienna Basins are located in Austria. The Vienna Basin lies on a fold belt at the junction of the Alps and Carpathian Mountains. The basement of the basin is Baikalian. The rocks of the lower structural stage are overlain transgressively by Neogene

Table 63.1 Major sedimentary basins

Country	Basin	Area (thousand km^2)	Major sedimentary rock Age	Major sedimentary rock Thickness (m)	Reservoir Age	Reservoir Lithology	Remarks
Austria	Vienna	8	C–Q	5000	T, E$_2$	Dolomite, sandstone	Extends into Czech, Slovakia
Italy	Po	50	Mz, R	9000	T, N$_2$	Dolomite, sandstone	
Italy	Preapennines	100	Mz, R–Q	2000	N$_2$	Sandstone	
Italy	Sardegna	16	R–Q	2000–3000			
Italy	Sicily	50	Mz, Kz	6000	T, J		
Albania	Korce	10	K–R	1500	R	Sandstone	Extends into Greece
Greece	Nestos	5	R	4000	N$_1$	Sandstone	
Greece	Salonica	15	R	3500	N$_1$	Sandstone	
Greece	Strumon	5	R	4000	N$_1$	Sandstone	

World Atlas of Oil And Gas Basins, First Edition. Li Guoyu.
© 2011 John Wiley & Sons, Ltd. Published 2011 by John Wiley & Sons, Ltd.

molasses. In the pre-molassic stage, the of Triassic, Jurassic, Cretaceous and Paleogene deposits are oil and gas saturated. Hydrocarbon accumulations in the molasses are associated mainly with anticlinal uplifts. Between 1934 and 1938, five small oil and gas fields were discovered. In the basin, 69 fields have now been identified, of which 45 are oil fields. By the end of 2008 the proven oil reserves were 6.8 million t, with annual oil production of 800,000 t. Proven gas reserves were 16 billion m^3 and annual gas production was 1.8 billion m^3.

Italy Italy is located in southern Europe, lying between the Adriatic and Mediterranean Seas, with an area of 301,333 km^2 and a population of 58 million. There are four sedimentary basins, namely, Po, Preapennines, Sicily and Sardegna. They are infilled by Triassic, Jurassic, Cretaceous and Tertiary sediments, with thicknesses ranging between 4000 m and 9000 m. About 80 per cent of the country is occupied by the Alps and the Apennines Mountains. The Preapennines Basin comprises pre-orogenic Permo-Mesozoic–Paleogene strata, essentially carbonates, 1–2 km in thickness, and an upper molassic Oligocene–Neogene strata 1–4 km thick. One large gas field has been discovered in the Po Basin in Quaternary sandstone. A total of 20 giant and 30 small oil and gas fields have been discovered in Italy. By the end of 2008 the proven oil reserves were 55.8 million t with oil production of 4.9 million t. Proven gas reserves stood at 94.1 billion m^3 with corresponding gas production of 9.6 billion m^3.

Albania Albania is located in southeastern Europe and has an area of 28,748 km^2 with a population of 3.13 million. There are two sedimentary basins to be found in the country: the Adriatic Sea Basin and the Korce Basin, filled by Permian, Triassic, Jurassic, Cretaceous and Tertiary sediments with thicknesses between 4000 m and 6000 m. Oil exploration started in 1918. By the end of 2008 proven oil reserves were 27.2 million t with oil production of 495,000 t. Proven gas reserves totalled 8 billion m^3.

Greece Greece is located at the southern tip of the Balkan Peninsula, with an area of 131,957 km^2 and a population of 11.1 million. Three small sedimentary basins are present in the country with areas ranging from 5000 to 15,000 km^2. Sedimentary cover is composed of Tertiary rocks with thicknesses in the 1500–4000 m range. In 1972 the South Kavala gas field was discovered. Daily production rate of this discovery was 580,000 m^3 of gas and 82 t of condensate oil. By the end of 2008 oil production was 6.5 million t.

San Marino

San Marino is the third smallest country in Europe and is an enclave in Italy. It has an area of 61 km^2 with a population of 30,000.

Malta Malta is a small island Mediterranean Sea, which has an area of 316 km^2 and a population of 400,000. Sedimentary covers consist of Triassic dolomite and Tertiary marine limestone and shale with thicknesses up to 5000 m. One exploration well was drilled in 1972 without any result.

SPAIN, PORTUGAL AND ANDORRA

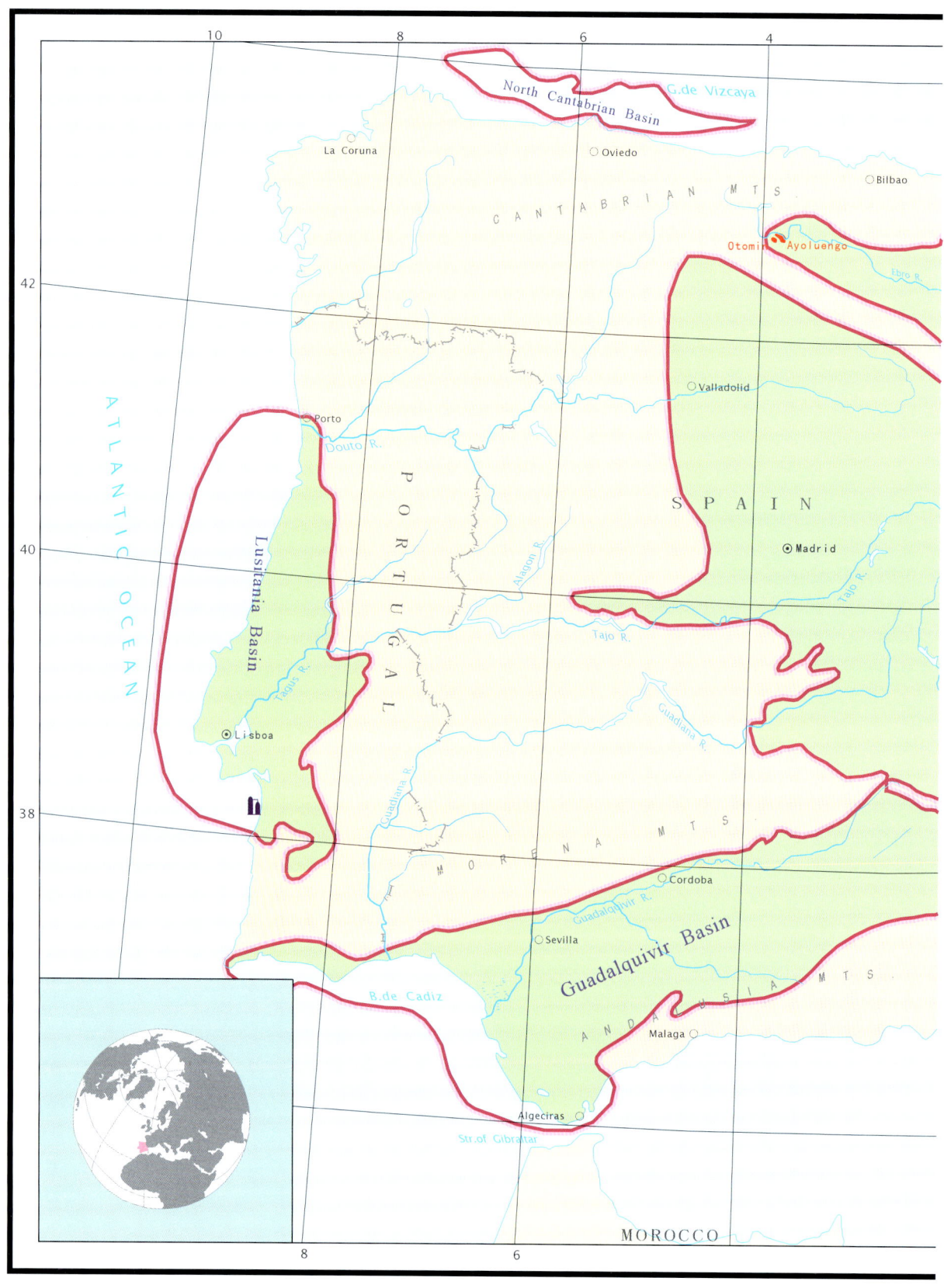

CHAPTER 64

64 Spain, Portugal and Andorra

Geography

Spain, Portugal and Andorra are located in the southwest of Europe, collectively occupying the Iberian peninsula.

History

The area has an exploration history that dates back to 1844 when, in Portugal, oil was discovered by digging surface oil sands. More modern exploration began in the 1950s, with Ayoluengo oil field discovered in the Ebro Basin after 169 dry wells had been drilled and abandoned. Spain is the only oil producing country in this group. The key data and indicators for these three countries are provided in Table 64.1.

Regional geology

The Iberian peninsula is a basement platform ringed by mountains. Five sedimentary basins are found in this area, including Castilian, Ebro, Guadalquivir, North Cantabrian and Lusitania (Table 64.2). The Castilian Basin is a very important oil producing basin, in which five offshore oil fields and one gas field have been discovered. The continental shelf off Valencia and Barcelona is 75 km wide in the central part and narrows to 30 km at both the northern and southern limits. The deep part of the Gulf de Valencia contains an ancient basement high around which the Mesozoic strata are pinched out due to depositional thinning and/or post-orogenic erosion. The Mesozoic strata are overlain unconformably by various Tertiary deposits, mostly of Miocene age. Oil is located in Cretaceous limestone below the unconformity, and the best production is related to karstic weathering of these formations. Oil production also comes from overlying Tertiary beds, known as the Amposta Chalk or the Tarraco Limestone. Some gas was also discovered in Miocene sandstone, known as the Castellon Sandstone. The Amposta Chalk reservoir is the first offshore discovery in Spain.

Structural features are Mesozoic buried mountains formed by Late Miocene or Late Pliocene movements. The first producing field was Amposta Marino where heavy oil was discovered in

Table 64.1 Key data for Spain, Portugal and Andorra

Index	Country		
	Spain	Portugal	Andorra
Area (km^2)	505,925	92,152	468
Population (thousand)	43,420	10,540	77
Proven oil reserves (million t)	20	—	—
Oil production (million t)	0.13	—	—
Proven gas reserves (billion m^3)	2.5	—	—
Gas production (billion m^3)	—	—	—

World Atlas of Oil And Gas Basins, First Edition. Li Guoyu.
© 2011 John Wiley & Sons, Ltd. Published 2011 by John Wiley & Sons, Ltd.

Table 64.2 Major sedimentary basins

Country	Basin	Area (thousand km²)	Major sedimentary rock		Reservoir	
			Age	Thickness (m)	Age	Lithology
Spain	Castilia	194	Mz, R	5000	J,K	Carbonate rock
	Ebro	52	Mz, R	4000	Mz	Carbonate rock
	Guadalquivir	76	R		N_1	Sandstone
	North Cantabrian	7	Mz, R		K	Carbonate rock
Portugal	Lusitania	48	Mz, R	5000	K	Limestone, sandstone

1970. The Lusitania basin corresponds to a Mesozoic depression in the body of the Palaeozoic Hesperian Mass. In the east, the basement is formed by a metamorphosed complex of the Hercynides, and in the west by oceanic crust. The cover of the basin is represented by Triassic evaporates, thick Jurassic clays and Cretaceous–Cenozoic sandstones and clays, with a total thickness up to 4 km.

Spain Spain has an area of 505,925 km² with a population of 43.4 million. About 60 per cent of this country is covered by sedimentary rocks. The Castilian Basin is a major oil-producing area infilled by Ceno-Mesozoic sediments and has a surface area of 194,000 km². Reservoirs are usually Mesozoic carbonate rocks. Cap rocks are Tertiary shales. The Amposta oil field was discovered in 1973. Oil production of this field declined from 690,000 t in 1977 to 380,000 t in 1978 due to a strong water-cone breakthrough.

Portugal Portugal is located on the western part of the Iberian Peninsula. As previously mentioned local inhabitants were known to have been exploiting surface oil sands in 1844. Between 1850 and 1865, several shallow wells were drilled without any results. Six shallow wells were drilled to depths of 300 m in the Lusitania Basin in the earlier part of this decade, but no appraisal of development drilling followed. This basin was recognized as a prospective basin, infilled by Mesozoic sediments with an area of 48,000 km². This country is classified as non-oil-producing.

Andorra Andorra is located in the Pyrenees Mountains lying between France and Spain. It has an area of 468 km² with a population of 77,000. Geological conditions are controlled by Alpine folding. The oil and gas prospect is unclear at present.

FRANCE, NETHERLANDS, BELGIUM AND MONACO

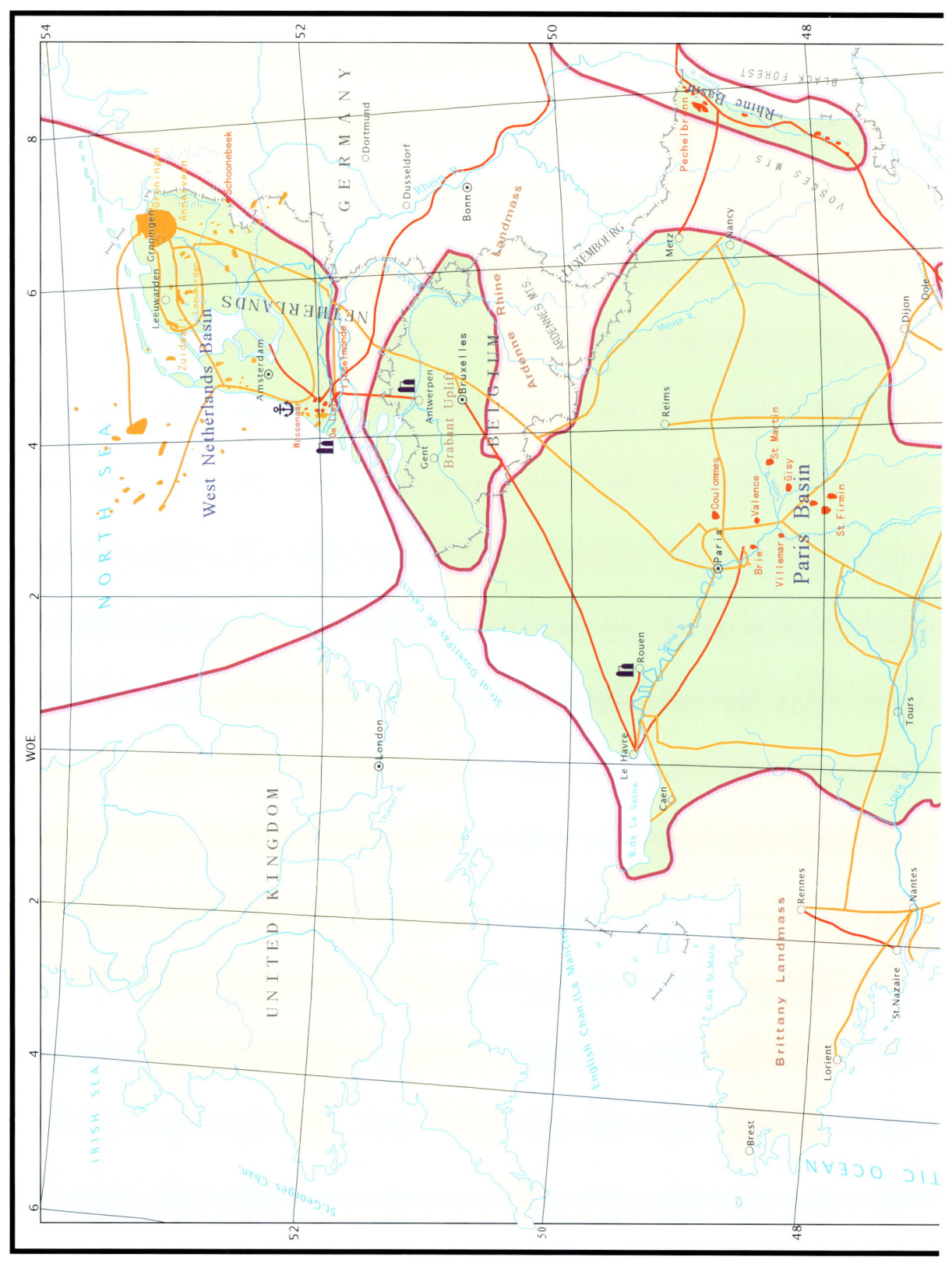

CHAPTER 65

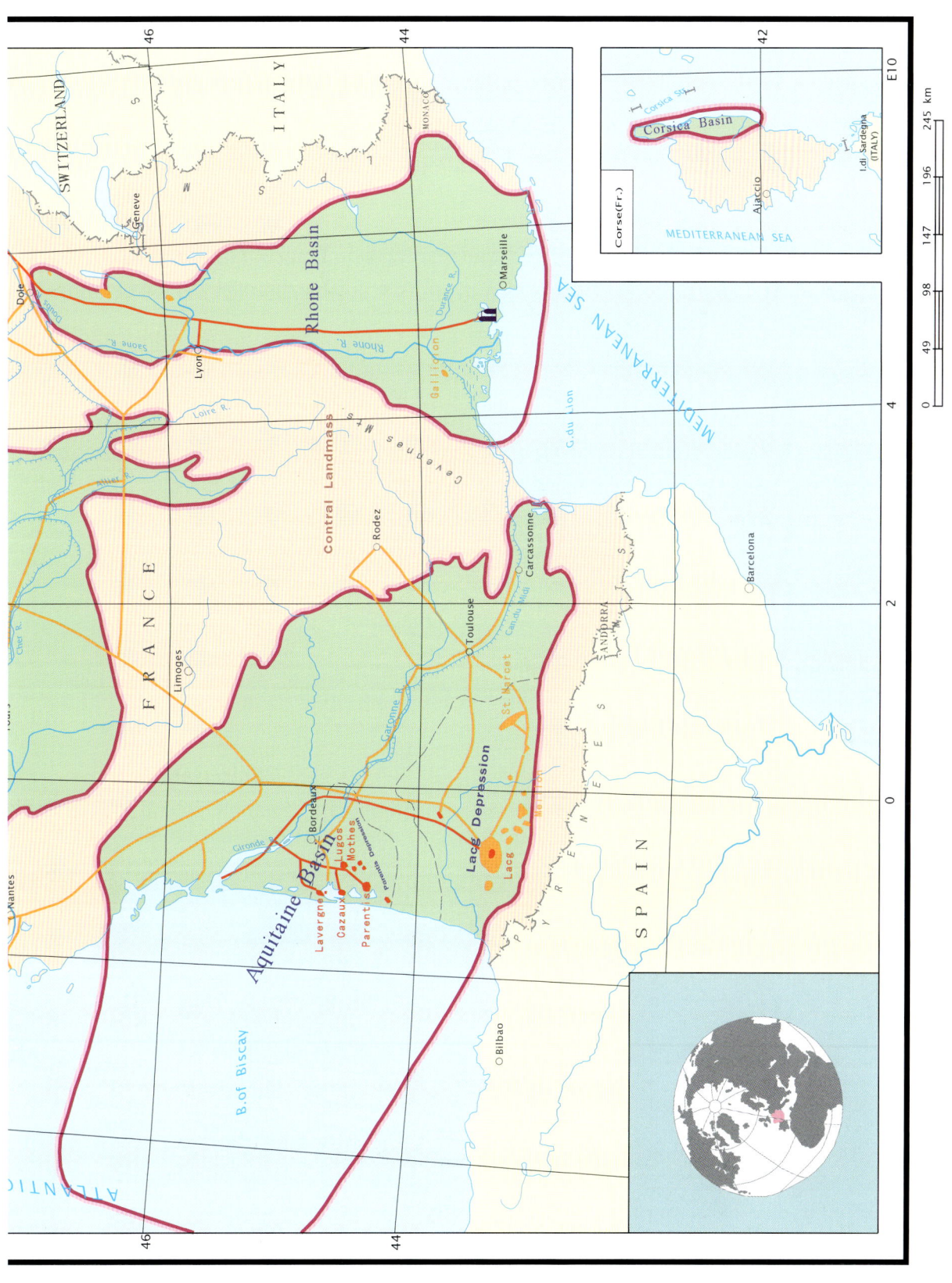

65 France, Netherlands, Belgium and Monaco

Geography

France, Netherlands, Belgium and Monaco are located in western Europe, bordering the North Sea, the Atlantic Ocean and the Mediterranean Sea (France only).

History

Oil exploration started as far back as 1735 in Aquitaine Basin in France. Further rapid development did not commence until after Second World War, with the notable discovery of the Groningen gas field (Fig. 65.1) in the Netherlands in 1957.

Regional geology

Generally this area is part of the European basement platform. Several sedimentary basins exist here (Table 65.1). The Pyrenees Mountains comprise Precambrian crystalline basement and Palaeozoic meta-sedimentary rocks. The basins are filled by Palaeozoic, Mesozoic and Cenozoic rocks with thicknesses ranging from 3500 m to 10,000 m. Reservoirs are sandstones and limestones.

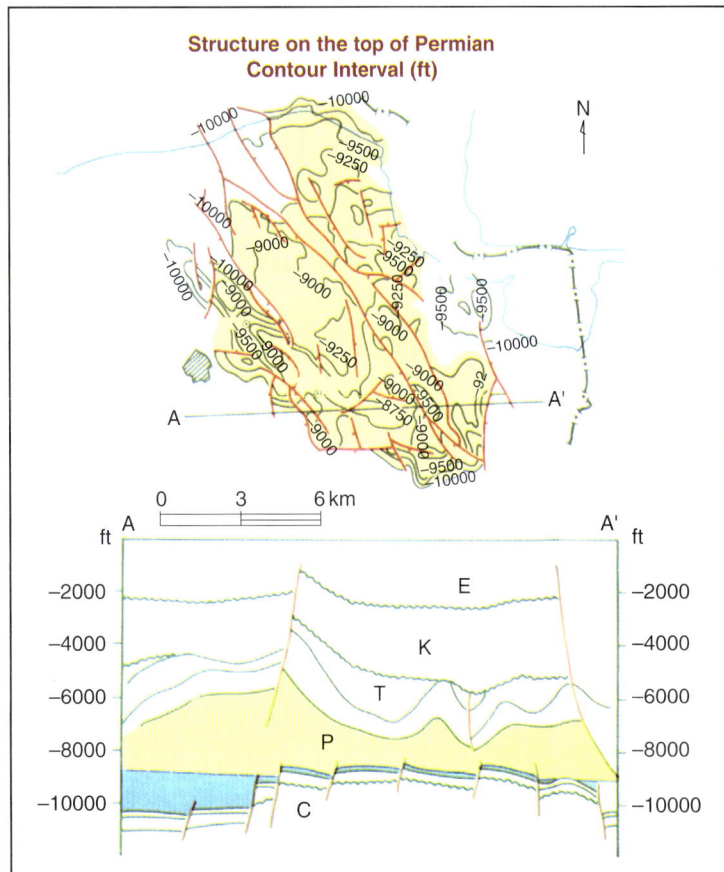

Groningen is the world's eighth supergiant gas field and is located in Netherlands.

Basic data

1. Name	Groningen
2. Discovery year	1957
3. Area (km^2)	800
4. Reservoir	Sandstone
5. Reservoir age	Permian
6. Depth (m)	2,700–3,600
7. Thickness (m)	158
8. Porosity (%)	15–20
9. Permeability (mD)	98–987
10. Recoverable reserves (trillion m^3)	2.8

Fig. 65.1 Details of the Groningen gas field.

World Atlas of Oil And Gas Basins, First Edition. Li Guoyu.
© 2011 John Wiley & Sons, Ltd. Published 2011 by John Wiley & Sons, Ltd.

Table 65.1 Major sedimentary basins

Country	Basin	Area (thousand km²)	Major Sedimentary rocks		Reservoir	
			Age	Thickness (m)	Age	Lithology
France	Aquitaine	180	Mz-Kz	10,000	J, K	Limestone
	Paris	130	T-R	3500	J, K	Sandstone
	Rhone	60	Mz-R	4000	R	Sandstone, limestone
Netherlands	West Netherlands	134	Mz	4100	J, K	Sandstone
Belgium	Brabant	24	K, R			

France France is located in western Europe and borders the Mediterranean Sea to the south, with an area of 551,602 km² and a population of 63.2 million. Four sedimentary basins are found in the country, namely, Aquitaine, Paris, Rhone and part of the Rhine Basin. The Aquitaine Basin occupies a syneclise of complicated structure, consisting of depressions and the Upper Cretaceous Flysch Trough. Oil fields are concentrated mainly in the Arkshaon Depression. The main source rocks are Upper Jurassic bituminous limestones. The Paris Basin has an area of 130,000 km². The basement occurs at a depth more than 3 km. The cover is formed by marine carbonate–terrigenous deposits ranging in age from Permo-Triassic to Quaternary. The basin is essentially oil-prone but out of 32 fields discovered, only one appears to be gas-producing. Petroleum potential of the basin is estimated as medium. By the end of 2008 the proven-oil reserves were 14 million t, with annual oil production of 990,000 t. Proven gas reserves stood at 9.6 billion m³ and annual gas production was 1 billion m³.

Netherlands The Netherlands is located in northwestern Europe and borders the North Sea to the north. It has an area of 338,145 km² and a population of 5.2 million. There are two sedimentary basins. The West Netherlands Basin is located both on land and offshore, with roughly equal portions lying in either part. It is filled by Palaeozoic and Mesozoic sediments, with a thickness up to 4100 m. The other basin is part of the Central Graben of the North Sea. A gigantic gas field was discovered in 1959, in the north of the country at Slochteren in Groningen Province, with reserves assessed at some 1.6 trillion m³. Most hydrocarbon fields are in country lying below sea level, and would be offshore fields but for the artificial drainage and sand filling of a significant part of the country in the past. The discovery of the Groningen giant gas field was not only significant from a point of view for the overall energy balance of the country, but also for Europe. It stimulated more aggressive exploration for oil and gas in the North Sea. By the end of 2008 the proven oil reserves stood at 13.6 million t with annual oil production of 1.75 million t. Proven gas reserves were 1.4 trillion m³ with gas production of 76.5 million m³.

Belgium Belgium is located on the coast of western Europe and occupies an area of 30,528 km² with a population of 10.4 million. In the country lies the Brabant Uplift, covered by thin Cretaceous and Tertiary sandstone and muddy limestone. Some exploration wells were drilled without any positive results. Belgium has one of Europe's largest petrochemical industries.

Monaco Monaco is located on the northern shore of the Mediterranean Sea between Italy and France, having an area of 1.95 km² and a population of 32,000. Its geology is dominated by intensely folded Cretaceous and Tertiary sediments.

UNITED KINGDOM AND IRELAND

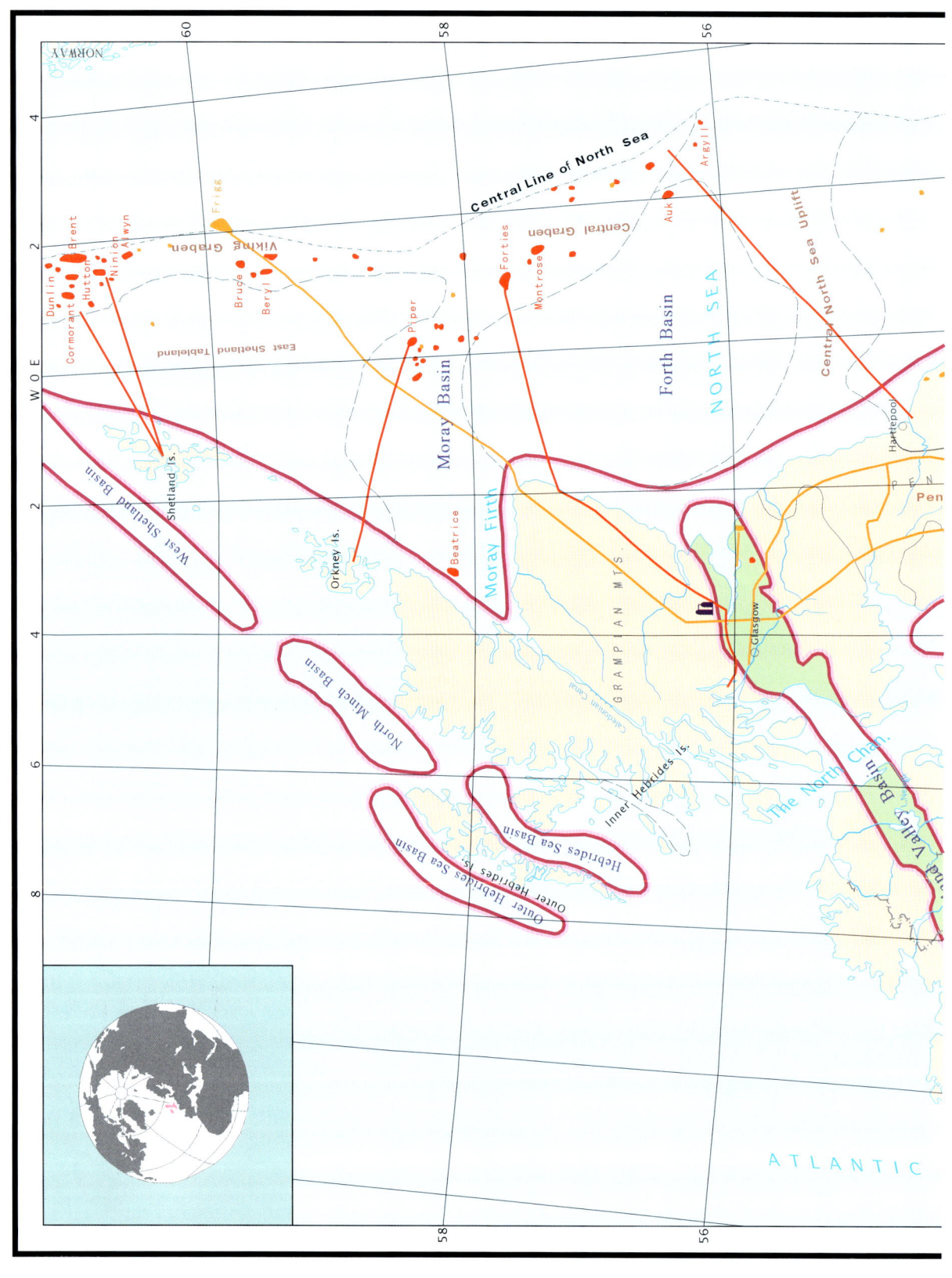

CHAPTER 66

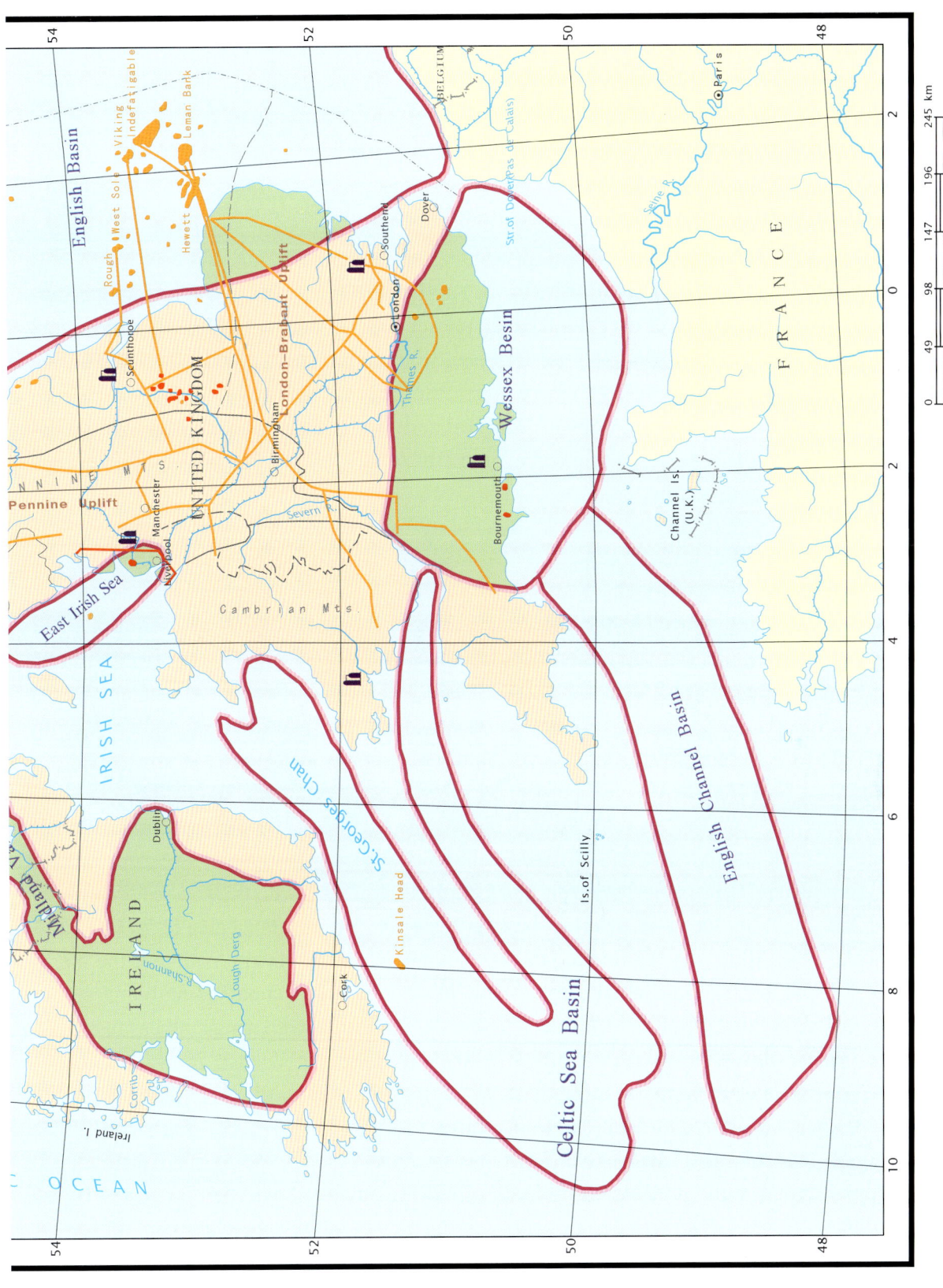

287

66 United Kingdom and Ireland

Geography

These two countries are located in Western Europe, consisting of several islands surrounded mainly by the Atlantic Ocean to the west and south.

History

Being the home of the Industrial Revolution in the 19th century, the UK's need for fuel was pressing and led to early exploration for oil, later for gas. Prior to that, there had been many occurrences of oil and gas seepages all over the islands, particularly associated with the nation's coal mines. Gas was discovered in 1896 in Cretaceous sandstone in the south of the country. Then in 1919, oil was discovered in Carboniferous Limestone. Prior to the historic opening up of the North Sea as a major oil and gas province, only 17 small oil and gas fields had been discovered in the country. Once North Sea production commenced, the UK would become one of the world's leading oil- and gas-producing nations.

Regional geology

The basement platform of the British Isles occupying central England and Ireland was formed over the continental basement produced by the Caledonian Orogenic Belt. It consists of a series of small platforms containing troughs and depressions (for instance, the West England Depression). Mobile orogenic belts are represented by Caledonian, Hercynian and Alpine. Twelve sedimentary basins (Table 66.1) are found in these regions, filled by Palaeozoic, Mesozoic and Cenozoic sediments with thicknesses ranging between 1200 m and 7500 m, comprising mainly sandstone and limestone.

United Kingdom

The United Kingdom is located on the western seaboard of Europe, comprising one main island and several smaller ones. It has a total area of 244,100 km^2 and a population of 60.2 million. Ten sedimentary basins are distributed all over the country ranging in size from 5000 to 60,000 km^2 and filled by Palaeozoic, Mesozoic and Cenozoic sediments with thicknesses of between 1200 m and 7500 m. By far the most important basin is North Sea Basin with an area of 570,000 km^2, of which about 244,000 km^2 lies within the UK. When the successful oil exploitation of the North Sea started in 1964, the UK transformed from being an oil-importing country to an oil exporting country. By the end of 2007 the proven oil reserves were 48.7 million t, with annual oil production of 70 million t. Proven gas reserves were 34.2 billion m^3 with annual gas production 76.3 billion m^3.

The first phase of deep exploration in England was commissioned by the government in 1918, with the first location being Hardstoft in Derbyshire. Small quantities were discovered, after which attention was shifted to the English Midlands in the hope of further similar discoveries. Geological predictions at the time were that further to the east there should be oil-bearing layers beneath the unconformable Mesozoic rocks, and also in east Yorkshire where Permian rocks was recognized as potentially prospective on the evidence of hydrocarbon shows in potash mines in Germany. Further prospective areas were thought to be the Lower Carboniferous of the Midland

World Atlas of Oil And Gas Basins, First Edition. Li Guoyu.
© 2011 John Wiley & Sons, Ltd. Published 2011 by John Wiley & Sons, Ltd.

Table 66.1 Major sedimentary basins

Country	Basin	Area (thousand km²)	Major sedimentary rock		Reservoir	
			Age	Thickness (m)	Age	Lithology
UK	East Irish Sea	8	P,T	2000	T	Sandstone
	English	52	D–R	5800	C	Limestone
	English Channel	60	Mz, R	1200		
	Forth	55	Mz, Kz			
	Hebrides Sea	5	Kz			
	Moray	26	Mz, Kz	1500	J	Sandstone
	North Minch	7	Kz			
	Outer Hebrides Sea	8				
	Wessex	43	Mz, R	3000	J	Limestone, sandstone
	West Shetland	8	Pz, Mz	7500	T	Sandstone
Ireland	Celtic sea	64	Mz, Kz	3000–6000		
	Midland valley	72	D–R	4500	C	Sandstone

Valley of Scotland. Each of these plays was met with modest success. In Scotland, a small gas field and a small oil field were established in 1937 east of Edinburgh, and in Yorkshire, gas in limited commercial quantities was found near Eskdale. Modest quantities of oil were also discovered at Eakring in Nottinghamshire in 1939, and three other finds were to be made close by thereafter. Although the Carboniferous Limestone had been the initial objective, this reservoir in fact proved to be relatively unimportant. The main oil reserves were in Lower Carboniferous sands (Gluyas and Hichens, 2003).

Ireland Ireland is the third largest island in Europe. It lies on the west side of the Irish Sea and on the east side of the Atlantic Ocean, with an area of 70,282 km² and a population of 4.13 million. It is a non-oil-producing country. Within this country lie two sedimentary basins. The Midland Valley Basin has an area of 72,000 km², filled by Mesozoic and Tertiary sediments with thicknesses between 3000 m and 4000 m in water depths of 1000 m. In 1978, two exploration wells were drilled in this basin with no shows. The other main basin is Celtic Sea Basin which has an area of 640,000 km², filled by Mesozoic and Tertiary sediments with thickness ranging between 3000 m and 6000 m in water depths of a few hundred metres. The Kinsale Head gas field was discovered at water depths of 100 m, with proven gas reserves of 28 billion m³.

NORWAY, SWEDEN, FINLAND, DENMARK AND ICELAND

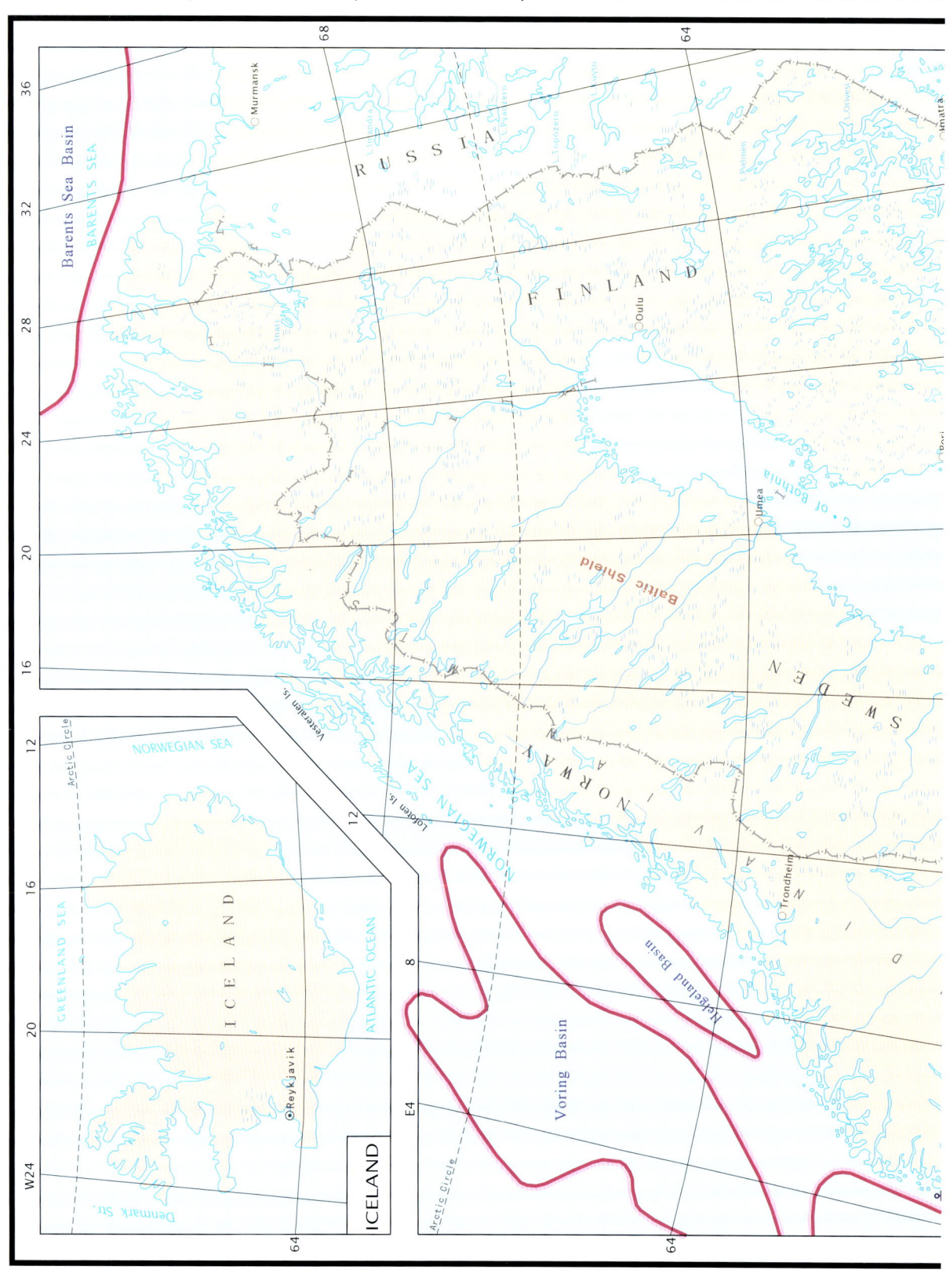

CHAPTER 67

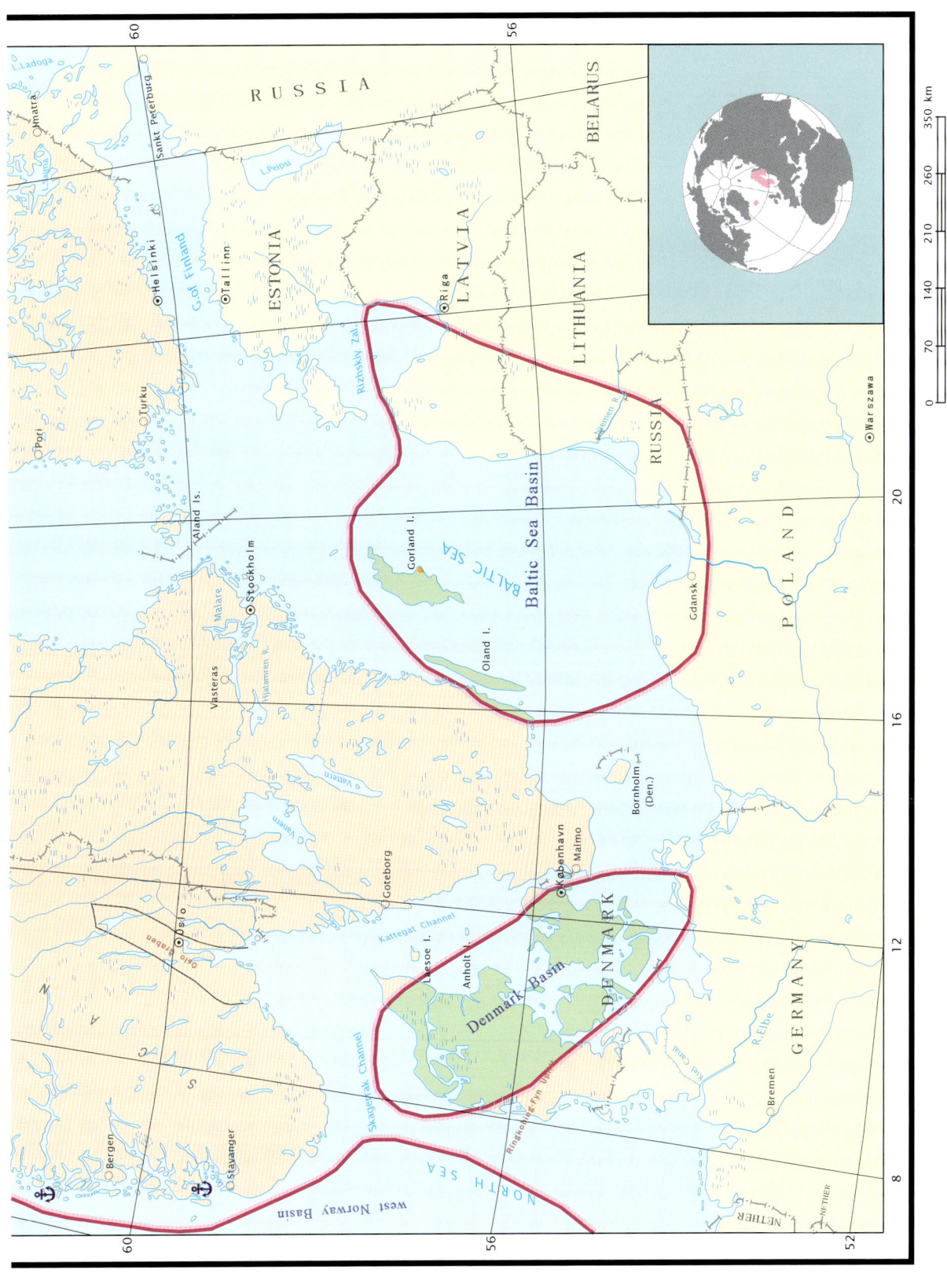

67 Norway, Sweden, Finland, Denmark and Iceland

Geography

Norway, Sweden, Finland, Denmark and Iceland are located in northern Europe, occupying the Baltic Shield and prominently featuring the Scandinavian Mountains.

History

Because no sedimentary basins are known to exist on land, active oil and gas exploration did not commence until after the North Sea discoveries. In 1965, Norway commenced oil exploration in the North Sea, and several giant oil and gas fields were rapidly discovered and developed. Norway thus became a major oil-producing country with a peak annual production of 160 million t in the year 2000.

Regional geology

As mentioned earlier in the mountainous regions on land, there are no sedimentary basins. However, both the North Sea and the Baltic Sea represent commercially viable oil provinces. Details of sedimentary basins in Norway and Denmark are listed in Table 67.1.

Norway Norway is located on the Scandinavian Peninsula, borders Finland to the east, and has an area of 385,155 km^2 and a population of 4.6 million. The Scandinavian Mountains run throughout Norway from the south to the north. This initially ruled Norway out in terms of oil and gas prospects. However, this was to change following the discovery of the large oil and gas fields in the British sector of the North Sea. Norway started exploration in its sector of the North Sea. Initial results arising out of continuous exploration and drilling between 1965 and 1969 were disappointing, with 32 dry holes found. The 33rd well led to the discovery of the huge Ekofisk oil field with recoverable reserves of 237 million t, producing out of Upper Cretaceous Chalk. A total of 55 oil and gas fields were discovered in the Norwegian sector of the North Sea, made up of 42 oil fields and 13 gas fields. Peak annual oil production reached 160 million t in 2000 and peak annual gas production reached 59.5 billion m^3 in 2007. By the end of 2008, the proven oil reserves were 915 million t with an annual oil production of 106 million t. Gas reserves were estimated at 23 trillion m^3. The Voring Basin occupies the continental shelf off mid-Norway. The sedimentary cover comprises Upper Palaeozoic carbonate deposits, Triassic and Jurassic terrigenous, essentially continental, rocks, and Cretaceous marine carbonate–terrigenous

Table 67.1 Major sedimentary basins Norway and Denmark

Country	Basin	Area (thousand km²)	Major sedimentary rock		Reservoir		Remarks
			Age	Thickness (m)	Age	Lithology	
Norway	Helgeland	12	K,R	4500			
	Voring	92	Mz, R				
	West Norway	39	P, Mz–Kz		J	Sandstone	
Denmark	Denmark	78	Pz, Mz	5000–6000	P	Dolomite	Extends into Germany

World Atlas of Oil And Gas Basins, First Edition. Li Guoyu.
© 2011 John Wiley & Sons, Ltd. Published 2011 by John Wiley & Sons, Ltd.

Norway, Sweden, Finland, Denmark and Iceland

rocks. The thickness of the Cretaceous sequence in the Mere Depression reaches 6 km. Ten fields have been discovered in the basin, 6 oil and 4 gas. The main productive sequence is represented by Lower and Middle Jurassic sandstones. The fields are confined, as a rule, to anticlines, which are often faulted. Source rocks are Upper Triassic coal bearing sequences and Upper Jurassic clays.

Sweden Sweden is a mountainous country located in northern Europe and has an area of 449,964 km^2 with a population of 9.1 million. The whole country consists of Precambrian Baltic Shield and Scandinavian Caledonides, with no sedimentary rocks on land, except Oland and Gotland where sedimentary rocks occur with thicknesses less than 10,000 m. Seven shallow wells were drilled on Gotland Island with one hydrocarbon occurrence found in Ordovician limestone and Silurian sandstone. In 1976, oil production was 3500 t. A small part of the Denmark Basin is found in the southwest part of Sweden. This basin is filled by Cambrian, Ordovician, Silurian and Mesozoic sediments.

Finland Finland is located in northern Europe, bordering Russia to the east and is covered by the Quaternary glacial deposits. The other key geological feature is the Baltic Precambrian Shield, on which there are grabens with thin sediments.

Denmark Denmark is located in northern Europe, bordering the North Sea to the west and the Baltic Sea to the east. It has an area of 43,094 km^2 with a population of 5.45 million. The Danish sector of the North Sea (56,000 km^2) represents about 10 per cent of the sea, and in this area 16 oil and gas fields have been discovered. Prior to that, oil exploration started onshore in 1936. The development of small oil fields in Germany just south of the Danish border stimulated post-war exploration in Denmark, using gravity and seismic methods. Oil shows were found particularly in Upper Cretaceous Chalk, but the results in terms of hydrocarbon discovery were disappointing – the main outcome was the location of salt deposits. Nevertheless knowledge of the deep structure and stratigraphy of Denmark was important in later assessment of the regional prospects of the North Sea Basin. Offshore oil and gas exploration was initiated in 1966 and the first oil field, Den, was discovered in 1971. The peak annual oil production reached 19.6 million t, while the peak annual gas production was 8.6 billion m^3 in 2004. By the end of 2007 proven oil reserves were 162 million t, with annual oil production of 15.6 million t. Proven gas reserves stood at 70.5 billion m^3, with annual gas production of 8.5 billion m^3.

Iceland Iceland is an island surrounded by the Atlantic Ocean, Denmark Straits, Greenland Sea and Norwegian Sea. It has an area of 103,000 km^2 with a population of 300,000. Iceland is covered by glaciers and comprises volcanic tuffs. Hot water geysers and pools have been utilized as a heat energy source by the local population. There are no sedimentary basins to be found on the island.

NORTH SEA OIL AND GAS REGION

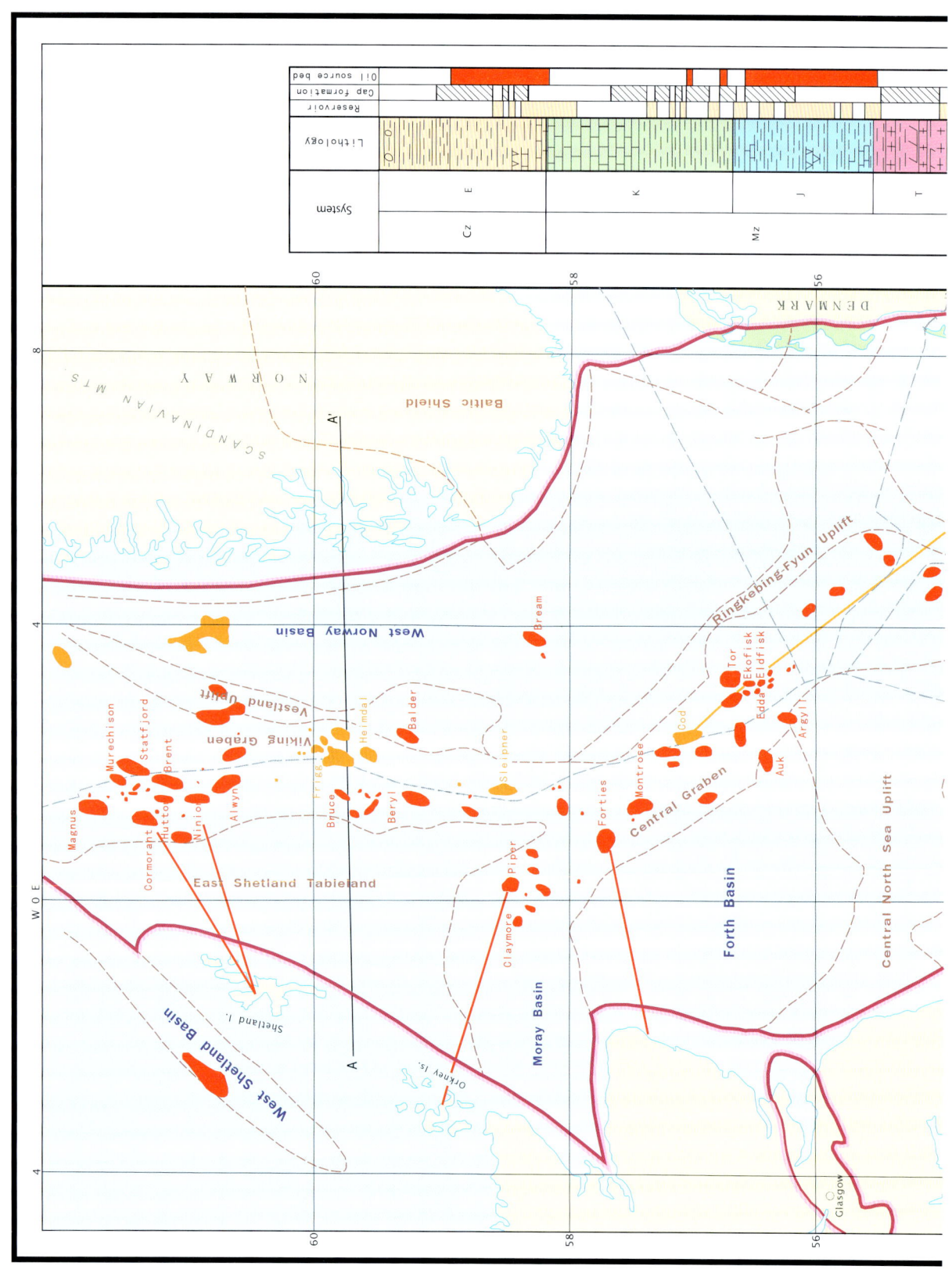

CHAPTER 68

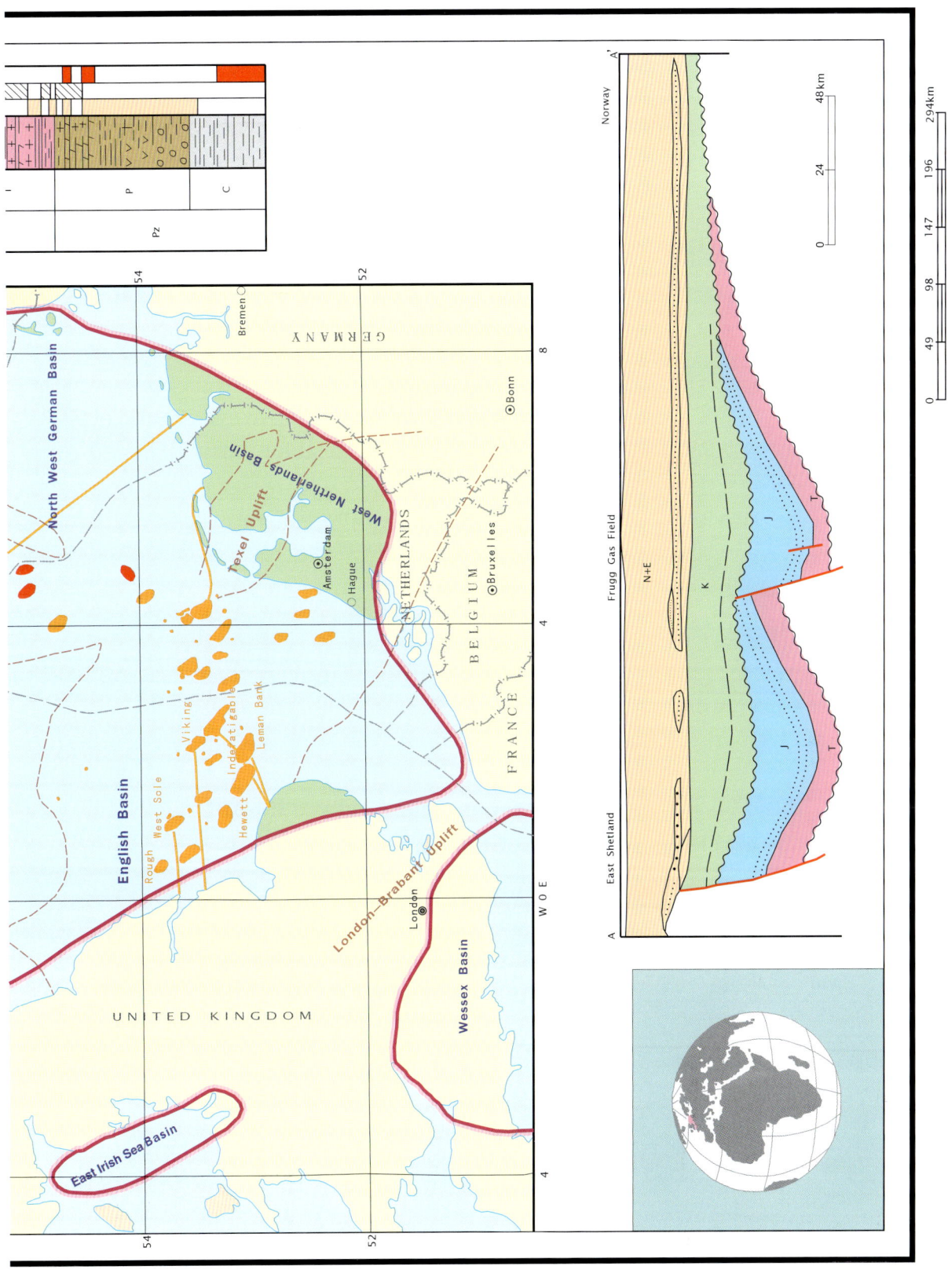

295

68 North Sea Oil and Gas Region

Geography
The North Sea Oil and Gas Region is located between the islands of Great Britain and the Scandinavia. The total area is 573,000 km^2 and parts of the basin lie in seven countries: the UK, Norway, Netherlands, Denmark, Germany, Belgium and France. The North Sea has very difficult weather and sea conditions, varying from relatively calm to hostile with high winds and waves, as well as frequent incidents of dense fog. Water depths in the UK sector are less than 200 m, reaching deeper levels than this in the Norwegian sector. At present, it is the largest oil producing area in Europe.

History
Since exploration began in the late 1960s, 40 dry wells were drilled. In 1970, the announcement of a wet gas discovery at Cod and of oil at Ekofisk provided a major fillip to hydrocarbon exploration in the northern North Sea, and exploration of the rich British sector followed. Some 62 per cent of this lies in the Lower and Middle Jurassic sandstone reservoirs, with a further 36 per cent shared approximately equally between the Upper Cretaceous–Lower Tertiary chalks and Tertiary sandstones. Details of the major oil and gas fields found up till 1979 are listed in Table 68.1. Many more have been found since then (Halbouty, 1992).

Regional geology
The northern part of the region is an ensemble of ensialic rift sedimentary basins that have undergone a polyphase history of deposition and tectonism (Glennie, 1990). The main geological evolution took place in a regional setting during Cretaceous times. The North Sea is now established as

Table 68.1 Some major oil and gas fields of the UK and Norway sectors of the North Sea Basin

Country	Name	Discovery year	Recoverable reserves		Depth (m)	Trap	Reservoir	
			Oil (million t)	Gas (billion m^3)			Age	Lithology
UK	Brent	1971	303	8.9	2500	Fault block	J	Sandstone
	Forties	1970	280	—	2100	Anticline	E	Sandstone
	Ninian	1974	154	—	3100	Fault block	J	Sandstone
	Piper	1973	117	—	2500	Fault block	J	Sandstone
Norway	Statfjord	1974	420	6.4	2400	Fault block	JT	Sandstone
	Ekofisk	1969	238	10.9	3000	Salt dome	E–K	Sandstone
	Oseberg	1979	140	6.7	2700	Fault block	J	Sandstone
	Gullfaks	1978	172	1.8	2000	Fault block	J	Sandstone
	Sonrre	1979	186	—	2400	Fault block	J–T	Sandstone
	Troll	1979	196	127.1	1400	Anticline	J	Sandstone

World Atlas of Oil And Gas Basins, First Edition. Li Guoyu.
© 2011 John Wiley & Sons, Ltd. Published 2011 by John Wiley & Sons, Ltd.

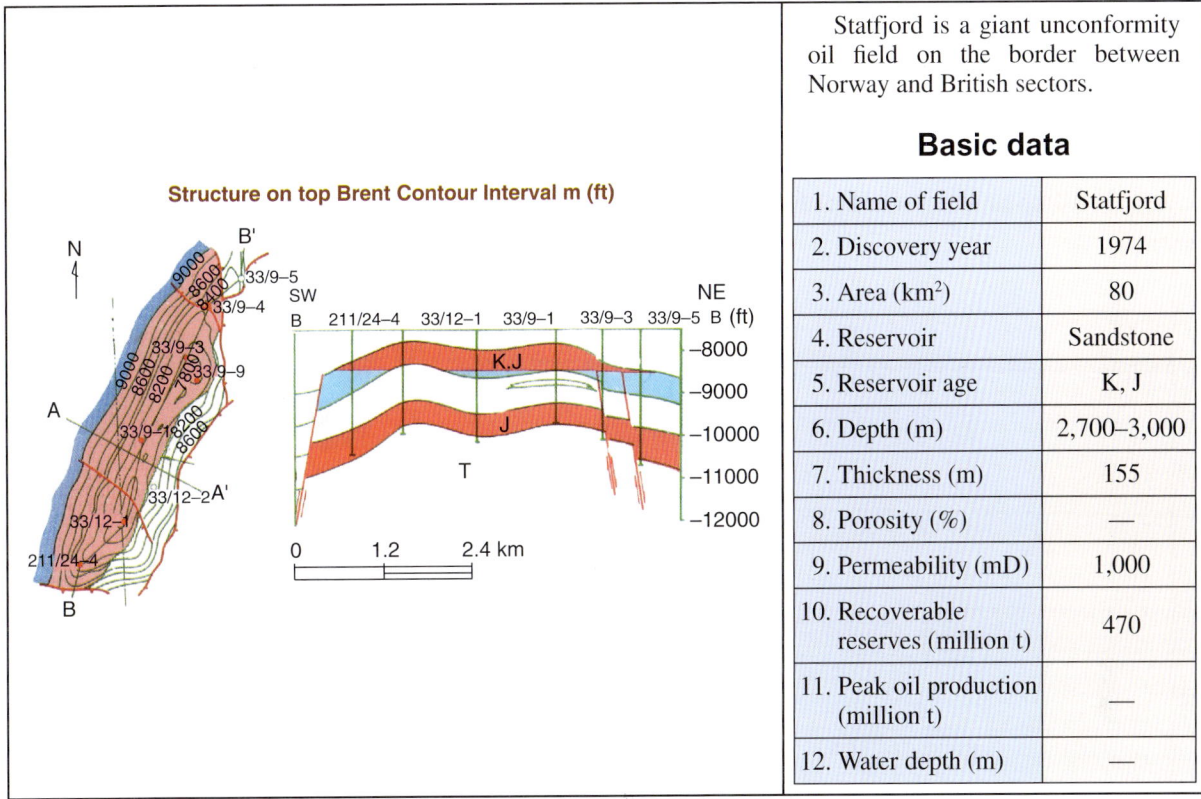

Fig. 68.1 Details of the Statfjord oil field.

a mature petroleum province. Since the commencement of exploration in the early 1960s, almost 2000 wells have been drilled, with over half these in the UK offshore sector. Reservoir horizons exist in rocks of virtually every geological period from the Carboniferous. Of importance are those in the Middle Jurassic of the Brent Province (e.g. Fig. 68.1) on the Shetland platform, the Tertiary of the Central and Viking Grabens and the Lower Permian of the Southern North Sea.

The most important source rocks are the Kimmeridgian Clay of the Central and Viking Grabens and the Carboniferous Coal Measures of the southern North Sea. Caprocks are provided by a variety of impervious strata, the most important being the basal thermal subsidence Lower Cretaceous shales draping the graben-fill Jurassic strata and the Zechstein evaporates and carbonates sealing the Rotliegend Sandstones.

The basin developed over Devonian sediments and crystalline basement. Clastic sediments were deposited in a rift setting during Permo-Triassic times. Middle Jurassic shallow marine sandstones are the principal reservoirs in the basin.

Future prospects

As a result of the extensive exploration activity of recent decades it is probable that all the giant fields south of 62°N have been discovered. However, further north of the region is still relatively unexplored and it is probable that giant fields await discovery. One of the most active areas of exploration in this region has been Haltenbanken, where the Mid-Upper Jurassic tilted fault block play has yielded successful results. Many of the other large graben-fill and thermal subsidence plays described from the areas to the south therefore may have hydrocarbon potential.

ESTONIA, LATVIA, LITHUANIA, BELARUS, UKRAINE

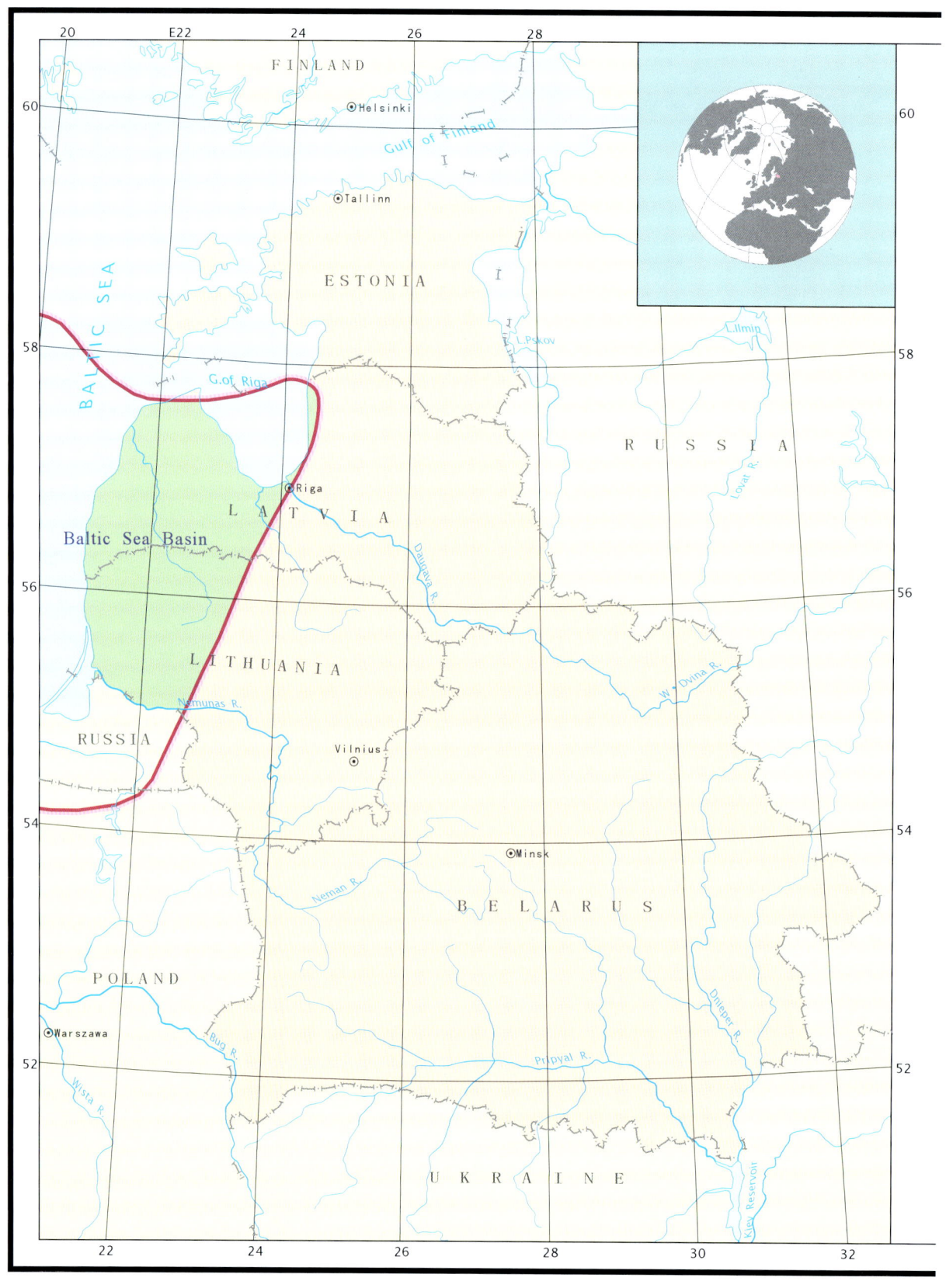

AND MOLDOVA CHAPTER 69

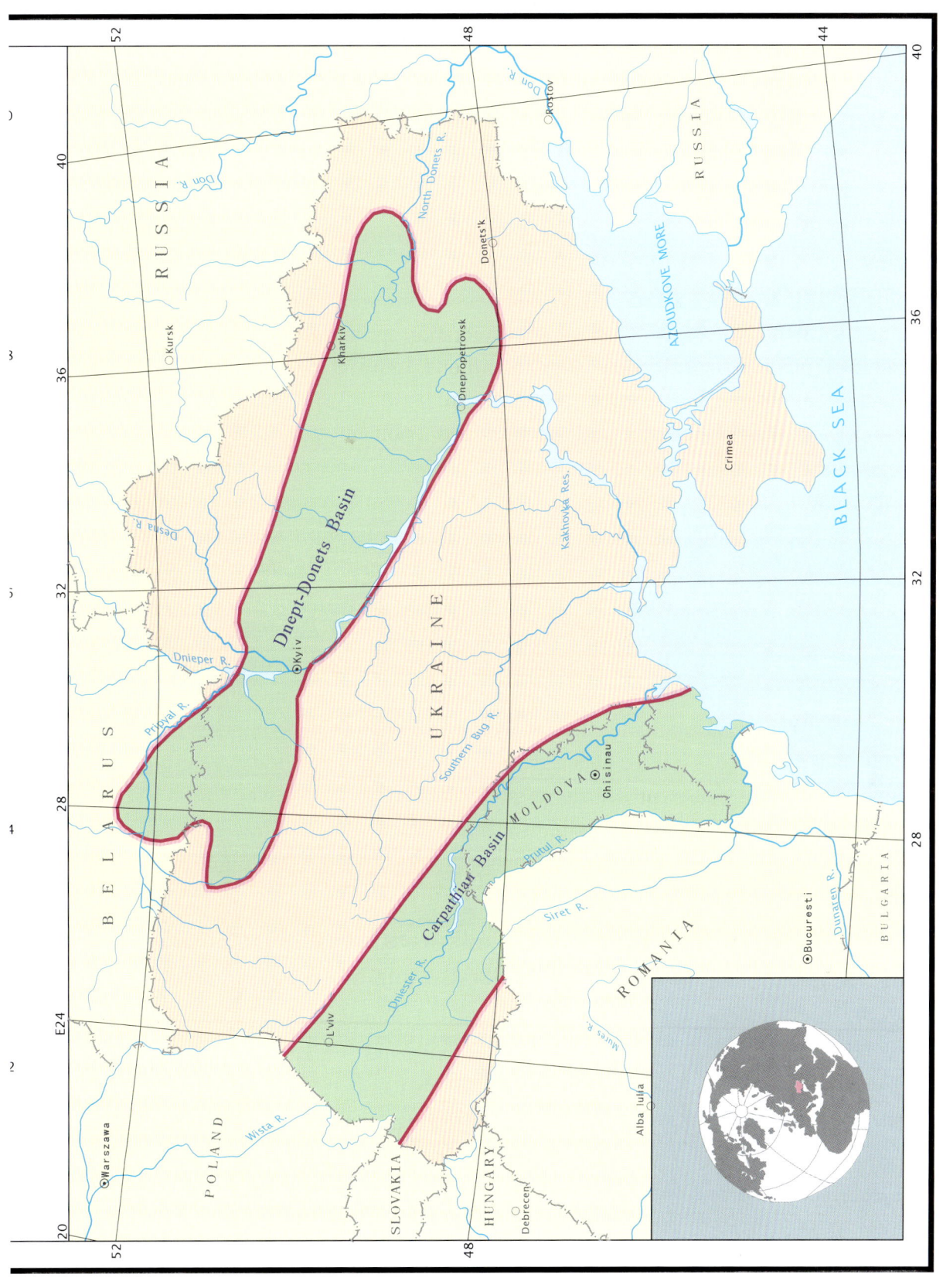

69 Estonia, Latvia, Lithuania, Belarus, Ukraine and Moldova

Geography

These countries are located in eastern Europe, with Lithuania, Latvia and Estonia bordered by the Baltic Sea to the west. Eastern Europe is an area of extensive plains.

History

Among these six countries, Ukraine is an old oil producing area, with the Dolina gas field being discovered in 1860. In the other countries, oil exploration and development started later. By the end of 2008 oil production of Ukraine was 3.75 million t, Lithuania was 2.8 million t and Belarus was 1.75 million t. The key data for these countries are listed in Table 69.1.

Estonia Estonia has an area of 45,200 km^2 and a population of 1.36 million. This country is rich in oil shale in the eastern section, which provides 75 per cent of energy consumption. Estonia is unique as one of the very few countries that have a developed shale oil industry; others are China, Russia, Germany and Israel. Shale oil is used for electricity generation and other energy and fuel requirements. Occasional fuel shortages are covered by imported Russian oil. One drawback in using shale oil is the pollution it generates.

Latvia Latvia has an area of 64,589 km^2 and a population of 2.28 million. This country is a non-producer of oil and gas. All oil and gas requirements are imported from Russia. The country has a large oil port, Ventspils, located on the Baltic Sea coast with an import handling capacity of 25 million t per year. Very importantly, it is an ice-free port thus making it a vital resource during the winter months. There is a large underground gas reservoir used for strategic fuel storage at Incukalns. The country is also used as a trans-shipment and transit gateway for oil and gas from Russia and Kazakhstan. This country has reasonable prospects of discovering offshore oil in the Baltic Sea, but no active exploration has commenced.

Table 69.1 The key data and indicators for Estonia, Latvia, Lithuania, Belarus, Ukraine and Moldova

Index	Country					
	Estonia	Latvia	Lithuania	Belarus	Ukraine	Moldova
Area (km^2)	45,200	64,589	65,300	307,600	603,700	33,800
Population (thousand)	1360	2280	3780	9750	46,890	3400
Proven oil reserves (million t)	—	—	1.6	2.7	54	—
Oil production (million t)	0.22	—	—	0.17	3.7	—
Proven gas reserves (billion m^3)	—	—	—	—	1104	—
Gas production (billion m^3)	—	—	—	—	119	—

World Atlas of Oil And Gas Basins, First Edition. Li Guoyu.
© 2011 John Wiley & Sons, Ltd. Published 2011 by John Wiley & Sons, Ltd.

Lithuania   Lithuania has an area of 65,300 km² and a population of 3.78 million. Oil consumption has reached 2.8 million t per annum. Oil reserves stand at only 1.6 million t. All oil and gas requirements are imported from Russia. There is a large refinery with an annual capacity of 13 million t. Estimated oil reserves are 31–60 million t primarily in reservoirs known to exist on the shelf of the Baltic Sea. By the end of 2008 the proven oil reserves stood at 16 million t.

Belarus   Belarus has an area of 307,600 km² and a population of 9.75 million. Oil exploration started in 1964. Some oil fields were discovered during the 1970s. Maximum oil production reached 7.9 million t in 1975. Because of a shortage of discovered recoverable reserves, production has been curtailed to 1.75 million t (2003), in an effort to conserve remaining reserves. The Pripyat Depression is a part of the Dnept-Donets Basin with a total area of 36,000 km². The sedimentary cover consists of Palaeozoic and Mesozoic sediments with total thickness reaching 8,000 m. Source rocks have been identified in the Upper Devonian dark shales. Reservoir rocks are associated with Devonian carbonate and terrigenous rocks. Traps are represented by faulted anticlines.

Ukraine   Ukraine has an area of 603,700 km² and a population of 9.7 million. This country is rich in many resources, including coal, oil and gas. It is mostly located in the eastern European plain. The Carpathian Mountains traverse the southwest part of Ukraine up to an altitude of 2061 m. Ukraine has a long history of discovering oil and gas. The Dolina gas field was discovered in 1860. More than 20 gas fields were discovered during the 1950s and the 1960s. In the years 1972–1975, peak oil production reached 14.5 million t and gas production reached 68.7 billion m³ annually. In 2008 annual oil production declined to 3.7 million t, with gas showing a similar downward trend to 11.9 billion m³. The large Dnept-Donets (Map 73) oil and gas basin is located in Ukraine.

Moldova   Moldova has an area of 33,300 km² and a population of 4.3 million. Exploration for oil started in 1946. The 20-year drilling programme led to the discovery of only one small oil field and four small gas fields. At the end of 1997, cumulative oil production was a mere 23,000 t. Recently, production in all of these fields has been suspended indefinitely.

DNEPT-DONETS BASIN

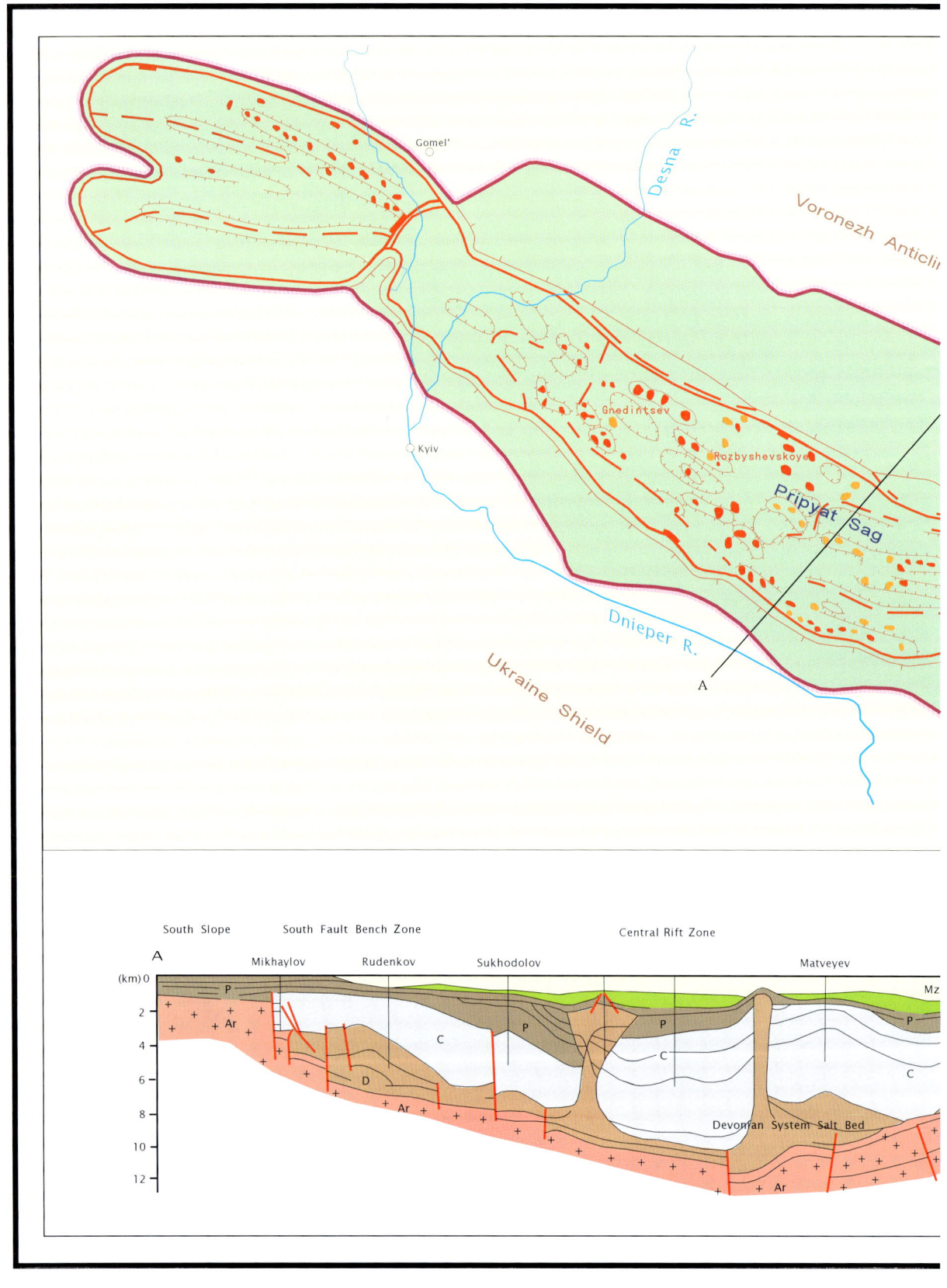

CHAPTER 70

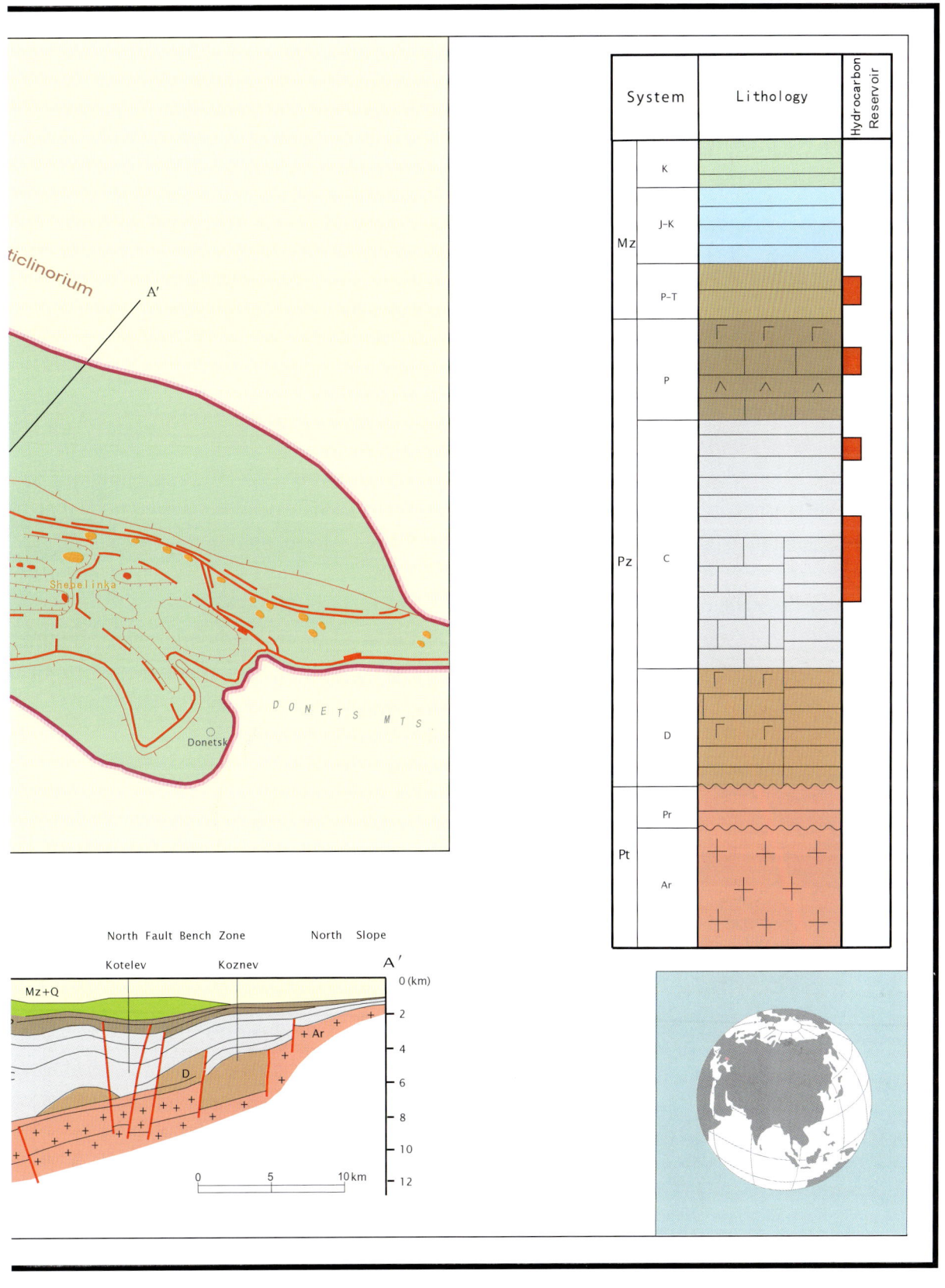

70 Dnept-Donets Basin

Geography

The Dnept-Donets Basin is located in eastern Ukraine, in the southwest part of the Russian platform and belongs to the foreland basin of the Dnept-Donets platform. It is bordered by the Ukraine Shield to the west, and has an area of 135,000 km².

History

In 1936, a small salt-dome oil field was discovered. After 1944, few large anticlines have been found in Permian and Mesozoic strata. The large gas field Shebelinka (Fig. 70.1) was discovered in a faulted anticlinal trap associated with a Permian salt dome. It has proven reserves of 527 billion m³ in a Permian sandstone reservoir at a depth of 1500 m, thickness of 450 m, porosity of 9.15 per cent and permeability of 5–10 mD. Methane content is 92–95 per cent During 1960–1975, a total of 50 oil and gas fields were discovered. Details of three gas fields are listed in Table 70.1. Annual oil production was 520,000 t in 1961, peaking at 19 million t in 1974, then dropping to 8.3 million t in 1980. Annual gas production was 14.1 billion m³ in 1961, peaking at 56.0 billion m³ in 1975. At present, in excess of 120 oil and gas fields have been discovered in this basin.

Regional geology

In this basin, reservoirs are associated with Devonian, Lower Carboniferous, Middle Carboniferous, Upper Carboniferous–Lower Permian and Upper Permian–Mesozoic strata at depths varying greatly between 400 m and 5920 m. Reservoir porosity also varies greatly at

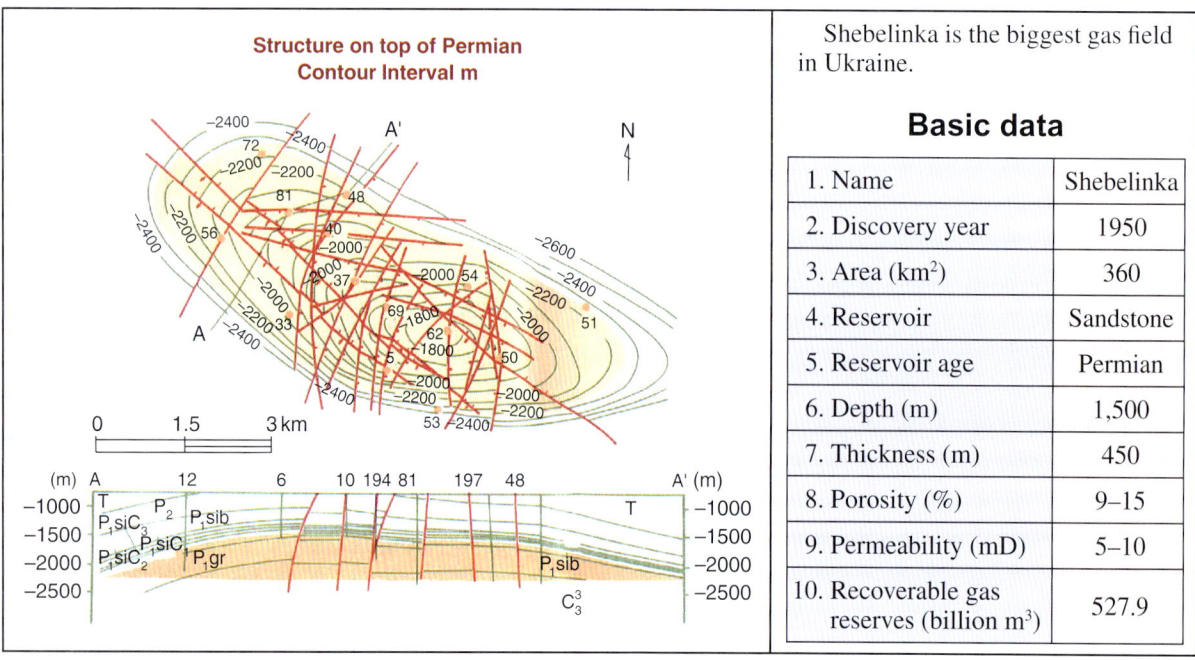

Fig. 70.1 Details of the Shebelinka gas field.

World Atlas of Oil And Gas Basins, First Edition. Li Guoyu.
© 2011 John Wiley & Sons, Ltd. Published 2011 by John Wiley & Sons, Ltd.

Table 70.1 Examples of major gas fields of the Dnept-Donets Basin

Name	Discovery year	Recoverable reserves		Depth (m)	Trap	Reservoir	
		Oil (million t)	Gas (billion m³)			Age	Lithology
Shebelinka	1950	—	527.9	1500	Salt dome, anticline	P	Sandstone
Zapadno krestischenskoye	1968	—	216.9	2900	Anticline	P	Sandstone
Yefremovskoye	1965	—	121.3	1800	Salt dome, anticline	P	Sandstone

7–28 per cent, as does permeability at 13–870 mD. The eastern European platform includes the Volga-Urals, Pre-Caspian, Dnept-Donets North Kavkaz and Moscow Basins. All, except the Dnept-Donets Basin which is confined to a rift structure of the same name, are associated with vast synclinal structural elements,. Within the synclinal basins the basement occurs at a depth of 2–3 km in the arched portions and up to 5–6 km in the zones of subsidence. In contrast, in the area of the Ural Foredeep and Dnept-Donets Trough basement occurs at ~ 10 km depth. The sedimentary cover comprises Proterozoic and Phanerozoic deposits. Most of it is formed by Riphean, Vendian and Palaeozoic terrigenous and carbonate strata, and also by Permian carbonate, sulphate and salt-bearing rocks. Mesozoic deposits occur but are not widespread. The Devonian salt-bearing sequence is 2–3 km or more in thickness and is typical of the Dnept-Donets Basin.

Commercial petroleum potential has been identified in Middle Devonian to Neogene deposits in the basins on the eastern European platform. In the Volga-Urals, Pre-Caspian and Dnept-Donets basins the productive horizons are Devonian, Carboniferous and Permian rocks, as well as Triassic and Jurassic strata in the Pre-Caspian and Dnept-Donets Basins. Each basin may contain from five to seven productive complexes formed by carbonates with frequent reef varieties, marine carbonate–terrigenous and terrigenous rocks. Lower Permian sulphate- and salt-bearing sequences, clayey carbonates and clayey bands within the Carboniferous and Devonian deposits, salt-bearing sequences in the Devonian rocks of the Dnept-Donets basin, Upper Permian, Triassic, Jurassic and Cretaceous clayey and clayey carbonate rocks serve as regional seals. Source rocks are represented by marine clayey and clayey carbonate bituminous strata found in the Upper Devonian (Frasnian Domanik Formation), Carboniferous and Lower Permian deposits, and also by clayey bituminous varieties occurring in the Mesozoic rocks. The majority of the fields are confined to anticlinal or brachyanticlinal folds (both faulted and unfaulted) and reef masses. In the Dnept-Donets Basin most of the fields are associated with salt domes.

RUSSIA

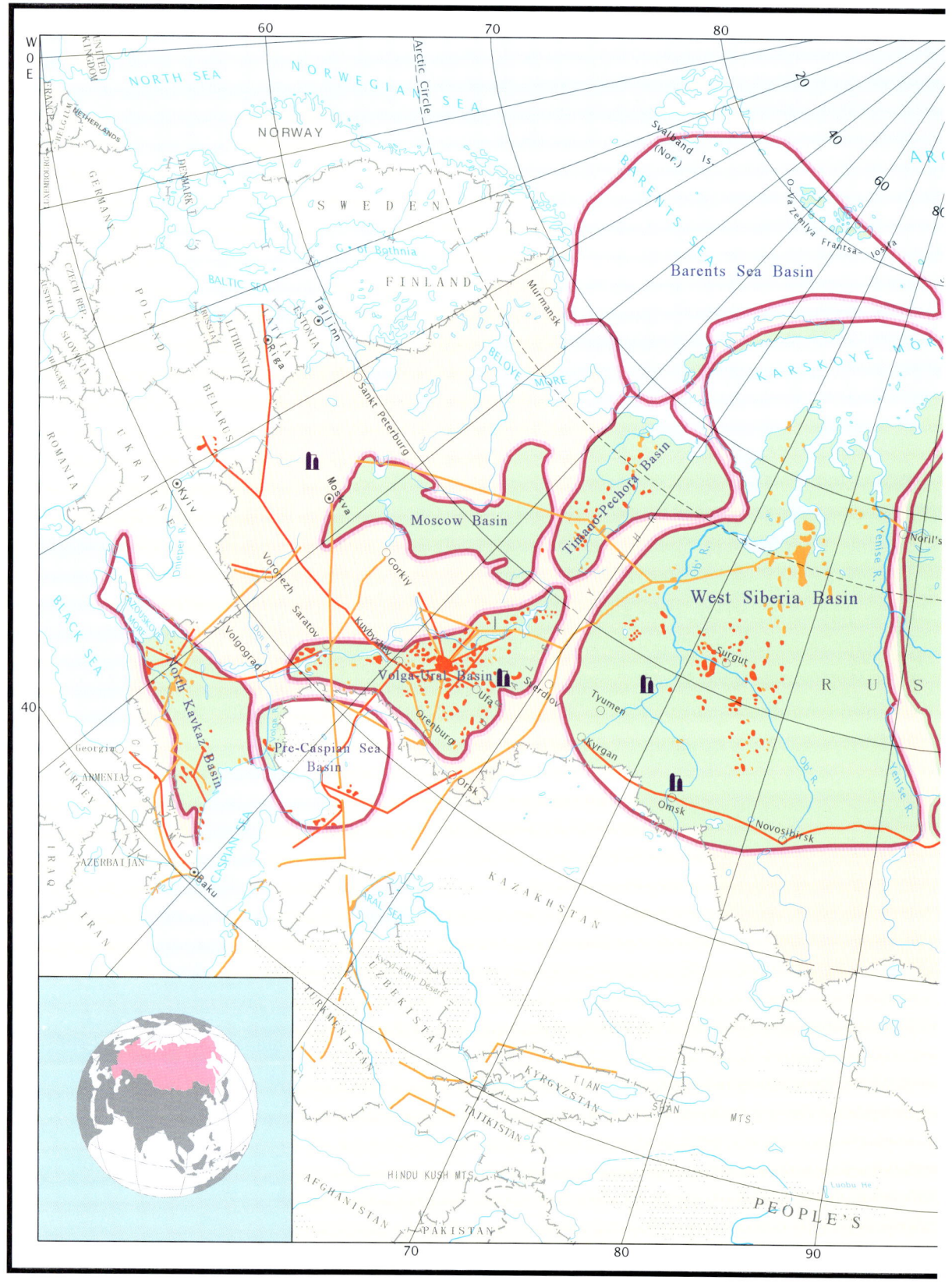

CHAPTER 71

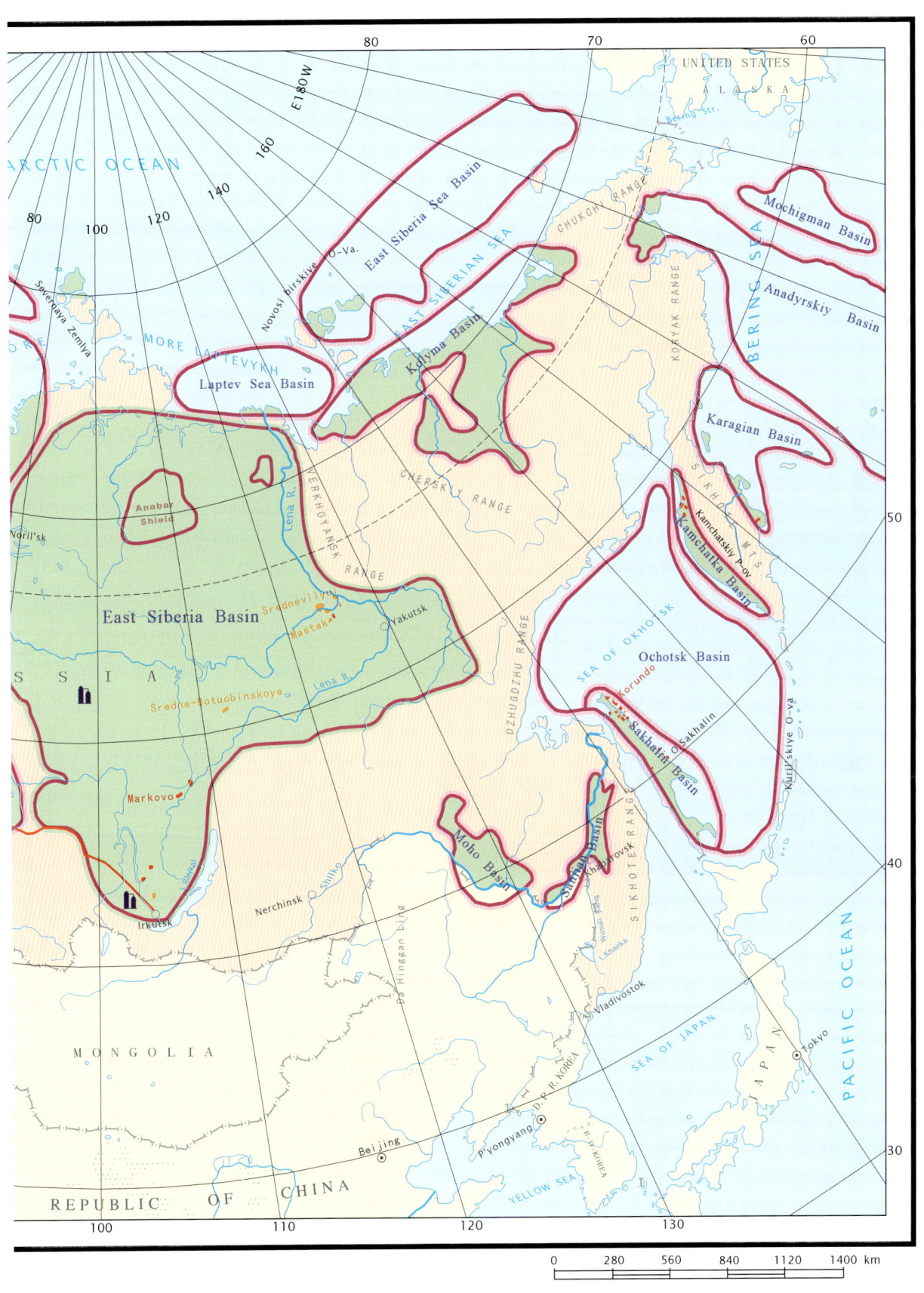

71 Russia

Geography

Russia is located in both Europe and Asia. It is the largest country in the world by area. Russia is a country that stretches from eastern Europe to the Pacific Ocean, with an area of 17,075,400 km^2. The vast territory of Russia has traditionally been divided into five geographical zones: tundra, taiga or forest, steppe or plain, arid and mountain. Most of Russia consists of two plains (the East European Plain and the West Siberian Plain), two lowlands (the North Siberian and the Kolyma, in far northeastern Siberia), two plateaus (the Central Siberian Plateau and the Lena Plateau in the east), and a series of mountainous areas concentrated mainly in the extreme northeast or extending intermittently along the southern border.

History

Russia holds abundant oil and gas resources, with a history of more than 140 years of development for the oil industry. However, the history of gas exploration is comparatively short, and began only in the 1950s. Since when, a substantial oil and gas industry system has been established. Following the end of the Second World War, thanks to the vigorous exploration of the two large oil and gas blocks in Volga-Urals and West Siberia, the oil and gas output of the Former Soviet Union generally has shown signs of high-speed growth. In the late 1980s, the oil and gas output of the Former Soviet Union ranked first in the world. In 1988 the annual output of the Former Soviet Union was 624 million t (Table 71.1), with Russia producing b568 million t of this, accounting for 91.1 per cent of the total. In 1991, the gas output of the Former Soviet Union was 766.4 billion m^3, with Russia producing 643 billion m^3 of this. As a result of the disintegration of the Former Soviet Union and the subsequent economic crisis, oil and gas output fell rapidly. However, that trend has been reversed particularly in the past 5–10 years. Recently, oil and gas output has been picking up gradually. In 2004, the remaining recoverable reserves of oil were 8.2 billion t (Table 71.2), with an annual output of 410 million t; gas reserves were 47.5 trillion m^3, with annual output of 637.5 billion m^3. The development of the oil and gas industry in Russia is closely related with that of other countries of the Commonwealth of Independent States (CIS), and is generally divided into the following periods.

The first period From the mid- and late 19th century to 1930, two old oil recovery bases, Baku and North Caucasus, were explored. At the turn of the 20th century, Baku Oil Block was the oil production centre in Russia. In 1880, annual oil output amounted to 516,000 t. In 1890, annual oil output increased to 4.0 million t, approximately accounting for 40 per cent of the world's total. In 1901, annual oil output stood at 12 million t (51 per cent of the world's total), from the two regions that include the Baku and North Caucasus oil production blocks.

The second period During the period from 1930 to 1960, the Volga-Urals (also named the second Baku) and Timano–Pechora Basins were vigorously explored to further expand the oil block exploration of North Caucasus. In 1929, the industrial oil flow of the Upper Palaeozoic Permian

World Atlas of Oil And Gas Basins, First Edition. Li Guoyu.
© 2011 John Wiley & Sons, Ltd. Published 2011 by John Wiley & Sons, Ltd.

Table 71.1 Oil and gas production trends of Russia (1981–2008)

Year	Remaining recoverable oil reserves (billion t)	Oil production (million t)	Remaining recoverable gas reserves (trillion m³)	Gas production (billion m³)	Oil export (million t)
1981	11.6	608	34.7	463	120
1982	8.6	612	39.6	500	125
1983	8.6	616	41.0	535	127
1984	8.6	612	42.5	586	125
1985	8.3	595		644	106
1986	8.0	614	43.8	685	129
1987	8.0	624	41.0	726	136
1988	7.9	624	42.4	769	
1989	7.9	607	42.4	796	160
1990	7.1	570	45.2	770	132
1991		459		643	93
1992		396			110
1993		350			121
1994		313			126
1995		303		596	136
1996		298		601	161
1997		302	48.1	571	170
1998		300	48.1	591	178
1999	6.6	302	48.1	591	200
2000	6.6	325	48.1	576	213
2001	6.6	345	47.5	568	233
2002	8.2	379	47.5	598	268
2003	8.2	410	47.5	637	209
2004	8.2	447	47.5	607	230
2005	8.2	470	47.5	580	240
2006	8.2	480	47.5	583	245
2007	8.2	491	47.5	592	257
2008	8.2	488	47.5	601	243

System on Upper Chusov of the Volga-Urals Basin was discovered. In 1930, a Palaeozoic oil field was discovered in the Timano-Pechora Basin, marking the discovery of new oil blocks and also marking the onset of the second period. During the 1930s and the 1940s, exploration of Volga-Urals region was increased. Carboniferous oil fields in the Kuybyshev, Perm and Bashkir regions were discovered. After the discovery of super-large oil fields, including the Durmaz (1944) and Romashkino (1948) oil fields, annual oil output increased from 38 million t in 1950 to the historic peak of 147 million t in 1960. Correspondingly the water injection volume increased from 90.01

Table 71.2 Remaining oil and gas reserves (proven and controlled) of Russia at the end of 2004

Region	Oil (billion t)	Gas (billion m³)
West Siberia	15.6	37,777
Volga-Urals	3.4	3936
North part of Europe	1.4	3497
North Kavkas	0.19	305
East Siberia and Far East	0.85	2651
Total	21.6*	48,167

* Proven oil reserves were 8.2 billion t, while controlled oil reserves were 12.4 billion t.

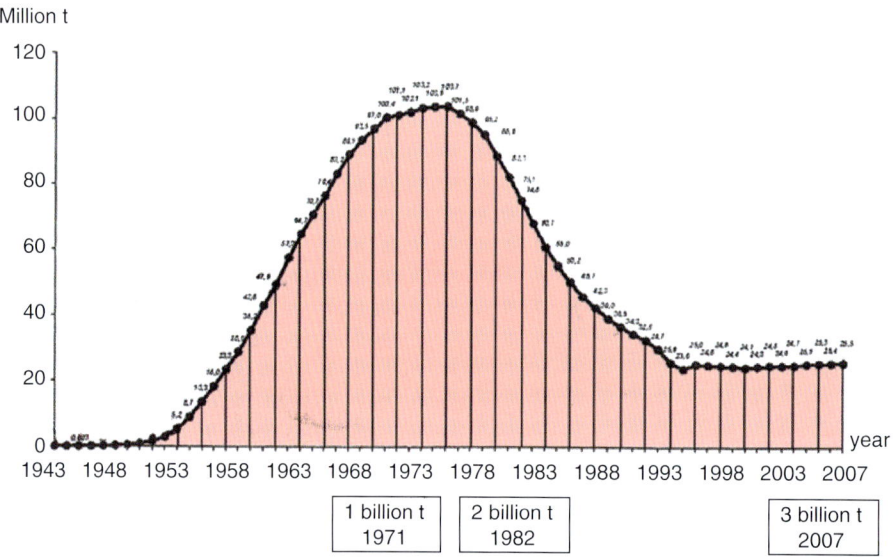

Fig. 71.1 Oil production of Tatarstan Republic of Russia during 1943–2007. This curve is a typical parabolic curve of the oil production profile of any oil region.

million m³ in 1950 to 329 million m³ in 1960, with the price of wellhead crude oil standing at 0.5–3.0 ruble/t. Figure 71.1 shows the temporal profile of oil production of a region that developed its hydrocarbon industry in this period

The third period From 1960 to 1980 great efforts were made to explore oil and gas basins in West Siberia. Since the Belezovo gas field in West Siberia region was discovered in 1960, the super-large Cretaceous oil fields such as Samotlor field (1965; Fig. 71.2) and Mamentovo field (1965) were discovered in the central area of the basin in the 1960s. At the same time, four giant gas fields including the Urengoy field were discovered in northern West Siberia. Thereafter, giant gas fields were discovered in the Yamal Peninsula, with cumulative proven and controlled oil reserves of 12.67 billion t and natural gas reserves of 24 trillion m³. All of these factors combined to make the annual oil production of the Former Soviet Union climb to 603 million t, with natural gas production of 435 billion m³. In these figures, Russian oil output was 547 million t and natural gas 250 billion m³, becoming one of the most important oil-producing countries in the world.

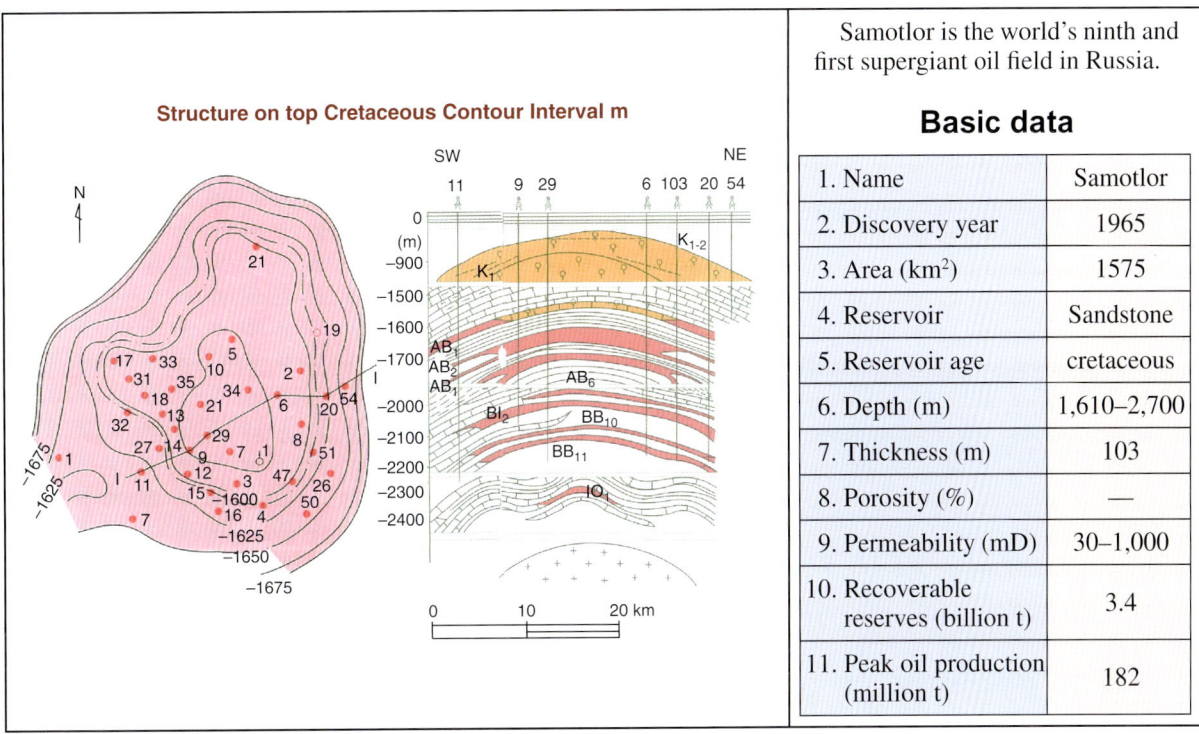

Fig. 71.2 Details of the Samotlor supergiant oil field.

The fourth period From 1981 to 1991 further progress was made in oil and gas exploration in the pre-Caspian Basin, Yamal Peninsula in the north of West Siberia, in the East Siberia Basin, including the coastal shelf, with a further series of large oil and gas fields being discovered. In 1985 the super-large Urobxin and Kovikgin oil and gas fields were discovered in the south of the East Siberia Basin. All the discoveries proved that the Russian coastal shelf and the East Siberia Basin enjoyed rich prospects in terms of oil and gas exploration and development. During the same period great efforts were made to develop large high-production oil and gas fields in West Siberia, ensuring that Russian annual output of oil and gas would increase further. Annual oil output stood at a record high of 607 million t in 1989, with the corresponding record high of natural gas production being 643 billion m³ in 1991, ranking the Former Soviet Union first in the world in terms of both oil and gas production.

The fifth period From 1992 to the present, due to the disintegration of the Former Soviet Union and the subsequent economic crisis, investment and workforce availability for exploration and development of oil and gas fell sharply. As a result addition of new oil and gas reserves could not meet annual production needs. In addition, fields matured rapidly and reservoir structures and integrity were adversely affected. The consequence of this was a steady decrease in annual oil and gas output. Annual output decreased from 396 million t in 1992 to 302 million t in 1999. Correspondingly, the annual output of natural gas also decreased from 643 billion m³ to 591 billion m³ during the same period. As at the time of compiling this atlas, the Russian economy appears to have taken a turn for the better. Investment in oil and gas exploration has increased and the old oil and gas fields are being reconstructed. This has resulted in the stabilization of oil and gas output, with a discernible upward trend. By the end of 2008, oil output amounted to 488 million t, and gas production was 601 billion m³. Tables 71.2 show details of remaining oil and gas

Table 71.3 Recent oil and gas development of Russia and predicted development for the coming decade

Index	Year						
	1990	1995	1999	2005	2010	2015	2020
Oil production (million t)	516	307	305	350	350	355	360
Refinery capacity (million t)	298	182	170	185	200	220	225
Oil export (million t)	286	179	134	165	180	185	170
Gas production (billion m^3)	641	596	589	600	645	670	700
Gas consumption (billion m^3)	426	403	462	360	395	410	435
Gas export (billion m^3)	215	193	126	235	250	260	265

reserves as at 2004, and Table 71.3 details of the development of the oil and gas industry over the past two decades as well as projections for the future (Shannon and Naylor, 1989).

Regional geology

Tectonically, Russia includes two old platforms (the Russian platform and the East Siberian platform), one young platform (the West Siberian platform) and five orogenic belts (Urals, Taimer, Altai, Alpes and Kamchatka). It is very difficult to provide detailed information regarding the huge oil and gas potential of Russia. The unique geological conditions, however, are characterized by following five features.

1. There are many large sedimentary basins (Table 71.4), including West Siberia basin which has an area of 3.3 million km^2, East Siberia Basin 3.5 million km^2, Barent Sea Basin 1.0 million km^2, Anadyrskiy Basin 740,000 km^2, Okhotsk Basin 850,000 km^2 and Volga-Urals Basin 700,000 km^2. Worldwide, there are only three basins larger than 3 million km^2 in area, and two of these (West Siberia and East Siberia) are located in Russia.

2. There are many large structures in Russia, mostly very simple, i.e. anticline types. Most of these structures have large areas, examples being the Romashkinskoe oil field with an area of 4000 km^2, the Urengoy gas field 4000 km^2, the Fedorov oil field 1800 km^2 and the Samotlor oil field 1600 km^2.

3. There are many large oil and gas fields. Urengoy is the second largest gas field in the world with proven gas reserves of 8 trillion m^3. Samotlor oil field is also megasized with proven oil reserves of 2.1 billion t.

4. There are many large oil and gas bearing formations. Urengoy gas field has gas-bearing strata of 2180 m in thickness while Fedorov oil field has oil bearing strata that are 1100 m in thickness, with porosity 25 per cent and permeability 500–1000 mD.

5. There are huge oil and gas resources. According to estimated data by the end of 2000, the total oil resources were 84.5 billion t and accumulated oil production only 15.6 billion t; total gas resources 236 trillion m^3 and accumulated gas production 46 trillion m^3.

Table 71.4 Major sedimentary basins of Russia

Basin	Area (thousand km²)	Major sedimentary rock		Reservoir		Remarks
		Age	Thickness (m)	Age	Lithology	
Anadarskiy	740	K–Kz	4000–6000	R	Sandstone	Extends into Norway
Barents Sea	1010	Pz–Kz		T,J	Sandstone	
East Siberia	3500	Pz, Mz	7000–12,000	Pt, E, C–J	Sandstone, limestone, dolomite	
East Siberia Sea	700	Kz				
Kamchatka	110	Kz	3000–10,000	R		
Karagian	235	Kz				
Kolyma	620	Mz		Mz		
Laptev Sea	240	Kz				
Mochigman	150	Kz				
Moho	100	J				Extends into China
Moscow	350	Pz, Mz				
North Kavkaz	350	Mz, Kz	12,000	T–N	Sandstone	Extends into Ukraine
Ochotsk	850	K–R	8000	R	Sandstone	
Sahalin	200	K, R	8000	N	Sandstone	
Sanjian	57	J–R	3500	J, K, R	Sandstone	Extends into China
Timano-Pechora	498	E–Kz	8000	D–P,T	Biohermal limestone, dolomite, mud volcano and sandstone	
Volga-Urals	700	D–P	6000	D,C,P	Sandstone, carbonate rock	
West Siberia	3300	Mz, Kz	8000	J,K	Sandstone	

OIL AND GAS PIPELINES OF RUSSIA AND

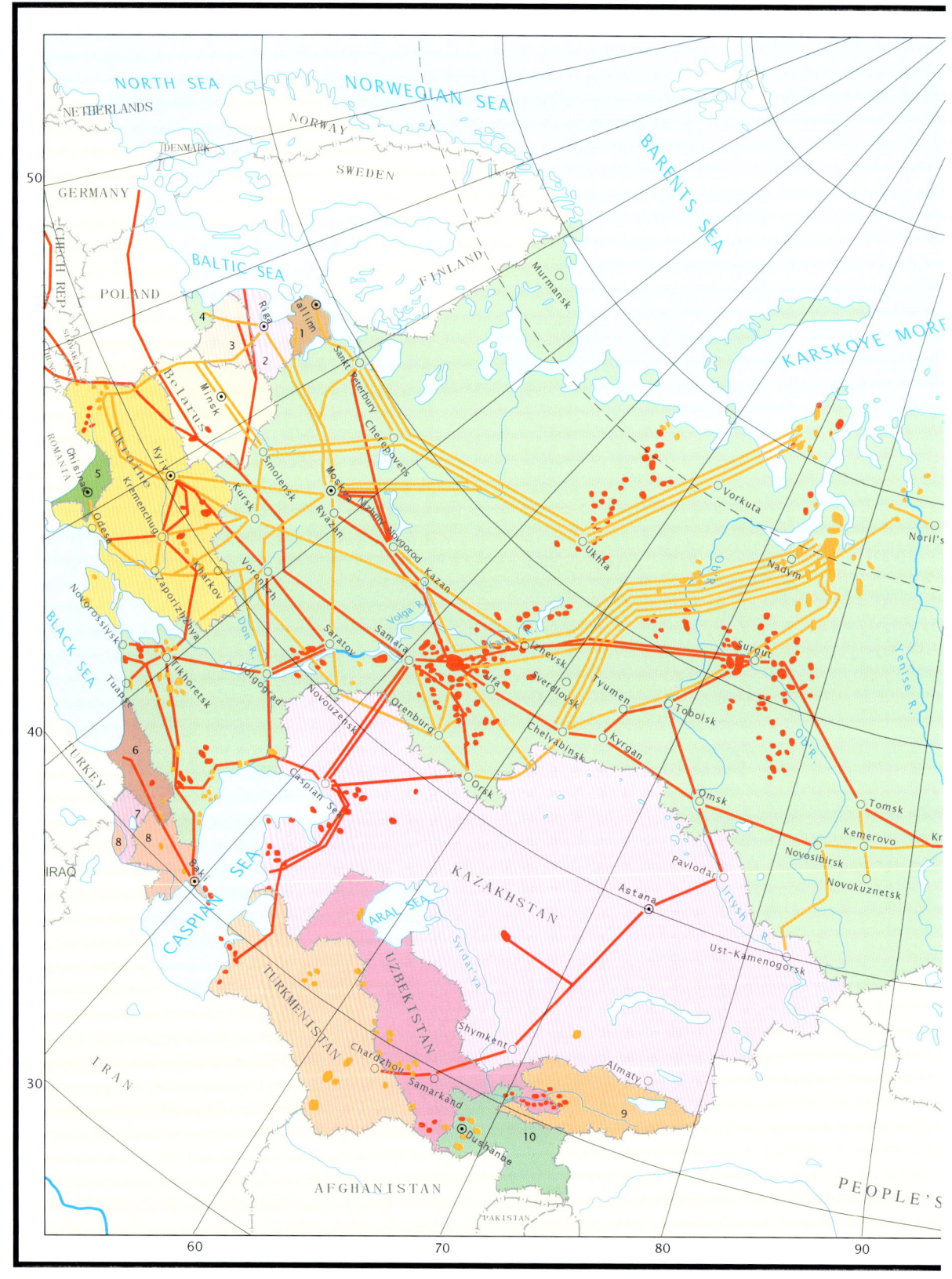

Country: 1 Estonia 2 Latvia 3 Lithuania 4 Russia 5 Moldova 6 Georgia 7 Armenia 8 Azerbajian 9 Kyrgyzstan 10 Tajikistan

NEIGHBOURING COUNTRIES   CHAPTER 72

72 Oil and Gas Pipelines of Russia and Neighbouring Countries

The uniqueness of the oil and gas industry in Russia arises from the fact that most of the output in Russia is transported via large diameter pipelines, whereas the prevailing mode of transportation in the rest of the world is by marine tankers. It is therefore important to discuss the pipeline networks of Russia in considering global oil and gas trade flows. Russia can be described as the most important of the oil- and gas-producing nations, with annual production of 488 million t of oil and 643 billion m^3 of gas in 2008. These production levels ranked the country as the second and fifth respectively in the world. Russian oil and gas production is either consumed domestically, or destined for the European markets, as well as, in the near future, to Asian countries such as China and Japan.

Oil pipelines have been constructed all over Russia over the past 120 years. Currently the oil and gas pipeline network is 219,000 km long in the country, consisting of 151,000 km of gas pipelines, 486,000 km of crude oil pipelines and 194,000 km of petroleum products pipeline. All natural gas production, 99 per cent of crude oil production and more than 50 per cent of petroleum products are transported by trunk pipelines. The transportation of gas, crude oil and petroleum products accounts for 55.4 per cent, 40.3 per cent and 4.3 per cent of the total trunk pipeline carrying capacity, respectively. In 2008 the total production of natural gas was 601 billion m^3 in the country. The total gas supply to its domestic market was 447.1 billion m^3 and exports totalled 191 billion m^3.

Large gas pipeline systems in Europe

During the 11th Five-Year Plan period from 1981 to 1985 of the Former Soviet Union, the giant Urengoy central station gas pipeline system was completed, with significant implications for global oil and gas pipeline transportation. The network starts in the giant Urengoy gas field in the Arctic Circle, traversing the north of the country all the way to the west, terminating in Uzhgorod in western Russia. It is a giant gas pipeline system, consisting of six gas pipelines.

The first gas pipeline was the Jarmar pipeline, dedicated solely to gas exports and with the largest quantity of construction materials. It has a diameter of 1.42 m. The total length is 4451 km within the Former Soviet Union, passing through 945 km of water area, crossing more than 700 rivers and running through 2128 ha of forest and permafrost areas. It still holds the world record for construction complexity, number of pumping stations and the highest gas compression rate. Construction of this pipeline was completed in four months, compelling proof of the advanced construction technology and logistical methods available.

The second pipeline was the Urengoy–Moscow gas pipeline with a length of 2976 km, most of which (2297 km) being 1420 mm in diameter. It was completed in May 1981, one year after construction commenced. The third pipeline was the Urengoy–Pskovsk gas pipeline. It is 3341 km long and 1420 mm in diameter. Construction started in 1981 and was completed in 1983, with 260 km passing through marshland. The fourth was the Urengoy–Petrov gas pipeline with a length of 2731 km and 1420 mm in diameter. Construction started in 1981 and the pipeline was brought

World Atlas of Oil And Gas Basins, First Edition. Li Guoyu.
© 2011 John Wiley & Sons, Ltd. Published 2011 by John Wiley & Sons, Ltd.

on stream in April 1982. The fifth and sixth pipelines in the system run parallel to each other, with individual lengths of 3429 km and 3384 km. They were both constructed from 1985 to 1989.

Construction of all six pipelines was effected in dedicated land corridors which ensured efficient centralized construction logistics along favourable basement and subsurface foundation conditions. The total length of this pipeline system is 20,000 km, most of which contains pipeline sections of 1420 mm diameter. The installed gas transmission capacity of the network is 180 billion m^3, supplying gas to both the domestic market and crucially to the markets of many European countries. The construction of this massive system had a very significant effect on Russian gas production. Gas production was 289.3 and 430 billion m^3 in 1975 and 1980, respectively. This rose to 644.8 billion in 1985 and hit 770 billion in 1990. By the end of 1989, 210,000 km of trunk gas pipelines had been completed in the Former Soviet Union.

The Friendship trunk oil pipeline

The launch of the Friendship oil pipeline project was as a result of the Cooperative Construction of Oil Pipelines Agreement entered into by the Former Soviet Union Government, the Czech Government, the Hungarian Government, the Polish Government, and the government of the then Germany Democratic Republic (East Germany) in 1959. Construction started in December 1960 and was concluded in October 1964. The total length of the pipeline is 4000 km. The pipeline is also referred to as the Druzhba Pipeline. The Friendship Trunk Oil Pipeline Company is an important part of the Transneft network, which is the largest oil transportation company in the world and operates the world's largest trunk oil pipeline system. Commencing in 2005, an ambitious programme of upgrading and enhancement of the Friendship pipeline commenced. These plans include construction of strategic links and extensions to the pipeline that will considerably increase its oil delivery capacity.

The First Russia to Asia pipeline

Russia recognizes the importance of China as a consumer of Russian oil and gas in the future. For this reason, discussions and negotiations have taken place over the past few years regarding the construction of a dedicated pipeline from Russian fields to China. Strategically, such a pipeline will improve the development of oil and gas resources in eastern Siberia, with additional markets to Japan and South Korea. Thus, the Russian government decided to build the Taishet-Nakhodka pipeline in 2005, from Taishet in Russia to Nakhodka in the north Pacific region. At the same time, one branch oil pipeline will be constructed from Skolovogino to China.

VOLGA-URALS BASIN

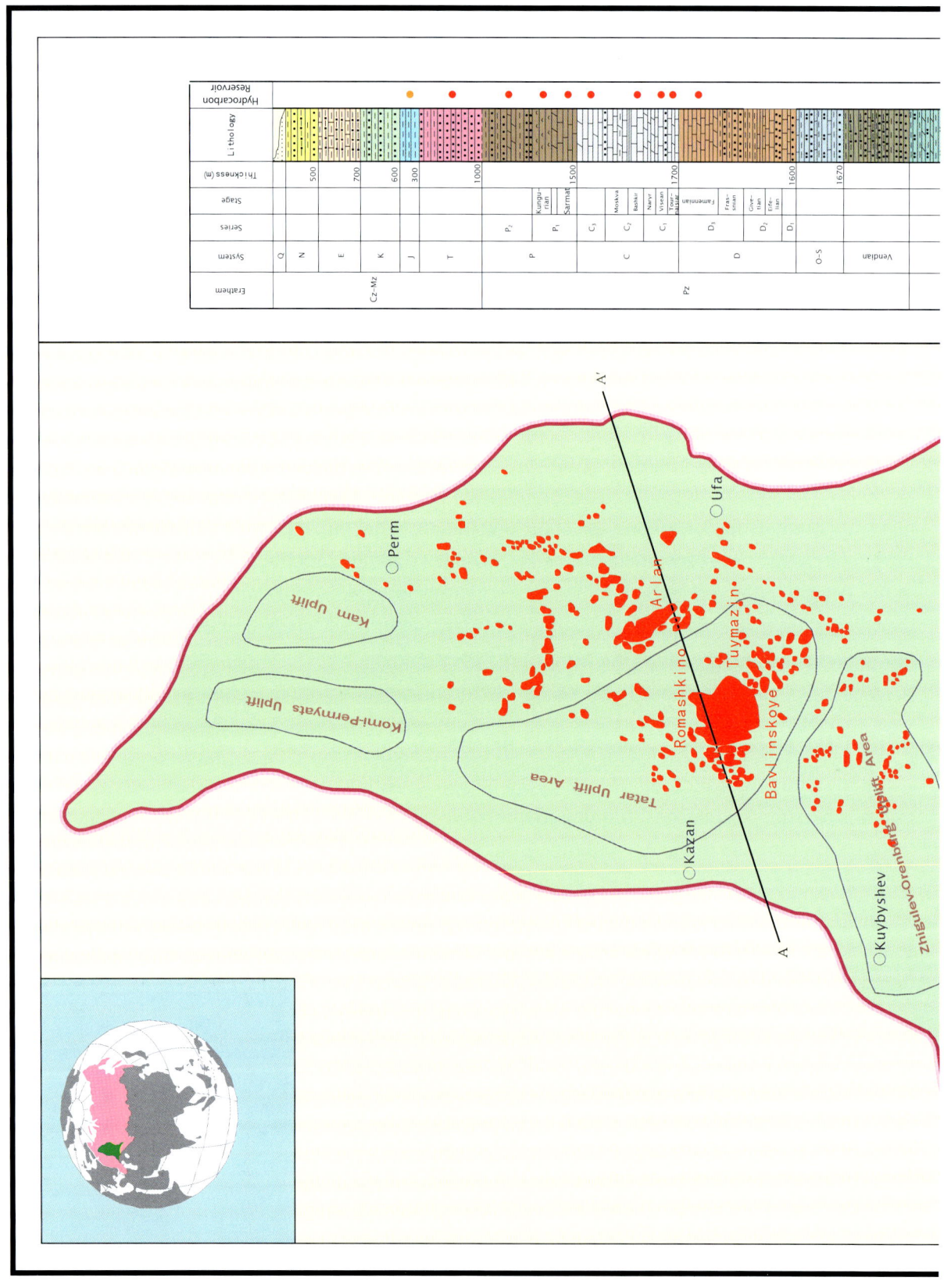

CHAPTER 73

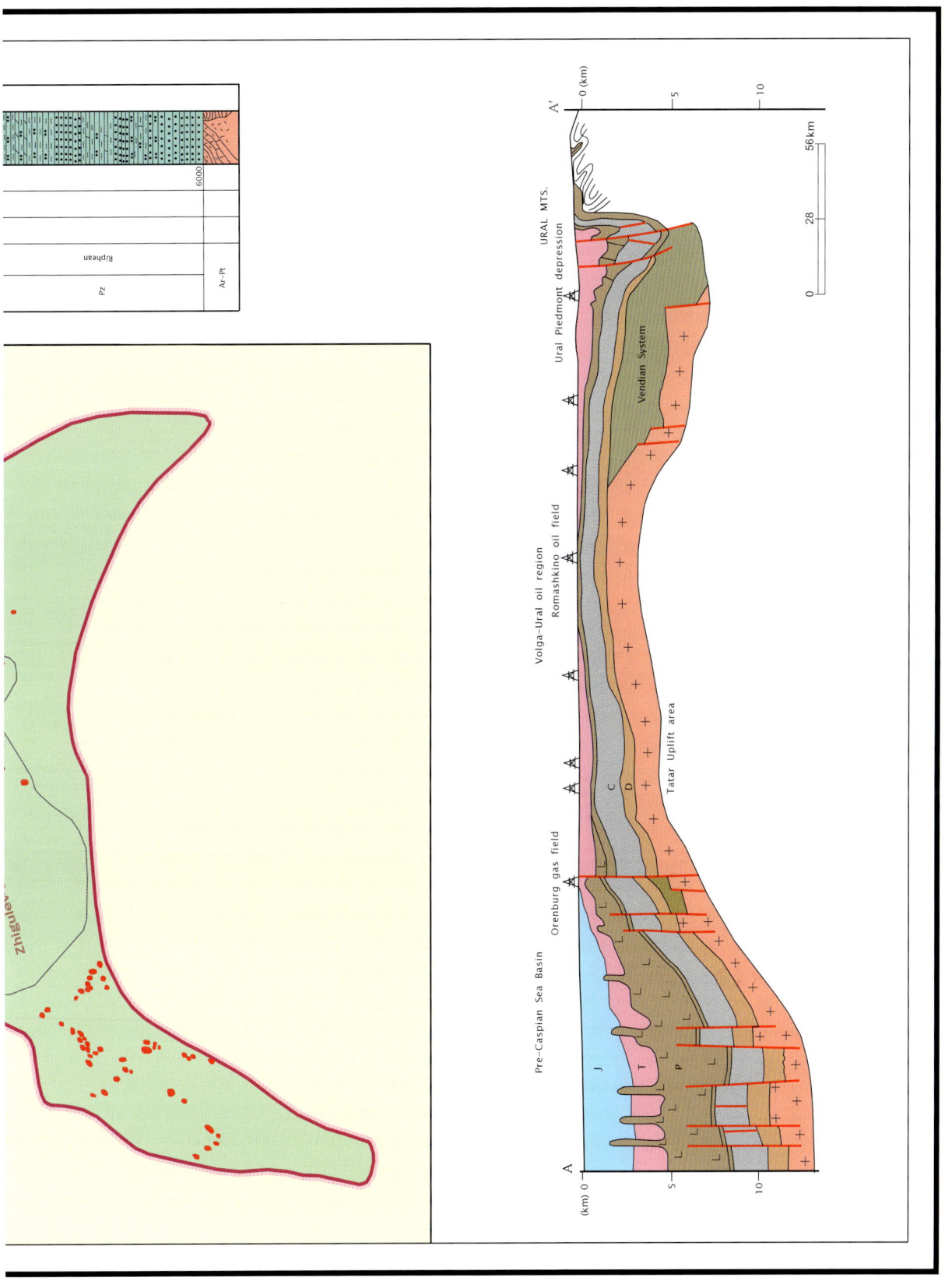

319

73 Volga-Urals Basin

Geography

The Volga-Urals Basin covers some 900,000 km² in eastern Europe, Russia and northern Kazakhstan. The province is bordered to the south by the Pre-Caspian (or North Caspian) Basin.

History

The first exploration activity in the basin was conducted in the 1920s. The first giant find was made in 1944, when oil was found by deeper drilling in the Tuymazy field in Middle Devonian clastics. Shortly afterwards, the supergiant Romashkino field (Fig. 73.1) was discovered in a similar play. In 1966, the supergiant Orenburg gas field was found in the southeast of the basin in Permian carbonates. The Volga-Urals Province is one of the richest Palaeozoic basins in the world and contains significant hydrocarbon reserves (Table 73.1), including two supergiant fields: Romashkino (2 billion t) and Orenburg (159.9 billion m³). Development drilling began in the early 1930s. Following the development of Romashkino in 1949, a large number of wells have been drilled, with numbers peaking at over 3500 wells a year in 1987. Production began in 1929 at low levels, and by 1946 output had reached 3.8 billion t. Romashkino was put on pilot production in 1949 and the province rapidly took over from the South Caspian Basin as the Former Soviet Union's major producing area, with production rates reaching 250 million t in 1965. In 1994 annual output was 78.5 million t, while annual oil production was about 30 million t in 2006.

The history of development of the Volga-Urals Basin is typical of any oil-producing area, and can be divided into five stages: initial, development, peak, decline and stable stages. Annual oil production was 2.8 million t in the initial stage in 1945, 105–200 million t in the development stage in 1960–1970, 226 million t in the peak stage in 1975, 110 million t in the decline stage in 1990 and 36.2 million t in the stable stage in 1999–2007. For example, see Fig. 71.1 showing the profile of Tatarstan oil production. The Tatarstan is a major oil region of the Volga-Urals Basin. By August of 2007 the cumulative oil production reached 3 billion t. Annual production was 2 million t in 1952, with peak production of 103 million t in 1975, but it had dropped to 25 million t in 2006. There are plans to keep this production level over the next 40–50 years when another 1 billion t of oil will be produced.

Regional geology

The Volga-Urals Basin is located in the eastern part of the Russian platform, having an area of 200,000 km². A few large uplifts of Precambrian crystalline basement controlled the distribution of many giant oil fields in the Volga-Urals Basin. For example the Romashkino and Tyimas oil fields are associated with the Tatar Uplift. In Riphean, Vendian and Cambrian times there was a period of rifting. The sedimentary cover comprises Palaeozoic, Mesozoic and Cenozoic sediments, but mainly Palaeozoic. Although source rocks have been recognized in the Riphean to Permian sediments, the major source rocks in the basin are the Domanik Formation shales and carbonates of Frasnian–Famennian

World Atlas of Oil And Gas Basins, First Edition. Li Guoyu.
© 2011 John Wiley & Sons, Ltd. Published 2011 by John Wiley & Sons, Ltd.

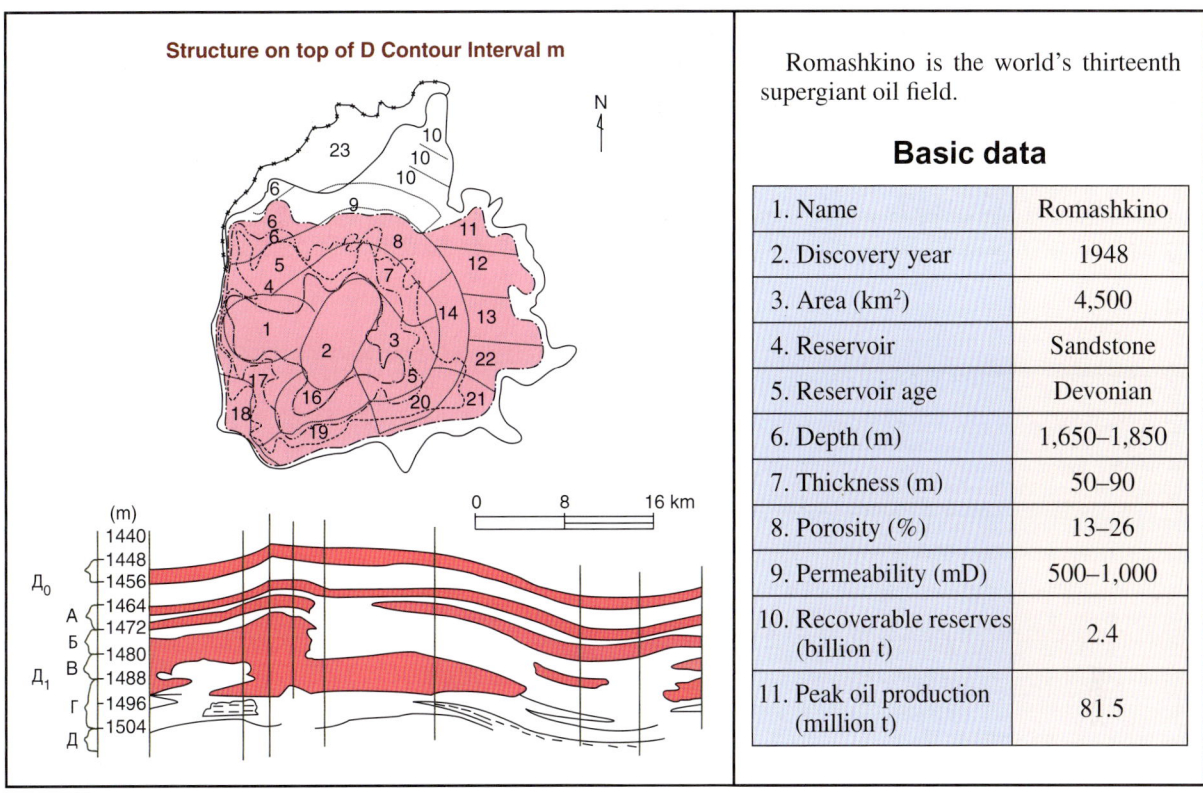

Fig. 73.1 Details of the Romashkino oil field.

Table 73.1 Major oil and gas fields of Volga-Urals Basin

Name	Discovery year	Recoverable reserves		Depth (m)	Trap	Reservoir	
		Oil (million t)	Gas (billion m³)			Age	Lithology
Romashkinskoye	1948	2,031	—	1700	Anticline	D	Sandstone
Mukhanovskoye	1945	214	—	2100	Anticline	C	Sandstone
Novo Elknov	1955	182	—	1660	Anticline	D	Sandstone
Shkapovskoye	1944	175	—	2000	Anticline	D	Sandstone
Tuymazy	1937	140	—	1200	Anticline	D	Sandstone
Kuleshovskoye	1958	106	—	1600	Anticline	C	Sandstone
Arlanskoye	1955	574	—	1200	Anticline	C	Sandstone
Orenburg	1966	490	1599	1600	Anticline	P	Carbonate rock

age, which have total organic carbon (TOC) values of up to 12.4 per cent and are generally mature for hydrocarbon generation throughout the basin. Reservoirs are Middle Devonian to Jurassic strata, but the most important are Middle Devonian sandstones. Five regional seal units have been recognized in the province, including Lower Frasnian, Tournisian, Visean and Lower Moscovian shales and carbonates and Kungurian salt. Traps are mainly Middle Devonian clastics.

TIMANO-PECHORA BASIN

CHAPTER 74

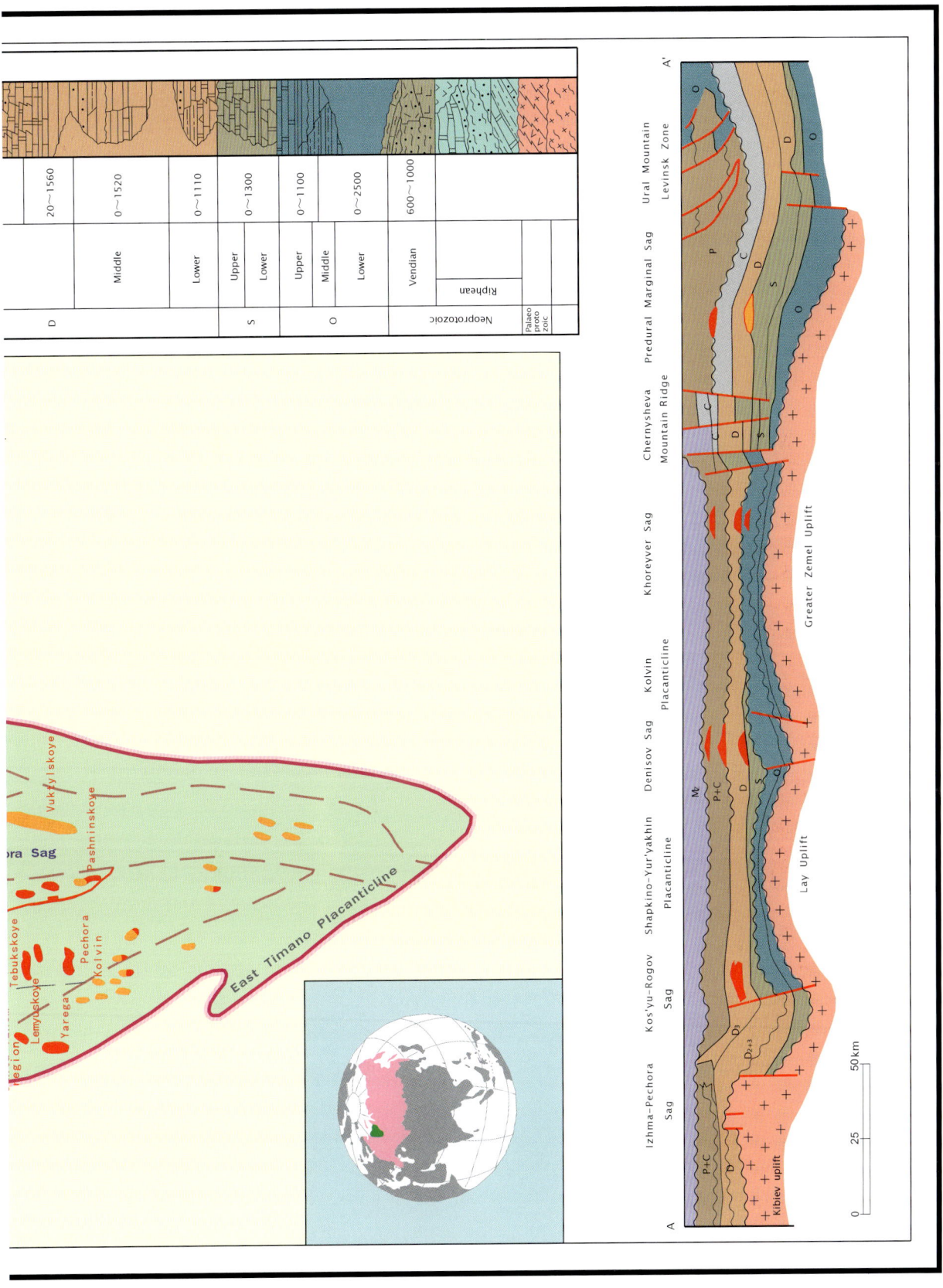

74 Timano-Pechora Basin

Geography

Timano-Pechora Basin is a large triangular basin in the northeast of European Russia with an area of 495,000 km^2. It is bounded to the east by the Ural Mountain Belt and the islands of Novaya Zemlya and to the southwest it is bordered by the Timano Ridge. The basin extends offshore into the Pechora Sea and is separated from the South Barents Sea Basin by the South Barents Fault Zone.

History

Oil seeps were first observed in the 1600s in the southwest of the basin near Ukhta. The Devonian surface oil sand led to exploitation in 1745 but eventually became too expensive to exploit. The first oil well was completed in 1890 in the Ukhta area and the first oil field was discovered in 1930. Thirteen oil and gas fields were discovered during 1959–1964, which stimulated further exploration. The total number of discovered oil and gas fields is 195, and peak annual production was 20.4 million t in 1980. Details of seven oil and gas are listed in Table 74.1. The first seismic surveys were undertaken in the early 1980s and the first wildcat well based on these surveys was drilled in 1981. The Pomor gas field was discovered in 1985 and was followed by successes at Gulyayev Severnyy (1986) and Prirazlomnoye (1989). Only ten wildcat wells have been drilled in the offshore sector.

Regional geology

The basin is located within the Timano-Pechora platform. It is confined to the Pechora Syneclise. The basin extends to the shelf of the Barents Sea. The basement occurs at depths of 1–3 km to 6–10 km. The sedimentary cover comprises Phanerozoic rocks in which oil- and gas-bearing complexes have been identified in Ordovician, Silurian, Devonian, Carboniferous, Permian and Triassic deposits. Among them, terrigenous and carbonate reservoirs are important. The distinctive feature of the basins is the presence of Mesozoic–Cenozoic rifts, troughs and horst–graben dislocations, which complicate the geology of the basins..

The sedimentary cover represents all stages of the Phanerozoic and attains up to 10–12 km in thickness. These are Palaeozoic and Upper Proterozoic deposits of marine origin; Mesozoic strata are represented by essentially continental red-coloured sediments with widely developed salt- and coal-bearing formations. Source rocks have been identified in the sediments of both marine and continental origin.

Reservoir rocks are associated with sandy deposits, mostly in Mesozoic strata, and also with Palaeozoic marine carbonate–terrigenous rocks. Clayey seals and sealing strata associated with gypsiferous rocks and rock salt are widespread. Regional seals are represented by Middle Devonian clays and a Lower Permian chemogenic sequence. The main source rocks relate to Upper Devonian bituminous clayey limestones and shales (Domanik Formation) as well as Lower Carboniferous clayey limestones. Most of the fields are confined to brachyanticlinal uplifts and

Table 74.1 Major oil and gas fields of Timano-Pechora Basin

Name	Discovery year	Recoverable reserves		Depth (m)	Trap	Reservoir		Remarks
		Oil (million t)	Gas (billion m³)			Age	Lithology	
Vozey	1972	140.0	—	1600	Anticline	D–C–P	Sandstone	
Usanovskoye	1963	91.0	—	1900	Anticline	C	Carbonate	
Layavozhskoye	1965	70.0	106	2000	Anticline	P	Carbonate	
Vuktylskoye	1964	—	483.6	2600	Anticline	C–P	Carbonate	Gas field
Vasilkovskoye	1970	—	89.2	1460	Anticline	T	Sandstone	Gas field
Vaneivisskoye	1973	—	85.4	2120	Anticline	C	Sandstone	Gas field
Kumzhinskoye	1974	—	99.1	1480	Anticline	T	Sandstone	Gas field

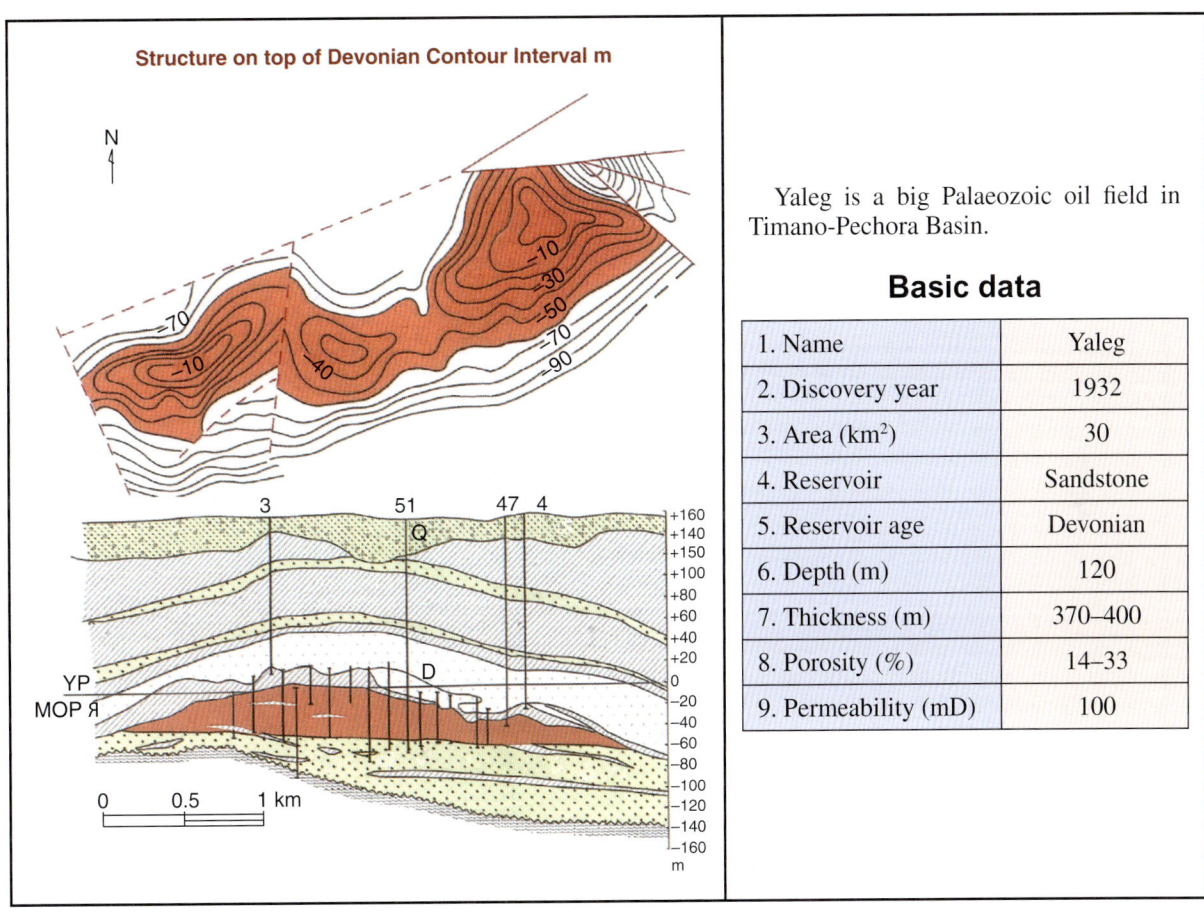

Fig. 74.1 Details of the Yaleg oil field.

reef masses. A number of large fields have been discovered in the basin, including the Usinskoye and Vozeiskoye oil fields, and the Vuktylskoye and Kyrtayelskoye gas fields. The amount of original in-place resources of the basin is classified as large. Traps are represented by structural and anticlinal types. Oil and gas accumulations are multilayered and mainly confined to Palaeozoic (e.g. Fig. 74.1) and Mesozoic deposits.

WEST SIBERIA BASIN

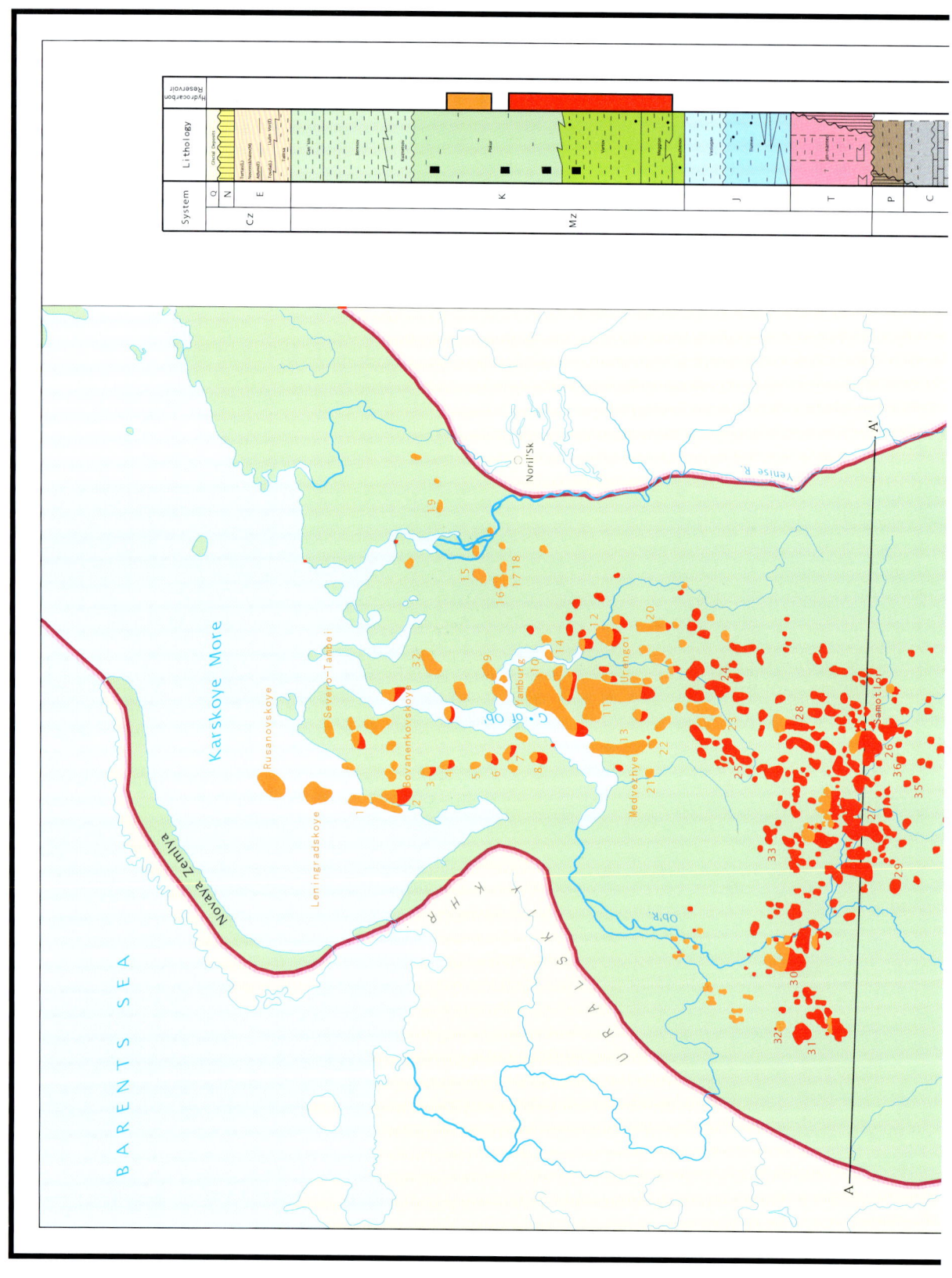

CHAPTER 75

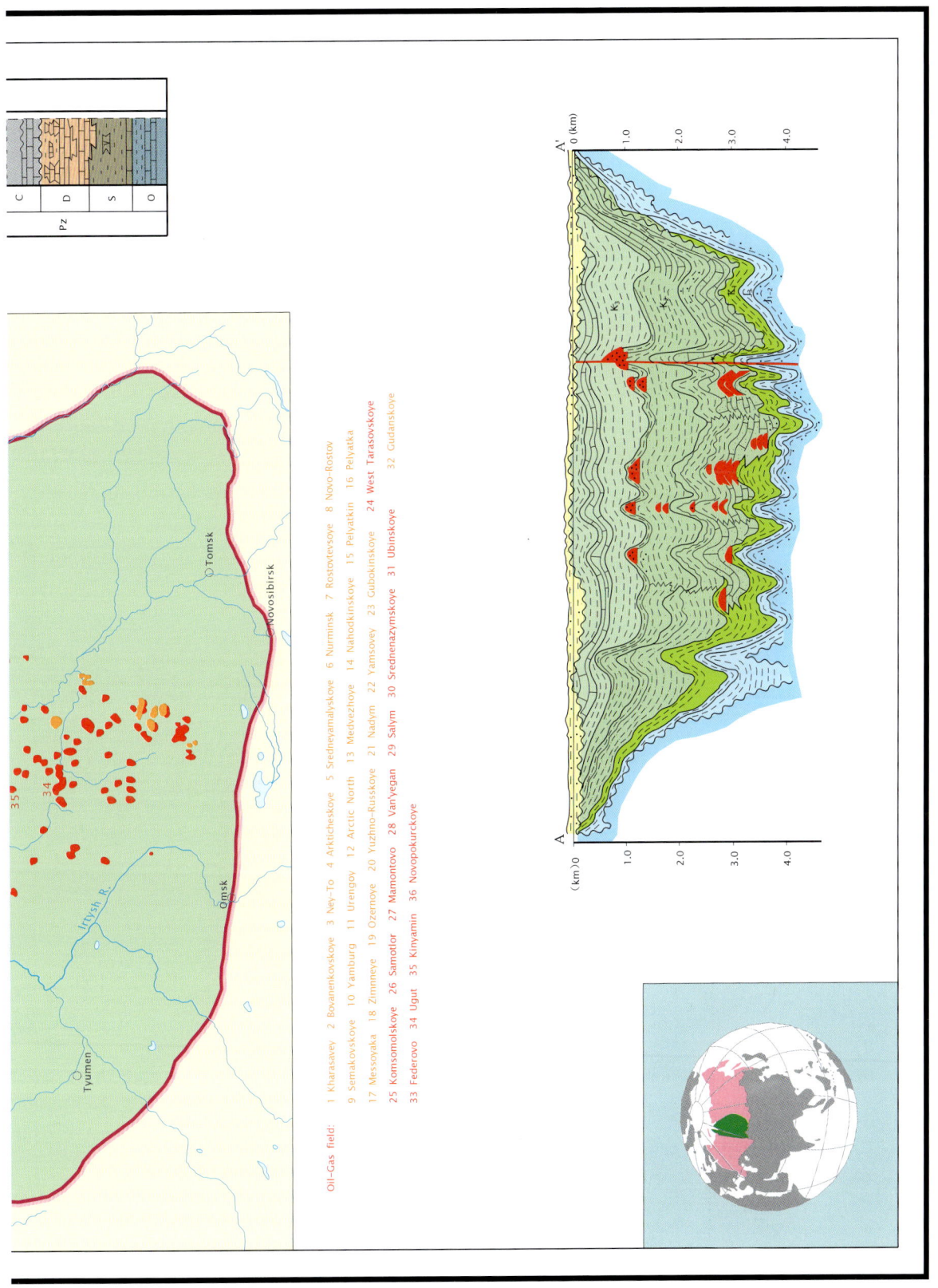

75   West Siberia Basin

Geography

The West Siberia Basin is part of the West Siberian Plain, which is the world's largest and extends east from the Urals to the Yenisey River.

History

Oil exploration started in 1930. Regional geological, geophysical and drilling works began in 1948. In the early stages exploration was concentrated along railway lines in the southern part of the West Siberia Basin. About 20 wells were drilled on 20 local structures, but this plan of exploration was unsuccessful as it did not lead to discovery of any significant oil and gas fields. Deep geological study, however, showed prospects in the central part of this basin, and since 1961 super giant oil and gas fields have been discovered almost continuously, such as Samotlor oil field in 1965 and Urengoy gas field in 1966 (Fig. 75.1). At the time of this compilation, a total of 417 oil and gas fields had been discovered in the West Siberia Basin. Annual oil production increased

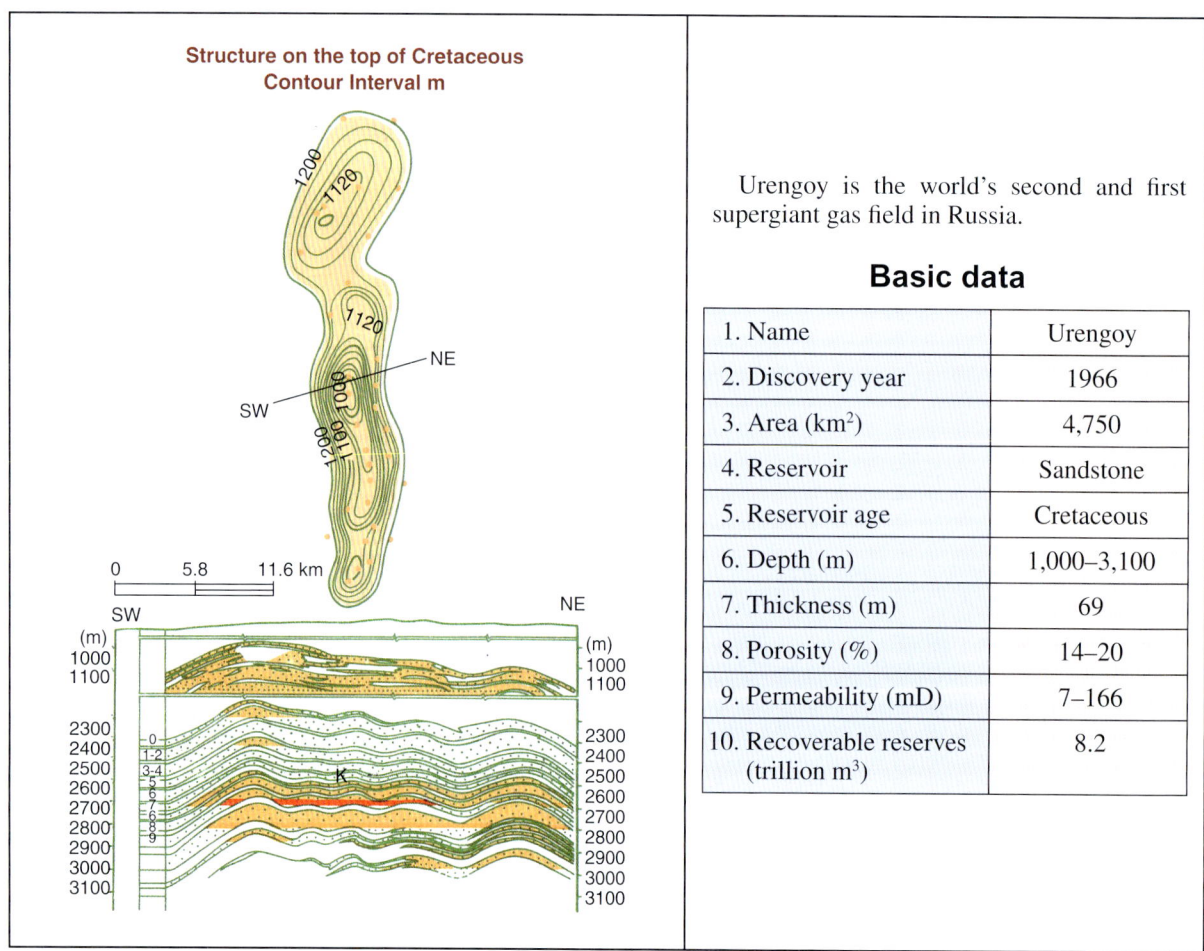

Urengoy is the world's second and first supergiant gas field in Russia.

Basic data

1. Name	Urengoy
2. Discovery year	1966
3. Area (km^2)	4,750
4. Reservoir	Sandstone
5. Reservoir age	Cretaceous
6. Depth (m)	1,000–3,100
7. Thickness (m)	69
8. Porosity (%)	14–20
9. Permeability (mD)	7–166
10. Recoverable reserves (trillion m^3)	8.2

Fig. 75.1 Details of the Urengoy gas field.

World Atlas of Oil And Gas Basins, First Edition. Li Guoyu.
© 2011 John Wiley & Sons, Ltd. Published 2011 by John Wiley & Sons, Ltd.

rapidly after the initial discoveries: 4 million t in 1965, 148 million t in 1974, 312 million t in 1980 and 414 million t in 1988 (peak production). After the disintegration of the Former Soviet Union, oil production declined to 200 million t in 2000, and then increased to more than 300 million t in 2006. The West Siberia Basin is rich in oil and gas resources, and is one of the world's largest oil and gas producing regions. Of the 417 hydrocarbon fields known at the time of this compilation, 307 are oil fields and 110 are gas–condensate fields.

Regional geology

The West Siberia Basin occupies an area of 3.3 million km^2. The heterogeneous basement is Hercynian in the west of the basin and Baikalian of the Hercynian in the south of the basin. The depth to basement is 1 km to 12 km. The sedimentary cover consists of a system of grabens filled with essentially Triassic and lowermost Jurassic deposits.

Commercial oil and gas accumulations are known from Palaeozoic carbonate deposits, Lower and Middle Jurassic sandy–clayey rocks(with thicknesses up to 1–5 km), Upper Jurassic marine and continental sandy–clayey sediments (being 200–300 m in thickness) and Cretaceous sandy-clayey deposits(up to 3 km in thickness). Triassic terrigenous rocks of are prospective for oil and gas in the northern areas of the basin. Details of 14 such oil/gas fields discovered between 1961 and 1976 are provided in Table 75.1. Other reservoirs are Jurassic sandstones with porosity varying from 14 per cent to 35 per cent, and Cretaceous sandstones with porosity of up to 25 per cent. Regional seals are Upper Jurassic–Lower Cretaceous Bazhenovsky clays (Vysotsky, 1995).

The huge potential for future discovery of new oil and gas accumulations cannot be overemphasized.

Table 75.1 Major oil and gas fields of the West Siberia Basin

Name	Discovery year	Recoverable reserves		Depth (m)	Trap	Reservoir		Remarks
		Oil (million t)	Gas (billion m^3)			Age	Lithology	
Samotlor	1965	2116	—	2200	Anticline	K	Sandstone	Oil
Fedorovskoye	1963	350	52.6	1700	Anticline	K	Sandstone	Oil
Sovetskoye	1962	588	—	3300	Anticline	K	Sandstone	Oil
Ust Balysk	1961	322	—	2600	Anticline	K	Sandstone	Oil
Zapadno Surugutskoye	1962	280	—	2300	Anticline	K	Sandstone	Oil
Mamontovskoye	1965	245	—	1900	Anticline	K	Sandstone	Oil
Pravdinskoye	1964	210	—	2100	Anticline	K	Sandstone	Oil
Yuzhno Cheremshanskoye	1969	202	—	1800	Anticline	K	Sandstone	Oil
Urengoy	1966	—	805.5	1200	Anticline	K	Sandstone	Gas
Yamburg	1969	—	475.6	1000	Anticline	K	Sandstone	Gas
Bovanenkovskoye	1971	—	415	1200	Anticline	K	Sandstone	Gas
Zapolyarnoye	1965	—	267	1200	Anticline	K	Sandstone	Gas
Kharasaveyskoye	1974	—	126.5	1500	Anticline	K	Sandstone	Gas
Kruzernshternovskoye	1976	—	112	660	Anticline	K	Sandstone	Gas

EAST SIBERIA BASIN

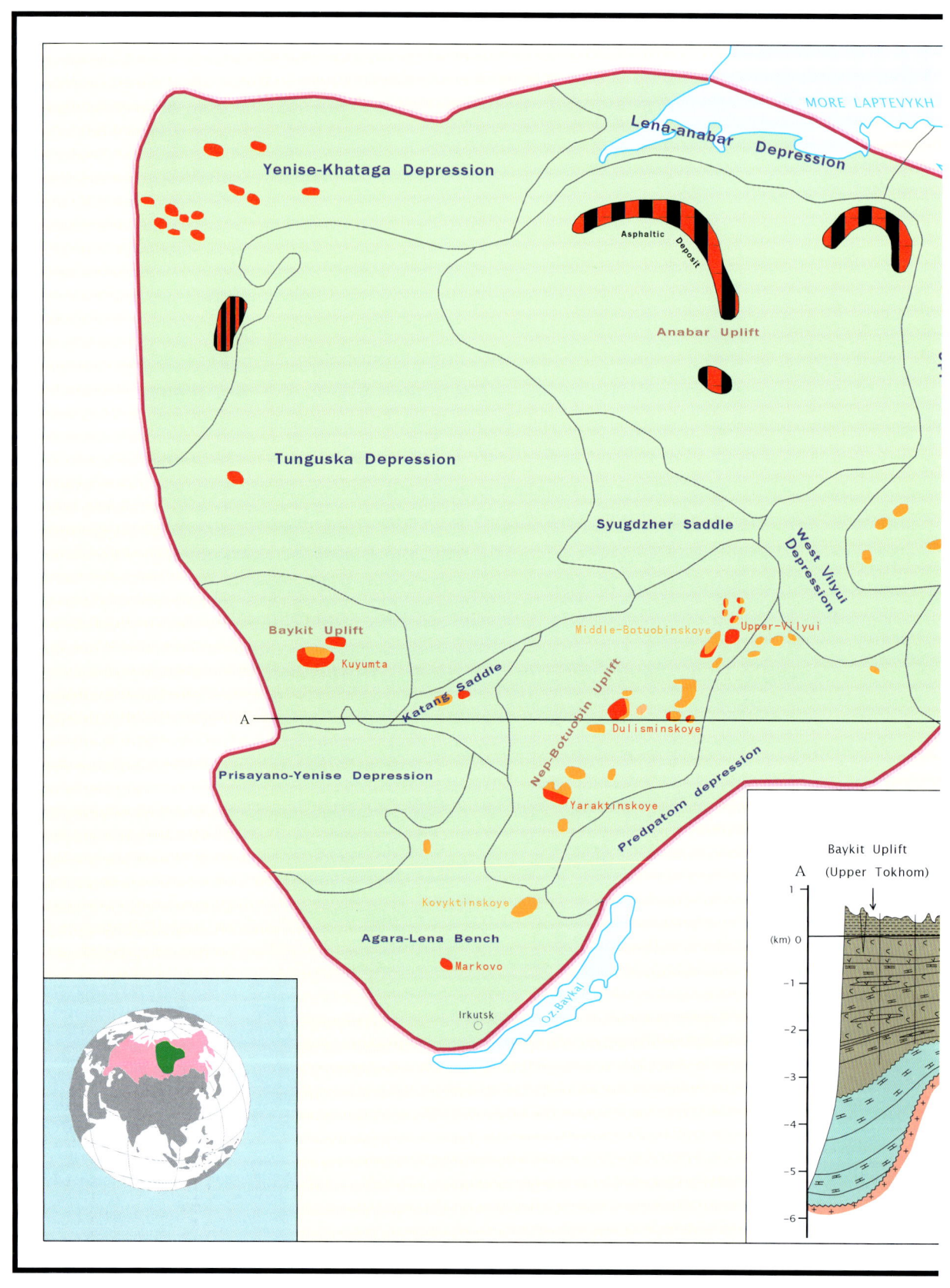

CHAPTER 76

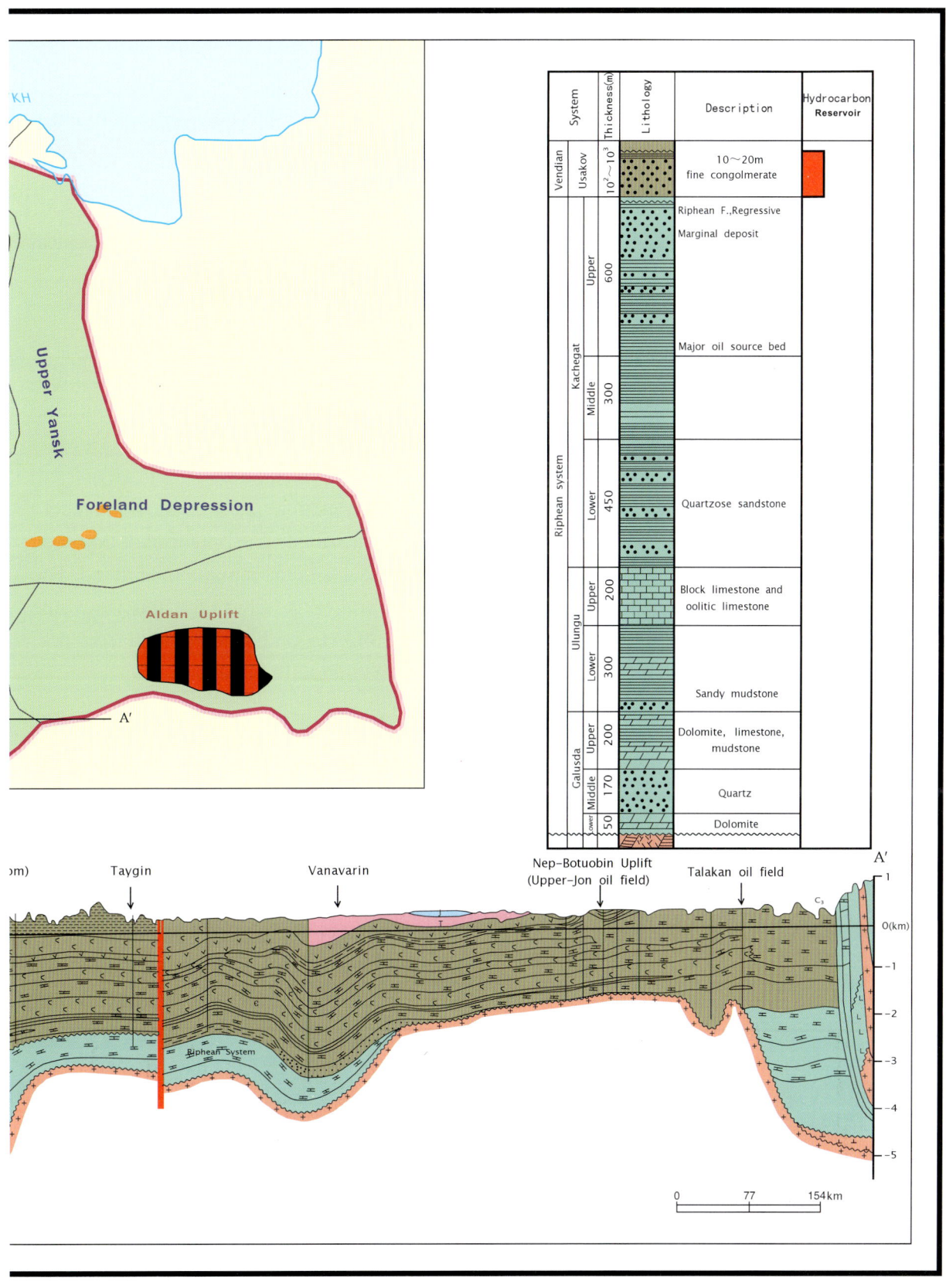

76 East Siberia Basin

Geography
The East Siberia Basin is located in the eastern part of Russia between the Lena and Yinist Rivers, to the north of Lake Baikal and bounded by the Leptev Sea to the north.

History
The first hydrocarbon field discovered in the basin was the Markov field in 1962, with proven and probable reserves of 17 billion m^3 of gas, 2.1 million t of condensate and 1.3 million t of oil. The second largest field to date is the giant Sredne-Botuobin gas–condensate field, found in 1970. Sredne-Botuobin is a structural trap in both Proterozoic and Lower Cambrian strata. Proven plus probable reserves are 168 billion m^3 of gas and 20 million t of condensate. In both fields, the major reserves are in Proterozoic strata. The largest field, Verkhnevilyuy was found in 1975, and has 285 billion m^3 of proven plus probable reserves of gas and about 35 million t of condensate. Yaraktin oil field was discovered in 1971 in Proterozoic strata. Proven, probable and potential reserves are about 34 million t. The field is a very large stratigraphic trap. The estimate of reserves is subject to upward and/or downward revision as stepout and infill wells are drilled. The potential of the basin is very high. Several hundred, perhaps several thousand, oil and gas fields remain to be found. The province appears to be gas prone, but the discovery of the Yaraktin oil field indicates that some areas will have oil production. In terms of ultimate recovery, this region has the potential for producing 100 billion barrels of oil and 200 trillion ft^3 of gas, together with condensate.

Regional geology
The East Siberia Basin is an attractive area for oil and gas exploration because it is one of the three largest sedimentary basins in the world, with an area of 3 million km^2. The other two are the Persian Gulf and West Siberia Basins. The main difference between these basins is the oil- and gas-producing strata, i.e. Cambrian in the East Siberia, Meso-Cenozoic in the Persian Gulf and Jurassic–Cretaceous in the West Siberia Basins.

The Siberia platform is characterized by the Archaean basement exposed on the Aldan Shield and the Anadyr Massif. The sedimentary cover comprises Middle Proterozoic Riphean, Palaeozoic and Mesozoic deposits. The largest structures here are the Angara-Lena, Tungus and Vilyuy Syneclises. The basement of the platform is Precambrian.

A minimum of nine commercial and 13 non-commercial discoveries were made from 1962 in the Lena-Tunguska petroleum province of the East Siberia Basin. However, the petroleum potential of this province has scarcely been tapped because the prospective area is greater than 1,737,000 km^2. Discovered reserves are in Proterozoic marine terrigenous clastic reservoirs and Early Cambrian fractured carbonates interbedded with thick evaporites. The Cambrian reservoirs are small and are associated with salt swells and salt pillows, whereas the Proterozoic discoveries are large and are in both stratigraphic and structural traps. For example, the Lena-Vilyuy Basin is 350,000 km^2 in

East Siberia Basin

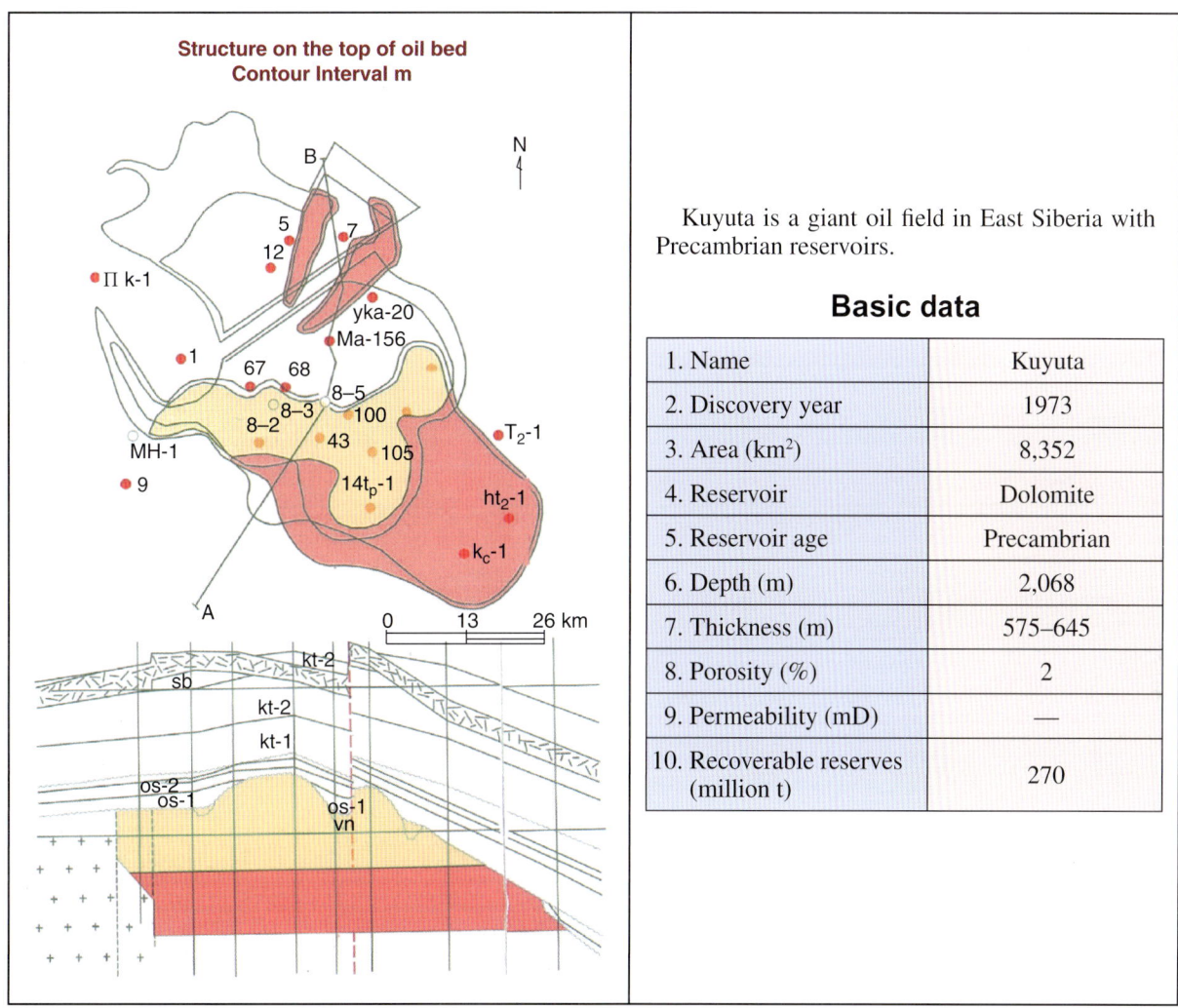

Fig. 76.1 Details of the Kuyuta oil field.

area and formed in the Verkhoyansky Foredeep, which overlies the Vilyuy Syneclise. The basement of the basin is comprises Archaean–Proterozoic crystalline rocks and occurs at depths ranging from 3–6 km to 10–16 km. Another example is the Kuyuta oil field (Fig. 76.1).

The sedimentary cover of the East Siberia Basin comprises platform and molassic sediments. At present, only the upper part of the cover dating back to the Permian Period has been studied. Nine gas and condensate fields have been discovered in anticlinal traps of the basin. The petroleum potential of the basin is estimated to be high. For example, in addition to the Lena-Vilyuy petroleum basin, the Yenisey-Khatanga petroleum basin occupies an area of about 250,000 km². The heterogeneous basement of this basin comprises metamorphosed Precambrian, Lower and Middle Palaeozoic rocks of. The depth to basement ranges from 3 km to 8–12 km. The sedimentary succession is represented by Palaeozoic terrigenous–carbonate and terrigenous deposits, reaching up to 5 km in thickness, and essentially terrigenous Mesozoic sequences up to 8 km in thickness. Petroleum potential is confined to Jurassic and Cretaceous sandstones with porosity of 25–32 per cent. Lower Cretaceous (Hauterivian Stage) clays serve as the regional seal. Fourteen gas and condensate fields and one oil–gas–condensate field have been discovered in the basin. The traps are of predominantly anticlinal type.

NORTH KAVKAZ BASIN

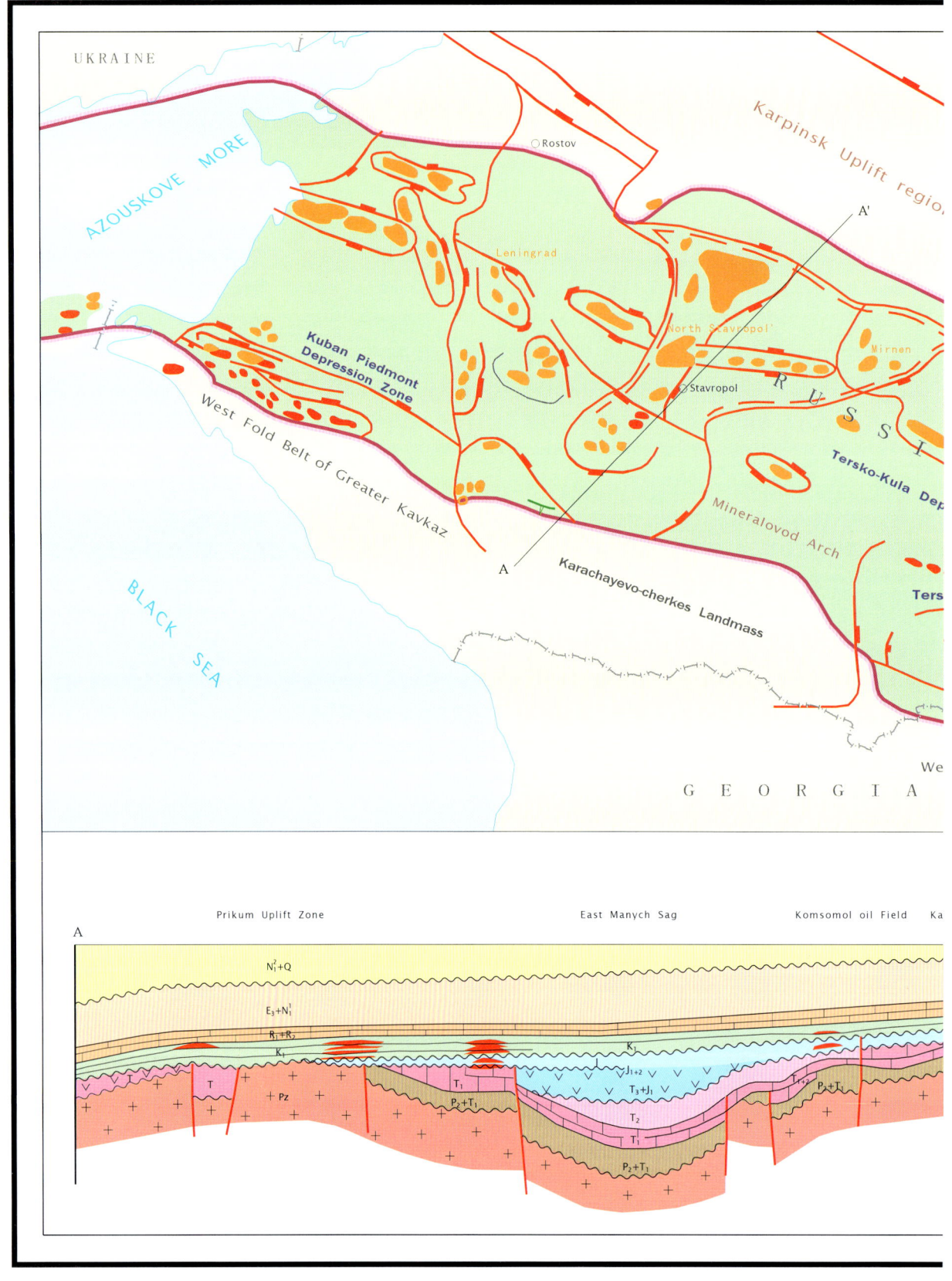

CHAPTER 77

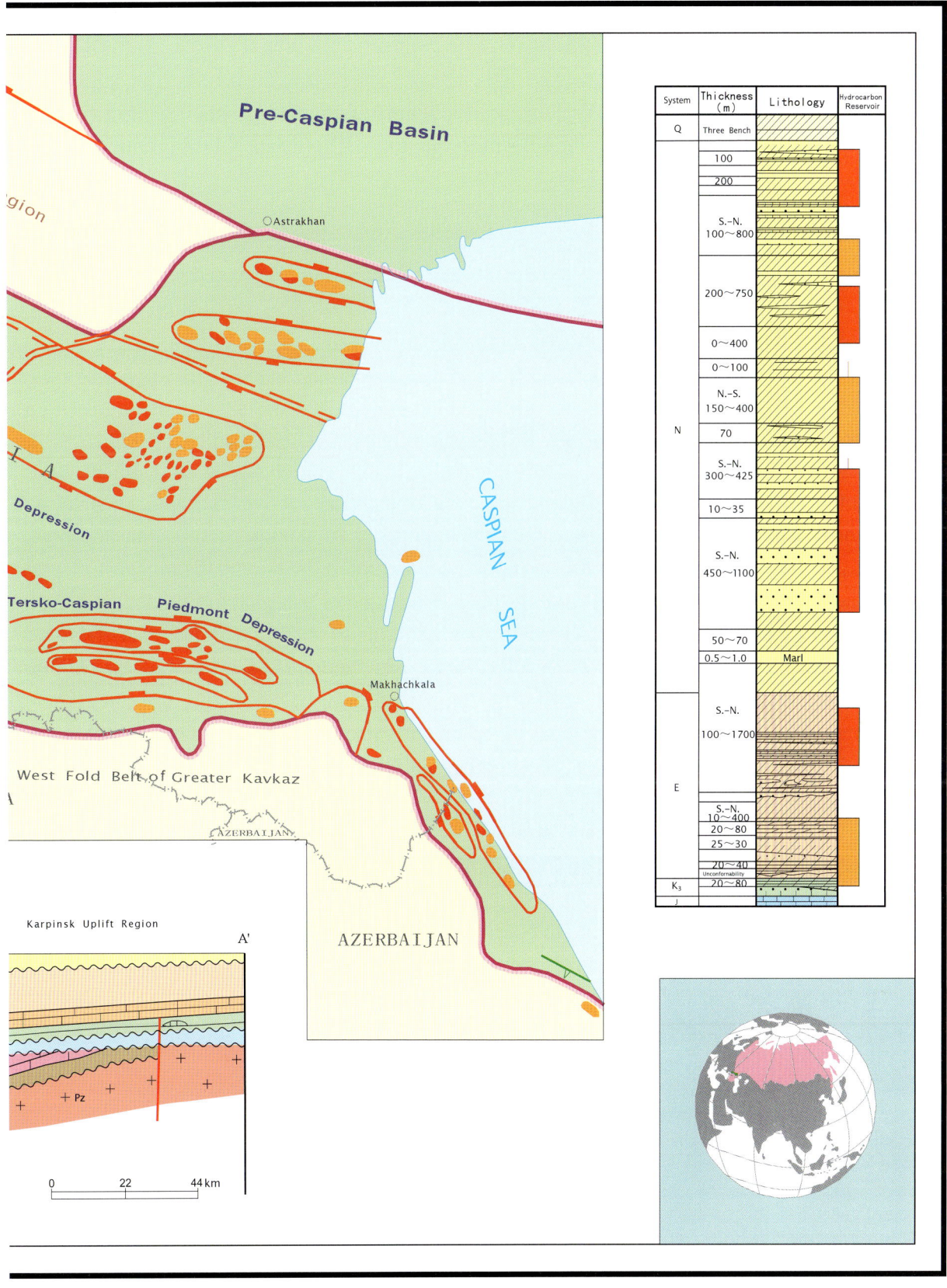

77 North Kavkaz Basin

Geography

The North Kavkaz Basin lies in the Prekavkaz Mountains, facing the Caspian Sea to the east, and is contiguous with the Pre-Caspian Sea Basin to the north. This is an old oil producing area of the world.

History

The first oil well was finished in 1864 and the Grozny oil field was discovered in 1893. Annual oil production reached 2 million t in 1920, 6.8 million t in 1960, with peak production of 35.5 million t in 1971, dropping sharply to 2.9 million t in 2000.

Regional geology

The North Kavkaz Basin occupies the Scythian platform, the South Mangyshlak Trough System of the Turonian platform and the Alpine marginal troughs (Indolo-Kubansky and Tersko-Caspian) of the Great Caucasus. The basin also includes the shelf of the Caspian Sea. Hercynian basement is present throughout a considerable area, with some portions of younger (Baikalian and pre-Baikalian) consolidation. The depth to the basement varies from 1–3 km to 6–14 km.

The sedimentary cover comprises two complexes: a lower one of pre-plate origin and an upper one coinciding with the plate formation. The lower complex consists of Upper Palaeozoic–Triassic terrigenous–marine carbonate and extrusive volcanic deposits, being 2–3 km in thickness (in the area lying in front of the Caucasus) and of terrigenous and grey-coloured rocks with carbonate bands. Its thickness is 4–5 km. The upper complex comprises essentially Jurassic and Upper Cretaceous–Paleogene terrigenous rocks. The thickness of Oligocene–Neogene molasses reaches 6–7 km.

Oil and gas accumulations have been identified in the Triassic to Neogene interval. A few petroleum complexes were identified in Permo-Triassic, Jurassic, Cretaceous, Paleogene and Neogene deposits. Source rocks are principally Jurassic and Cretaceous clayey carbonate deposits, and Paleogene and Neogene clayey rocks. In the North Kavkaz Basin, 290 fields have been identified made up of 200 oil fields and 90 gas fields, controlled predominantly by anticlinal uplifts (e.g. Fig. 77.1). Among the largest of these are the Malgobek-Voznesenskoye, and Starogtoznenskoye, oil fields, the Anastasiyevskoye gas–oil field and the Severo-Stavropolskoye and Mailopskoye gas fields (Table 77.1).

World Atlas of Oil And Gas Basins, First Edition. Li Guoyu.
© 2011 John Wiley & Sons, Ltd. Published 2011 by John Wiley & Sons, Ltd.

North Kavkaz Basin

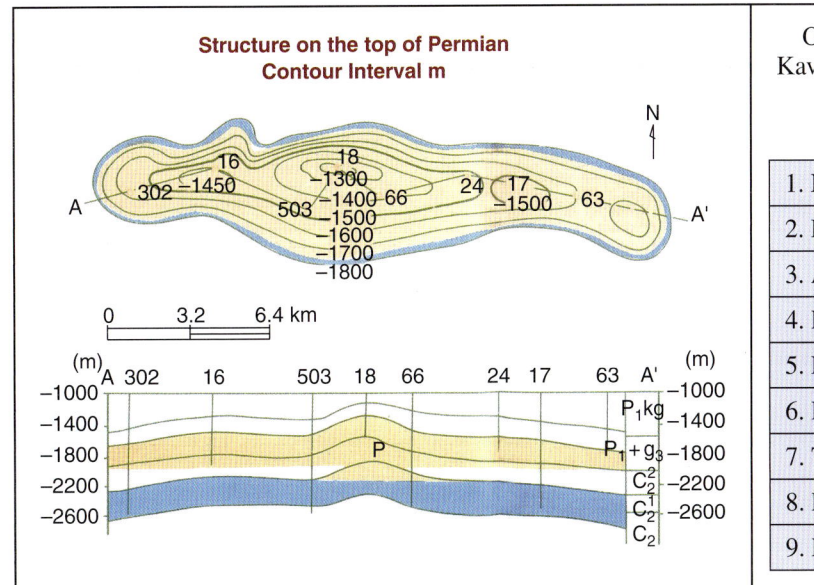

Orenburg is a giant gas field in North Kavkaz Basin.

Basic data

1. Name	Orenburg
2. Discovery year	1966
3. Area (km²)	2,354
4. Reservoir	Limestone
5. Reservoir age	Permian
6. Depth (m)	1,700
7. Thickness (m)	89–253
8. Porosity (%)	11.3
9. Permeability (mD)	1–34.6

Fig. 77.1 Details of the Orenburg gas field.

Table 77.1 Major oil and gas fields of North Kavkaz Basin

| Name | Discovery year | Recoverable reserves | | Depth (m) | Trap | Reservoir | | Remarks |
		Oil (million t)	Gas (billion m³)			Age	Lithology	
Malgobek Voznesenskoye	1915	420.0	—	500	Anticline	N	Sandstone	Oil field
Anastasiyevskoye	1953	35.0	23.4	2100	Anticline	N	Sandstone	Oil–gas field
Starogtoznenskoye	1893	91.0	—	4100	Anticline	K	Sandstone	Oil field
Severo Stavropolskoye	1950	—	22.8	900	Anticline	N–E	Sandstone	Gas field
Mailopskoye	1958	—	8.5	2435	Anticline	K	Sandstone	Gas field

SAKHALIN BASIN

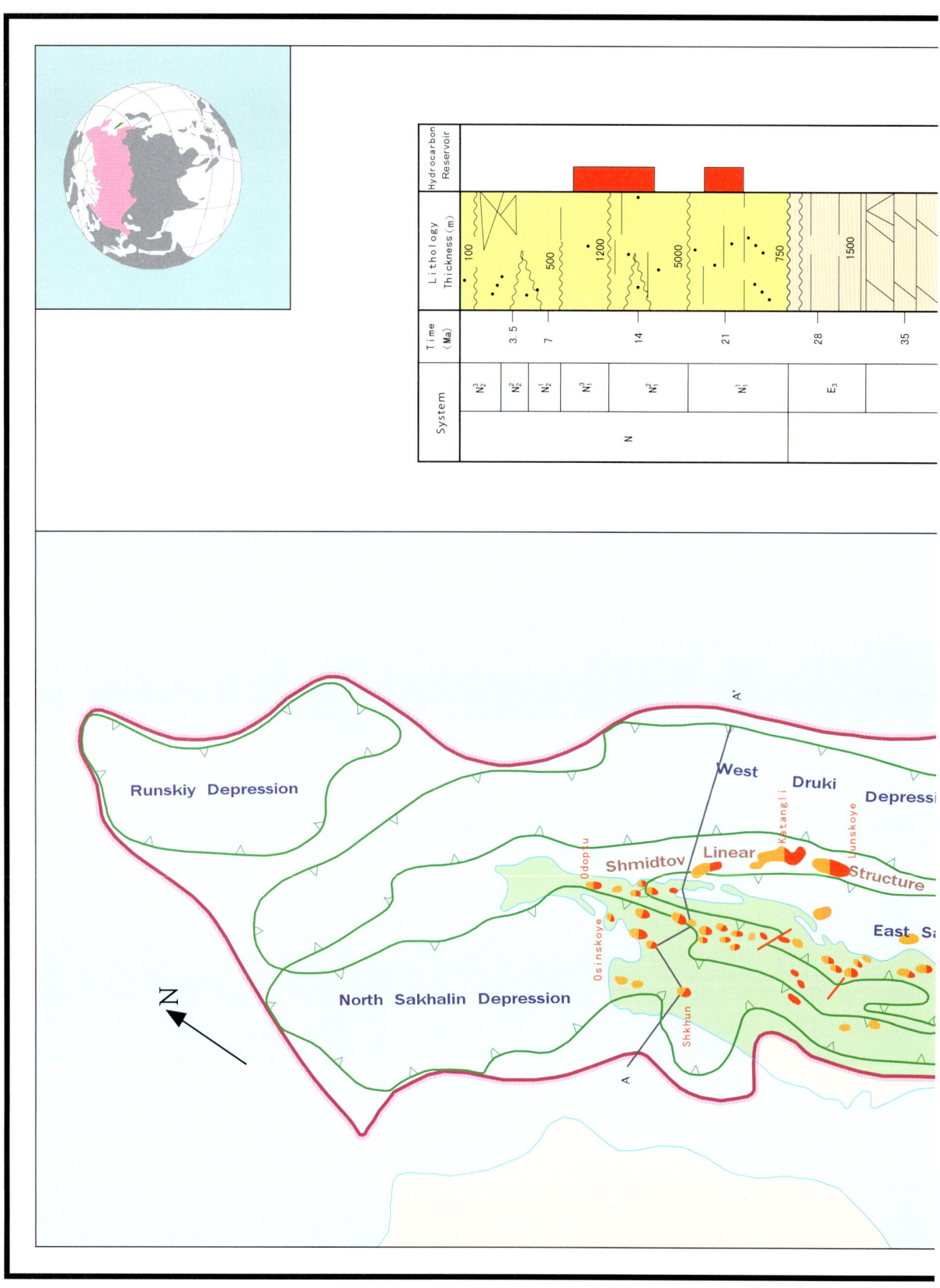

CHAPTER 78

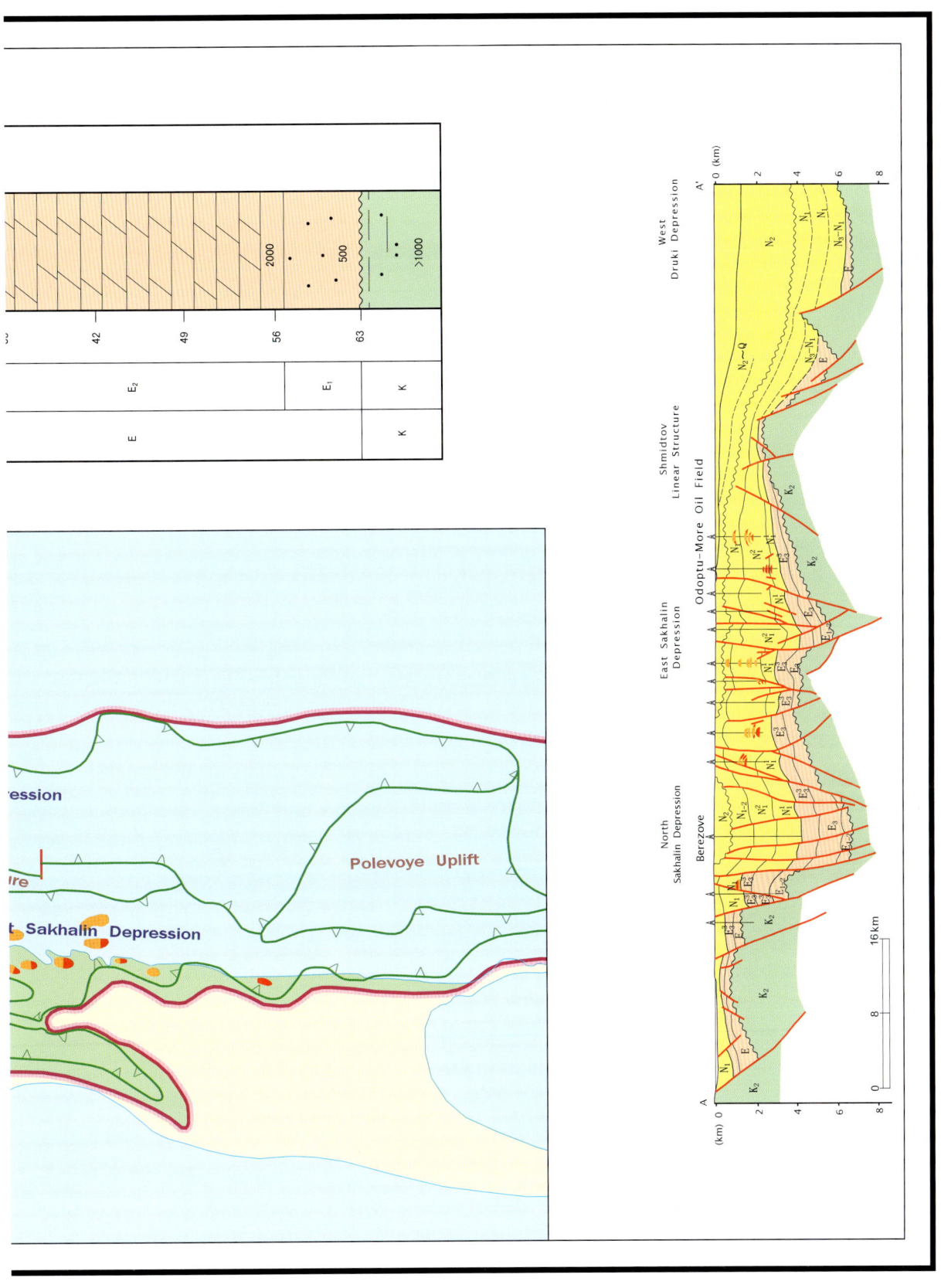

78 Sakhalin Basin

Geography

Sakhalin Basin is located on the Russian island of Sakhalin, with an area of 200,000 km^2, including 70,000 km^2 offshore. Sakhalin lies off the Pacific coast of Russia and to the north of Japan.

History

Oil sands were first discovered in the onshore Sakhalin Basin in 1880. The Okha oil field was discovered in 1923. During 1950–1960 oil and gas exploration was undertaken on a large scale, mainly onshore. At present a total of 70 oil and gas fields have been discovered and more than 30 fields are producing. More than 4000 development wells were drilled in the Okha, Katangli, Ekhabi and Vostochnoye fields (each containing more than 500 development wells). Onshore oil production commenced in 1923, when the Okha field was brought onstream. Until the early 1960s the associated gas was mostly flared.

Regional geology

The late Tertiary sedimentary succession overlies an accretionary complex of Mesozoic island arcs and microcontinents welded to the eastern margin of Eurasia. Paleocene to Oligocene sediments deposited in north–south intermontane grabens, including continental clastics and coal measures, are overlain by marine sediments that include probable source rocks. Late Oligocene to Early Pliocene deposition was in a dominantly transtensional setting. All comprise multiple interbedded sandstones and shales, with coals in part; the sandstones include most of the basin's reservoirs, while the shales and coals may have local significance as gas-prone source rocks. Lateral facies changes provide a stratigraphic component to many hydrocarbon traps. In the Late Pliocene, the motion on the shear zone became transpressional, and basin inversion began. Deep marine siliceous mudstones of the Late Oligocene–Early Miocene strata are the principal source rocks, for the onshore sector of the basin at least. The total organic carbon values range up to 1.5 per cent.

The Late Miocene to Pliocene Okobykay and Lower Nutovo Formations are probably gas-prone, but are largely immature in the onshore sector; they may nevertheless have contributed significantly to the offshore accumulations where they will have been buried to greater depths. The principal reservoirs are the stacked sandstones and silty sandstones of the Miocene Dagi and Okobykay Formations (onshore), and the Pliocene Nutovo Formation (offshore), as well as sandstone reservoirs. Seals are provided by the shales interbedded with the multiple stacked sandstone reservoirs and are thus mainly intraformational. Traps are typically anticline and faulted anticline, probably dominated by the hanging-wall type with respect to inversion-related thrust tectonics. Fault-related compartmentalization of anticlines is characteristic of many structural prospects.

The Nutovo Formation (Pliocene) plays are the most significant, accounting for over half the basin's ultimately recoverable hydrocarbons, mainly in structural and combination traps. Offshore reserves are dominantly in this group of plays. All the Oligocene through Pliocene strata are source rocks, and all the Upper Oligocene to Lower Pliocene strata are reservoirs. Differences

World Atlas of Oil And Gas Basins, First Edition. Li Guoyu.
© 2011 John Wiley & Sons, Ltd. Published 2011 by John Wiley & Sons, Ltd.

Sakhalin Basin

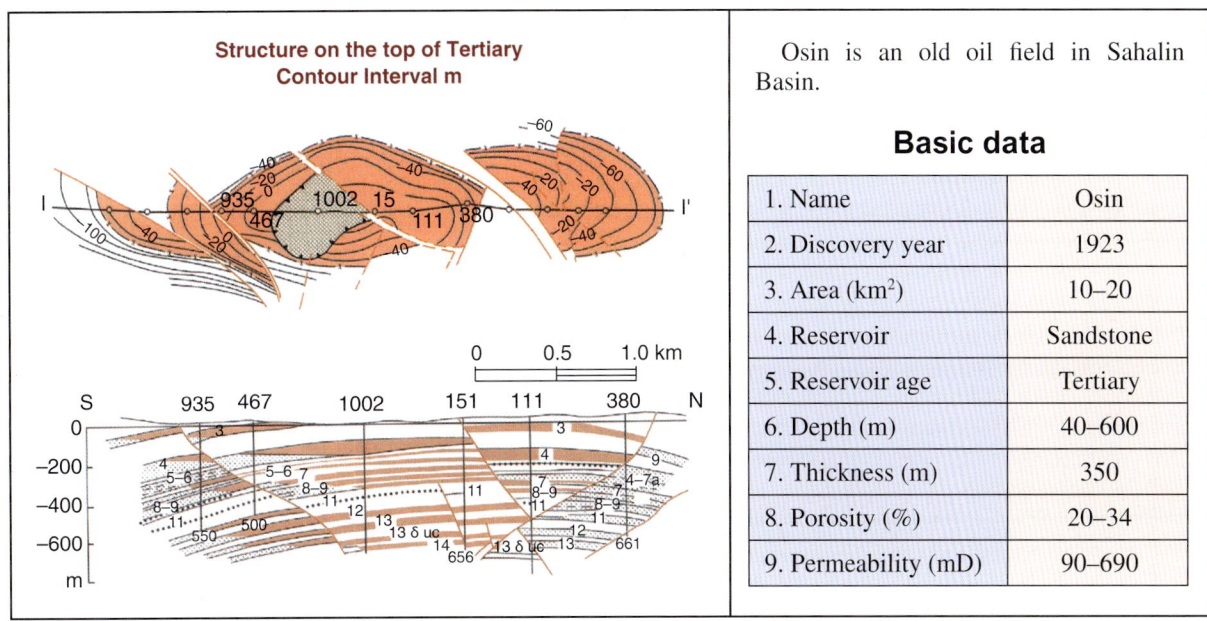

Osin is an old oil field in Sahalin Basin.

Basic data

1. Name	Osin
2. Discovery year	1923
3. Area (km²)	10–20
4. Reservoir	Sandstone
5. Reservoir age	Tertiary
6. Depth (m)	40–600
7. Thickness (m)	350
8. Porosity (%)	20–34
9. Permeability (mD)	90–690

Fig. 78.1 Details of the Osin oil field.

in oil composition appear to correlate mainly with reservoir depth, and it is thus thought likely that oils from different source rocks have become mixed in the migration process. The petroleum system was active from about 23 million years ago to the present, with peak expulsion and migration relating to the Late Pliocene to recent inversion event. The larger structures offshore remain incompletely charged. The onshore sector of the basin is mature, although some Okobykay and Dagi Formation prospects remain undrilled. Offshore, the largest structures have been drilled but a number of moderate sized structures remain untested. Improved understanding of the stratigraphy and sedimentology of the reservoir-bearing units could lead to the identification of prospects with a significant component of stratigraphic trapping. The deeper offshore part of the basin remains almost completely unexplored.

NORTH AMERICAN OIL AND GAS BASINS

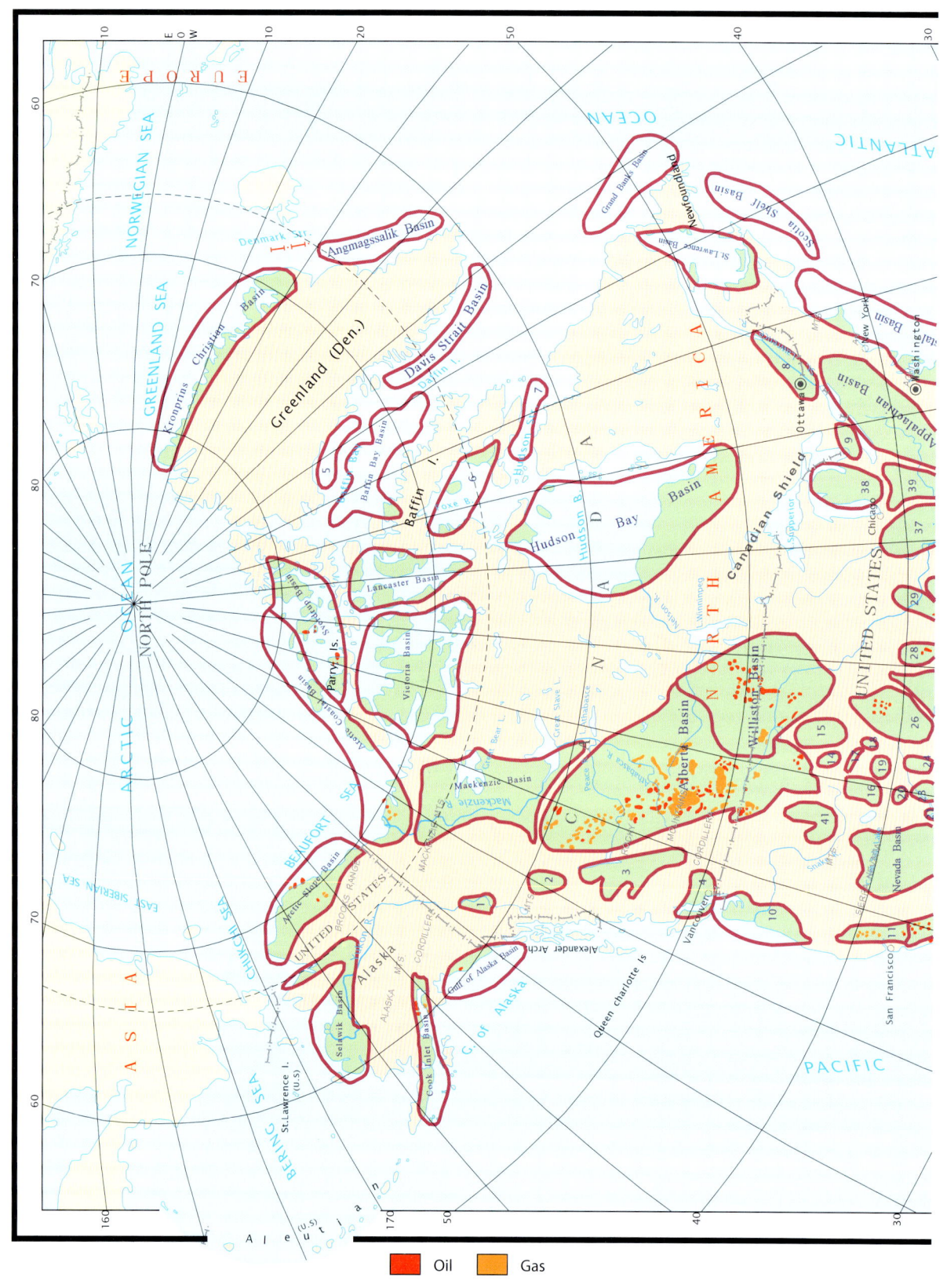

PART V

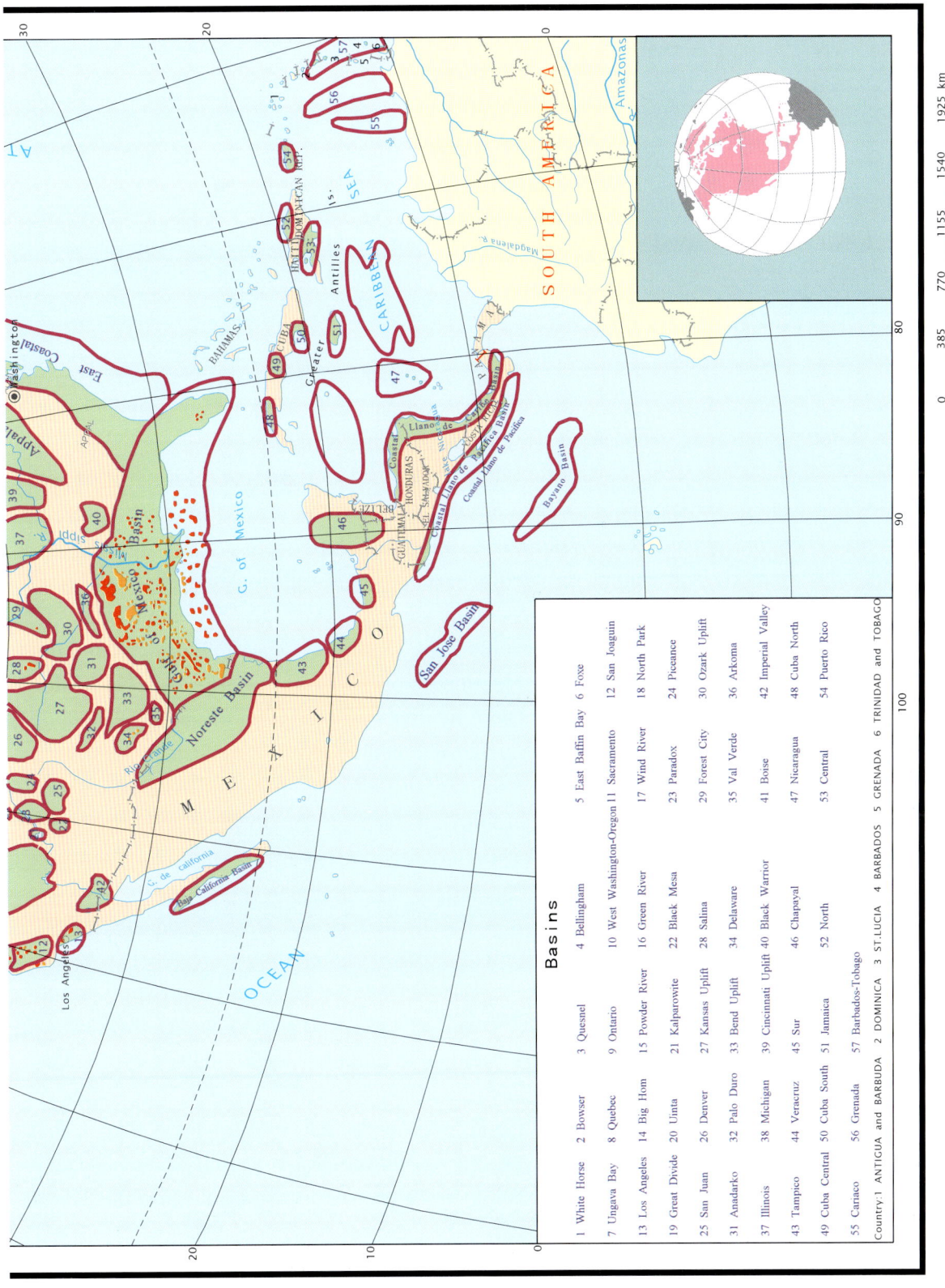

343

Part V

North American Oil and Gas Basins

Geography

The North American Region includes the continent of the North America and large islands (Greenland, Canadian Arctic Archipelago, Aleutian Islands), together with the adjacent deep-water areas of the oceans and seas including the Gulf of Mexico and Caribbean region. The region covers an area of 42.2 million km^2, with a population of 528 million.

History

North America has a long history of oil and gas exploration. Early exploration commenced in the foreland Appalachian Mountain Basin of the USA. The first gas field was discovered in 1776, while the first oil field was discovered almost a century later, in 1859. Annual oil production was 67,000 t in 1860, which at that time was 98 per cent of the known oil production worldwide. In 1910 a total of 450 oil and gas fields were discovered. In North America the increasing trend of oil production was dramatic and annual production reached 28.36 million t in 1910, reaching 121 million t in 1930, most of which was produced in the USA. After the Second World War, the petroleum industry witnessed more rapid development. Discovery of several new oil and gas fields resulted in the continental oil and gas production leaping to 410 million t and 375 billion m^3 respectively in the decade after the war. In 1960, the discovery of giant oil and gas fields in Alaska, Gulf of Mexico, western Canada and Mexico allowed even more rapid increases in oil production. By 1970 annual crude oil production was 600 million t, while gas production reached 677 billion m^3. Significantly, after this year, oil production started to decline in continental North America. Annual oil production in the USA showed a reduction from 530 million t in that year down to 245 million t by 2008 (Table V.1), with the total continental production declining to just 542 million t.

Regional geology

The North American Region occupies a considerable portion of the lithospheric plate of the same name. To all intents and purposes this is a continental area, but locally can be typified by oceanic crust. A small portion of the Pacific Plate underlies the coastal regions of the State of California, forming a conterminous presence of both continental and oceanic crust types. This region can be divided into: a central stable shield; the Cordillera Orogenic Belt in the west; the Appalachian Orogenic Belt in the east; the Franklin Geosyncline in the north; and the North America platform in the south.

The contact between the North American and Pacific Plates runs along the Pacific coast with little lithospheric variation. This contact also runs partially along a spreading axis in the Gulf of California area. In the context of geodynamic evolution, most of the North American Region

World Atlas of Oil And Gas Basins, First Edition. Li Guoyu.
© 2011 John Wiley & Sons, Ltd. Published 2011 by John Wiley & Sons, Ltd.

Table V.1 Proven reserves and production of oil and gas of North America (by the end of 2008)

Country	Proven reserves		Production	
	Oil (million t)	Gas (billion m³)	Oil (million t)	Gas (billion m³)
Total North America	28,880	9352	542	834.8
Barbados	0.29	0.1	0.04	—
Belize	0.9	—	—	—
Canada	24,396	1648	128	158
Cuba	16	70	2	—
Guatemala	11	—	0.7	—
Mexico	1438	372	140	71.6
Trinidad and Tobago	99	531	5	39
USA	2920	6731	245	607

represents a passive continental margin. The Gulf of Mexico was formerly an inner deep-sea basin, but is part of a complex area of microplate collision and subduction. As a result the deep-water portion of the Gulf of Mexico with oceanic crust is now separated from the Atlantic Ocean by Florida and its shelf.

Ninety-three sedimentary basins have been discovered within the limits of the North American Region. These basins can be divided generally into two simple categories, continental and foreland, but the features of some basins may be complex, combining several types and phases of formation. In such cases, they are classified by reference to the predominant tectonic factor. A total of 41,000 oil and gas fields have been discovered in this region, and more than 16,000 fields are associated with the Gulf of Mexico Basin (Vysotsky *et al.*, 1995).

CANADA

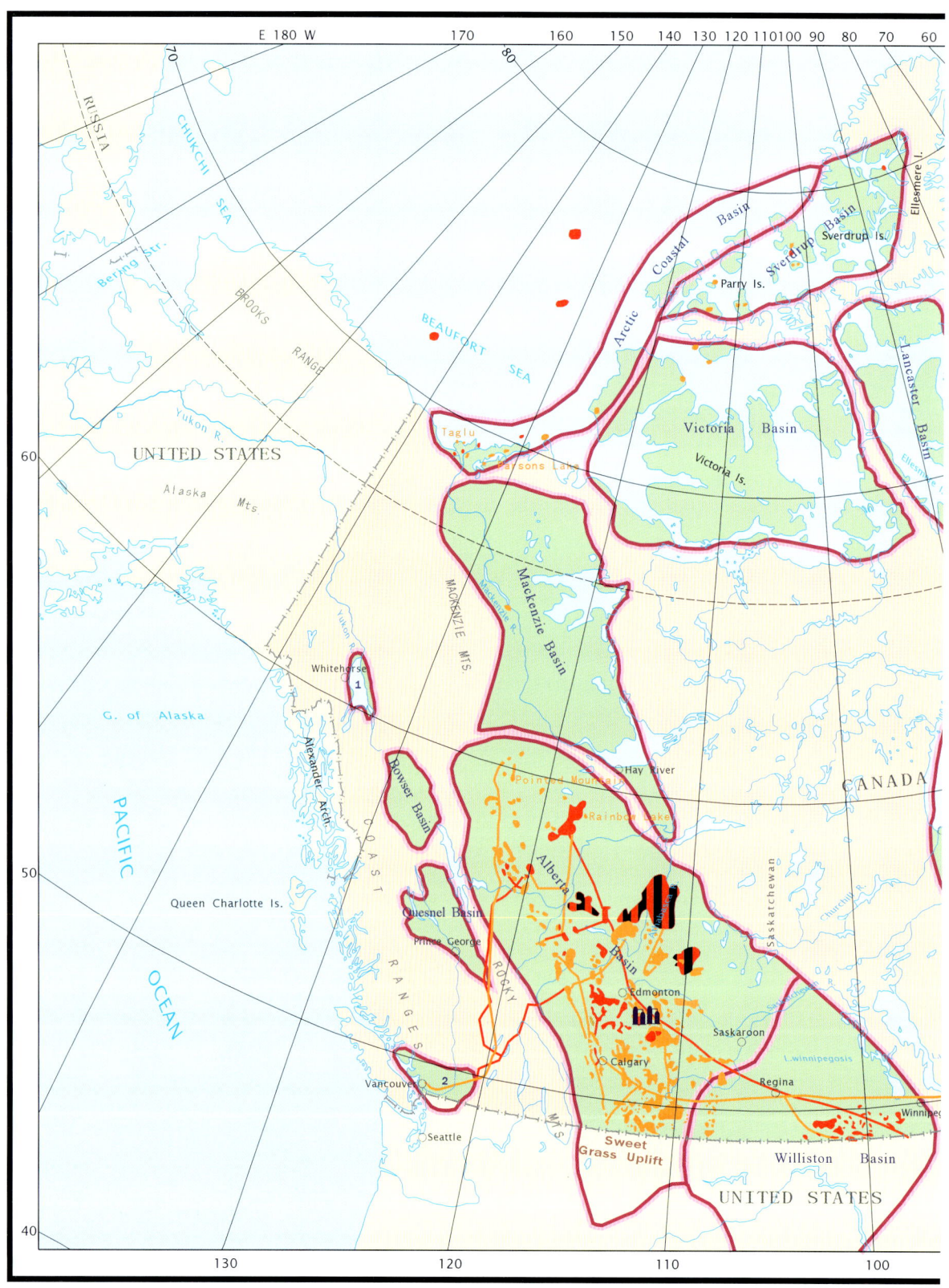

Basin: 1 White Horse 2 Bellingham 3 Ontario 4 Quebec 5 Ungava Bay

CHAPTER 79

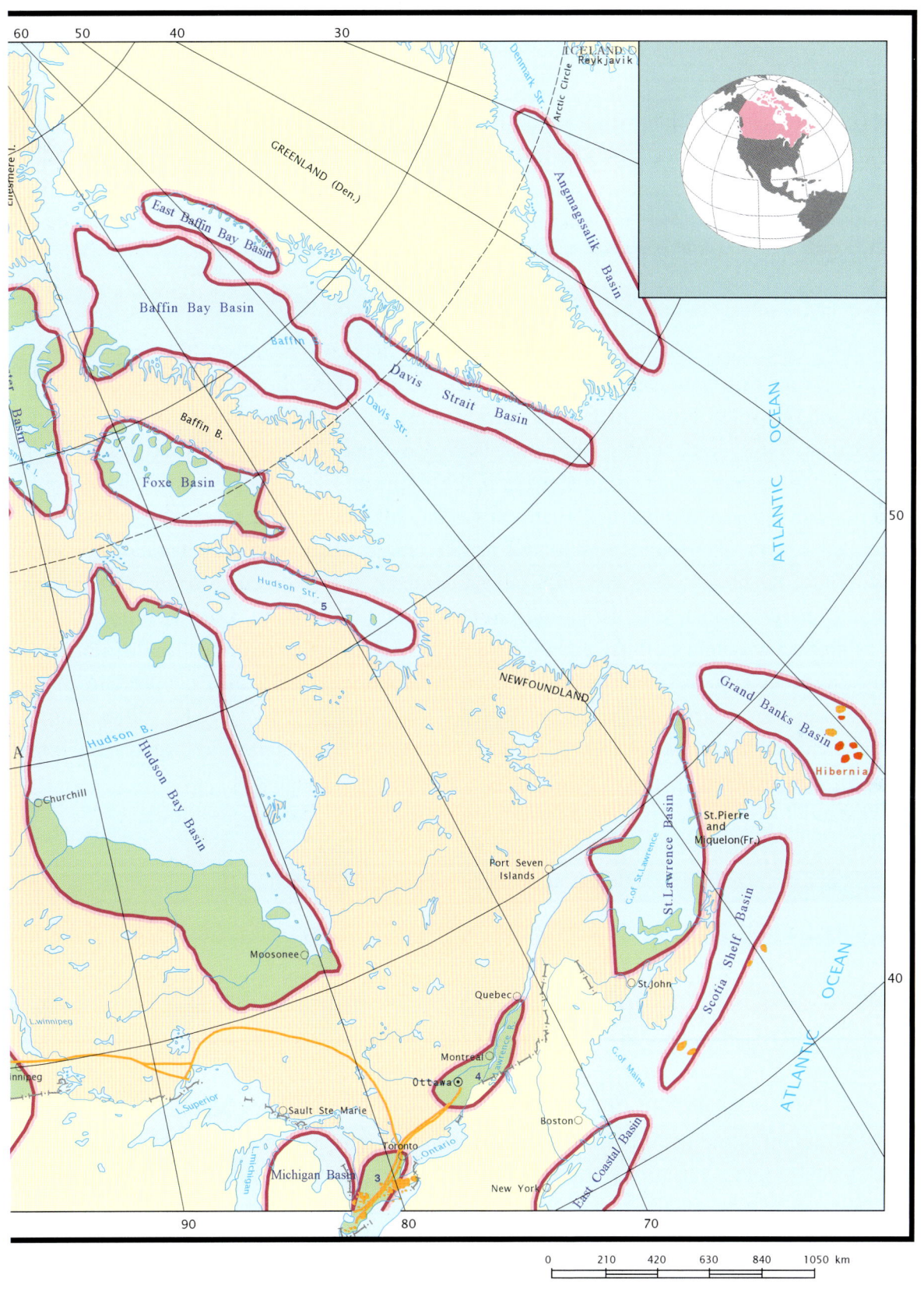

79 Canada

Geography

Canada is located in the northern part of North America, surrounded by the Atlantic Ocean to the east, the Pacific Ocean to the west and the Arctic Ocean to the north. The Rocky Mountains Range lies in the west of the country, while the northerly limit of the Appalachian Mountains is located in the east, separated from the Laurentian Plateau by the St Lawrence River. Canada has an area of approximately 9,984,670 km^2, and a total population of 33.6 million.

History

As with many other oil producing countries in the world, the early records of the development of the hydrocarbon industry are uncertain. Williams No.1 oil well in Oil Springs, Ontario, is considered to be the first oil well in Canada (indeed, it is recorded as pre-dating the more celebrated Drake wildcat in Pennsylvania of the following year). It is generally agreed that oil production commenced in the western provinces of Canada in 1900. In early 1947, Imperial Oil discovered light gravity oil in Upper Devonian carbonates at Leduc, Alberta. Significant amongst these are the Pembina oil field in 1953, Swan Hills oil field in 1957, Hibernia oil field in 1979 and Elmworth gas field in 1976. By the end of 2008, 24.3 billion t (mostly oil sand reservoirs) in proven oil reserves had been discovered, with annual oil production of 128 million t. Canada is also a significant gas-producing country. By the end of 2008 the proven gas reserves were 1.64 trillion m^3, with annual gas production of 158 billion m^3. An event of historical significance occurred in 2002 when recorded proven reserves increased dramatically from 665 million t to 24.6 billion t as a result of the inclusion of the massive oil sand reserves in the total proven reserves figure. The key features of the development history of the petroleum industry in Canada are:

1 the substantial Alberta Basin which is exceptionally rich in oil and gas reservoirs;
2 the oil sand reserves are the largest in the world;
3 the presence of huge basins in the Atlantic Ocean regions to the east.

Regional geology

Twenty sedimentary basins are distributed in the Canada landmass with areas ranging from 20,000 km^2 to 1.05 million km^2 (Hudson Bay). Details of 12 of these are provided in Table 79.1. The largest oil- and gas- producing basin, the Alberta Basin, has an area of approximately 980,000 km^2. The post-peat reservoirs are confined to the intermontane depressions located in some portions of the eastern zone of the Rocky Mountains. The zone occupies the western margin of the North American Platform which was active during Cenozoic times. Individual basins (within the Alberta Basin) with hydrocarbon potential are filled with Meso-Cenozoic terrigenous strata, overlying relatively thin Palaeozoic carbonate rocks that rest on Precambrian basement. These basins are characterized by the extended linear anticlinal zones of submeridional folds located in the marginal parts of the basins, while locally anticlinal folds occur in the central part of the basin that, as a rule, are extensively faulted. The basins are not considered to be large,

World Atlas of Oil And Gas Basins, First Edition. Li Guoyu.
© 2011 John Wiley & Sons, Ltd. Published 2011 by John Wiley & Sons, Ltd.

Table 79.1 Major sedimentary basins of Canada

Basin	Area (thousand km²)	Major sedimentary rock		Reservoir	
		Age	Thickness (m)	Age	Lithology
Alberta	980	E–R	5700	D, K	Carbonate rock, sandstone
Arctic Coastal	240	Mz, Kz	9450	K, R	Clastic rock
Baffin Bay	350	Mz, Kz			
Foxe	320	Pz			
Hudson Bay	1050	Pz			
Lancaster	240	Mz, Kz			
Mackenzie	520	E–D	4500	D	Reef
St Lawrence	260	E–C		C, O	Sandstone, limestone
Sverdrup	240	C–R	18,000	T, J	Sandstone
Victoria	740	Pz			
Williston	520	E	3000	C, J, K	Sandstone, limestone

either for size or the original oil and gas resources in place, and they are characterized by the predominance of oil resources over gas (approximately 2:1). Productive belts are generally in the Mesozoic and Cenozoic rocks, but sometimes are present in the Palaeozoic strata.

Most of the accumulations are associated with structural traps. Major accumulations are concentrated in the subthrust zone and are confined to the arched parts of the anticlines. The main productive horizons within the thrust belt are Jurassic sandstones and limestones, and also Cretaceous sandstones, Triassic and Permian carbonates, Pennsylvanian sandstones and Mississippian, Devonian and Ordovician carbonates. About 2600 fields have been discovered in all these basins, among which 900 are gas fields. In these basins the gas-producing sequences are Paleocene–Eocene, Upper Cretaceous and Upper–Middle Jurassic sandstones.

The western Canada sedimentary basin (comprising the Alberta and Williston Basins) is infilled by Cambrian to recent sediments that form a wedge, thickening westward from zero at the Canadian Shield to in excess of 6000 m in its deepest part. The structure is simple, with only the Sweetgrass Uplift in the south and the Peace River Arch in the centre of the region to break the monotony of the west-dipping monocline. Outcrops are rare, occurring mainly near the edge of the Canadian Shield and in some of the more deeply incised river valleys. At the western limit of the basin the complex structure of the Foothills and Rocky Mountains is in sharp contrast with the remainder of the basin (Perry et al., 1984).

WESTERN CANADA OIL AND GAS REGION

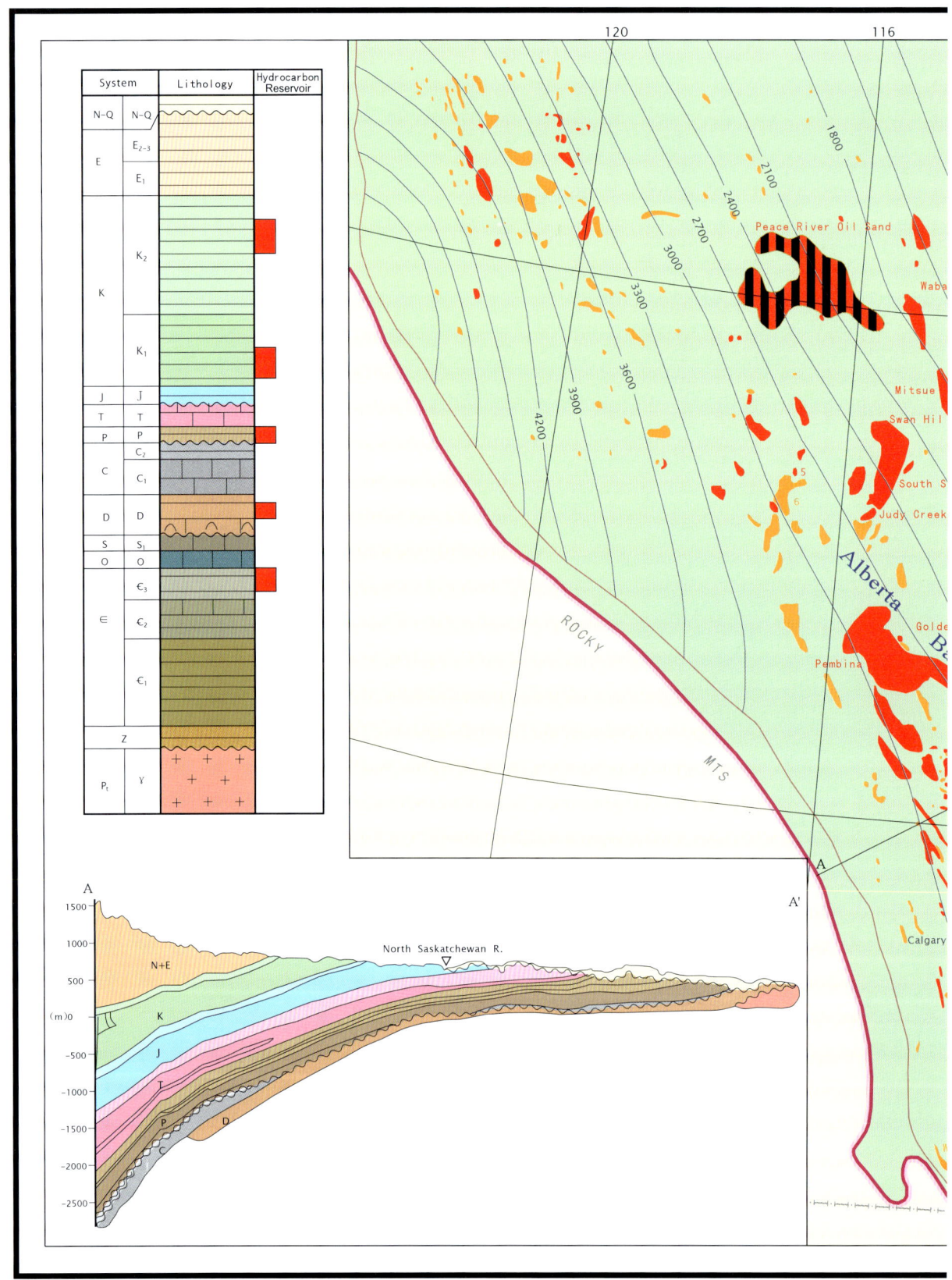

Oli-gas field: 1 Crossfield   2 Leduc-Woodbend   3 Wizard Lake   4 Bonnie Glen   5 Kaybob   6 Kaybob South

CHAPTER 80

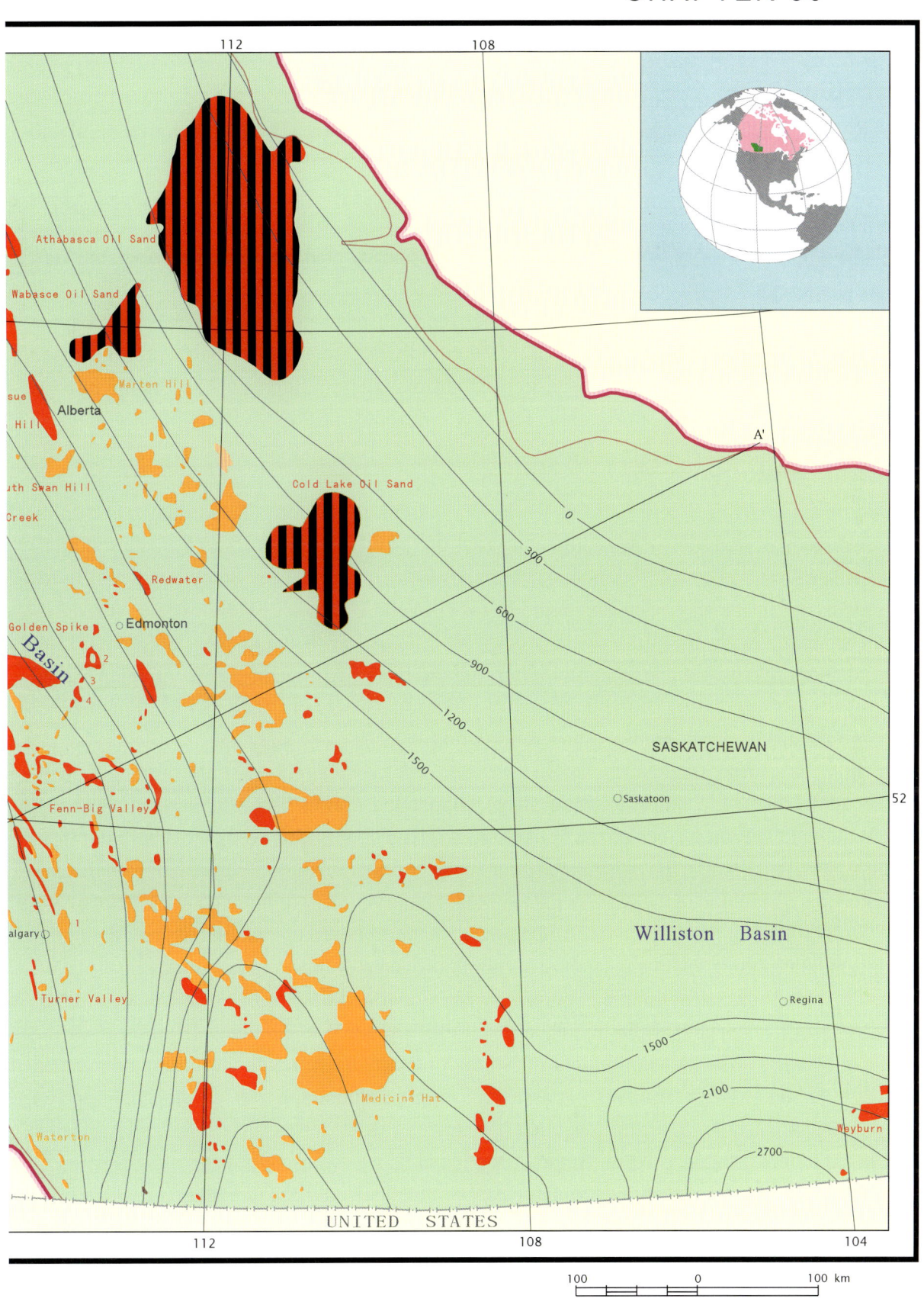

80 Western Canada Oil and Gas Region

Geography

The western Canada oil and gas region has an area of 980,000 km². The region is rich in oil and gas resources and is renowned for its huge tar sand resources.

History

In 1904 a commercial gas field was discovered in Upper Cretaceous rocks at Medicine Hat. A further discovery was made in 1910 when oil was produced from thrust folds in the Upper Cretaceous Turner Valley field in the Foothills. These early discoveries can be described retrospectively as surprising, given the importance of production from Devonian strata in western Canada today. The western Canada sedimentary basin is a multiplay region. Surges of exploration interest have concentrated on different stratigraphic levels at different times for a variety of reasons.

Regional geology

Between the Canadian Shield and the Rocky Mountains in western Canada there is a thick wedge of Palaeozoic and Mesozoic sedimentary rocks referred to as the western Canada sedimentary basin. The western Canada sedimentary basin has for a long time been known particularly for its Devonian reef production. The Precambrian rocks of the Canadian Shield extend westwards beneath the basin. At the foothills of the Rocky Mountains, basin thicknesses exceed 6000 m. The sedimentary pile then thickens abruptly beneath the Rocky Mountains. The Canadian Shield thus forms the eastern margin of the basin, whilst the thrust sheets of the Rocky Mountain Front Ranges form a striking and abrupt western limit (Miall, 1980).

The cratonic core of western North America was constructed by a process of Precambrian nucleation. By mid-Proterozoic times, a single stable continental craton was established in Canada. Throughout the Late Proterozoic and the Palaeozoic, a terrigenous clastic wedge was built out from the faulted western margin of the crystalline craton in the cordilleran zone. Sedimentation on the cratonic platform in the region of the western Canada sedimentary basin was initiated in the Palaeozoic. During the Mesozoic and Tertiary, a foreland basin was superimposed on both the western continental wedge and the cratonic platform, leading to further orogenic events in the Cordilleran (orogeny) Province. In Late Proterozoic time, the North American craton was surrounded by passive trailing margins. The mio-geosynclinal clastic sequences are of the passive Atlantic margin type in the Cordilleran Province. The western margin of North America collided with a volcanic arc system (product of the Sonoma orogeny) as the Triassic east-dipping subduction zone developed. Followed by the volcanic arc accretion, further accretion of extracontinental terranes occurred and reached a climax in the Late Mesozoic as North America continued its westward motion, now as the overriding plate. Specific regions of accreted terranes have been recognized in the Cordillera Province from Alaska to southern California, as subduction and collision continued. The process is continuing to the present day, although it was largely completed in British Columbia by the end of the Cretaceous.

World Atlas of Oil And Gas Basins, First Edition. Li Guoyu.
© 2011 John Wiley & Sons, Ltd. Published 2011 by John Wiley & Sons, Ltd.

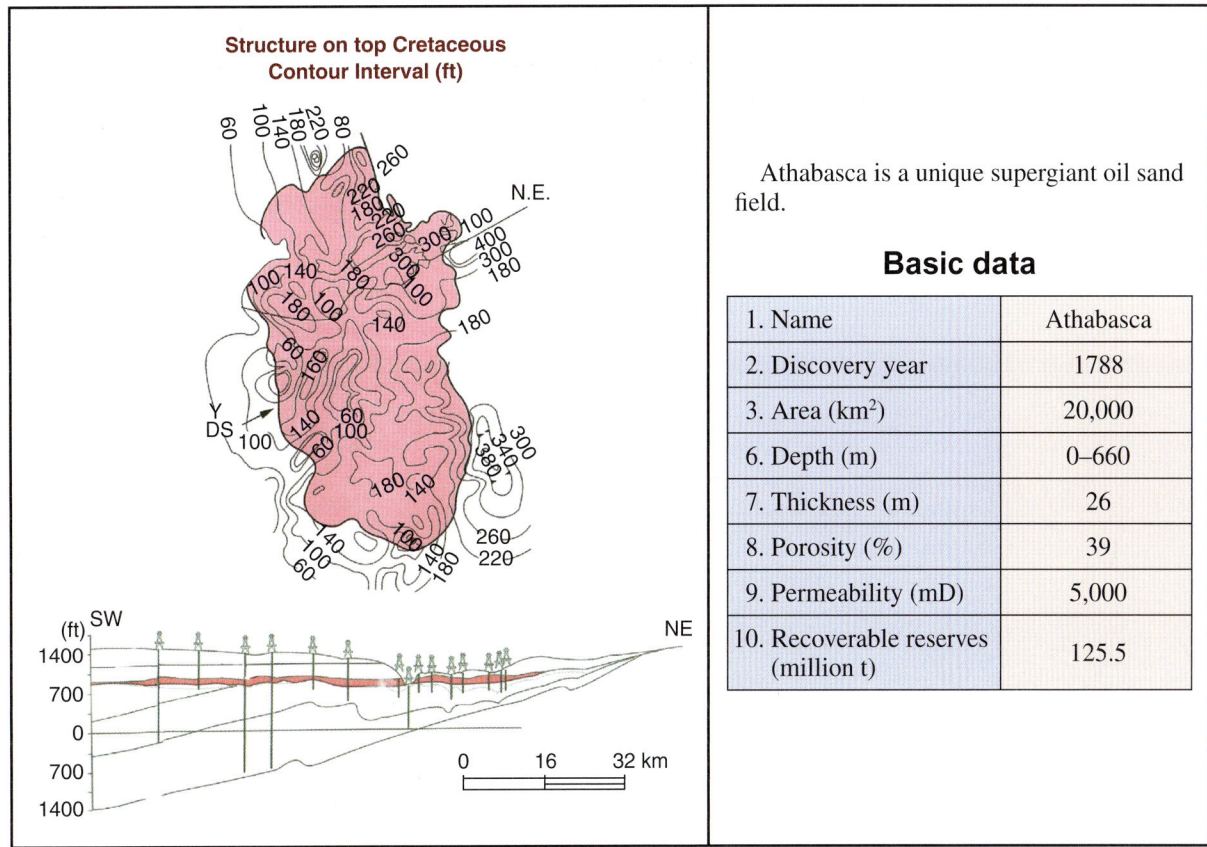

Fig. 80.1 Details of the Athabasca oil field: YDS, Y-delta starter.

The Cordilleran orogeny began to form at the end of the Palaeozoic. Following a major phase of accretion in Late Cretaceous time, there were changes in the directions and rates of movement of the interactive plates at the western seaboard of North America. Multiple deformations and strike-slip movements have produced a complex structural zone.

The Alberta Basin contains a variety of petroleum types ranging from dry gas to heavy oil, with a range of geological settings from Devonian to Cretaceous in age. The majority of the conventional oil reserves are confined to the Devonian and the Upper Cretaceous (Upper Colorado) strata, but oil is distributed throughout the stratigraphic column. It must be noted that heavy bitumen within the Mannville Group (lowermost Cretaceous), notably the Athabasca Tar sands (Fig. 80.1), represents a considerable resource that is not amenable to conventional reserve estimates. Gas is more evenly distributed (Hills, 1974).

USA

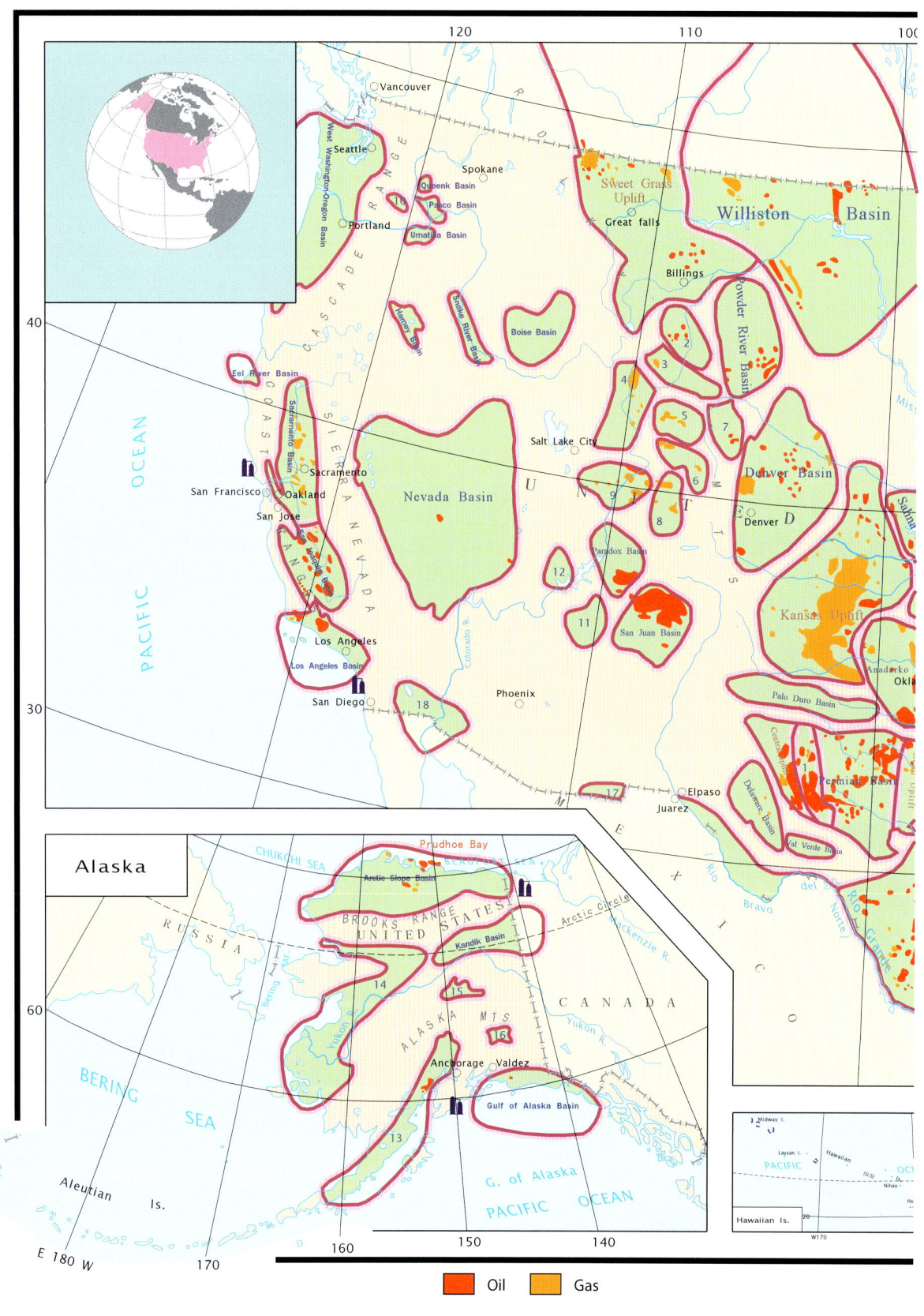

CHAPTER 81

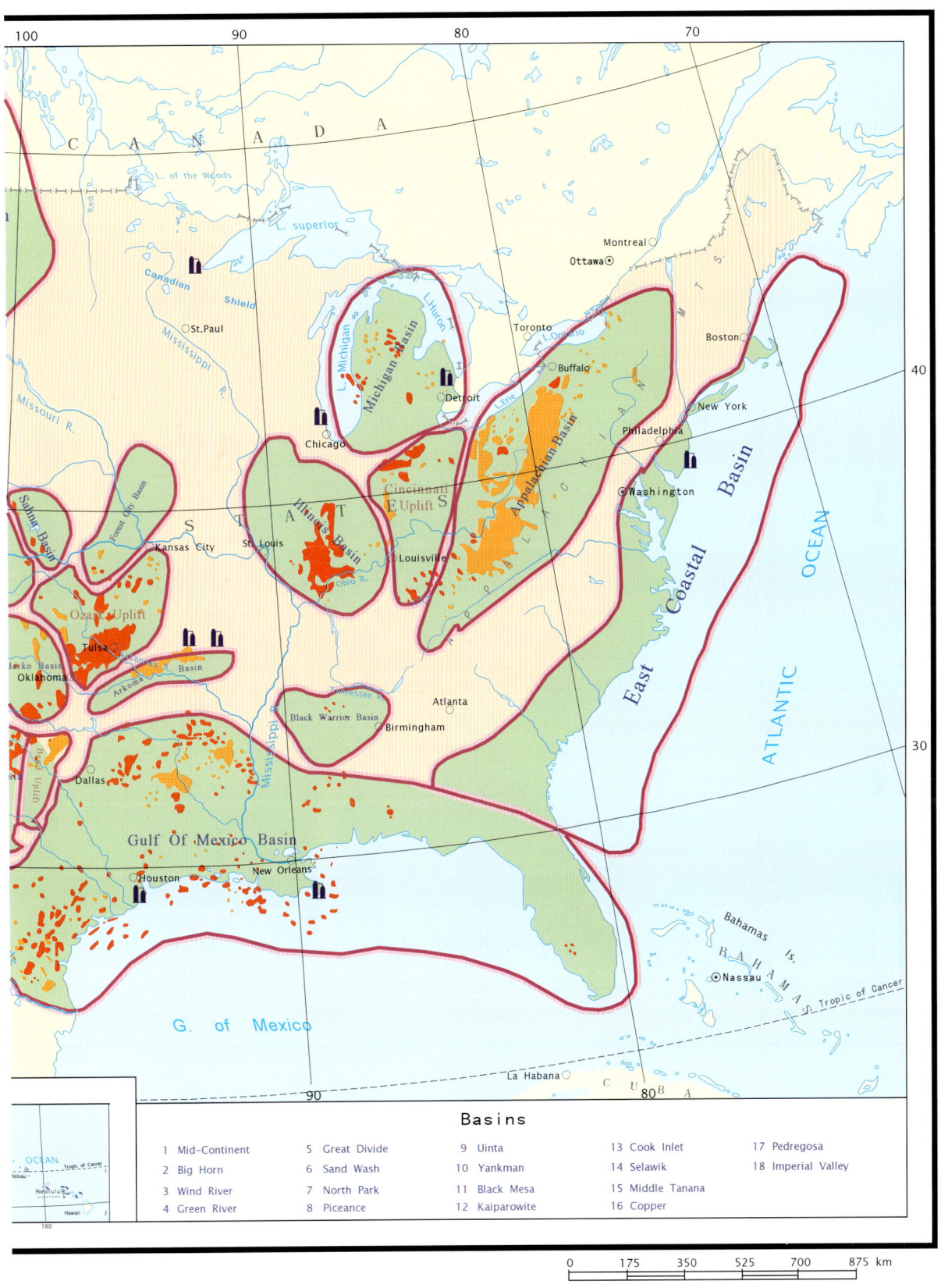

81 USA

Geography

The USA is located on the North American continent, facing the Pacific Ocean to the west and the Atlantic Ocean to the east. Within the USA, the Appalachian Mountains lie in the eastern part, while the Rocky Mountains lie in the western part. The country has an area of 9.37 million km^2, ranking fourth in the world after Russia, Canada and China. Today the population of the USA is estimated at about 304 million.

History

As one of the most important oil producers in the world, the USA's oil industry has a long history that can be dated back to the 1850s. The first oil well, the famous Drake wildcat in Pennsylvania, was drilled in 1859. Prior to 1954, the country's production of crude oil accounted for over 60 per cent of the world's total. Since that year the proportion of the USA oil production against the world's total has fallen continuously, accounting for about 25 per cent of the world's total in the 1980s, 10 per cent in 1994, and even as low as 9.8 per cent in 1997. The history of the USA oil industry can generally be divided into five periods.

Initial period (1859–1900) Oil and gas exploration during this period was initially in the form of anticlinal exploration related to oil and gas seepages. The first oil well was successfully drilled close to the oil seepage of the Appalachia region. In 1862 the first oil field, the Florence-Canon oil field, was discovered in the Rocky Mountains. In 1873 the annual crude oil production exceeded 1 million t for the first time. Before the 20th century, the Appalachian Basin and the Cincinnati Uplift were the major oil basins, with the annual output accounting for about 75 per cent of the nation's total.

Early period (1901–1930) This period witnessed a large-scale increase in the number of oil and gas fields discovered in the USA. It is estimated that 50 per cent of large oil and gas fields located in the USA were discovered between 1859 and 1930. In 1902 the annual output exceeded 10 million t for the first time, with remaining recoverable reserves estimated at 430 million t. After 1907 the annual output of crude oil exceeded 20 million t, with recorded estimated reserves increasing by 100 million t. From 1911 to 1920 a total of 38 large oil and gas fields were discovered in various regions. In 1923 the annual output of crude oil increased to 100 million t. During the five years from 1925 to 1930, 51 large-scale oil and gas fields were discovered. By 1930 the landmark statistic for the USA petroleum industry of 121 million t in annual output of crude oil was reached.

High-speed development period (1930–1966) After 1930 the pace of exploration quickened significantly. About 300,000 wells were drilled, with oil output and reserves rising rapidly. Thirty-five per cent of oil resources were proven. From 1936 to 1940, 31 large-scale fields were discovered. The annual output of crude oil in the USA was 200 million t in 1943 (Table 81.1). From the mid-1940s to the mid-1950s, offshore oil and gas exploration was increased, and onshore exploration was expanded to the Alaska region. Forty-three large-scale oil and gas fields were discovered

World Atlas of Oil And Gas Basins, First Edition. Li Guoyu.
© 2011 John Wiley & Sons, Ltd. Published 2011 by John Wiley & Sons, Ltd.

Table 81.1 Oil and gas reserves, production and imports in the USA

Year	Proven oil reserves (billion t)	Oil production (million t)	Proven gas reserves (billion m³)	Gas production (billion m³)	Oil import (million t)	Total oil consumption (million t)
1900	0.3	8.5				
1914	0.7	35.8				
1919	0.9	51	424			
1925	1.1	104	651	34		
1943	2.7	202	3,115	99		
1951	3.5	323	5488	200	25	
1963	4.1	409	7819	398	56	
1968	4.2	500	8136	524	64	671
1969	4.0	508	7790	561	70	706
1970	4.0	530	8233	603	66	734
1988	3.6	408	5297	495		796
1990	3.5	359	4703	519	401	781
1991	3.5	369	4791	528	389	765
1992	3.3	357	4730	527	394	782
1993	3.2	341	4672	544	431	787
1994	3.1	331	4596	560	446	885
1995	3.0	326	4636	556	441	886
1996	3.0	321	4673	566	470	836
1997	3.0	318	4711	563	495	848
1998	3.0	318	4732	563	519	863
1999	2.8	298	4642	556	527	888
2000	2.9	291	4737	567	554	897
2001	3.0	290	5024	575	580	896
2002	3.0	288	5195	570	567	897
2003	3.0	286	5293	557	612	912
2004	2.9	270	5353	552	630	948
2005	2.9	256	5451	511	635	944
2006	2.9	256	5980	523	640	943
2007	2.9	250	5980	545	636	943
2008	2.9	245	4839	607	636	884

in total, including 15 offshore oil fields. In the 1960s domestic oil consumption increased rapidly (Fig. 81.1), and outpaced the capacity of domestic production, so that the world's hitherto largest exporter of oil became the largest importer (Beebe, 1968).

Steady development period (1968–1985) By the end of the 1960s, oil production in the USA entered a steady period of high production, with annual output exceeding 400 million t for

North American Oil and Gas Basins

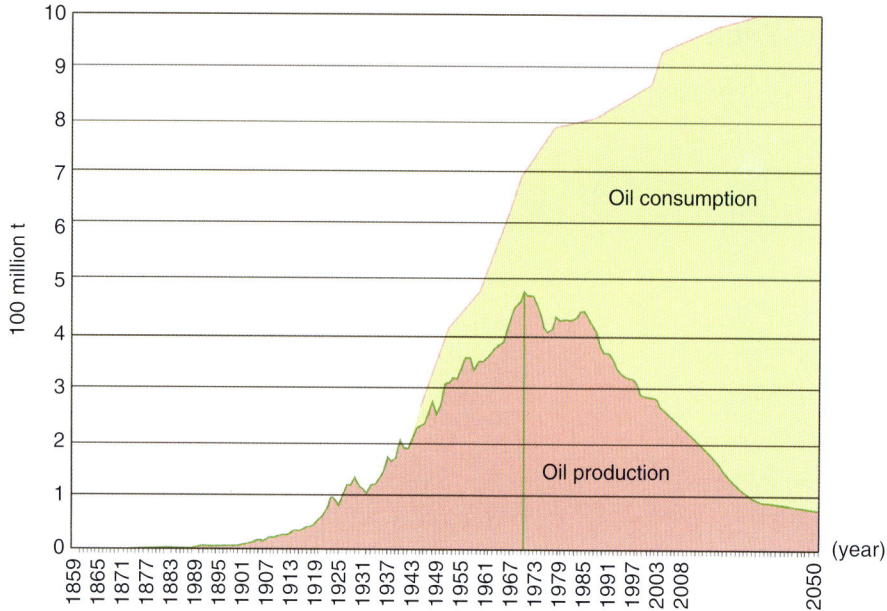

Fig. 81.1 Oil production and total oil consumption of USA.

18 consecutive years. In 1968, the super-large oil and gas field Prudhoe Bay was found in Alaska, which enabled the remaining proven reserves of crude oil and natural gas to reach a historic peak (5.27 billion t of crude oil, 8.23 trillion m³ of natural gas). The annual output of oil and gas reached a peak of 530 million t of oil and 620.4 billion m³ of gas. In this period, the discovered oil fields were generally smaller in size and not able to match the increased consumption despite the high production rates over several years, and consequently the recoverable reserves displayed a negative trend.

Declining production period (1986-Present) By 1986 the period of steady and high production was truly over. Crude oil production and reserves started to decrease. In 2004 the remaining recoverable reserves of oil were 2.99 billion t, with annual output of 245 million t. The reserves of natural gas were 6.7 trillion m³, with the annual production of 607 billion m³. By the end of 2008, the proven oil reserves were 2.9 billion t (Table 81.2), with annual oil production of 245 million t, while the proven gas reserves reached 6.7 trillion m³, with annual gas production of 607 billion m³.

Features of the USA oil industry The remarkable development of the oil and gas industry in the USA has substantially contributed to the economic advancement of the developed nations.

1. From 1859 to the end of 2008, approximately 2 million wells had been drilled, with more than 40,000 oil and gas fields discovered and over 20 billion t of oil recovered, making the USA the largest accumulated oil producer in the world. The complex range of geological conditions encountered alongside the total of 8.03 million km² sedimentary rocks and the huge amounts of recovered oil, tends to confirm the potential for abundant of oil and gas resources on a worldwide basis.

2. Leading oil technology: the USA oil technology has responded to the pressure of demand for oil and gas by reaching and maintaining a global leadership position. This is particularly true in areas such as geological surveys, seismic exploration, drilling, recovery, enhanced recovery techniques and refining technology.

Table 81.2 Petroleum industry of the USA and the world by the end of 2008

Index	USA	World	USA/World (per cent)
Sedimentary rock area (km²)	830	10,000	8.3
Population (million t)	305	6700	4.5
Oil consumption (million t)	943	3600	26
Cumulative oil production (billion t)	24.2	166	14
Total oil and gas fields	51,635	67,720	76
Total wells	448,619	667,853	74
Daily production per well (t)	1.5	10.4	7
Oil reserves (billion t)	2.9	183	1.5
Oil production (million t)	245	3648	6
Gas reserves (trillion t)	6.7	1771	3
Gas production (billion t)	607	3050	15

3 Advances in theory and developments: the USA is at the forefront of scientific and technological development related to hydrocarbon resources. These include theories related to organic origin of oil and gas, detection of anticlinal structures and 'hidden' reservoirs, and seismic stratigraphy. These and many other contributions have enhanced the development of the petroleum industry in several countries worldwide.

4 Over reliance on oil and gas. In 2008, the consumption of oil and natural gas were over 900 million t and 600 billion m^3, respectively, which accounts for about 25 per cent of the world's total in each commodity. For country with a population of about 300 million, this is clearly out of proportion to the global demand for oil and gas resources. Arguably such avaricious demand for oil has led to the USA being involved in several wars since the 1960s in a seemingly unquenchable desire to control oil and gas resources worldwide.

In this context the USA with 4.5 per cent population uses 26 per cent (900 million t) oil of the world's annual oil output. The USA uses 1.5 per cent of its oil reserves (2.9 billion t) annually to produce 6 per cent of the world's total oil output, and 3 per cent (6.7 trillion m^3) of its gas reserves to produce 15 per cent of the world's total gas output. This is a unhealthy economic situation for any country and petroleum industry, and it will lead to serious problems in the future that it will be necessary to solve.

Regional geology

The USA is an apparent oil platform, comprising three regions, the Cordillera orogenic belt of the western part, basement platform of the central part and the Appalachian orogenic belt of the eastern part. Details of 25 of the major sedimentary basins are provided in Table 81.3. In the Rocky Mountains and the central stable platform in the west, a series of petroleum basins are distributed from New Mexico to Montana, including the Colorado Plateau. The petroleum formations range in age from Palaeozoic to Cenozoic. The Cordillera thrust fault belt has had great effect on the oil reserves in the Uinta, Piceance and San Juan Basins, in the western USA, many of the Cenozoic

Table 81.3 Major sedimentary basins of the USA

Basin	Area (thousand km²)	Major sedimentary rock		Reservoir	
		Age	Thickness (m)	Age	Lithology
Anadarko	60	E–P	>12,000	P, C	Carbonate rock, sandstone
Appalachian	530	E–C	12,000	D, C, S	Sandstone, carbonate rock, shale
Arctic Slope	100	D–R	9150	C, P, T, K	Limestone, sandstone
Bend Uplift	50	Pz–Mz			
Black Warrior	75	Pz, Mz			
Cincinnati Uplift	120	E–C	2300	O	Limestone
Cook Inlet	75	E_3–N_2	8230	E_3–N_2	Sandstone, siltstone
Denver	150	E–R	4600	K	Sandstone
East Coastal	1110	J–R	7620	K	Limestone, dolomite
Forest City	83	E–P	1220	C	Sandstone
Gulf of Alaska	70	R	3048–4572	R	Sandstone
Illinois	180	O–C	4800	C	Sandstone, carbonate rock
Kansas Uplift	73	E–P	1500	O	Carbonate rock
Michigan	190	E–J	5000	D, C	Carbonate rock, sandstone
Nevada	240	K–R			
Ozark Uplift	120	Pz			
Palo Duro	50	Pz	7500	P	Carbonate rock
Permian	130	E–P	8000	P, C	Carbonate rock, sandstone
Powder River	60	E–R	5150	K, C	Sandstone
Sacramento	50	J–R	18,290	K, E_2	Sandstone
Selawik	60	Pz–Mz			
Sweet Grass	100	E–R	2900	K, C	Sandstone, carbonate rock
West Washington–Oregon	150				
Williston	240	E–R	4600	D, C	Carbonate rock
Gulf of Mexico	1300	R–J	13,000–15,000	R, K, J	Sandstone

basins are hydrocarbon-bearing. In Alaska, there are two oil producing zones. One is the Cook Inlet Basin with Cenozoic strata. The other is North Slope Basin, where the famous Prudhoe Bay oil field is located. In the Prudhoe Bay oil field, the plays are Palaeozoic, Mesozoic and Cenozoic in age (Cram, 1971).

In the central part of the USA, the major hydrocarbon-bearing rocks are Palaeozoic in age. In the western USA, the hydrocarbon is found in intermontane basins and foreland basins. The strata are Palaeozoic, Mesozoic and Cenozoic in age. The Mid-continent region covers a large area of the central stable platform. There are two deep basins in the region. One is Anadarko Basin in the west; the other is Arkoma Basin in the east. Both of the basins have thick Palaeozoic sequences, which are hydrocarbon-bearing. The hydrocarbons in that area are concentrated in Kansas and Oklahoma. The structures are of various types, including structural, stratigraphic and composite traps. Along the coast of the Gulf of Mexico there are Mesozoic and Cenozoic composite basins on the southern margin of the North American Continent. The basin structural units in western Texas and eastern New Mexico include the Permian Basin, Bend–Fort Worth Region and Marathon-Ouachita thrust fault belt. Most of the Palaeozoic strata in this area can produce hydrocarbons. The Permian Basin is very rich in hydrocarbon resources. To the east of Appalachian orogenic belt the reservoirs are Palaeozoic rocks.

Based on the oil and gas production, the region around the Gulf of Mexico is ranked first, the second largest region is western Texas and eastern New Mexico, where the Permian Basin is located, and the third largest is the Mid-Continent region.

LOS ANGELES BASIN

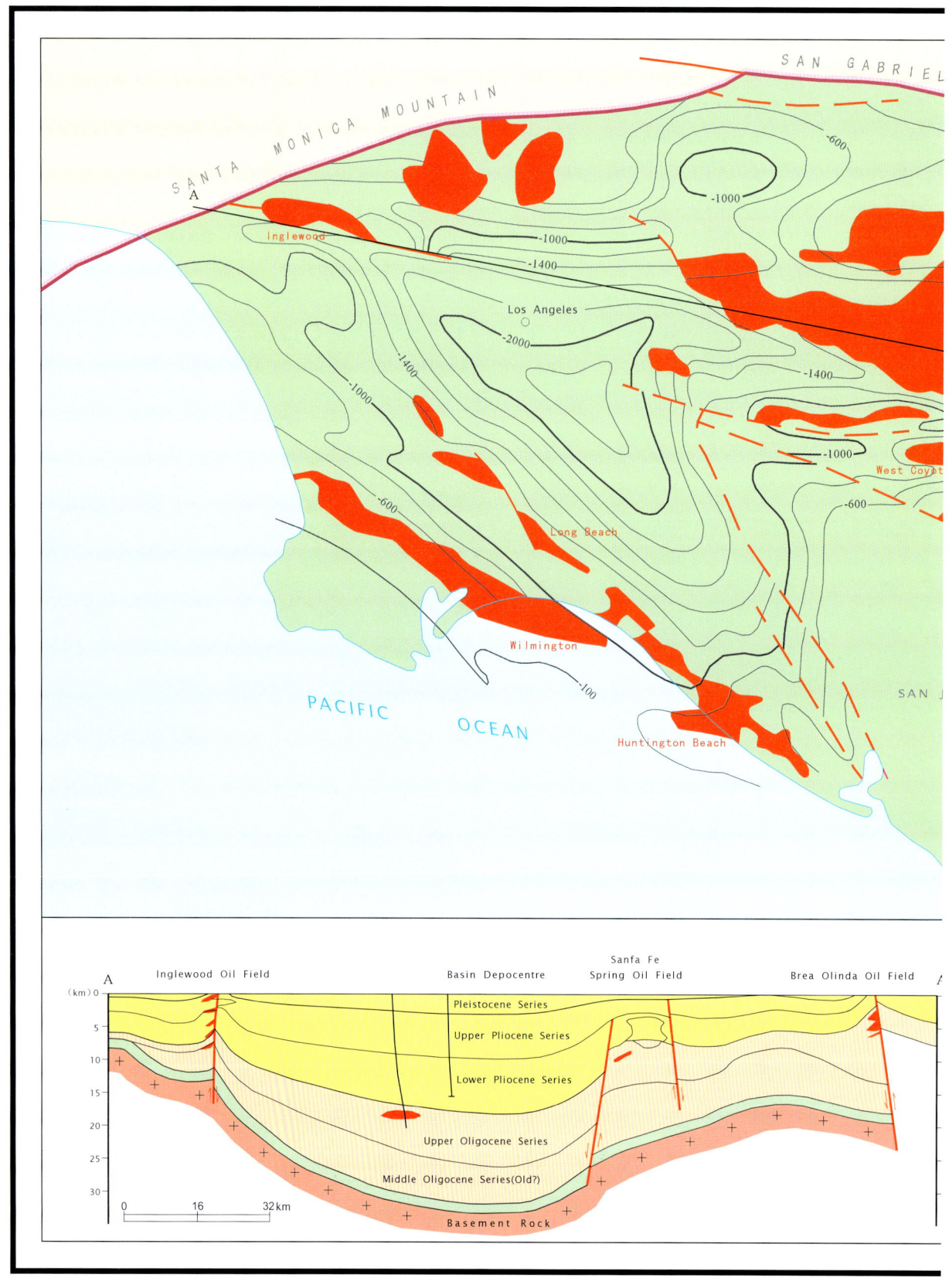

CHAPTER 82

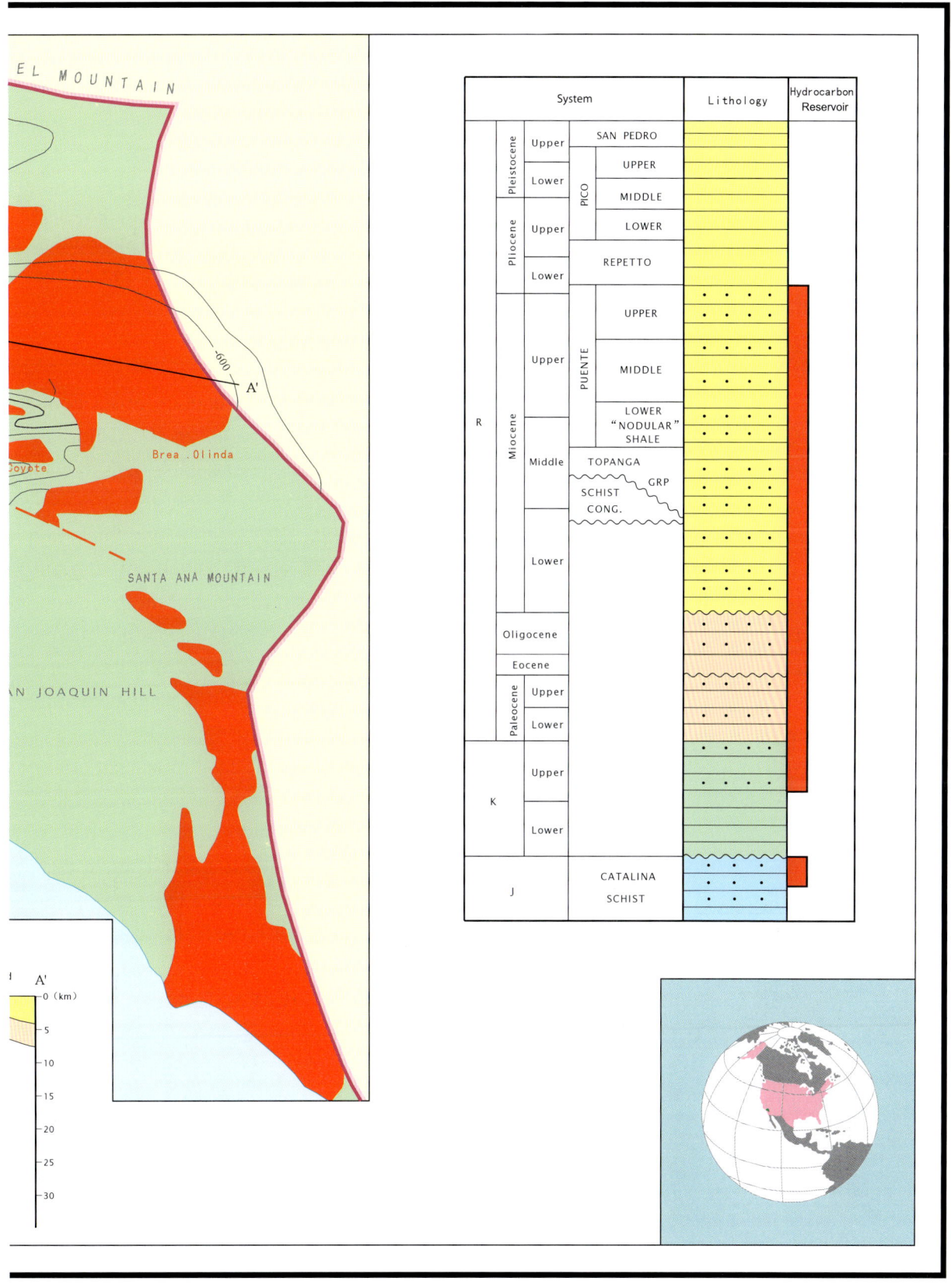

82 Los Angeles Basin

Geography
The Los Angeles Basin is located in the western coastal state of California.

History
Oil exploration commenced in 1860. The first oil field was discovered in 1875. Wilmington oil field was discovered in 1932, with reserves of 306 million t (Table 82.1 and Fig. 82.1). Since then, there have been more discoveries of significant oil and gas fields, making this basin one of the most important in continental USA

Regional geology
The basin is located at the junction of two lithospheric plates. The sedimentary cover comprises extensive Quaternary deposits that overlie Pliocene deltaic sediments of the Colorado River, which in turn overlie deeper-water terrigenous rocks with basalts and conglomerates at the base.

This is a prolific petroleum basin that has been producing oil since 1989, with a total thickness of cover up to 5–6 km. Up to 50 fields have been discovered to date, among which are 12 giant fields. Unlike in the Gulf of California Basin, gas fields have not been discovered in this basin. Virtually all the production has come from the Upper Miocene and Lower Pliocene strata where organic-rich shales are primarily developed. Traces of subcommercial hydrocarbons have also been reported from Middle Miocene, Cretaceous and Jurassic rocks. Some oil is produced from Jurassic and Cretaceous Franciscan Formation fractured schists. Folds and faults dominate the offshore structures. The results of oil exploration in this basin indicate that the Tertiary anticlines are oil-producing. Most oil fields are associated with faulted anticlines, and a few are associated with stratigraphic traps.

In the west the oil fields discovered lie along a southeast–northwest widening trend, and a variety of plays occur. The main structures lying to the northeast of the Palos Verdes Hills are simple broad anticlinal drapes over basement highs formed of schist. A trend of faulted anticlines occurs

Table 82.1 Major oil and gas fields of the Los Angeles Basin

Name	Discovery year	Recoverable reserves		Depth (m)	Reservoir	
		Oil (million t)	Gas (billion m^3)		Age	Lithology
Wilmington	1932	306	3.7	600	N	Sandstone
Huntington Beach	1920	153	2.3	1500	N	Sandstone
Long Beach	1921	131	3.0	2200	N	Sandstone
Santa Fe Springs	1919	87	7.1	300	N	Sandstone
Brea Olinda	1884	62	1.5	1200	N	Sandstone

World Atlas of Oil And Gas Basins, First Edition. Li Guoyu.
© 2011 John Wiley & Sons, Ltd. Published 2011 by John Wiley & Sons, Ltd.

Los Angeles Basin

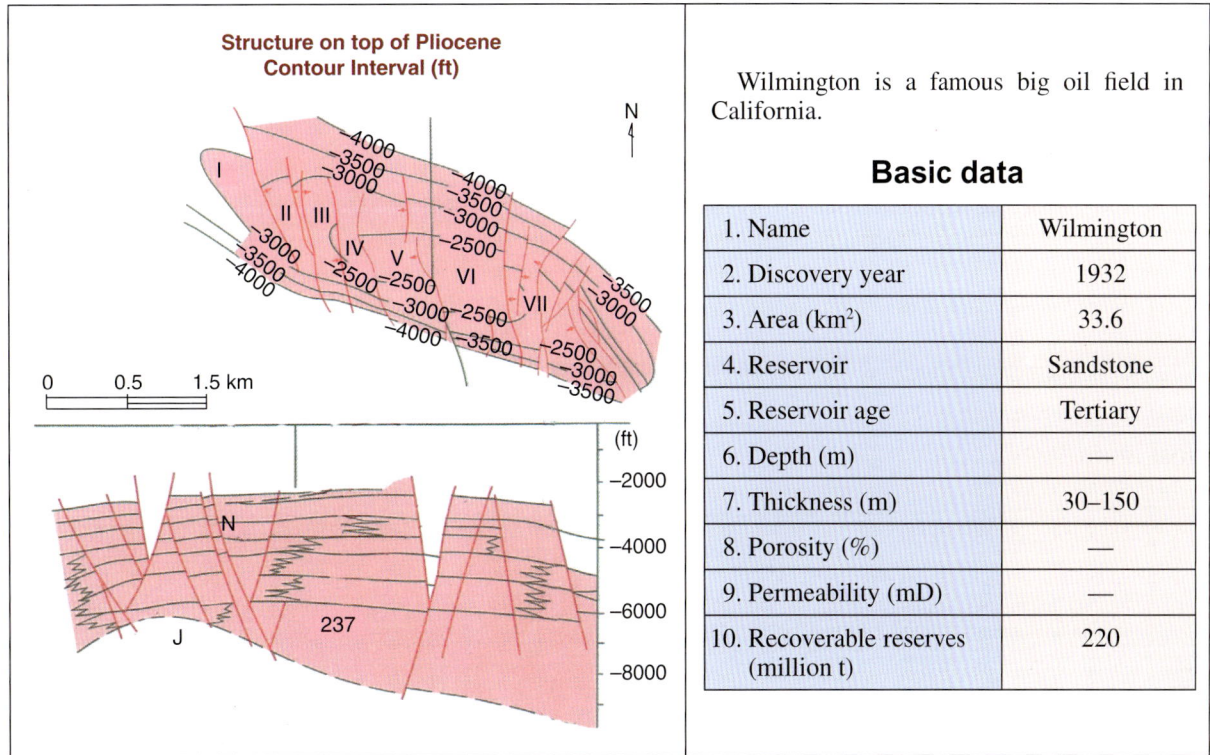

Fig. 82.1 Details of the Wilmington oil field.

further inland. Oil was first discovered in 1920. There are now six major producing fields, each with reserves in excess of 50 million barrels, plus several smaller fields. Total cumulative production from these oil fields is in excess of 2.5 billion barrels. The multiple sandstone reservoirs are Miocene and Pliocene in age. In addition, a number of broad simple domal buckle fold structures exist in the east.

In the vicinity of the Puente Hills further east, the structures are fault-controlled and associated with the Whittier Fault Zone. Montebello oil field is an en échelon WSW-plunging anticlinal field, closed by the fault. Whittier oil field is an elongated structure aligned parallel with the main fault zone. Sansinena oil field is located at a strike swing in the fault and en échelon drag folds are preserved under the hanging wall of the fault. Brea-Olinda oil field is a regional homocline closed updip by the fault, with a stratigraphic component that facilitates up-dip pinchout of some of the reservoir sandstones.

In the offshore parts of the Los Angels Basin the major plays are also Tertiary in age and comparable in style to those in the onshore basin. In particular, the Upper Miocene and Pliocene strata contain both source rocks and thick, generally turbiditic, sandstone reservoirs. Although the Los Angeles Basin is relatively small, it contains not only a thick sequence of Neogene hydrocarbon-bearing sediments, but also plentiful potential oil-producing horizons. The basin, both onshore and offshore, is likely to hold significant potential for plays, comparable in style to the discoveries described, especially at Upper Miocene to Pliocene levels. These are most likely to occur in the relatively deep central part of the onshore basin and in the offshore areas.

ROCKY MOUNTAIN BASINS

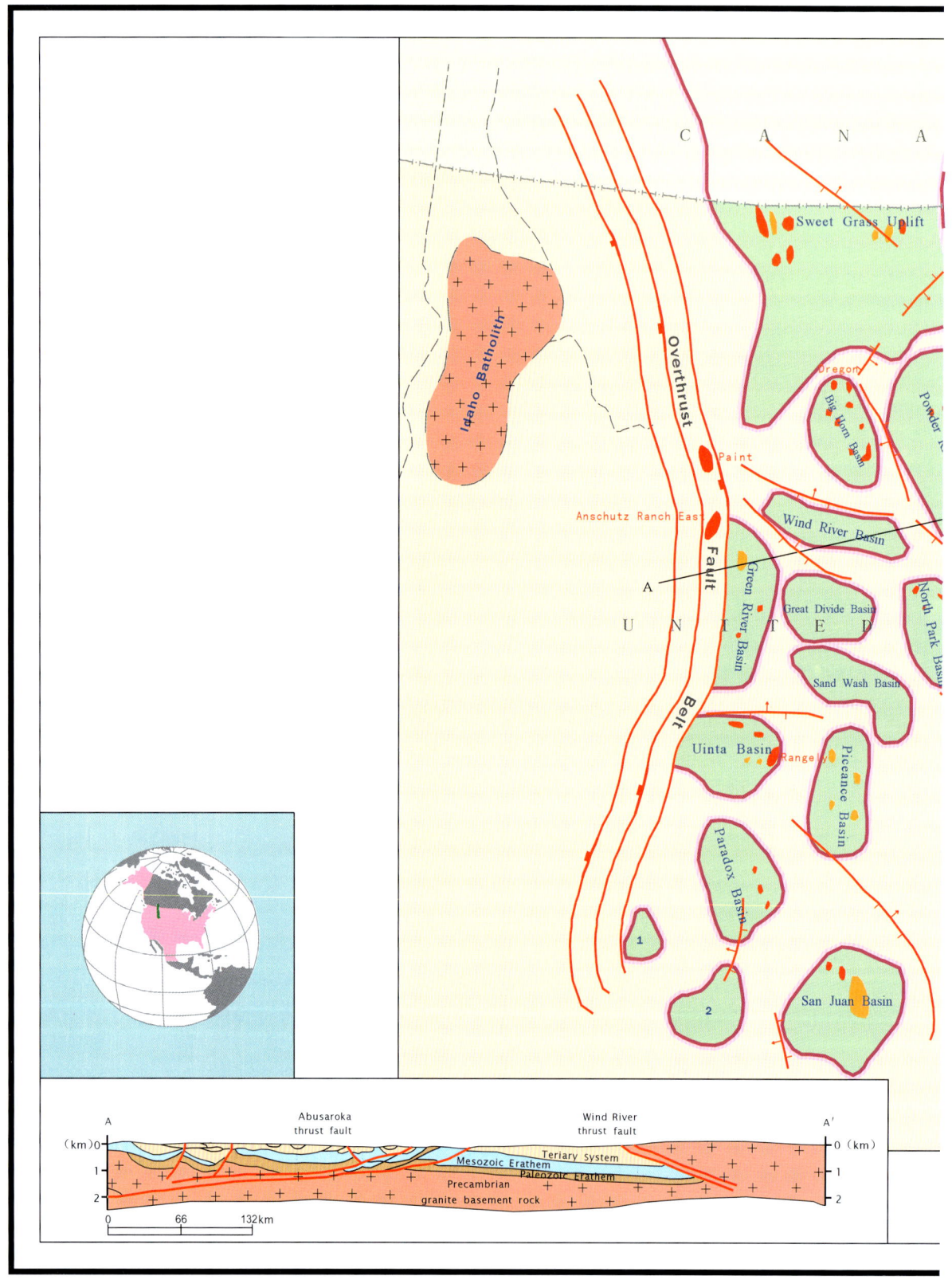

Basin: 1 Kalparowite 2 Black Mesa

CHAPTER 83

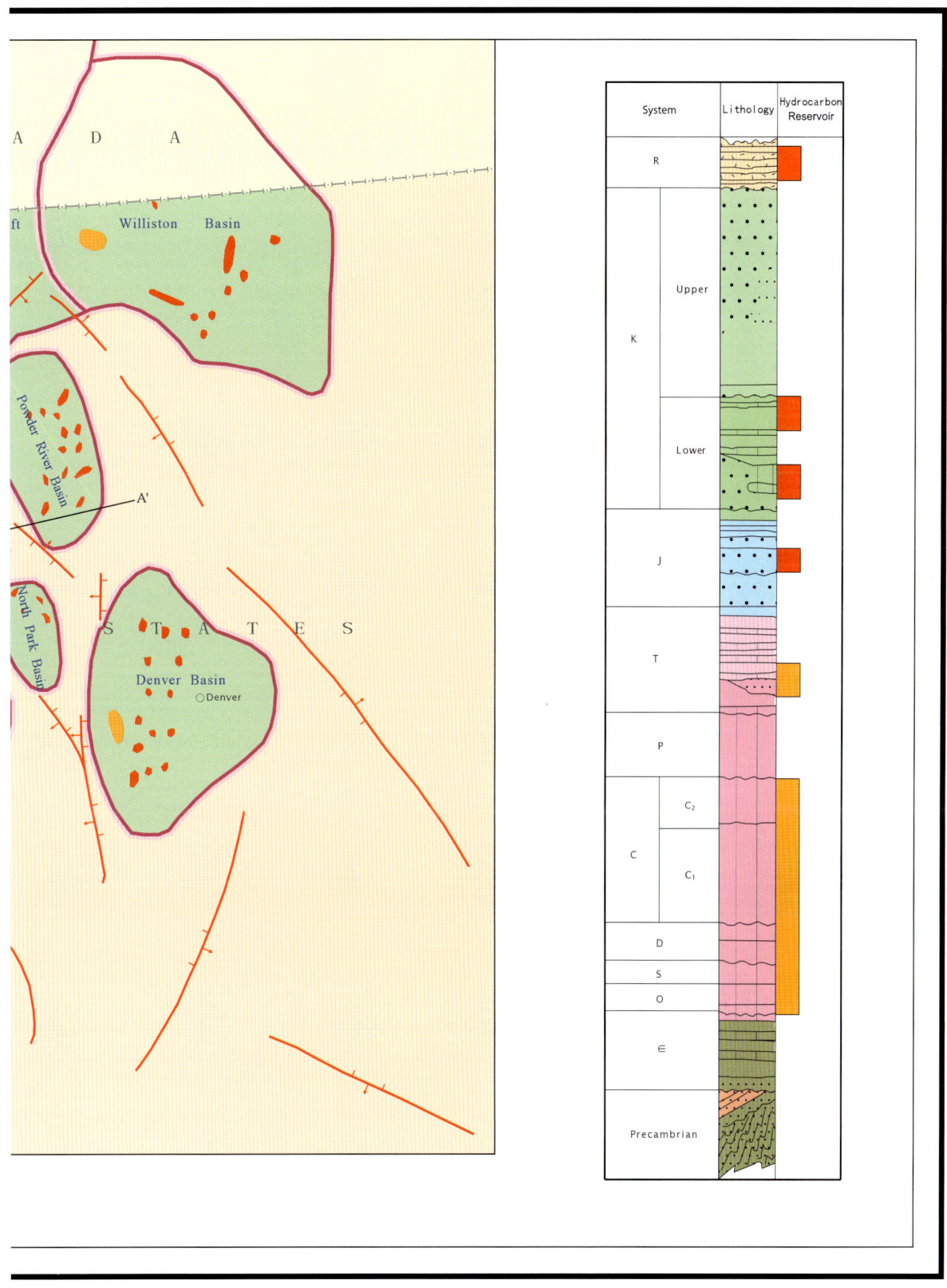

83 Rocky Mountain Basins

Geography

The Rocky Mountains petroleum province, in the west central part of the USA, is a NNW-trending series of Palaeozoic and Mesozoic hydrocarbon-bearing basins. They encompass parts of the States of Idaho, Wyoming, Nevada, Utah, Colorado and Arizona.

History

Oil exploration commenced in 1861 and shortly thereafter the Florence oil field was discovered. This is considered to be the second most important discovery after that in the Appalachian Basin in the eastern USA. Oil production reached 32 million t in 1975. A total of more than 1000 oil and gas fields have been discovered in this basin.

Regional geology

The Rocky Mountains petroleum province encompasses the western overthrust belt, the eastern foreland thrust belt and a series of intermontane or foreland thrust sheets. The major hydrocarbon production in the basins is associated with Devonian carbonate reefs, Cretaceous sandstone drapes and stratigraphic and structural (e.g. Fig. 83.1) traps. The Rocky Mountains Petroleum Province is one of the most prolific hydrocarbon-producing regions in North America, with an annual production close to 47 million t.

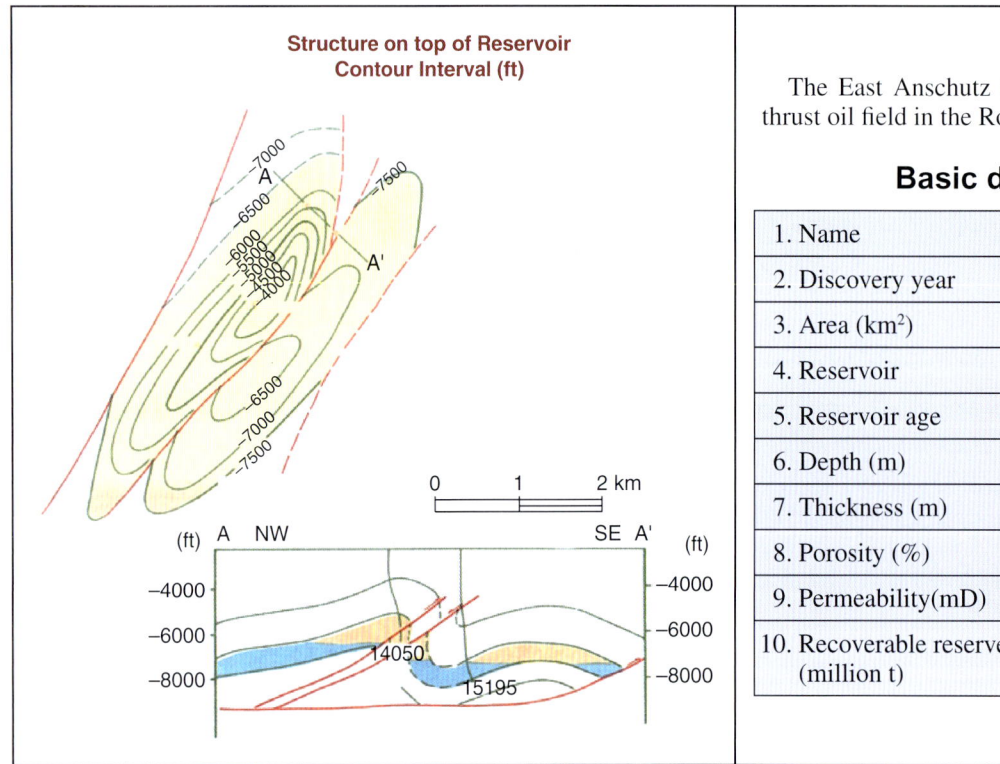

The East Anschutz is a typical overthrust oil field in the Rocky Mountains.

Basic data

1. Name	East Anschutz
2. Discovery year	1979
3. Area (km^2)	53.7
4. Reservoir	Sandstone
5. Reservoir age	Jurassic
6. Depth (m)	3,900
7. Thickness (m)	335
8. Porosity (%)	12
9. Permeability (mD)	0.01–460
10. Recoverable reserves (million t)	100–150 (in place)

Fig. 83.1 Details of the East Anschutz oil field.

World Atlas of Oil And Gas Basins, First Edition. Li Guoyu.
© 2011 John Wiley & Sons, Ltd. Published 2011 by John Wiley & Sons, Ltd.

The Rocky Mountains area has a complex history extending from Palaeozoic to Tertiary times. At the end of the Devonian, folding and eastward thrusting of the order of 30 km occurred behind the geanticline, while eastward-dipping oceanic subduction was continuing at the continental margin. Late Cretaceous compression produced a foreland thrust region of steep thrusts where the crustal shortening and thickening resulted in the thrusting of generally Precambrian basement across Palaeozoic and Mesozoic sediments of the retroarc basin, and the underlying Palaeozoic and early Mesozoic strata of the miogeosyncline or proto-retroarc basin. The occurrence of metamorphism and deformation at this significant distance from the continental plate edge is thought to reflect a shallow dipping subduction zone beneath the region. During the Middle Miocene, rocks of the Basin-and-Range Province were uplifted and faulted. An effect instigated through regional uplift caused by subduction beneath the continental plate of part of the East Pacific Rise. Alternatively, the stretching and extensional faulting may result from heating and expansion of the subducted mantle material beneath the continental edge.

In the Rocky Mountains petroleum province of the sedimentary cover comprises Precambrian to recent sediments, which rest upon Precambrian crystalline basement. The region has experienced a complex tectonic history, undergoing various phases of compression in a subduction-associated plate tectonic setting. This has resulted in a complex stratigraphic sequence characterized by interdigitating facies punctuated by eustatic and tectonic unconformities. The sedimentation regime remained regionally and relatively uniform during Palaeozoic times, becoming more variable in Mesozoic times and complex and laterally variable from one basin to the next in Cenozoic times, when local facies patterns were greatly influenced by their proximity to local and regional orogenic centres and uplifted sediment source areas. The Palaeozoic and Lower Mesozoic strata of the Rocky Mountains are predominantly a marine sequence, with tectonic effects inducing transgression/regression cycles from Upper Palaeozoic times.

The basement Precambrian rocks in the region are subdivided into the metamorphic and igneous rocks of Montana, Wyoming, Colorado and northern Utah and Arizona, which have radiometric ages in the range of 1.3–2.7 Ga. The Cambrian strata typically comprise diachronous sequences of quartzites, siltstones and carbonates, in a pattern of eastward onlap of the continental platform that is continued through the Lower Ordovician.

The Rocky Mountains petroleum province contains a total of 18 giant oil/gas fields and a large number of smaller fields, with a huge potential in petroleum. Hydrocarbon-bearing basins include the Sweet Grass Uplift (100,000 km^2), Wind River (26,000 km^2), Big Horn (30,000 km^2), Paradox (45,000 km^2), Williston (240,000 km^2), San Juan (40,000 km^2), Powder River (60,000 km^2), Denver (150,000 km^2) and Green River (40,000 km^2) (Cowan *et al.*, 1986).

WILLISTON BASIN

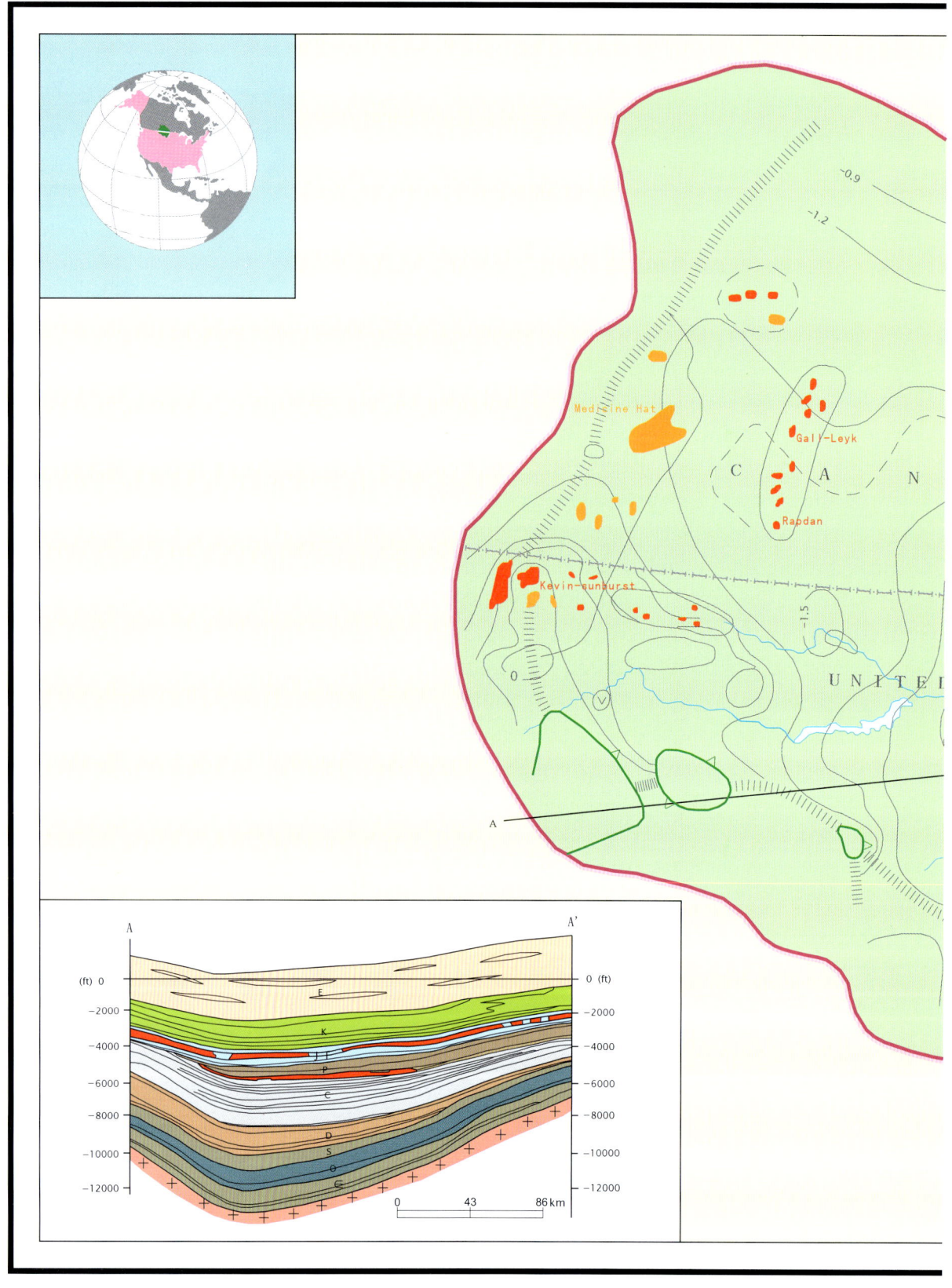

CHAPTER 84

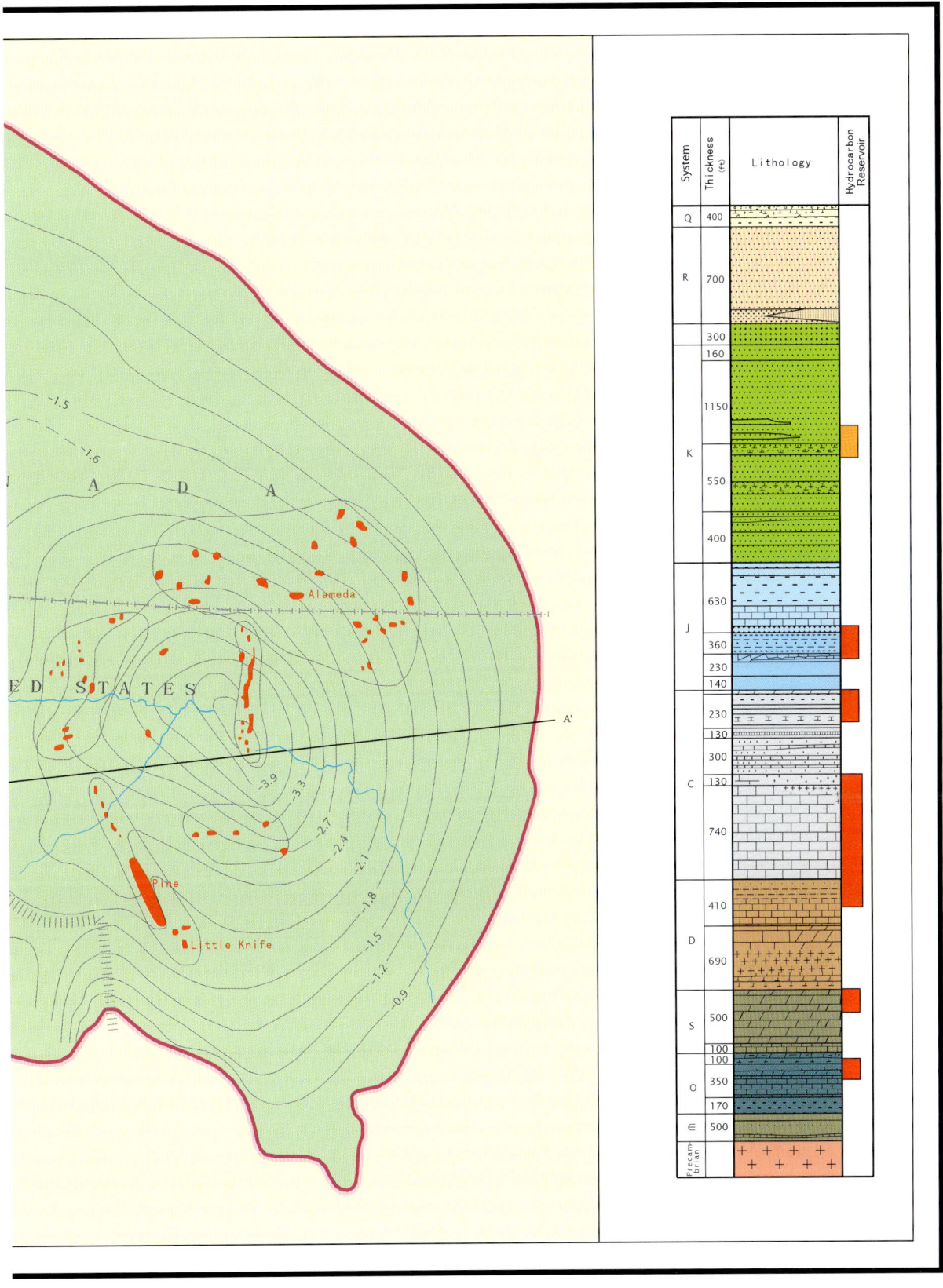

84 Williston Basin

Geography

The Williston Basin, which covers an area of 335,660 km², lies in north-central USA and extends northwards into Canada.

History

The first gas discoveries were made in 1914 on the Cedar Creek Anticline in the USA. In Canada, the drilling of structural anomalies at the Mississippian–Mesozoic unconformity led to discoveries of large stratigraphically trapped oil accumulations. Six of the largest fields were discovered in the early 1950s, including the Pine and Beaver Lodge oil fields (Table 84.1).

Regional geology

The Williston Basin is centred on the states of North Dakota and Montana in the USA and Saskatchewan in Canada. The sedimentary cover comprises Upper Palaeozoic, Mesozoic and Tertiary strata. The entire sedimentary succession reaches a maximum thickness in excess of 4.6 km in the depocentre of North Dakota. As with other typical interior basins, the succession in the Williston Basin thins gradually towards the basin margins and displays a concentric outcrop pattern, with progressively older strata occurring towards the basin edge. Regional dips in the basin are typically less than 10°. As a result, many of the traps are rather subtle stratigraphic features. Basement in the region is the Precambrian igneous and metamorphic complex of the Canadian Shield. It is overlain by a Lower Palaeozoic succession of up to 1 km thickness.

Although the Williston Basin contains no giant fields and, like the other interior basins is not a major producer by world standards, it produces oil and gas from a large number of small and often interesting accumulations. As mentioned above, the stratigraphic sequence in the Williston Basin extends from the Cambrian through to the Tertiary. However, more than 90 per cent of the prospective reservoirs lie in the Palaeozoic with over 90 per cent of these being carbonates. Commercial production has been established from 13 or so separate rock formations ranging in age from the Cambrian to the Cretaceous. These are mostly in the Upper Devonian and Mississippian. Between 1952 and 1957 oil fields were discovered in Mississippian and younger strata, while subsequent exploration has encountered success in older, especially Devonian, strata. The major successful plays to date are the

Table 84.1 Examples of oil fields in Williston Basin

Name	Discovery year	Age	Lithology	Area (km²)	Depth (m)	Residual recoverable reserves (thousand t)
Pine	1951	O	Carbonate rock	53.9	2560	349
Beaver Lodge	1951	C, D, O	Dolomite, limestone	87.1	2590–3870	694
Little Knife	1977	C	Carbonate rock	95	800	7560

World Atlas of Oil And Gas Basins, First Edition. Li Guoyu.
© 2011 John Wiley & Sons, Ltd. Published 2011 by John Wiley & Sons, Ltd.

Williston Basin

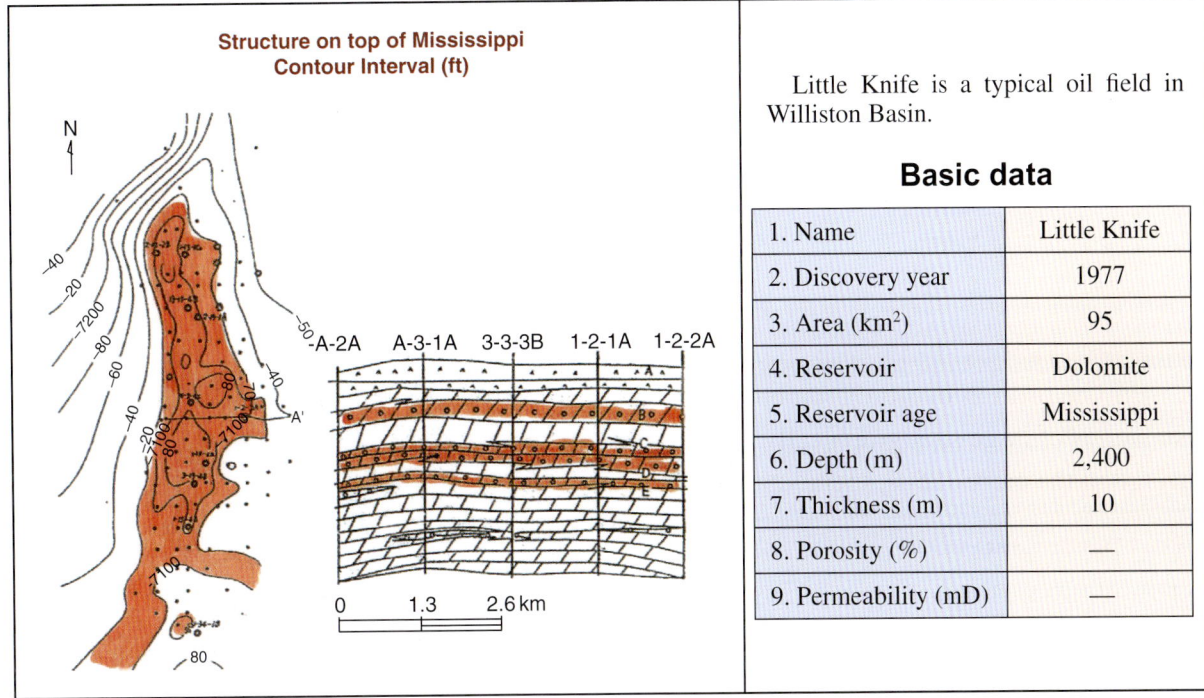

Fig. 84.1 Details of the Little Knife oil field.

Lower Devonian carbonates, which occur along the western margin of the basin as a long and prolific trend extending along the line of the Rocky Mountains into western Alberta and northeast British Columbia, and the Ordovician Red River Formation. The plays associated with the Mississippian Madison Group structural/stratigraphic traps are anticlinal and domal closures.

The cratonic Williston Basin is filled by predominantly marine sediments ranging in age from Cambrian to Tertiary. Cambrian to Mississippian sediments are predominantly carbonates which were formed under shallow marine conditions, frequently as carbonate–evaporite cycles. Tertiary compression causing deep-seated lateral movements are thought to be responsible for the folding and faulting along two extensive structural trends that had shown some positive structural tendency during earlier phases of basin development.

The most important Palaeozoic oil source is the Devonian–Mississippian Bakken Formation. Shales in the Ordovician Winnipeg Formation are the likely source of the oil in the Red River Formation reservoirs of the same age. So far the most important reservoirs are the dolomites and limestones of the Mississippian Madison Group (e.g. Fig. 84.1). Seals are provided by the alternation of carbonates, shales and evaporitic sediments within the reservoir sequences.

MICHIGAN BASIN

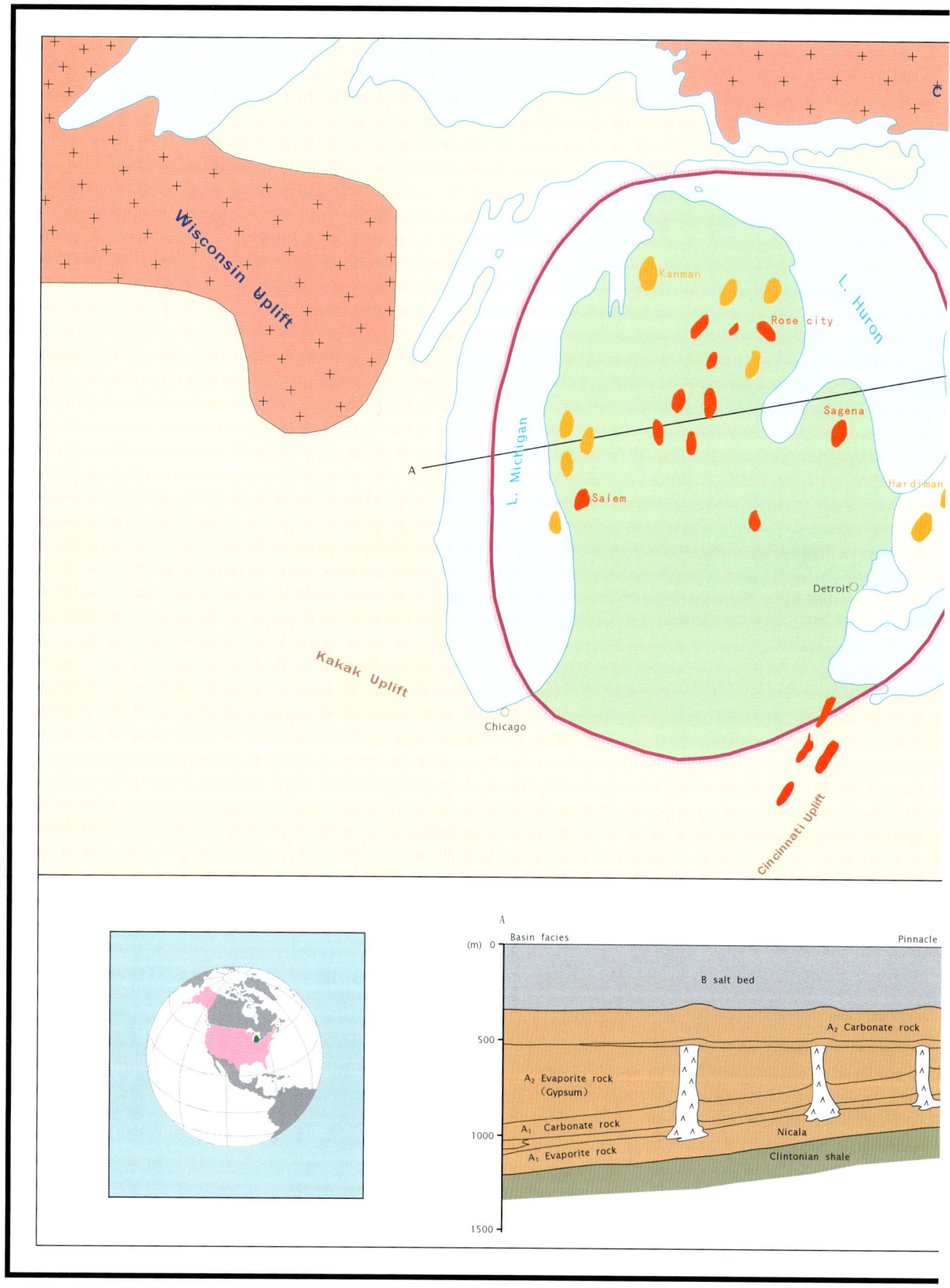

CHAPTER 85

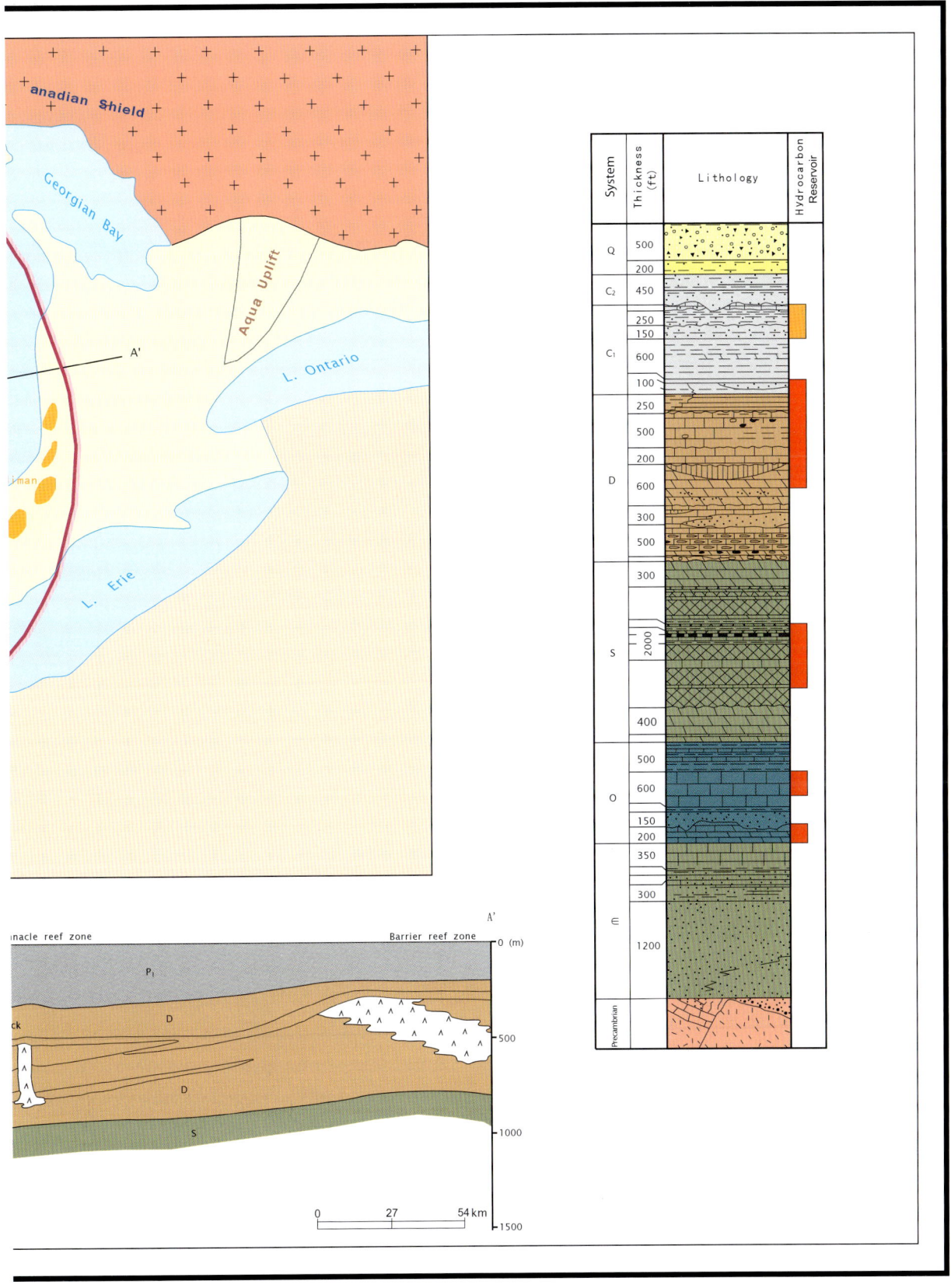

85 Michigan Basin

Geography

Michigan Basin lies in the Great Lakes region of the USA, to the southeast of Williston Basin and the northeast of Illinois Basin. It lies chiefly in the State of Michigan with extension into Indiana and Ohio to the south and Wisconsin to the west. The basin has an area of 6000 km^2 and is bordered to the west by the Wisconsin Arch, to the south by the Kankakee and Findlay Arches, to the east by the Algonquin Arch and to the north by the Precambrian igneous and metamorphic rocks of the Canadian Shield. It appears to conform to the classic broadly circular morphology of an interior hydrocarbon-bearing basin.

History

In common with many of the USA hydrocarbon-bearing basins, the Michigan Basin has a history of exploration stretching back into the 19th century. Production has been established in rocks ranging from Ordovician to Quaternary in age, but the most important reservoirs are in the Ordovician, Silurian and Devonian strata. In the 1880s, the first fields were discovered in the southern part of the basin, in Indiana and Ohio, and produced from Middle Devonian carbonate reefs. In the mid-1920s, significant production was established firstly from Mississippian and then from Upper Devonian strata in the centre of the basin. By the mid- to late 1930s, oil production had been established from Ordovician and Lower Devonian rocks (e.g. Fig. 85.1), while the first commercial oil discoveries in the Silurian were made in the 1950s. Gas fields were discovered first in the Carboniferous strata, followed by discoveries in Devonian and Silurian strata.

Regional geology

The sedimentary cover consists largely of a Lower Palaeozoic succession with a thinner sequence of Upper Palaeozoic and Mesozoic strata. The facies are largely shallow marine and commence with basal Cambrian sandstone overlain by shales, carbonates and evaporites. The latter consists of basinal carbonates rimmed by pinnacle reefs on a platform slope, a barrier reef and platform back-reef carbonates. The depositional topography has been infilled with Upper Silurian evaporites and minor amounts of carbonates. The Devonian and Mississippian are dominated by marine carbonates which give a way to Pennsylvanian non-marine clastics.

Structurally the basin is rather simple. A series of post-Mississippian northwest–southwest trending folds, slightly asymmetrical, dominate the central and southeastern parts of the basin, while in the southwest a group of folds plunge northeastwards, and others appear to have a random orientation and are probably controlled by movement of the Palaeozoic salt.

The mature Silurian Niagaran reef play is now relatively well understood. Estimated primary recoverable reserves from the known reefs are in the order of 300–400 million barrels. Although exceeded by the Devonian and Mississippian in terms of oil and gas production, it is probably the most famous and best documented play in the basin and information from the exploration has

Michigan Basin

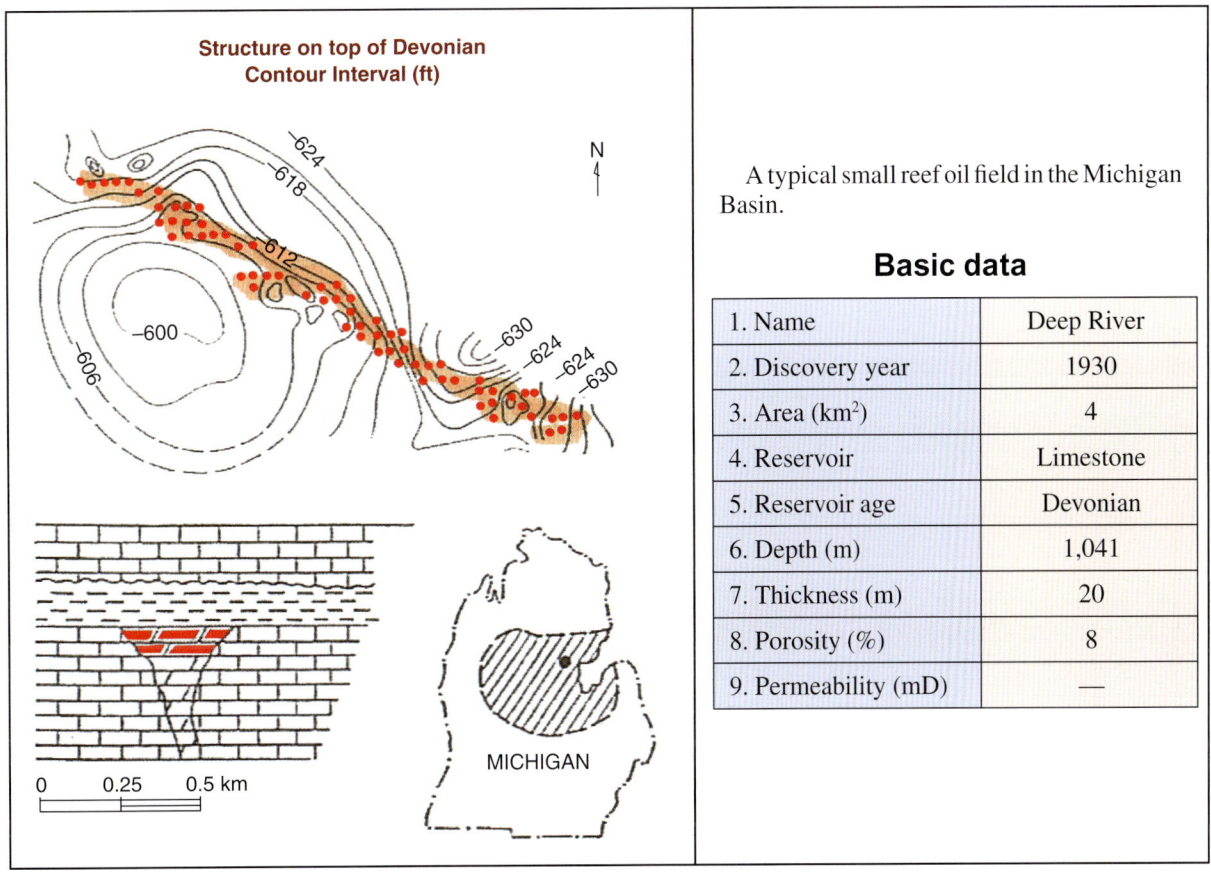

Fig. 85.1 Details of a typical small reef oil field in the Michigan Basin.

been instrumental in refining models of differential migration and entrapment. Although the first discoveries were made in the 1950s, extensive exploration of the reef play did not commence until 1966. This has concentrated around the northern zones.

To date more than 30,000 wells have been drilled in the basin and have resulted in the discovery of approximately 600 oil and gas accumulations. As is typical of interior basins, the Michigan Basin is not a major petroleum province and the majority of the fields are small, each holding less than 140,000 t of oil or less than 170 million m^3 of gas. One giant oil field, Albion-Scipio (200 million barrels recoverable) has been found in the basin and produces from a dolomitized fault zone in the Ordovician Trenton–Black River Formation. To the southeast of the Michigan Basin, the Findlay Arch, separating the basin from the Illinois Basin, contains the prolific Lima-Indiana trend, with major reservoirs in the Ordovician Trenton Formation. Reserves in this trend are of the order of 60 million t.

The central part of the basin is the most hydrocarbon-rich, with Devonian and Mississippian carbonate reefs, dolomitized fault zones and sandstones as major producers. There is also significant production from Silurian reefs and Ordovician strata on the northern and southern shelves of the basin respectively, while smaller reserves occur in the Pennsylvanian, and some gas even occurs in Quaternary strata.

MEXICO GULF OIL AND GAS REGION

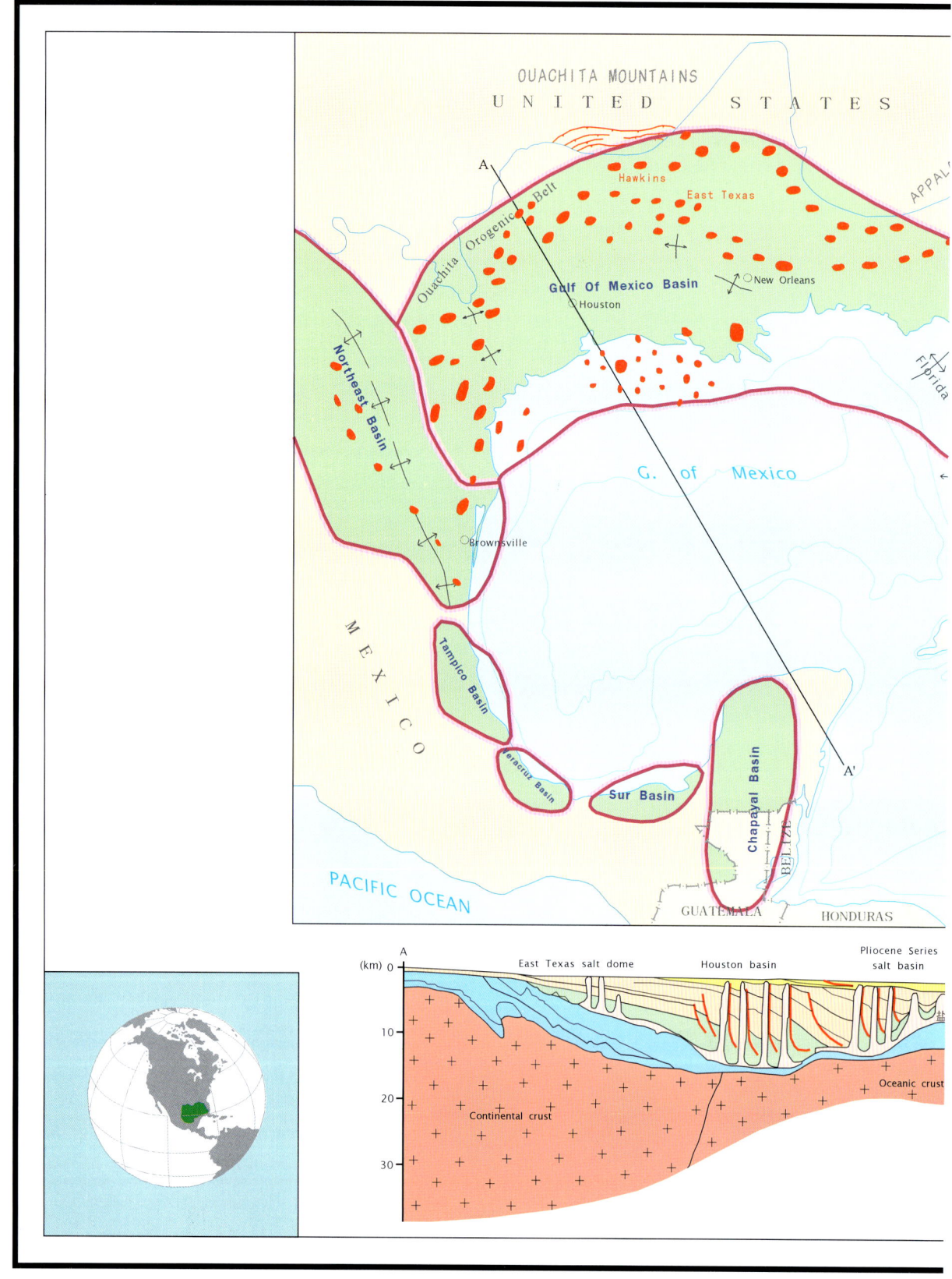

CHAPTER 86

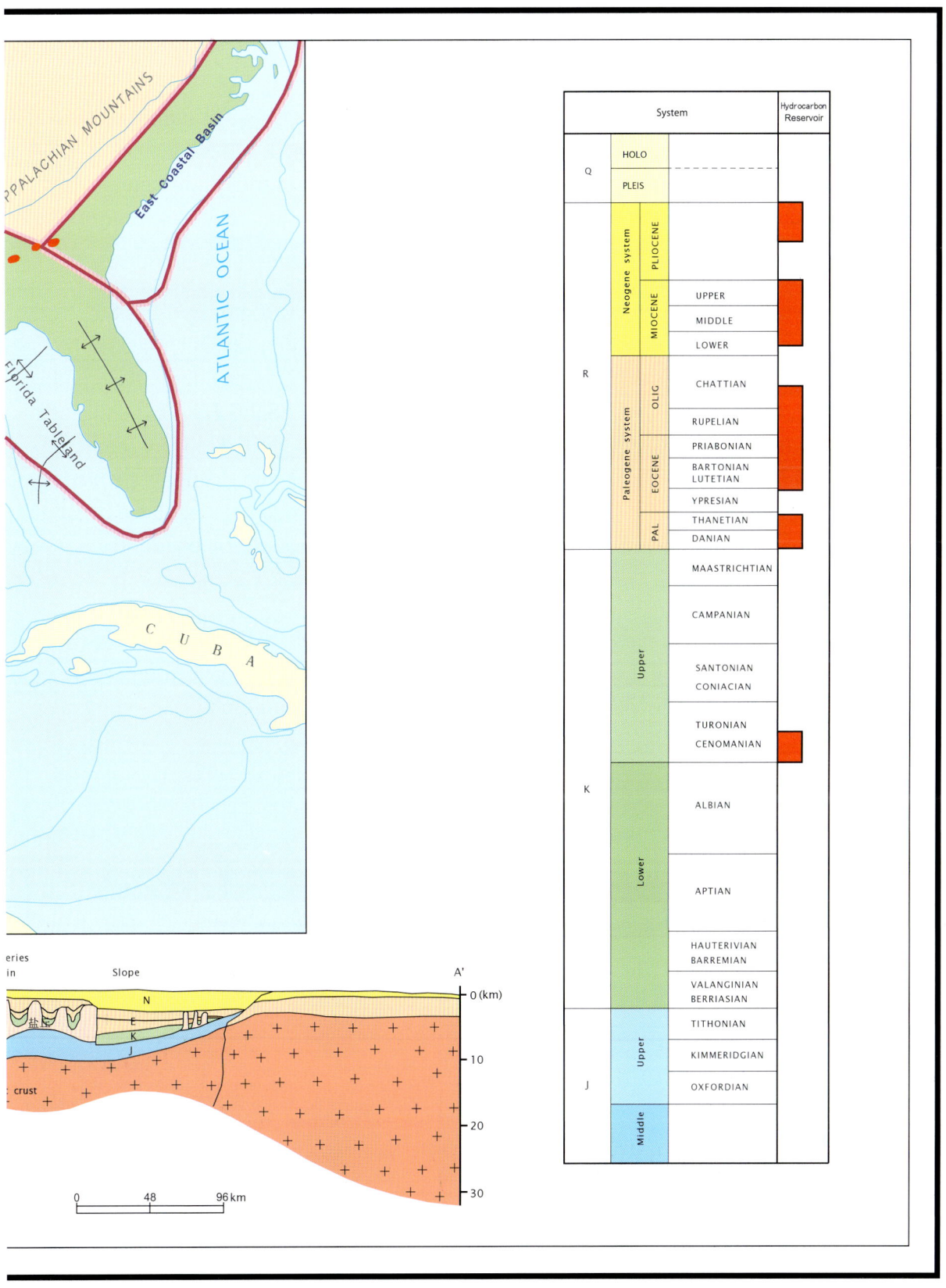

86 Mexico Gulf Oil and Gas Region

Geography

The Mexico Gulf oil and gas region is located in the south of the USA, including the States of Texas, Louisiana, Arkansas, Mississippi, Alabama and Florida, and also includes the shelf of the Gulf of Mexico with a water depth of 60 m. It has an area of 1.3 million km² and is a very renowned oil-producing area in the USA

History

Oil exploration commenced before 1860 with the first oil field was discovered in 1865. One exploration well near surface seepages in Spinto flowed oil at a rate of 14,000 t from cave limestone with depths of 340 m. This event strongly stimulated oil exploration on a large scale in this area. In particular, after the discovery of the mammoth East Texas oil field (Fig. 86.1), this area quickly became a major oil producing region. A total of 4087 oil and gas fields were found (offshore containing 177 oil and gas fields). Details of eight of the major oil and gas fields are listed in Table 86.1. By the end of 1972, 1900 platforms for oil drilling and production had been installed with annual oil production of 46 million t. Oil production reached 71.5 million t in 2000.

Regional geology

The Mexico Gulf oil and gas region occupies some parts of the Pre-Atlantic platform, continental slope, deep-water basin of the Gulf of Mexico with an oceanic crust and a foredeep of the Antilles Orogenic Belt. The presence of an active oceanic margin is inferred. The sedimentary

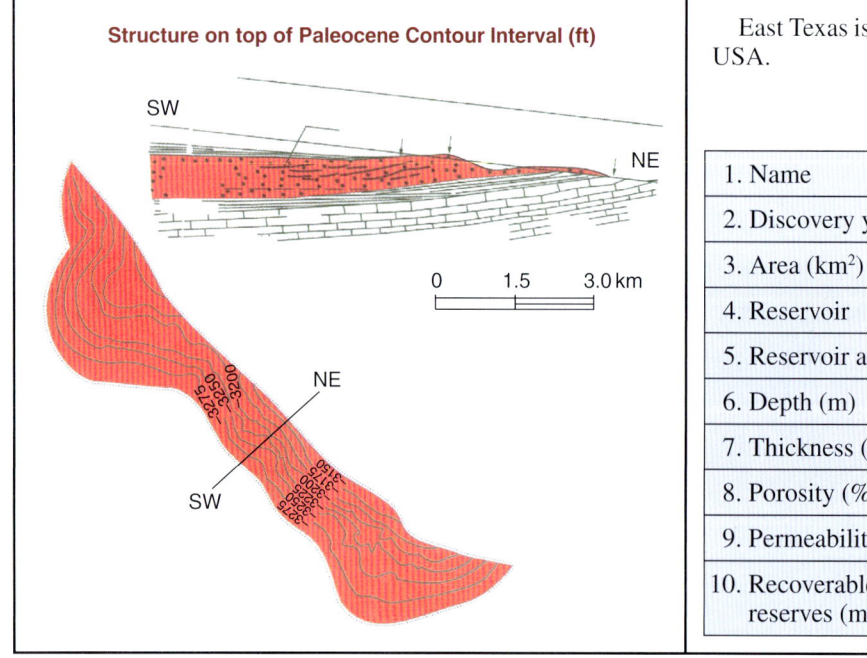

East Texas is the second giant oil field in the USA.

Basic data

1. Name	East Texas
2. Discovery year	1930
3. Area (km²)	566
4. Reservoir	Sandstone
5. Reservoir age	Cretaceous
6. Depth (m)	1,100
7. Thickness (m)	11.3
8. Porosity (%)	25
9. Permeability (mD)	2,098
10. Recoverable reserves (million t)	784

Fig. 86.1 Details of the East Texas giant oil field.

World Atlas of Oil And Gas Basins, First Edition. Li Guoyu.
© 2011 John Wiley & Sons, Ltd. Published 2011 by John Wiley & Sons, Ltd.

Table 86.1 Major oil and gas fields of Mexico Gulf oil–gas region discovered between 1930 and 1940

Name	Discovery year	Recoverable reserves		Depth (m)	Trap	Reservoir	
		Oil (million t)	Gas (billion m^3)			Age	Lithology
East Texas	1930	784	—	1100	Stratigraphic	K	Sandstone
Hawkins	1940	119	—	1400	Anticline	K	Sandstone
Hastings	1934	106	—	1800	Anticline	E	Sandstone
Greta	1934	134	—	1600	Anticline	E	Sandstone
Conroe	1931	104	—	1500	Anticline	E	Sandstone
Carthage	1936	21	50.9	1800	Anticline	K	Limestone
Katy	1934	3	16.8	2000	Anticline	E	Sandstone
Old Ocean	1934	18	14.0	3240	Anticline	E	Sandstone

cover comprises Cenozoic and Mesozoic strata. The base of the successions is formed by a thick evaporite sequence of predominantly Jurassic age. This sequence predetermined the wide development of salt-dome tectonics, with which a large number of oil and gas fields are associated. The upper part of the Jurassic strata is represented by terrigenous and marine carbonate rocks. The Cretaceous sequence is essentially comprised of carbonates. The Cenozoic part of the succession, which accounts for the most of the sedimentary cover volume, is composed mainly of terrigenous rocks. Albian–Senomanian, as well as Upper Jurassic and Paleogene, reef formations are widely developed throughout the succession. Maximum thickness of the sedimentary cover is up to 13–15 km (Dietz and Holden, 1970).

The reservoirs are represented by Pleistocene, Neogene, Paleogene, Upper Cretaceous, Middle and Lower Jurassic and Permo-Triassic, and also by Paleocene, Lower Cretaceous, Upper Jurassic and, probably, Mississippian limestones. In total 15,925 oil and gas fields have been discovered in the basin. Oil and gas accumulations are confined to various structural forms, such as anticlines, salt domes, reef masses, draping folds, zones of thinning and so on. The source rocks are Paleogene, Cretaceous, Jurassic and Triassic shales. In the outer part of the basin, oil accumulations are revealed in the anticlinal folds and zones of thinning. The oils are, as a rule, light and low in sulphur. In the accumulations associated with caprocks, sulphur content increases.

APPALACHIAN BASIN

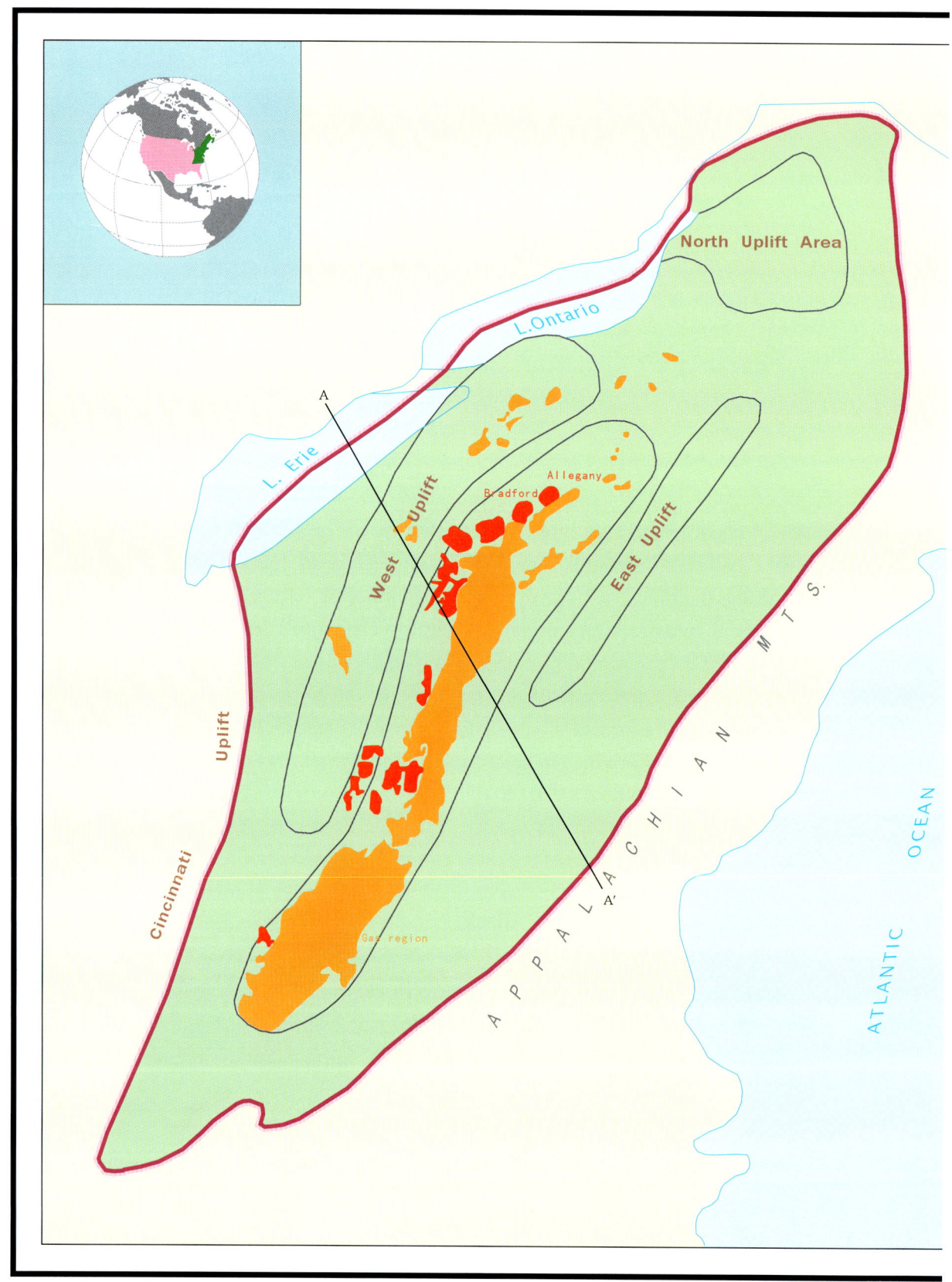

CHAPTER 87

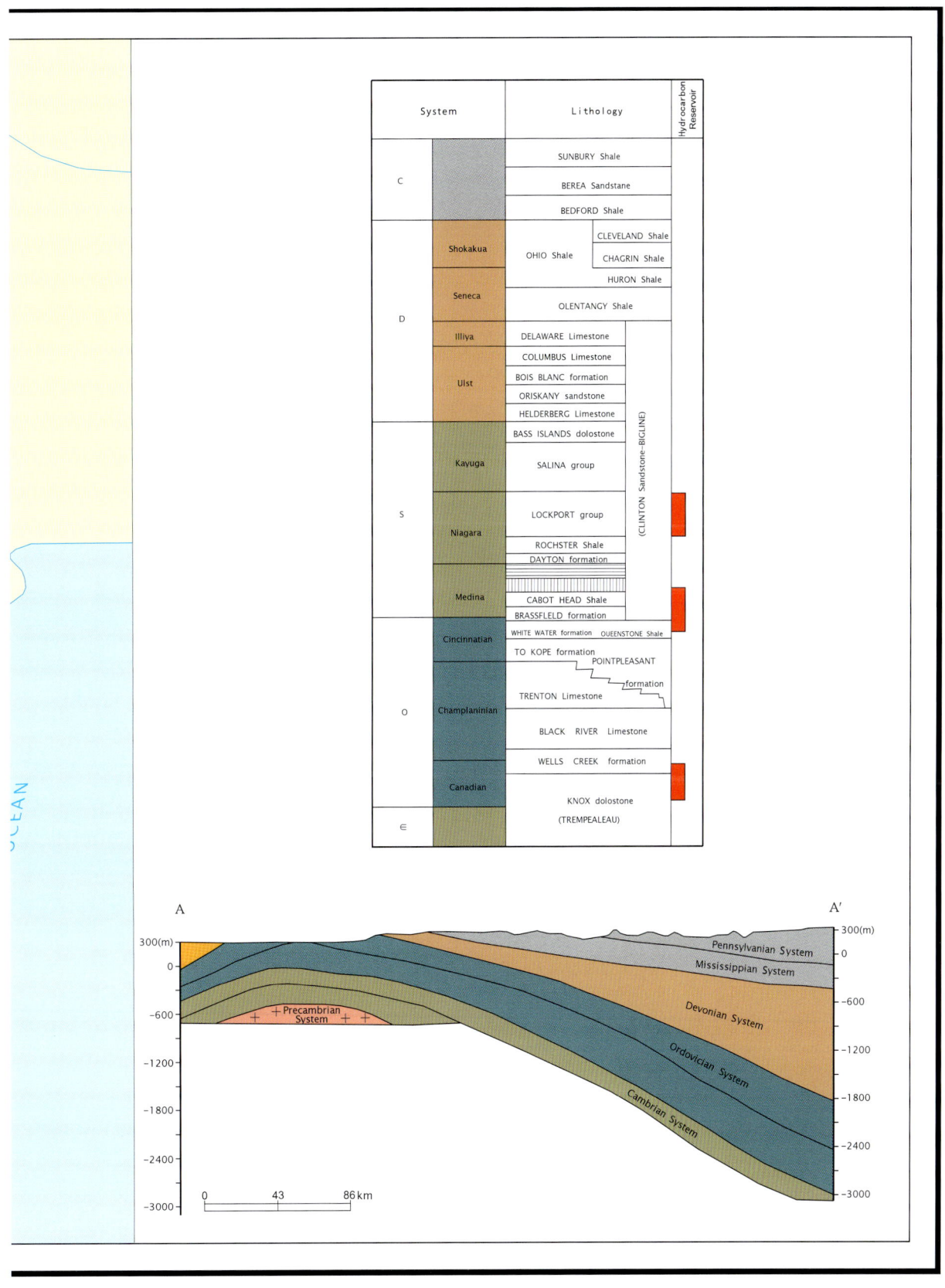

87 Appalachian Basin

Geography

This Appalachian Basin is located in the eastern part of the USA, including the states of New York, Pennsylvania, New Jersey, Ohio, Kentucky and Tennessee. Comprising an area of 530,000 km^2 this considered to be the birthplace of the establishment of the petroleum industry of the USA

History

Oil seepages were well known as far back as 1627. The first gas well was completed in 1821. The region is famed for being the home of the first wildcat well drilled by Edwin Drake in Pennsylvania. This well produced oil at rate of 274 t with a depth more than 20 m from Devonian sandstone in that year. Americans considered this event as marking the beginning of the world's petroleum industry. The Appalachian Basin is predominantly a gas-producing area. In 1978 annual oil production was 2.3 million t, with gas production of 12.5 billion m^3. At present, oil exploration is being carried out in the overthrust region of this basin and some oil and gas fields have been discovered.

Regional geology

The Appalachian Orogenic Belt separates the North American and the Pre-Atlantic platforms, and extends for a considerable distance (1600 km long) along the eastern seaboard of the USA, with a width of 300 km and area of 256,000 km^2, The belt consists of three segments – northern, central and southern. The age of folding in the first two segments is determined as Early Palaeozoic (Caledonian). The southern segment is considered to be a Late Palaeozoic (late Hercynidian) folded structure. Along the northern margin of the belt there exists a foredeep filled with the Ordovician to Lower Permian rocks. The Appalachian Belt is located on the basement of the ancient North America craton. The sedimentary cover is represented by Cambrian to Quaternary deposits. Mainly, the rocks that fill the basin are Palaeozoic, essentially carbonates. Mesozoic–Cenozoic strata are relatively limited and represented by terrigenous rocks. Productive sequences in the succession are Cambrian to Cretaceous inclusive. Reservoirs are represented by limestones, dolomites and sandstones. Oil and gas accumulations are, as a rule, confined to anticlinal folds, zones of thinning, erosional protrusions of the basement, reef masses and draping folds. Source rocks are represented by Mississippian, Devonian, Silurian and Ordovician shales and claystones, and by Lower Cretaceous shales.

Basins of this type are considerable in size (about 4100 million km^2) and possess large original in-place hydrocarbon resources. They have been producing for more than 60 years so that much of the resources have been extracted. Remaining prospective reserves are subthrust zones, located mainly within the limits of the Cordilleras Belt, and also oil and gas accumulations occurring at great depths in the foredeeps of the Cordilleras Belt and Appalachian Mountains.

World Atlas of Oil And Gas Basins, First Edition. Li Guoyu.
© 2011 John Wiley & Sons, Ltd. Published 2011 by John Wiley & Sons, Ltd.

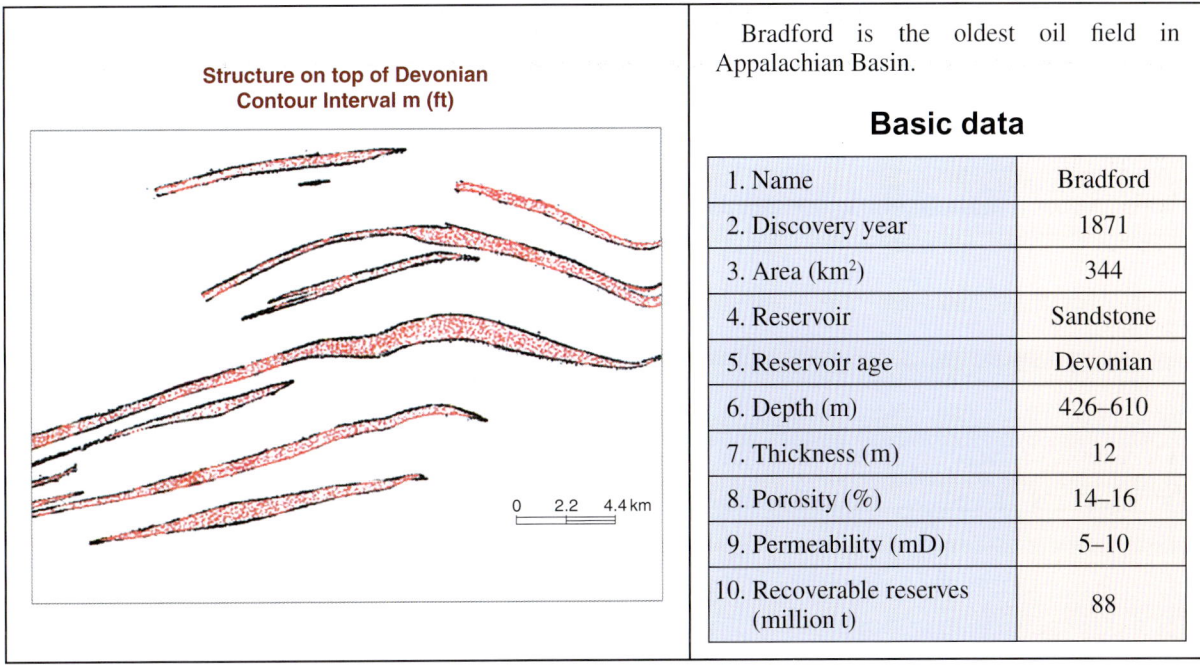

Fig. 87.1 Details of the Bradford oil field.

A thin-skinned fold model of Appalachian Plateau fold is considered to be impossible geometrically. However, a two-layer model consisting of a gently folded competent layer overlying a deformed zone, in which material flows from the synclines to the anticlines, is both reasonable geometrically and supported by geological evidence from the Plateau Province of Pennsylvania and south-central New York. The deformed zone is composed of weak rocks – evaporites and shale – which have been deformed much more than the competent rocks above; the décollement marks the lower boundary of this deformed zone. Care is required when reconstructing the depocentres of the deformed zone because the original thicknesses have been altered by later deformation.

The Appalachian Plateau of Pennsylvania and New York may be divided into two regions on the basis of fold mechanics. In the northwestern and far western plateau of Pennsylvania and south-central New York, Region I, the salt-bearing part of the Silurian Salina Group is the deformed zone; no other part of the stratigraphic section has flowed significantly. Synclines are broader than anticlines in the higher relief folds of Region I because much material has been removed from beneath the synclines and flanks to fill anticlines. In southwestern and central Pennsylvania, Region II, the anhydrite-rich Tonoloway Formation and/or the Upper Ordovician Reedsville Shale flowed as well as the salt.

Complex thrust faulting and disharmonic folding at the level of the Oriskany Sandstone contributes significantly to the structural relief, but it cannot provide the major part of the relief. Balancing the shortening present at the Oriskany level requires either stretching at this level in the synclines or significant layer-parallel shortening at shallower stratigraphic levels.

The Bradford oil field (Fig. 87.1) was discovered in 1871. It has an area of 344 km² and is the most densely developed oil field ever discovered in the USA with a total of 60,000 drilled wells. Peak annual oil production reached 3.6 million t in 1880, which, at the time, accounted for 75 per cent of oil consumption in the USA

ALASKA OIL AND GAS REGION

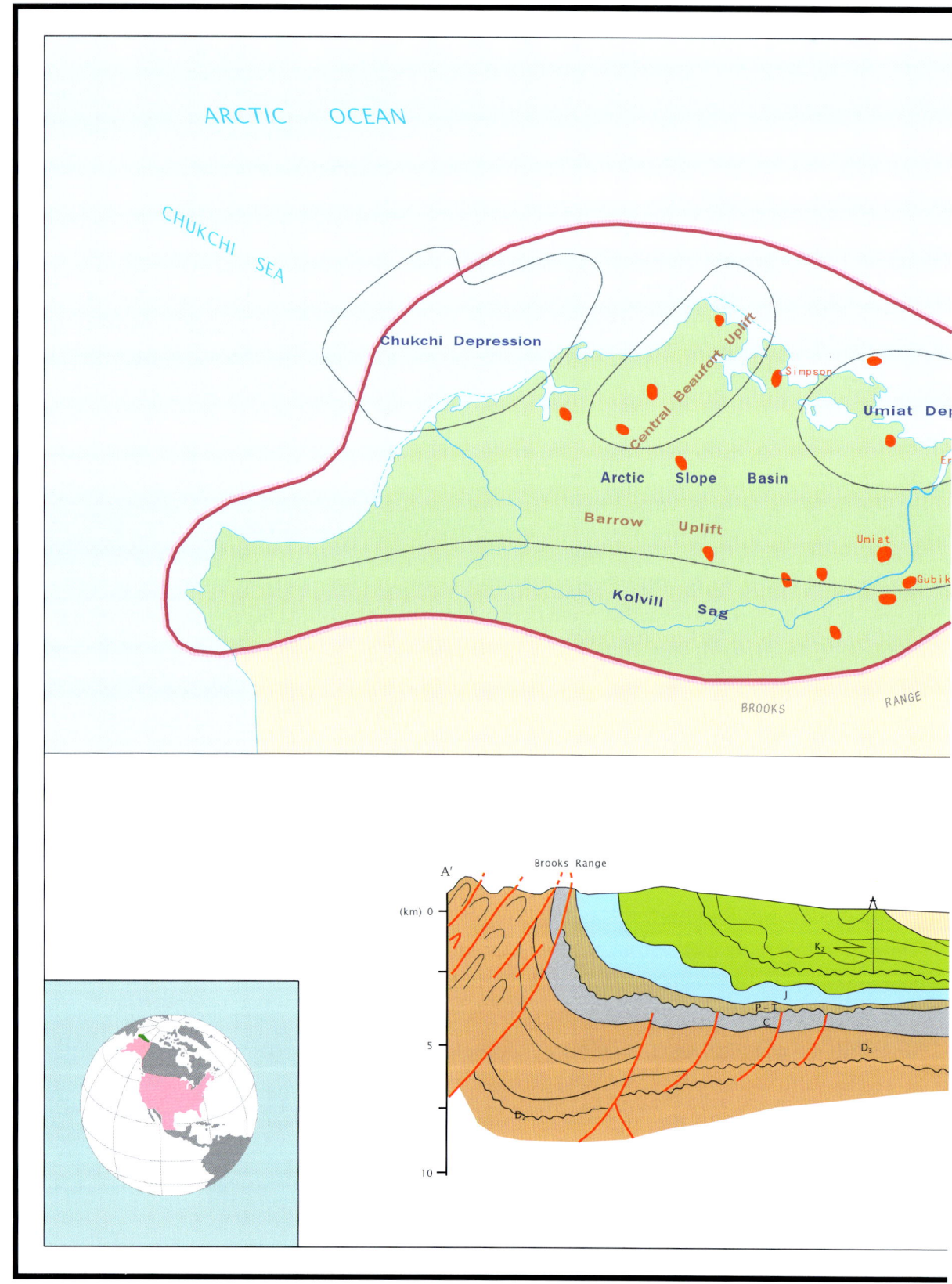

CHAPTER 88

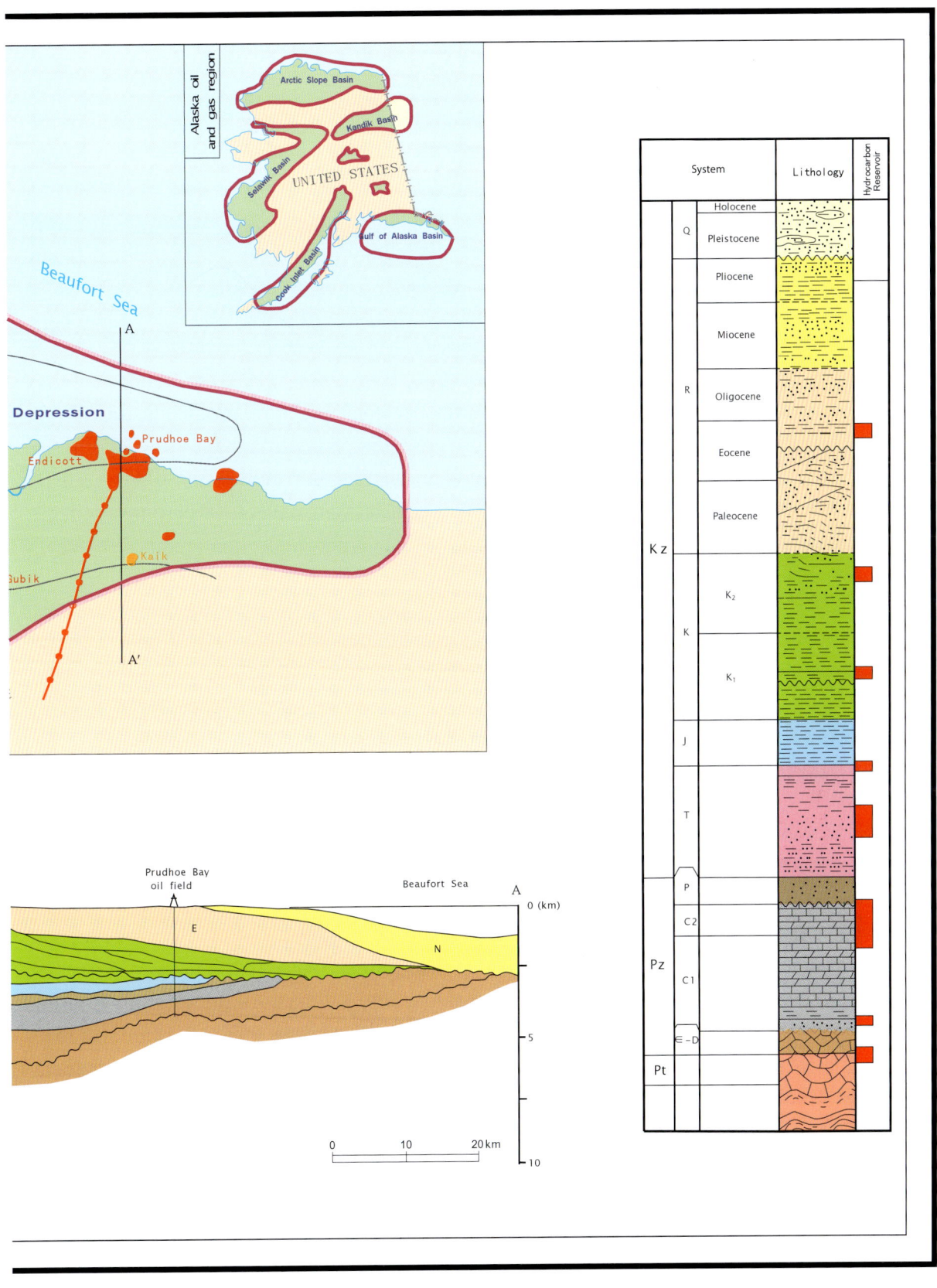

88 Alaska Oil and Gas Region

Geography

Alaska is located in the northwest of North America. The Alaska Range lies as an arch in the south, while the Brooks Range lies to the north. There are seven sedimentary basins located within the curtilage of Alaska, including the Cook Inlet and Arctic Slope Basins. Alaska has an area of 1.52 million km^2, of which the Arctic Slope Basin occupies an area of 100,000 km^2. At present, the Arctic Slope Basin is major oil producing region since the discovery of Prudhoe Bay oil field with proven oil reserves of 1.3 billion t.

History

The first oil field in 1957 was discovered in Cook Inlet Basin. Followed by the important discovery of the Prudhoe Bay oil field (Fig. 88.1), this area became a major oil producing area (Table 88.1). The giant Prudhoe Bay oil field was discovered in 1968. This is the largest oil field in the USA, with recoverable reserves of 1.3 billion t and peak production of 81 million t in 1980. Alaska's oil and gas reserves continue to play a central role. The focus has remained on developing a scheme to transport natural gas from the Alaskan North Slope to the markets in the contiguous 48 States.

Regional geology

Hydrocarbon accumulations of the Prudhoe Bay Complex generally are located on or near its crest. Two cross-sections (Map 92 and Fig. 88.1) depict the general structural and stratigraphic aspects of the complex and illustrate the trapping effect of the Lower Cretaceous unconformity, especially where it is combines with a favourable structural position. The Permo-Triassic reservoirs of the Prudhoe Bay oil field are important factors because they contain the bulk of the hydrocarbons in place as well as the only reserves currently being produced. Other reservoirs include the Mississippian–Pennsylvanian carbonates of the Lisburne Group, Lower Cretaceous sandstones of the Kuparuk River Formation and the Upper Cretaceous West Sak Sands.

The various reservoirs of the Prudhoe Bay Complex are estimated to contain at least 23.5 billion barrels of oil in place. Approximately 20 billion barrels of oil occur in the Permo-Triassic of the Prudhoe Bay oil field. An additional 3.5 billion barrels of oil in place are attributed to the Lower Cretaceous Kuparuk River Formation. Inadequate data exist to make reliable estimates of the volume of oil for other accumulations. Commercial production of hydrocarbons trapped in the accumulations other than the Permo-Triassic of the Prudhoe Bay oil field that will depend primarily on production development economics, and thus will be extremely sensitive to the production rate as well as reserves available.

The most prolific reservoirs of the Prudhoe Bay Complex occur within the Permo-Triassic strata. They are the Sadlerochit Group, the Shublik Formation and the Sag River Formation. The Sadlerochit Group has been subdivided, in ascending order, into the Echooka Formation, Kavik Shale and Ivishak Sandstone. Put River Sandstone of Lower Cretaceous age lies unconformably on the Sag River Formation in a limited area of the Prudhoe Bay oil field, but is considered part of the 'Permo-Triassic Reservoir'.

World Atlas of Oil And Gas Basins, First Edition. Li Guoyu.
© 2011 John Wiley & Sons, Ltd. Published 2011 by John Wiley & Sons, Ltd.

Alaska Oil and Gas Region

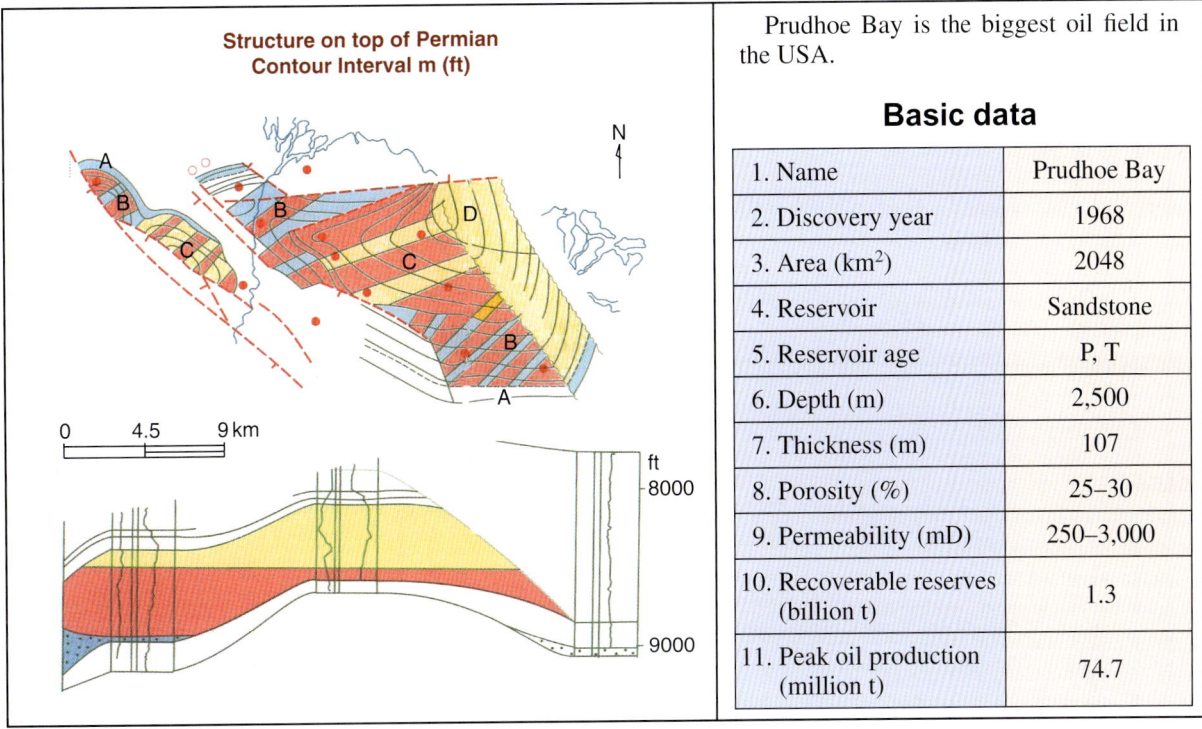

Fig. 88.1 Details of the Prudhoe Bay oil field.

Prudhoe Bay is the biggest oil field in the USA.

Basic data

1. Name	Prudhoe Bay
2. Discovery year	1968
3. Area (km^2)	2048
4. Reservoir	Sandstone
5. Reservoir age	P, T
6. Depth (m)	2,500
7. Thickness (m)	107
8. Porosity (%)	25–30
9. Permeability (mD)	250–3,000
10. Recoverable reserves (billion t)	1.3
11. Peak oil production (million t)	74.7

Table 88.1 Major oil and gas fields of the Alaska oil–gas region

Name	Discovery year	Recoverable reserves		Basin	Depth (m)	Trap	Reservoir	
		Oil (million t)	Gas (billion m^3)				Age	Lithology
Prudhoe Bay	1968	1,329.	72.8	Arctic slope	2500	Anticline	P, T	Sandstone
Kuparuk	1969	175	—	Arctic slope	1900	Stratigraphic	J	Sandstone
Point Thomson	1975	59	14.0	Arctic slope	3500	Anticline	E	Sandstone
McArthur River	1965	77	2.0	Cook Inlet	1600	Anticline	N	Sandstone
Kenai	1959	—	8.8	Cook Inlet	2000	Anticline	N	Sandstone

MEXICO

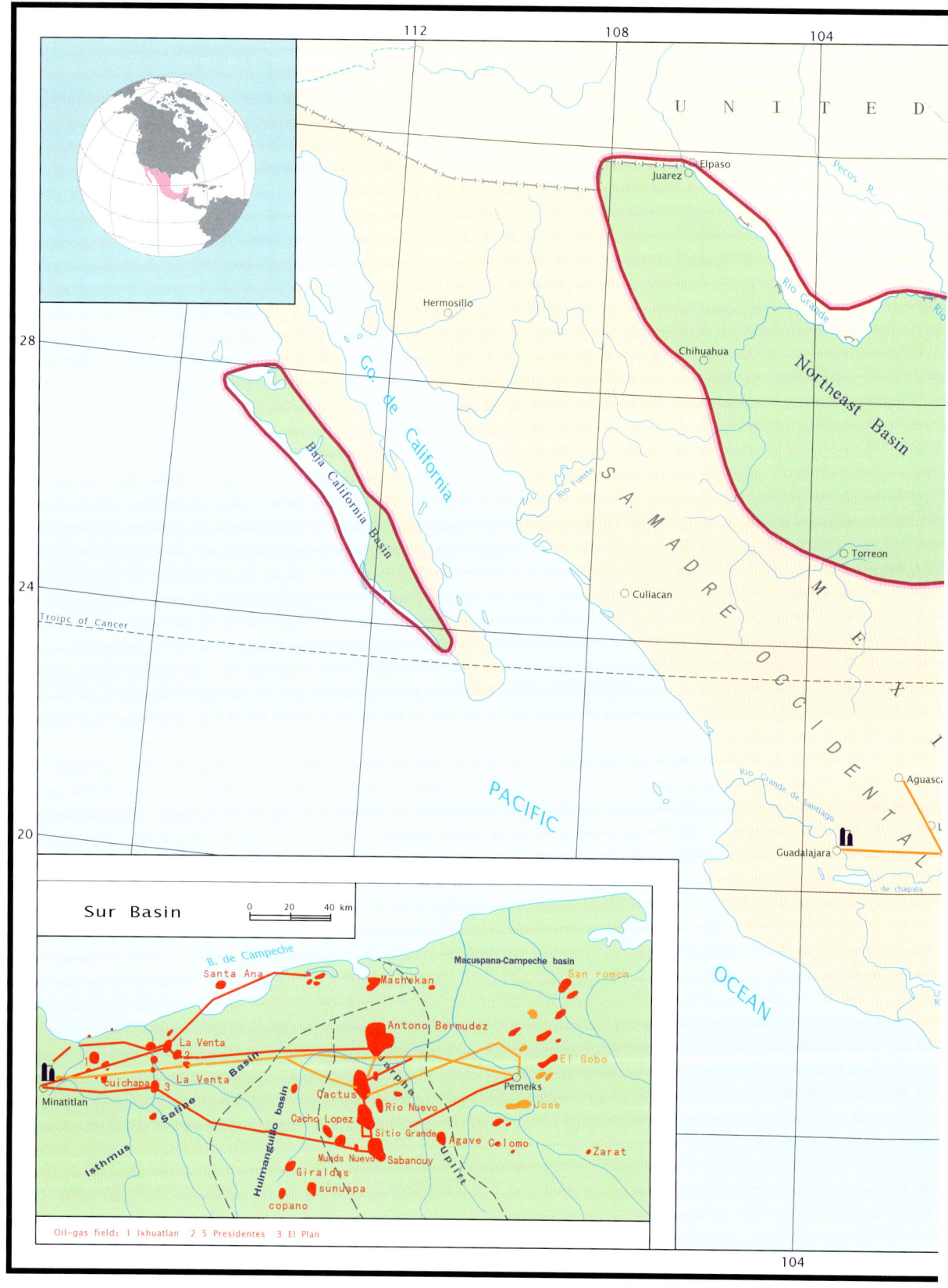

CHAPTER 89

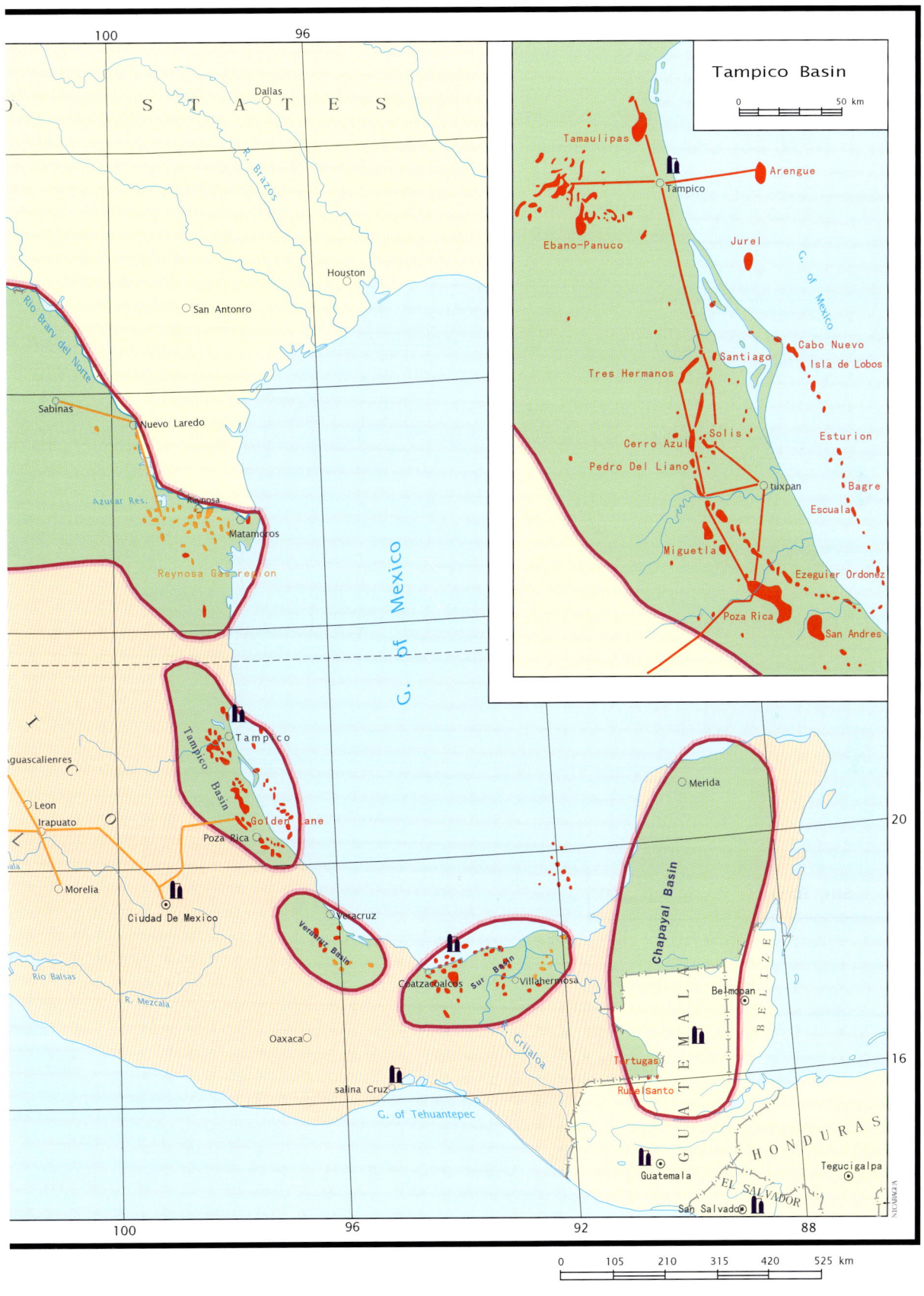

89 Mexico

Geography

Mexico curves from northwest to southeast, is generally shaped as a narrowing peninsula, has an area of 1.96 million km² and a population of 105 million. The Pacific Ocean lies to the west and the Gulf of Mexico to the east. The northernmost part of Mexico is covered by two different deserts. The northern border is the USA and Guatemala forms most of the southern border.

History

Mexico is an important oil-producing country. Oil seepages were well known as long ago as 1864. The first oil field was discovered in Tampico in 1901 with a depth of 490 m. As a result of high production from many giant limestone oil fields, annual oil production during 1930–1976 increased rapidly year by year from 1000 t in 1901, 12.2 million t in 1919 to 53 million t in 1977, rising to 119 million t in 1981 and 150 million t in 2000. By the end of 2008 the proven oil reserves were 1.43 billion t and annual oil production was 140 million t. There were also proven gas reserves of 372 billion m³, with 71.6 billion m³ of gas production.

Regional geology

Mexico has six main sedimentary basins (Table 89.1) that are filled by Mesozoic and Cenozoic sediments. The country lies in the middle portion of the Cordillera Mountains and comprised eight tectonic elements:

1 the lower California Ceno-Mesozoic fold belt;
2 the east California coast fold belt;
3 the South Madera fold belt;
4 the broken volcanic trend;
5 the West Madera Mountains;
6 the East Madera Mountains;

Table 89.1 Major sedimentary basins of Mexico

Basin	Area (thousand km²)	Major sedimentary rock		Reservoir	
		Age	Thickness (m)	Age	Lithology
Baja California	60	K, Kz		K	Sandstone
Northeast	410	Mz, Kz		E_2, E_3	Sandstone
Sur	60	Mz, Kz	8250	K, R	Sandstone, limestone
Tampico	73	J–R	3000	K, J	Carbonate rock
Veracruz	30	J–R	7000	K, E_3	Sandstone, limestone
Chapayal	168	J–R	8000	K, R	Sandstone, limestone

World Atlas of Oil And Gas Basins, First Edition. Li Guoyu.
© 2011 John Wiley & Sons, Ltd. Published 2011 by John Wiley & Sons, Ltd.

7 the Chiapas region;

8 the Gulf of Mexico plains.

Southern Mexico has received great attention from the oil industry in the past five years. Undoubtedly, some of the most prolific new fields in the world are being developed in the provinces of Chiapas and Tabasco in Mexico. The Cactus field has 33 wells and is producing 18,300 m^3 of gas per day. The Samaraia–Iride–Cunduacan complex, originally believed to be five separate fields and now known as the A.J. Bermudez field, is producing over 79,000 m^3 from 62 wells. Recently four individual discoveries – Chac, Bacab, Akal and Tunich hydrocarbon fields – were completed offshore. Thick play horizons of a lithology very similar to that exploited onshore Cretaceous trend were encountered. The production horizon marks the crest of what is referred to as the Chiapas Anticlinorium. It extends from south of Villahermosa to the north end of the Yucatan Peninsula. The trend of this anticlinorium at the northern end is somewhat in doubt and the Ixchel 1 hydrocarbon field well may indicate a slightly more northerly trend of a widening of the anticlinorium. The sedimentary succession comprises a very thick bank deposit consisting of calcarenites and dolarenites. Underlying these is a detrital limestone with excellent vuggy porosity. Recent work indicates some of these rocks to be as old as Kimmeridgian, although the majority are Turonian to Neocomian. This sedimentary succession includes the productive horizon of the onshore fields, and marks the zone of high-energy carbonates.

West of these bank deposits the sedimentary succession changes to a deep water facies, comprising marlstone and pelagic deposits, and fractured limestone. This succession is typical of the Isthmus Saline Basin. To the east the sedimentary succession changes to platform type carbonates, which continue into Guatemala.

The very thick accumulation of limestones and dolomites of the bank-type deposit is located between the impervious deep-water sediments to the west and the platform-type anhydrite deposits to the east, provides an ideal geological occurrence for exploitation of hydrocarbons. The structure controlling oil accumulation in the immediate area is closed anticlines, usually with normal faults on the flanks. The regional structure is not known in detail, although the west flank is bounded by the Comalcalco Fault, which is a normal fault of large displacement. The eastern flank is bounded in part by the Frontera Fault, which is also a normal fault. Probably these faults, or en échelon faults, bound the entire length of the area underlain by the bank carbonate deposits. The southern boundary of the bank deposits is also limited by a normal fault. The relationship of the locus of the bank deposits and the thickness of the bank deposits to the normal faulting remains unknown. It is believed that the bank deposits mark the hinge line between the platform sediments and the deep-water sediments. This hinge line was faulted, leading to rifting and spreading, which in effect formed a graben that was sinking contemporaneously with deposition. The settlement was uniformly slow and the shallow water resulted in very thick bank-type sediments. The structure of the area became complicated by the movement of the Louann salt. The exact effect of the salt movement is not known. Salt intrusions are known to the east and west of the Chiapas Anticlinorium but are not found within the belt of high-energy carbonates. Thrust faults that have been mapped are probably caused by salt movement.

Mexico is well know as an important oil-producing country in North America. In 2001 the daily oil production rate per well reached 139 t, compared with 1.5 t in the USA and 3 t in Canada. In particular the T-Lama oil well flowed at a rate of 31,800 m^3 per day, because the reservoir is thick limestone with high porosity and permeability.

TAMPICO BASIN

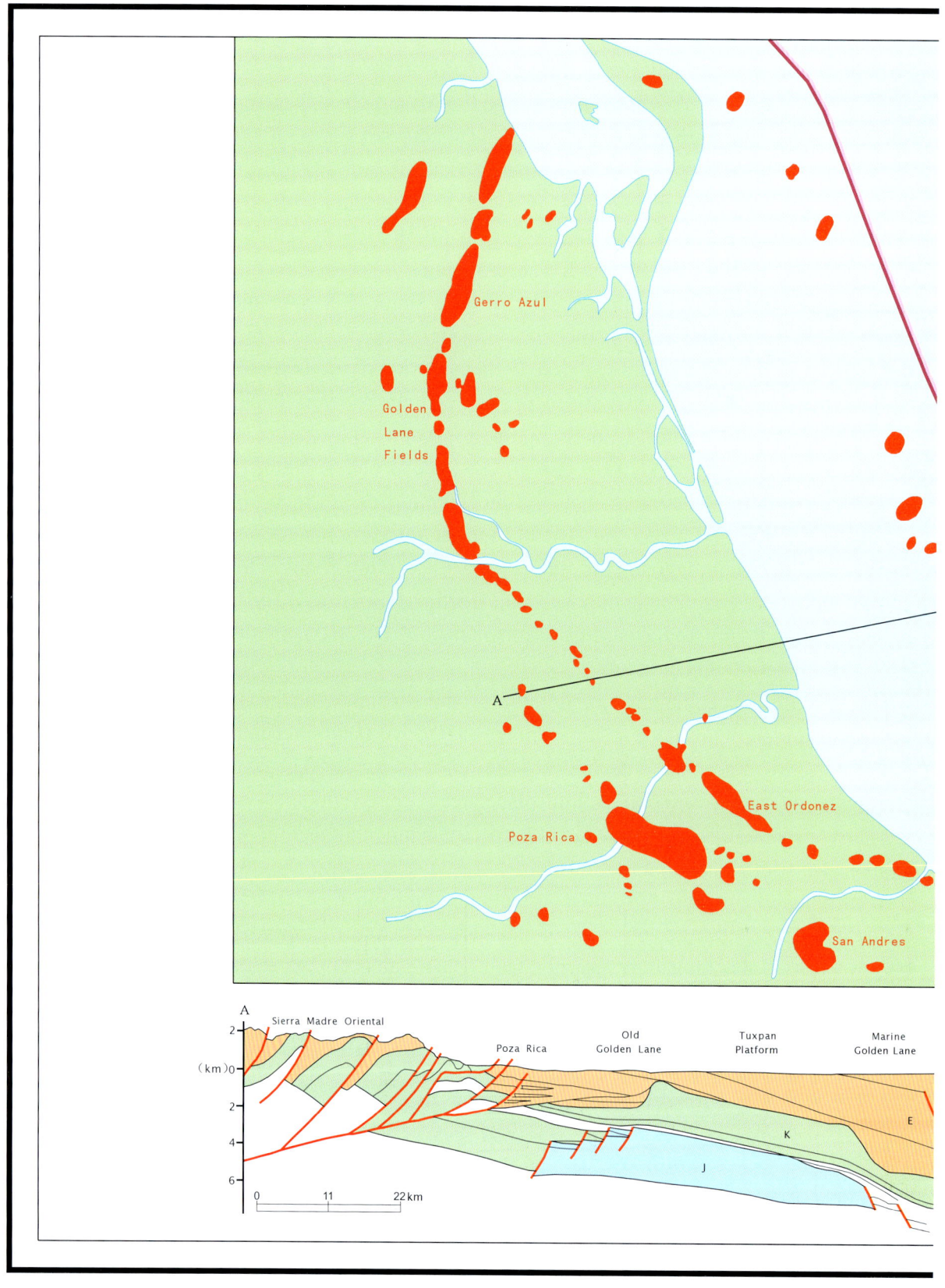

CHAPTER 90

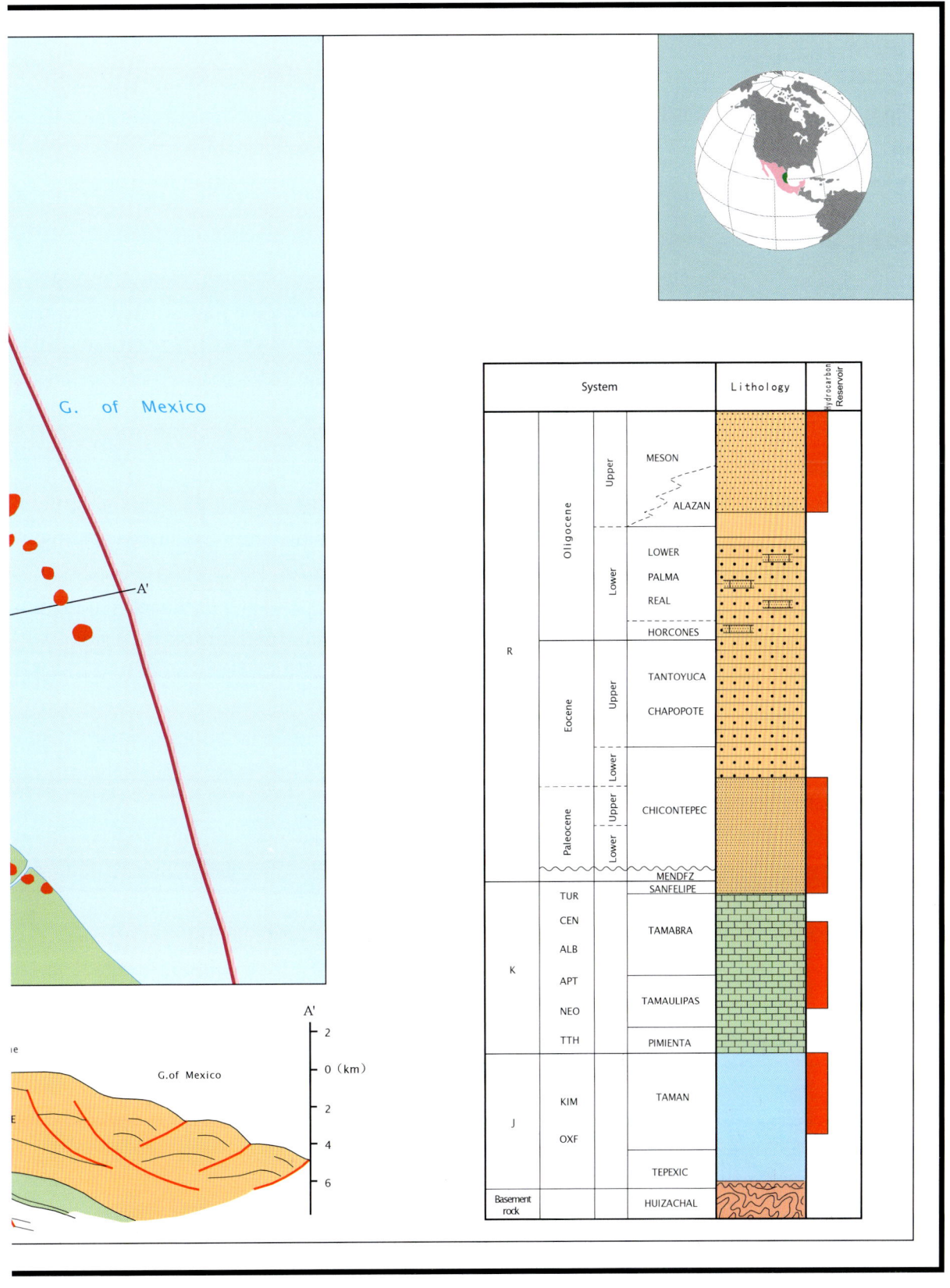

90 Tampico Basin

Geography

The Tampico Basin straddles the Gulf Coast of central Mexico, between Northeast Basin to the north and Veracruz Basin to the south. It is a mature basin, having been exploited by some 3000 exploration wells, and its cumulative production exceeds 700 million t of oil and 269 billion m^3 of gas.

History

Although being the second most important producing area in Mexico today, it contributes only 3.5 per cent of daily production. The oil exploration was initiated in 1901, the first oil field, Ebano, was discovered in 1904. This is a trap with a fractured carbonate reservoir. The Golden Lane oil field was discovered in 1908. The oil well T-Lama-3 flowed at a rate of 31,800 m^3 per day. This was a very important event in the history of world petroleum industry.

Regional geology

The key events are the Jurassic rifting related to the break-up of Pangea and the opening of the Gulf of Mexico, when continental sediments were deposited in north–south trending graben. During the subsequent Cretaceous passive margin development, marine deposition was predominant with thick carbonate sequences developing on the platform highs, and deeper water carbonates in more basinal areas. In the Tertiary, the basin was affected by the compressional tectonics of the Laramide Orogeny which led to the formation of many of the traps. In addition, a foreland basin developed in front of the rising Sierra Madre Oriental and caused the maturation of the Late Jurassic source rocks. A variety of plays have been identified ranging from the Jurassic to Miocene in age, but the mid-Cretaceous carbonate plays of the Poza Rica and Golden Lane trends are of the most significance as they account for more than 90 per cent of the basin's production.

Poza Rica trend The Tamabra (Table 90.1) structural–stratigraphic or Cretaceous breccia play was discovered in the 1930s. The Poza Rica structure was identified as gravity high due to uplifted basement beneath, to the west of the Golden Lane. In the Poza Rica 2 well the gas cap was discovered in 1930, and the oil leg of the field came on line in 1932. Recoverable reserves were estimated at 201 billion barrels of oil making this a giant field, and the most important play in the

Table 90.1 Major oil and gas trends in Tampico Basin

Major trend/Group (age)	Lithology	Recoverable oil reserves (million t)
Tamabra (K)	Conglomerate	400
El Abra (K)	Limestone	289
Tamaulipas (K)	Limestone	165
San Andres (J)	Limestone	121
Chicontecpec (E)	Turbidite	52

World Atlas of Oil And Gas Basins, First Edition. Li Guoyu.
© 2011 John Wiley & Sons, Ltd. Published 2011 by John Wiley & Sons, Ltd.

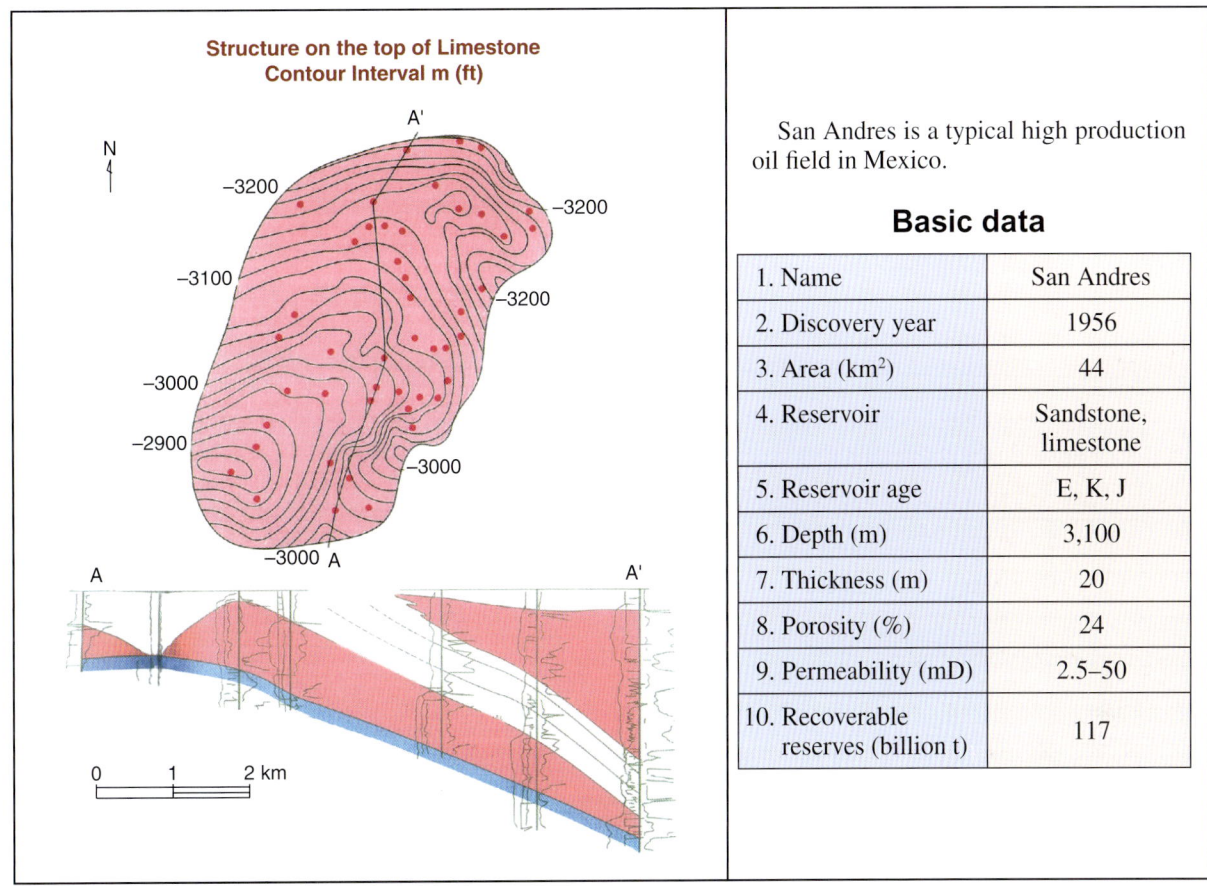

Fig. 90.1 Details of the San Andres oil field.

basin. There is also some production from oolites of the Upper Jurassic San Andrés structural–stratigraphic play (Fig. 90.1), and turbidites of the overlying Chicontepec stratigraphic play.

Golden Lane trend The El Abra stratigraphic–unconformity play is in karstified Cretaceous platform carbonates of the Tuxpan platform. The Golden Lane trend is a 180 km long arc. The individual closures were recognized as being karst pinnacles during later drilling, but were originally thought to be due to cross-faults or en échelon folds. By 1969 over 1000 wells had been drilled on the Old Golden Lane (onshore), of which over half were producers.

Chicontepec sub-basin The Chicontepec stratigraphic play was discovered during infill drilling of the Presidente Alemán field in the 1950s, when a Paleogene sandstone interval (Chicontepec Formation) was tested and resulted in significant production.

Ebano-Pánuco area This producing region is located at the southeast end of a major anticlinorium known as the Tamulipas arch or platform. The Tamaulipas structural play in fractured basinal carbonates was the first to be discovered in Mexico (in 1904), with the discovery La Pez 1 well drilled on surface seeps. The play has produced an estimated 1 billion barrels of oil and is now nearing depletion. The Pimienta–Tamabra petroleum system accounts for the bulk of the discoveries. Maturation of the Late Jurassic source rocks occurred in two stages. The first was in the Late Eocene, when the sources beneath the Chicontepec foreland in-fill expelled oil up-dip to the east to the Poza Rica and Golden Lane traps. The second followed basinward tilting in the Neogene, when a thick clastic package to the east of the Tuxpan platform helped mature the underlying Jurassic sources; this time migration was in the reverse direction, easterly up-dip towards the platform.

CARIBBEAN SEA REGION

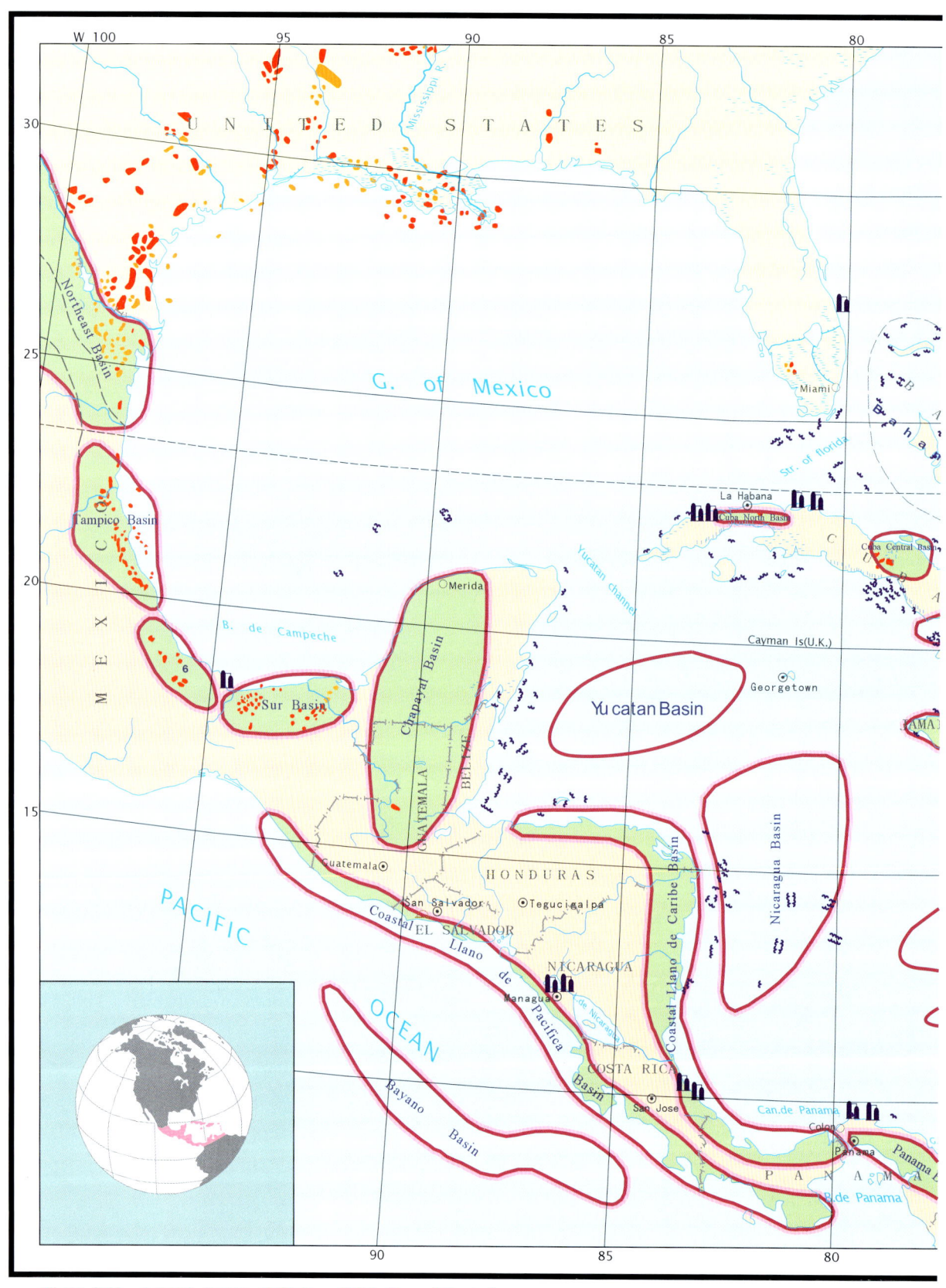

Oil-gas field: 1 Forest Reserve 2 Brighton 3 Palo Seco 4 Fyzabad
Basin: 1 Cuba South 2 Jamaica 3 Magdalena Medio 4 Falcon 5 Puerto Rico 6 Veracruz 7 Venezuela Gulf

CHAPTER 91

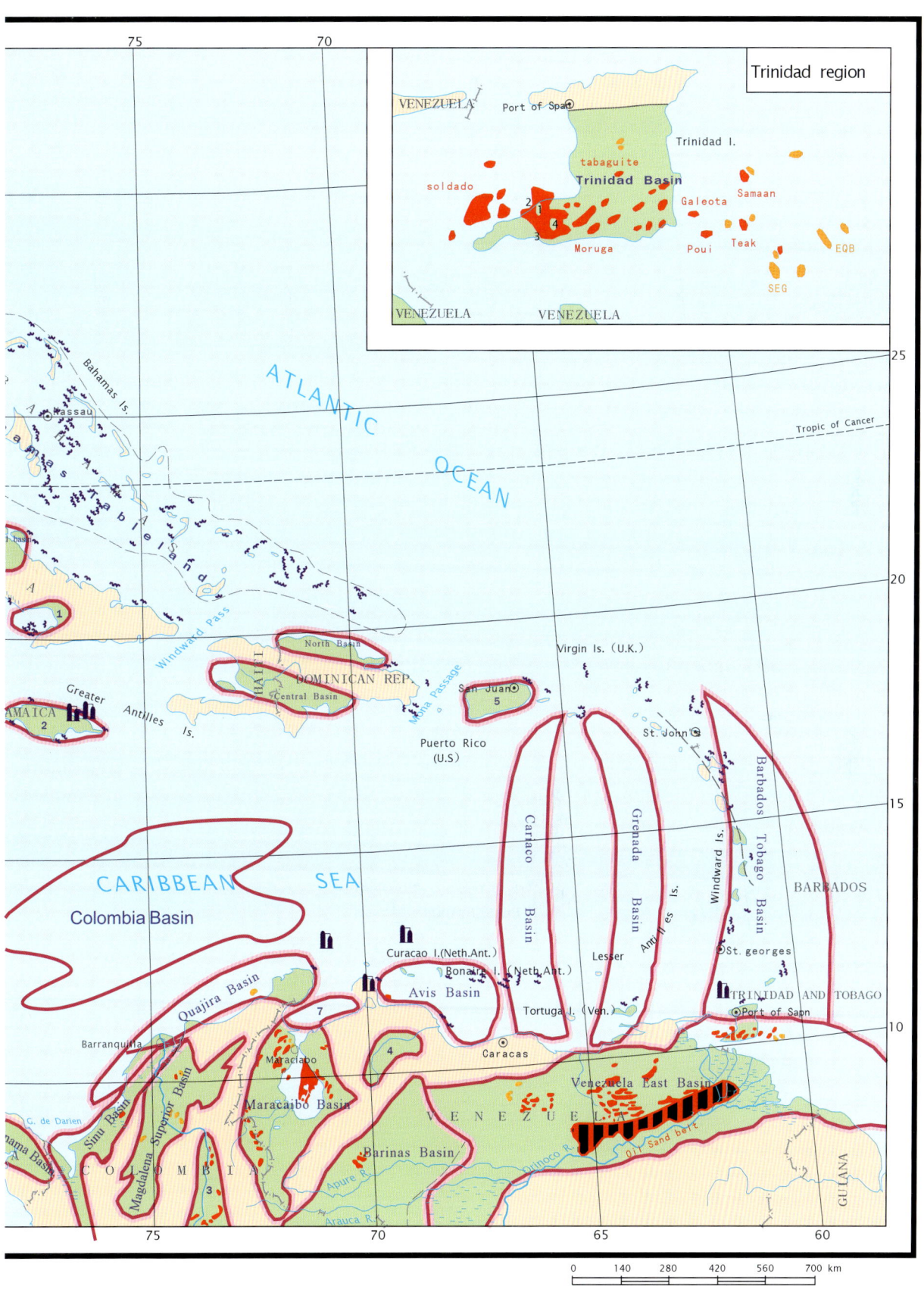

91 Caribbean Sea Region

Geography

The region of Central America and the West Indies includes the Caribbean Sea with the surrounding islands as well as Central America with the adjacent areas of the Pacific and Atlantic Oceans. This region includes more than 30 countries with a total area of 4.7 million km^2 and aggregate population of approximately 140 million.

History

Oil was originally discovered in this region during the late 18th and early 19th centuries. Out of the 30 countries, the oil fields were discovered only in four countries: Trinidad and Tobago, Cuba, Guatemala and Barbados. By the end of 2008, the oil production of Trinidad and Tobago was 5.6 million t, Cuba 2.5 million t, Guatemala 700,000 t and Barbados 40,000 t. Other countries are still essentially non-producing countries. The data in Table 91.1 show the basic situation of Trinidad and Tobago, Cuba and Guatemala at present. As a whole, the region has been poorly studied. However, there are 18 sedimentary basins (Table 91.2) recognized in the region with undoubted potential in petroleum resources.

Regional geology

The region of Central America and West Indies mainly occupies the Caribbean Plate where the crust relates to predominantly oceanic type, and partially, the North and South America Plates, with continental and oceanic crust. There are also regions of the oceanic Cocos Plate. The region represents an area of Alpine folding development which is expressed here by the elements of the regional Pacific Orogenic Belt, being a link between the Cordilleras of North and South America. An important part of the structure of the region is occupied by the Central American Median Mass. Within the structure of the region the following elements have been distinguished:

1 massifs (of Honduras, Guatemala and Maya known as the Central American Mass);
2 orogenic belts (of Antilles, Guatemala, together with volcanic ridges and plateaux);

Table 91.1 Major data of some oil-producing countries in the Caribbean Sea region

Index	Country		
	Trinidad and Tobago	Cuba	Guatemala
Area (km^2)	5128	110,860	108,889
Population (thousand)	1300	11,260	12,900
Proven oil reserves (million t)	99	16	11
Oil production (million t)	5.6	2.1	0.7
Proven gas reserves (billion m^3)	531	70	—

World Atlas of Oil And Gas Basins, First Edition. Li Guoyu.
© 2011 John Wiley & Sons, Ltd. Published 2011 by John Wiley & Sons, Ltd.

Table 91.2 Major sedimentary basins of the Caribbean Sea region

Basin	Area (thousand km²)	Major sedimentary rock		Reservoir		Remarks
		Age	Thickness (m)	Age	Lithology	
Bayano	80				Sandstone	Extends into inshore of Pacific ocean
Barbados–Tobago	200	K–E$_2$	2100	E$_2$		Extends into Trinidad and Tobago, Barbados, Dominican Republic
Cariaco	50	K, R				Venezuela North
Central	30	Mz, Kz				Extends into Haiti, Dominican Republic
Coastal Llano de Caribe	100	Mz, Kz				Extends into Honduras, Nicaragua, Costa Rica, Panama
Coastal Llano de Pacifica	120	K–R				Extends into Mexico Guatemala, Honduras, Nicaragua, Costa Rica, Panama
Cuba Central	25	K–R	5000–7000	K	Tuffaceous sandstone, limestone	Extends into Cuba
Cuba North	25	J–R	>4000	J, K	Serpentinite, tuff, limestone	Extends into Cuba
Cuba South	15					Extends into Cuba
Grenada	80					Venezuela North
Jamaica	23	K–Kz	4900			Extends into Jamaica
Nicaragua	160					Extends into Nicaragua
North	20	K–R		N$_1$	Conglomerate, gritstone	Extends into Nicaragua
Panama	22	K–R	500			Extends into Colombia
Puerto Rico	12	K–E$_2$	2100	K	Limestone	Puerto Rico (USA)
Chapayal	168	Pz–Kz	10,000	K	Carbonate rock	Extends into Mexico
Colombian	368					
Yucatan	300					

3 island arcs (of Antilles);
4 deep-water trenches and troughs (Central American, Cayman and Puerto Rico);
5 deep-water basins (Yucatan, Colombian and Venezuelan).

The Central American Mass has a basement of Palaeozoic age. The sedimentary cover predominantly consists of Lower Cretaceous carbonate deposits. In some areas, Triassic, Jurassic, Permian and, probably, Upper Carboniferous deposits have been identified. The Orogenic Belt of the Antilles occupies a part of Cuba Island. The Antilles island arc might have been formed during the interval from the middle of the Senomanian to the end of the Eocene. The sedimentary cover of deep-water basins consists of essentially Neogene–Quaternary and Upper Cretaceous–Cenozoic rocks of up to 5 km in thickness.

SOUTH AMERICAN OIL AND GAS BASINS

PART VI

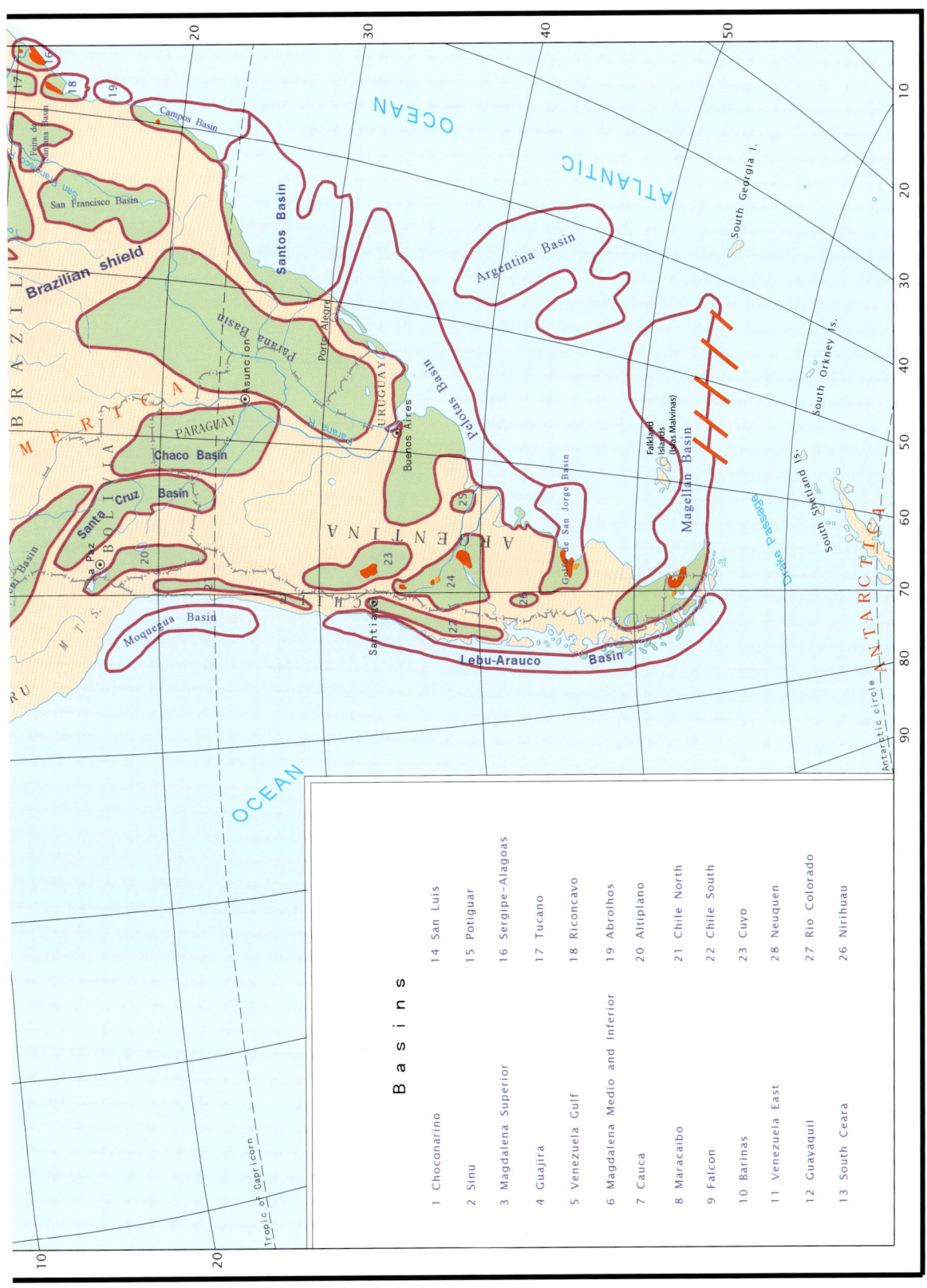

Part VI
South American Oil and Gas Basins

Geography

The South American region includes the continent of the same name, the adjacent deep-water areas of the Pacific and Atlantic Oceans. The region also includes the Falkland (Malvinas) Islands and the Tierra del Fuego Archipelago.

History

South America is an early oil producing continent. In 1869 the La brea-parinas oil field was discovered in Peru as a result of nearby surface oil seepages. Peru was therefore the first oil-producing country in South America. The country's annual oil production was 40,000 t in 1890 and 710,000 t in 1910. The next important oil field was discovered in the Maracaibo Lake area of Venezuela in 1924. South America was the leading offshore oil production region by 1970 with annual production of 120 million t. At present, Argentina, Bolivia, Brazil, Chile, Colombia, Ecuador, Peru, Surinam and Venezuela are oil-producing countries (Table VI.1). Amongst these, as of 2008, Venezuela is the largest producing country with annual oil production of 117 million t. Brazil is the second largest with annual production of 90 million t. Within the South American region, 59 petroleum and possibly petroleum basins have been identified, and the petroleum potential is therefore estimated to be very high.

Table VI.1 Key data for the countries of the South America region

Country	Proven reserves		Production	
	Oil (million t)	Gas (billion m^3)	Oil (million t)	Gas (billion m^3)
Total	16,652	6933	272	94.2
Argentina	354	441	30	42.9
Bolivia	63	750	2	14.1
Brazil	1729	365	90	12.4
Chile	20	97.9	1.1	—
Colombia	185	105.8	28	—
Ecuador	638	—	25	0.3
Peru	56	335	3.7	3.3
Surinam	10	—	0.8	—
Venezuela	13,613	4839	117	25.1

World Atlas of Oil And Gas Basins, First Edition. Li Guoyu.
© 2011 John Wiley & Sons, Ltd. Published 2011 by John Wiley & Sons, Ltd.

Regional geology

The South American region occupies mainly the western half of the lithospheric plate of the same name, which includes the coastal area, shelf, continental slope and a piedmont corresponding to ancient continental crust. The sedimentary cover comprises Ceno-Mesozoic strata with thicknesses of 2000–10,000 m. Reservoirs are sandstones, limestones, tuffs and conglomerates. The sedimentary basins are small- to medium-sized with areas of 12,000–200,000 km^2. Among the piedmont platforms we can distinguish between the ancient Brazilian (craton) and the young Patagonian (cratogene) platforms. The Brazilian Platform with Precambrian basement occupies most of the continent. The Guiana Shield being the northernmost uplift of the basement is formed by Archaean rocks. The Amazon Syneclise separates the West Brazilian and the Guiana Shields.

The Andes Orogenic Belt fringes the Brazilian and Patagonian platforms in the west and stretches along the Pacific Coast. The western (Pre-Pacific) zone was formed during the Cenozoic and coincides with the late Alpine cycle of tectogenesis. It comprises Palaeozoic and Mesozoic deposits intruded by Late Cretaceous and Cenozoic rocks. The zone is represented by mountainous build-ups of the cordilleras and troughs that partly occupy the continental slope, slopes of deep-water trenches and basins. The eastern zone is composed of the rocks of pre-Alpine and Alpine cycles. A zonal pattern is represented as individual segments of the orogenic belt – South (Patagonian), Central and North Andes – are aligned transverse to each other. It is traced laterally by changes in the structural complexes that form the cores of the ancient tectonic elements, and by the timing of the associated orogenic elements. Boundaries between the segments are determined by the position of regional fault zones.

The southeastern part of the Andes Mobile Belt in the area of the Sierra Pampa massif was formed as a result of post-platform orogeny, which in recent times involved the Brazilian platform margins. The mass is composed of basement pre-Late Precambrian rocks overlain by Upper Palaeozoic and Triassic continental clastic–volcanic cover. Faults have divided the massif into horsts and grabens filled with Neogene and Pleistocene deposits. The depression of La Rioja is associated with a system of lowered blocks. The zone of troughs traced from the northeastern extremity of the continent to the Pampasa Mass is considered to be a simple Andes foredeep. It includes eastern Venezuelian (Orinoco).

The Andes foredeep developed on Caledonian–Variscian basement in the south, whereas in the north it evolved on Precambrian basement. The marginal troughs comprise mainly Cenozoic molasses and are underlain by Mesozoic and Palaeozoic platform rocks that are frequently crumpled, forming linear folds. The thalassocraton includes a number of deep-water basins filled with Upper Cretaceous–Cenozoic deposits (Vysotsky *et al.*, 1995).

COLOMBIA AND ECUADOR

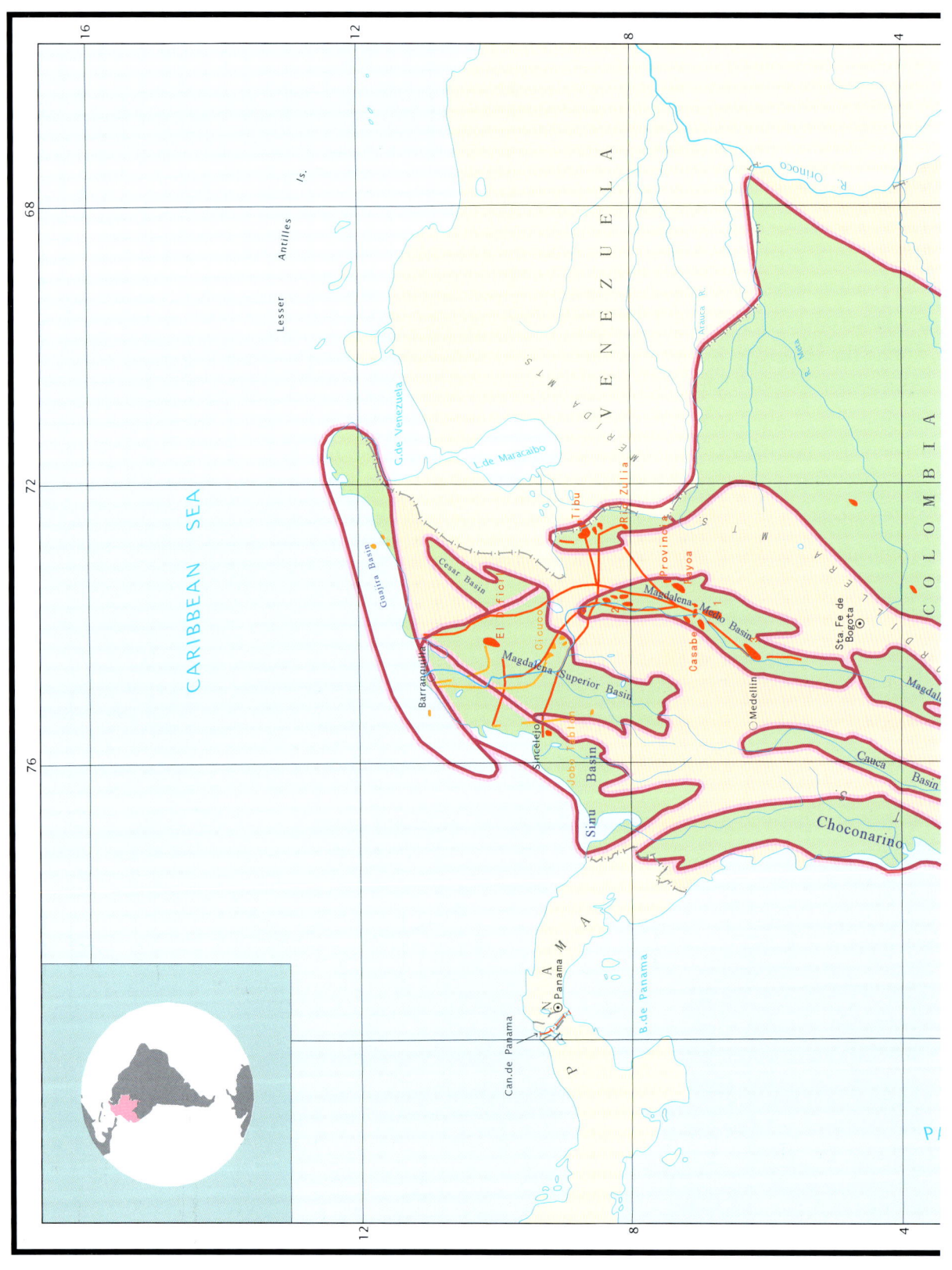

CHAPTER 92

92 Colombia and Ecuador

Geography

Colombia and Ecuador are located in the northwest of South America, facing the Caribbean Sea to the north and the Pacific Ocean to the west. The Andes Mountain Range lies in the western part of this region, occupying about 65 per cent of the combined area of the two countries. There are ten sedimentary basins recognised in the two countries (Table 92.1).

History

In these two countries oil exploration started in the early 1900s. This exploration effort resulted in the discovery of some large fields, thereafter commercial oil production commenced and increased rapidly. The key data relating to these two countries are reported in Table 92.2.

Regional geology

Colombia and Ecuador are located in northern part of Andes Mountains. The sedimentary basins trend northeast–southwest, filled by Jurassic, Cretaceous and Tertiary rocks with thicknesses of 6000–10,000 m. The traps are mainly anticline, reservoirs are sandstone and limestone, and seal rocks are shale.

Colombia Colombia is located in northern South America, bordering the Caribbean Sea and occupying an area of 1.14 million km^2, with a population of 46 million. Nine sedimentary basins lie in Colombia with individual areas between 12,000 km^2 and 450,000 km^2. The largest

Table 92.1 Major sedimentary basins

Country	Basin	Area (km$^{\pm 2}$)	Major sedimentary rock		Reservoir		Remark
			Age	Thickness (m)	Age	Lithology	
Colombia	Cauca	20	Kz				
	Cesar	15	Kz				
	Choconarino	73	Kz				Extends into Ecuador
	Guajira	12	J, K, R		R	Sandstone	
	Llanos	450	Mz, Kz	6,000	K	Sandstone	Extends into Peru
	Magdelena inferior	33	Mz, R	10,000	E$_3$, E$_2$	Limestone, sandstone	
	Magdalena Medio	40	Mz, Kz		K. R	Limestone, sandstone	
	Magdalena Superior	50	Mz, Mz		R	Sandstone	
	Sinu	50	K, R	6,600	R	Clastic rock	
Ecuador	Guayaquil	150	K, R	7,500	R	Sandstone	Extends into Peru

World Atlas of Oil And Gas Basins, First Edition. Li Guoyu.
© 2011 John Wiley & Sons, Ltd. Published 2011 by John Wiley & Sons, Ltd.

Table 92.2 Major data for Columbia and Ecuador by the end of 2008

Index	Country	
	Colombia	Ecuador
Area (km^2)	1,141,748	256,370
Population (thousand)	46,040	13,220
Proven oil reserves (million t)	185	638
Oil production (million t)	28	25
Proven gas reserves (billion m^3)	105	8.9
Gas production (billion m^3)	6.6	3.4

basin is Lianos Basin, with a size of 450,000 km^2. Oil and gas fields have been discovered in five of these basins.

Three coastal basins open towards the Caribbean Sea, and are bounded by the Andean Cordilleras in the south, by the spurs of the Cordilleras (Serrania del Interior) in the west, and by the buried uplift of the basement in the east. They have a total area of 500,000 km^2, of which 441,000 km^2 is occupied by deep waters. The three basins are filled with rocks ranging from Upper Cretaceous to recent; their total thickness exceeds 9 km. Cretaceous carbonate rocks at the base of the sedimentary succession are overlain by a Cenozoic carbonate–clayey sequence containing rare interbeds of sandstones and conglomerates. In the basins, two oil fields and seven gas fields have been discovered. Regionally oil- and gas-bearing rocks are of Oligocene and Miocene ages. Reservoirs are represented by sandstones and limestones. Source rocks are the Oligocene and Miocene clayey intervals of the succession as well as Upper Cretaceous limestones. The original in-place hydrocarbon resources are estimated as medium. The resources have been explored only to an extent of approximately 5 per cent for gas, while oil exploration can be regarded as negligible in terms of its potential. In 1950 a total of 1312 oil wells were completed with combined oil production of 4.71 million t. The peak annual oil production reached 41.25 million t. By the end of 2008 the proven oil reserves were 185 million t with annual oil production of 28.9 million t, while proven gas reserves were estimated at 105 billion m^3, with annual gas production of 6.6 billion m^3.

Ecuador Ecuador is located on the Equator and occupies an area of 254,000 km^2, of which about 68 per cent is covered by forest. The population is estimated at 13.2 million. Exploration in Ecuador commenced in the late 1940s, with initial wildcats recording oil and gas shows. In the Peruvian part of the Putumayo Basin, three large fields (Corrientes, Capahuari Sur and Shiviyacu) were discovered in the early 1970s. Since then, a number of small to medium sized discoveries have been made in Ecuador, the last one being Petroperu's 30 million barrel Chambira Este discovery in 1989. Oil production was 8000 t in 1917, 4 million t in 1972 and 20 million t in 2000. By the end of 2008 Ecuador's proven oil reserves were 638 million t with oil production being 25 million t annually.

VENEZUELA, GUYANA AND SURINAM

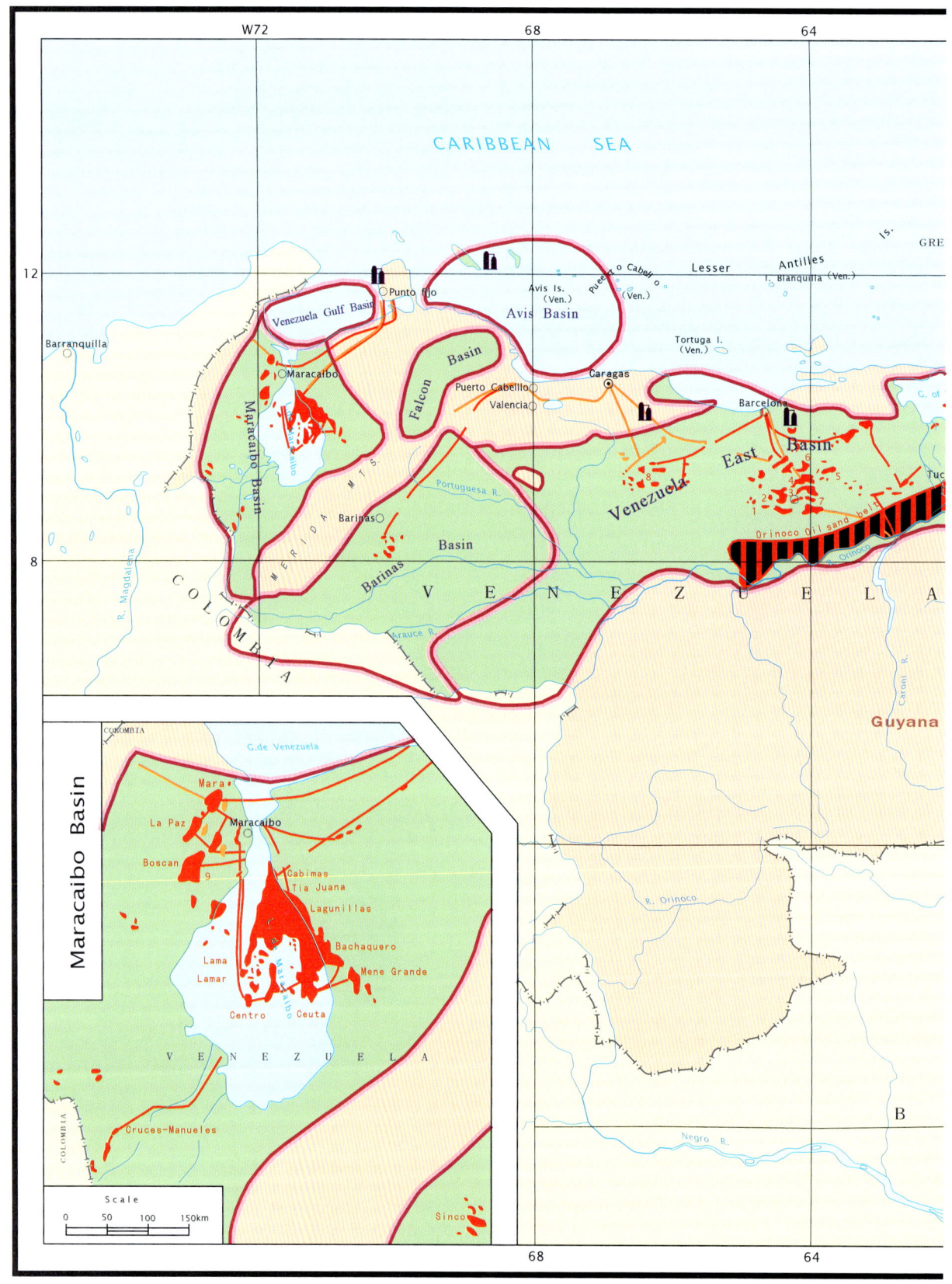

Oil-gas field: 1 Chimire 2 Oficina 3 Guara 4 Nipa 5 Oscurote 6 Leona 7 Dacion 8 La Mercedes 9 Concepcion

CHAPTER 93

93 Venezuela, Guyana and Surinam

Geography

Venezuela, Guyana and Surinam are located on the northern coast of South America, facing the Caribbean Sea and the Atlantic Ocean to the north, and incorporating the large Lake Maracaibo.

History

In ancient times local inhabitants of the region were reportedly familiar with surface oil seepages. Spanish colonialists often presented some of these collected oil samples to the King of Spain. Oil exploration started in 1787. Due to the discovery of many giant oil fields, Venezuela became the largest oil-producing country in the world by the year of 1955, with annual production of 110 million t. The key data for the three countries are presented in Table 93.1.

Regional geology

This area is divided into four tectonic elements: Guiana Shield, Venezuela sedimentary belt, Maracaibo Lake and Caribbean Sea, and the Atlantic Ocean continental shelf. Within the limits of these countries eight sedimentary basins have been identified, of which 4 (Maracaibo, Barinas, Venezuela East and Marajo) have proven commercial oil and gas accumulations. In these basins, 195 oil fields and 27 gas fields have been discovered. is the sedimentary basins are associated with the Maracaibo Intermontane Depression, which formed on Precambrian crustal blocks during the Palaeozoic and Early Mesozoic times, the South Andes Marginal Trough and the Caribbean Ridge. The sedimentary infill comprises essentially Mesozoic and Cenozoic deposits and attains 15 km in thickness in the most subsided parts of the basins.

Venezuela

The country occupies most of the northern coast of South America on the Caribbean Sea. It is bordered by Colombia to the west, and has an area of 916,700 km^2 with a population of 26.5 million. The national oil company, PDVSA, estimates that 54,000 m^2 of the Orinoco area (Venezuela East Basin) contain 1.2 trillion barrels of oil in place, of which 100–300 billion bbl might ultimately be recovered. Venezuela also contains Latin America's largest natural gas reserves, mostly in Maracaibo Basin. Venezuela has a long history of oil exploration and development. In ancient times local people were familiar with the use of oil in daily life. Spanish colonialists presented a bottle of oil as a sample to the King of Spain in 1539. Four shallow wells were completed close to nearby oil seepages on the shores of Lake Maracaibo in 1890. Heavy oil was discovered in the Orinoco area in 1913. In Venezuela, two prolific oil and gas bearing basins are known as the Maracaibo and the Venezuela East, with an area of 80,000–200,000 km^2 and a thickness of 5000–13,000 m Tertiary strata reservoirs. Oil production was 18,000t in 1917. By the end of 2008 the proven oil reserves were 13.6 billion t with annual oil production of 117 million t. The figures for gas were: proven reserves of 4.7 trillion m^3 and annual gas production of 24.6 billion m^3. These figures emphasize Venezuela's position as a leading oil and gas producer with potential for even higher production capacity. Details of Venezuela's major oil- and gas-bearing basins are provided in Table 93.2 (Shannon and Naylor, 1989).

World Atlas of Oil And Gas Basins, First Edition. Li Guoyu.
© 2011 John Wiley & Sons, Ltd. Published 2011 by John Wiley & Sons, Ltd.

Table 93.1 Key data Venezuela, Guyana and Surinam by the end of 2008

Index	Country		
	Venezuela	Guyana	Surinam
Area (km²)	916,700	214,969	163,826
Population (thousand)	26,560	750	493
Proven oil reserves (million t)	13,613	—	10
Oil production (million t)	117	—	0.8
Proven gas reserves (billion m³)	4839	—	—
Gas production (billion m³)	25	—	—

Table 93.2 Major oil and gas basins of Venezuela

Basin	Area (thousand km²)	Major sedimentary rock		Reservoir		Remark
		Age	Thickness (m)	Age	Lithology	
Avis	40	Kz				
Barinas	100	K, Kz	5000	E_2	Sandstone	Extends into Colombia
Falcon	13	Kz		N_1	Sandstone	
Maracaibo	80	K, R	5500–9,000	$E_2, E_3–N_1$	Sandstone	
Venezuela East	200	K, R	10,000–13,000	E_3	Sandstone	

Guyana Guyana is located in the northern part of South America, bordering the North Atlantic Ocean, between Surinam and Venezuela. It has an area of 214,969 km² and a population of 750,000. Geologically this country is located on the Guyana Shield, consisting of Precambrian metamorphic rocks. Part of Takutu Basin and part of offshore Marajo Basin are located within Guyana. During 1940–1941, one offshore well was completed with a depth of 1840 m in Tertiary strata with occurrence of heavy oil. At present, Guyana is a non oil-producing country.

Surinam Surinam is located on the northern coast of South America and has an area of 163,826 km² with a population of 493,000. As for Guyana, Surinam is located on the Guiana Shield, and contains the southern part of Marajo Basin, which is infilled by Jurassic, Cretaceous and Tertiary strata. The first well was drilled onshore without any commercial shows. An offshore exploration well was completed in 1965. By the end of 2008 the proven oil reserves were 12 million t with a modest annual oil production of 740,000 t.

MARACAIBO BASIN

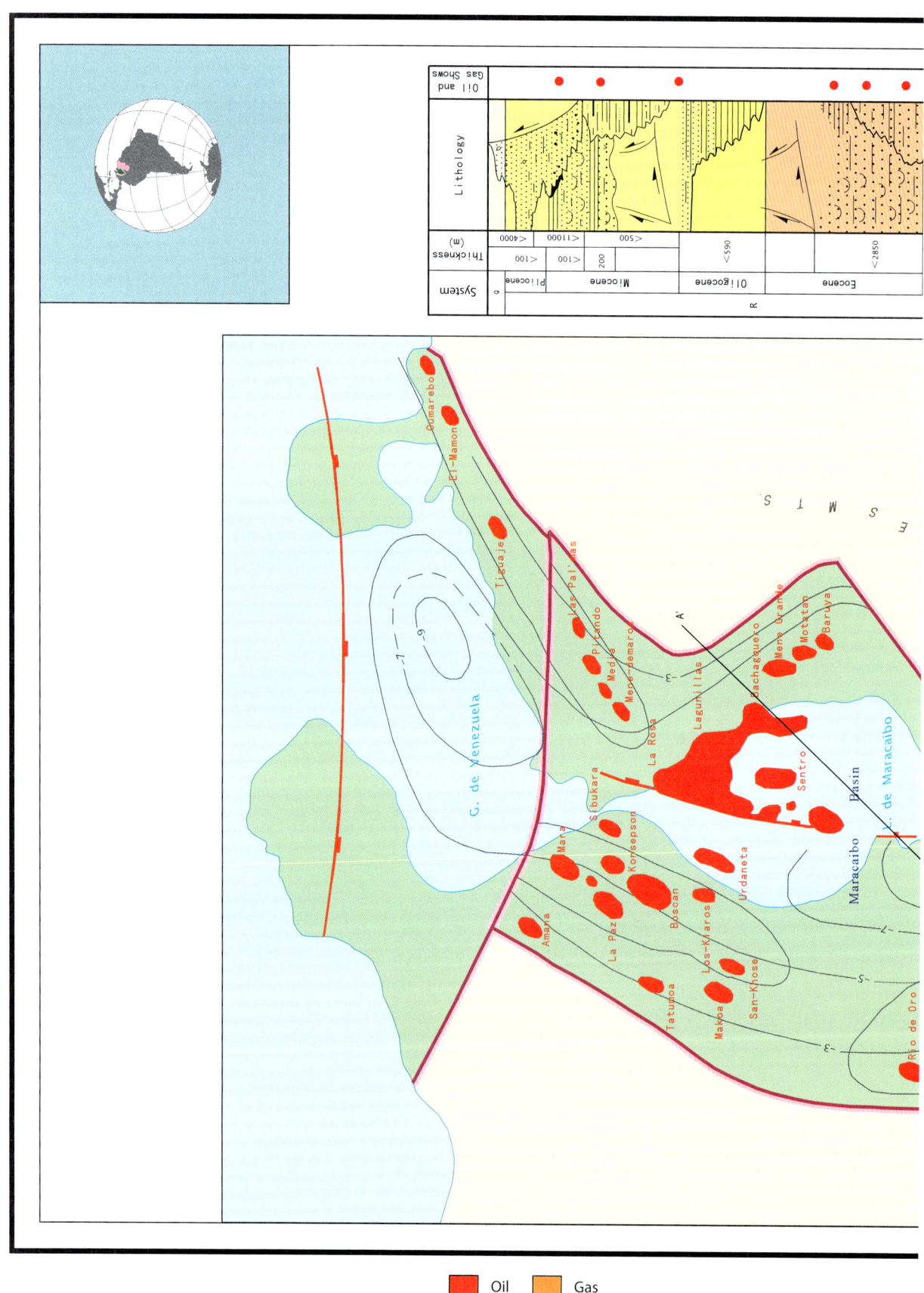

Oil Gas

CHAPTER 94

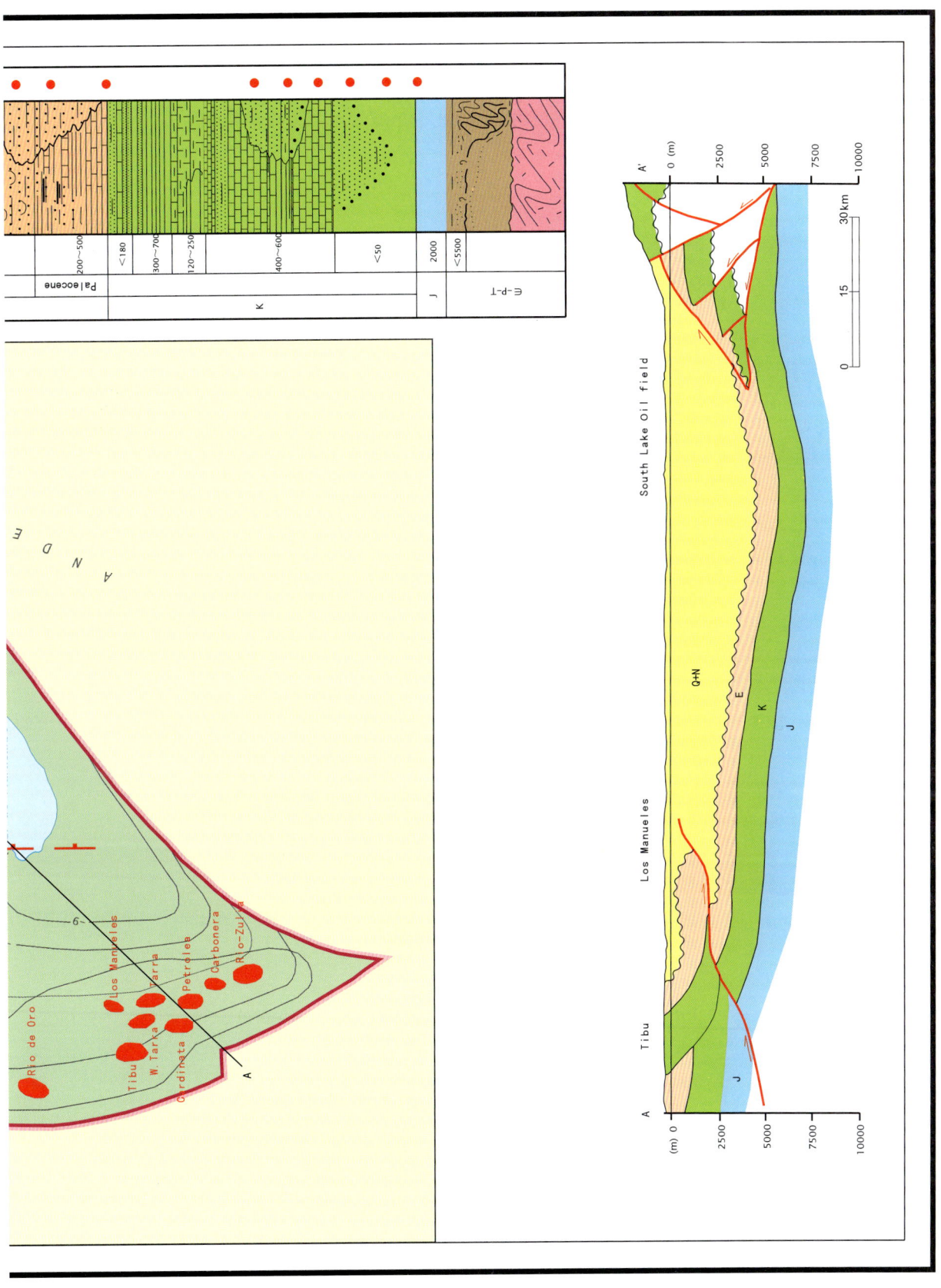

417

94 Maracaibo Basin

Geography

The Maracaibo Basin lies mainly in northwestern Venezuela. Lake Maracaibo lies entirely within the basin.

History

The level of exploration activity remained low until the early 1920s. The first oil of the Bolivar Coastal oil field (Fig. 94.1) was discovered in 1917. The level of exploratory drilling work peaked in the 1950s and began to decline in the 1960s. Production was 10,000 t in the late 1920s and increased to a peak of around 40,000 t in the late 1950s. In the early 1970s, basin production halved. Details of the major oil and gas fields of the Maracaibo Basin are listed in Table 94.1.

Regional geology

The basement comprises Late Precambrian metamorphics, and folded and partially metamorphosed Lower Palaeozoic rocks. Sedimentary cover comprises Ceno-Mesozoic deposits with thicknesses of 1700–420 m. Source rocks are the Cretaceous Laluna Formation bituminous limestones with an organic content of 10 per cent. Reservoirs are mainly in fractured basement rocks. Seals are Cretaceous mudstone and traps are various structural types, including faulted anticlines. Syn-rift continental sediments and volcanics of Triassic to Jurassic ages underlie the main prospective horizon in a series of extensional troughs that represent early Atlantic rifting. The Cretaceous to Paleocene post-rift section comprises passive (Atlantic) margin sediments, in which platform limestones associated with organic-rich clastics give way periodically to regressive deltaic clastics that are derived from the eroding Guyana Shield to the south. Maximal Cretaceous transgression, with development of widespread reducing conditions, occurred in the Cenomanian to Santonian interval. Late Cretaceous regression was accomplished by some fault reactivation and was followed by mainly Paleocene non-marine clastic deposition. In the Late Miocene, and continuing to the present, compressive uplift of the Merida Andes was accompanied by structural reactivation within the basin, and by deposition of large volumes of clastic sediments.

The Late Cretaceous bituminous limestones of the La Luna Formation are the source for most of the huge volumes of hydrocarbons that have been generated in the Maracaibo Basin. Significant volumes of hydrocarbons are found in reservoirs in the fractured basement. Various Cretaceous limestone comprise the reservoirs. Lower and Middle Eocene sandstone reservoirs are volumetrically the most important reservoirs. Paleocene and Oligocene reservoirs also contain significant quantities of reserves, but represent less than 2 per cent of the basin's total. The main regional seal is the Late Cretaceous Colon Formation mudstone overlying the La Luna source rock and associated reservoirs. Structural and combined structural–stratigraphic plays exist at all levels from Cretaceous to Pliocene but are of very variable volumetric importance.

World Atlas of Oil And Gas Basins, First Edition. Li Guoyu.
© 2011 John Wiley & Sons, Ltd. Published 2011 by John Wiley & Sons, Ltd.

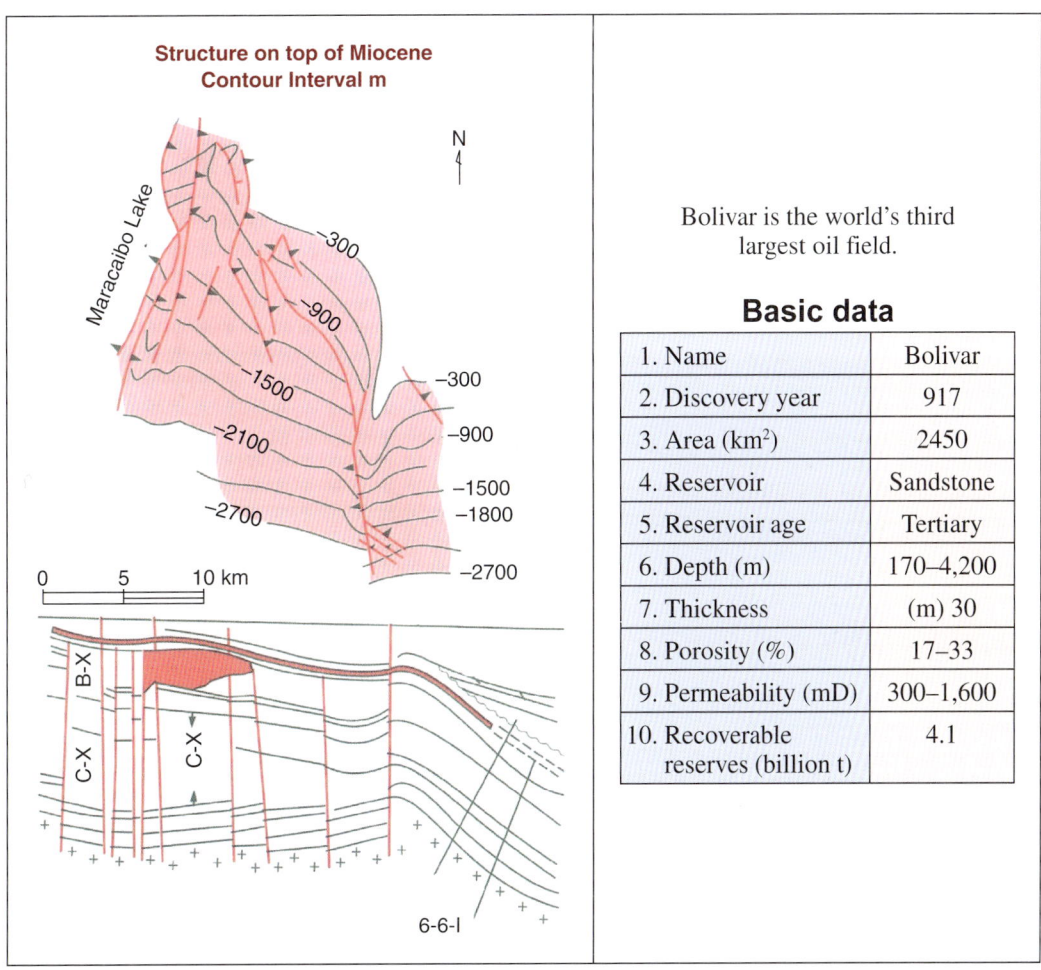

Fig. 94.1 Details of the Bolivar Coastal oil field.

Table 94.1 Major oil and gas fields of Maracaibo Basin

Name	Discovery year	Depth(m)	Producing formation	Lithology	Oil recoverable reserves (million t)
Boscan	1946	2100	E_2	Sandstone	137
Centro	1957	3000	E_2	Sandstone	137
La Paz	1925	1200	K	Carbonate rock	123
Bolivar Coastal	1917	170~4,200	N_1	Sandstone	4123
Mara	1945	1500			205
Lamar	1957	3900	E_3	Sandstone	205

Potential remains at basement and Cretaceous levels, although exploration will require improved methods for predicting fracture porosity. Paleocene plays offer good prospects in the southwest of the basin in subthrust situations. The Eocene and Oligocene plays are the least well explored in the mountain-front locations.

VENEZUELA EAST BASIN

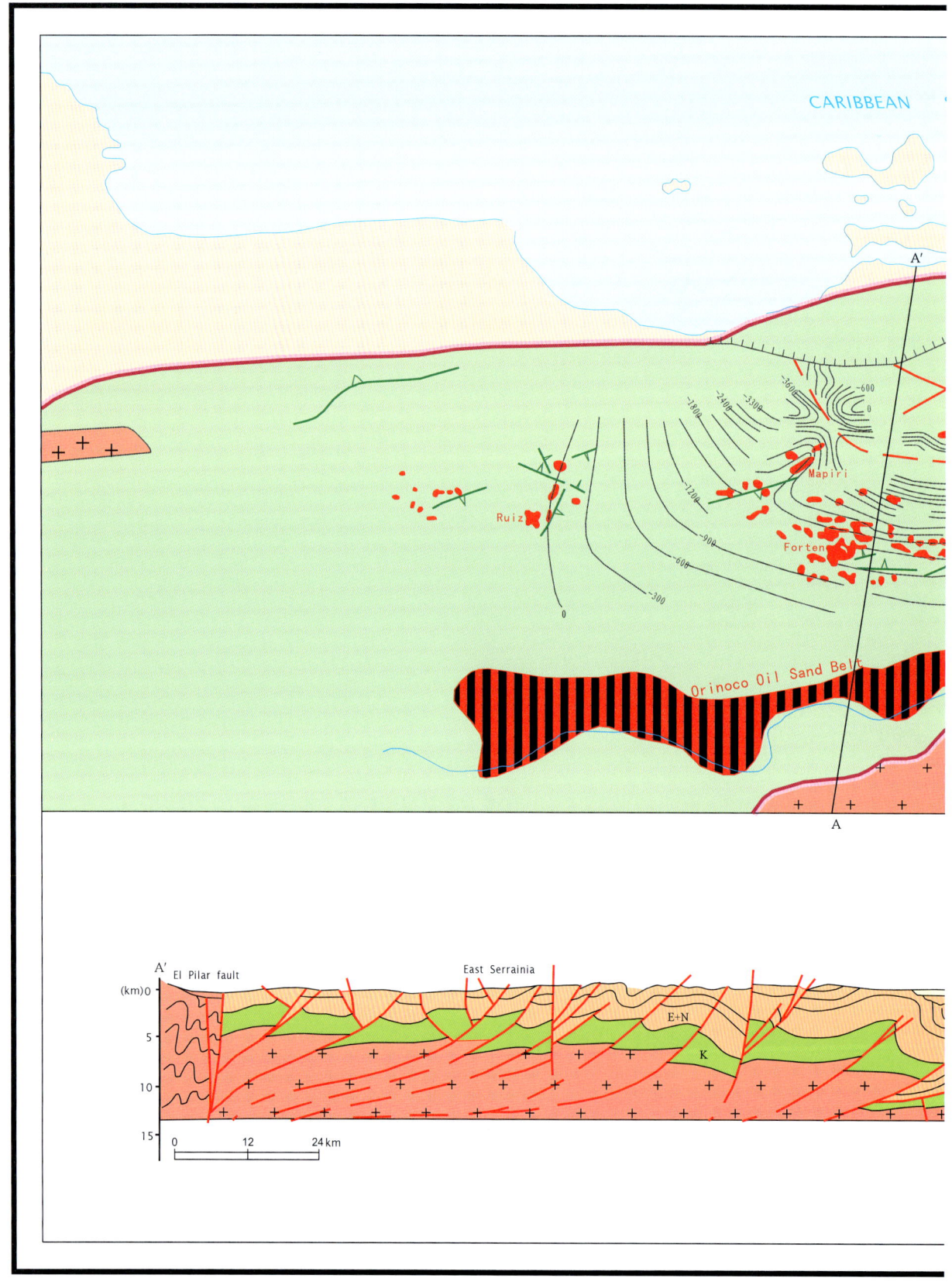

CHAPTER 95

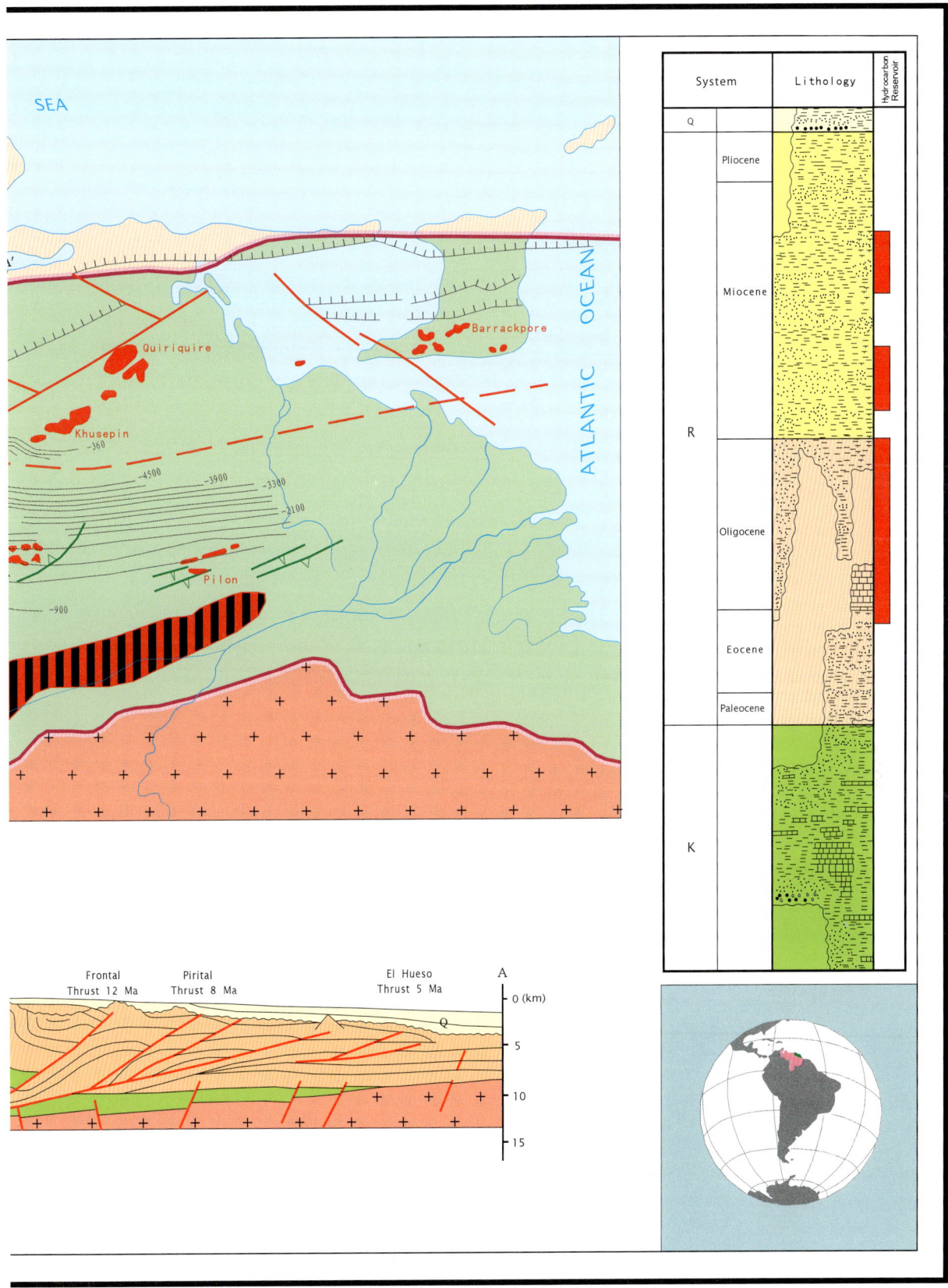

95 Venezuela East Basin

Geography

The Venezuela East Basin is a large asymmetric basin which covers most of central and eastern Venezuela and the island of Trinidad. Its eastward limit is the 1000 m submarine contour in the Atlantic Ocean which extends into Guyanan waters.

History

Onshore geological survey started in the 1850s. The seismic acquisition density is low due to poor terrain, and also because of land being used for farming or for industrial purposes. The first successful well was finished in 1867, with production of 20,000 barrels of asphaltic oil. The Venezuela Basin is world renowned for its occurrences of asphalt, particularly the Oficina oil field with recoverable reserves of 3.9 billion t. This oil field was discovered in 1937 with an area of 54,000 km^2 at a depth of 1580 m and oil density of 7.7–17.4 API. Offshore, seismic surveying began in the 1950s as exploration scientists searched for the seaward extensions of onshore fields.

More than 749 development wells have been drilled since 1948. Only two wells were drilled offshore, both in the Posa Field. Over 600 wells were completed after 1981, in almost 70 fields. Before the 1980s, approximately half the known reserves had been exploited and production was declining. The addition of the Carito and Furrial-Musipan reserves has resulted in an increase in production, which rose again in 1990 and 1991. Total cumulative production to date is approximately 8500 million barrels.

Regional geology

Following a phase of Late Triassic to Early Cretaceous rifting related to the opening of the Central Atlantic, an Early to Late Cretaceous passive margin sequence was developed as the South Atlantic opened. Clastic sediment supply was from the Guyana Shield in the south. Sediments were provenanced largely from the rising mountain belts to the north, and deformation migrated progressively to the south and east. By Late Miocene times, sedimentation had ceased in the west of the basin and a major delta system developed in the east of the basin, sourced by the proto-Orinoco River. Generation seems to have begun in mid-Oligocene to Middle Miocene times in the north of the basin, including areas now located in the Serranfa Interior Oriental. In the southern parts of the basin, Cretaceous source rocks are either absent or immature.

All oil reservoirs in the basin are sandstones (Table 95.1 and Fig. 95.1), with the exception of the Tertiary and Aptian–Albian carbonates of the EI Canto Formation in Venezuela. Seals are interbedded shales, lignites and clays. Altogether, seven groups of plays are recognized within the basin. The Oficina structural and structural–stratigraphic plays contain 41 per cent of the basin´s oil and 38 per cent of its gas, largely in the Oficina and Merecure Formations of Venezuela. The Oficina stratigraphic (self-sealing) play is the sole play in the heavy oil belt. It is restricted to the southern

World Atlas of Oil And Gas Basins, First Edition. Li Guoyu.
© 2011 John Wiley & Sons, Ltd. Published 2011 by John Wiley & Sons, Ltd.

Table 95.1 Major oil and gas fields of Venezuela East Basin

Name	Discovery year	Depth (m)	Producing formation	Lithology	Ultimate recoverable Oil reserves (million t)
Quiriquire	1928	500	N_2	Sandstone	137
Oficina	1937	1500	E_3	Sandstone	132
East Guara	1942	2100	N_1	Sandstone	86.3
Nipa	1945	2200	N_1	Sandstone	79.5
Mata	1954	2800	N_1	Sandstone	68.5

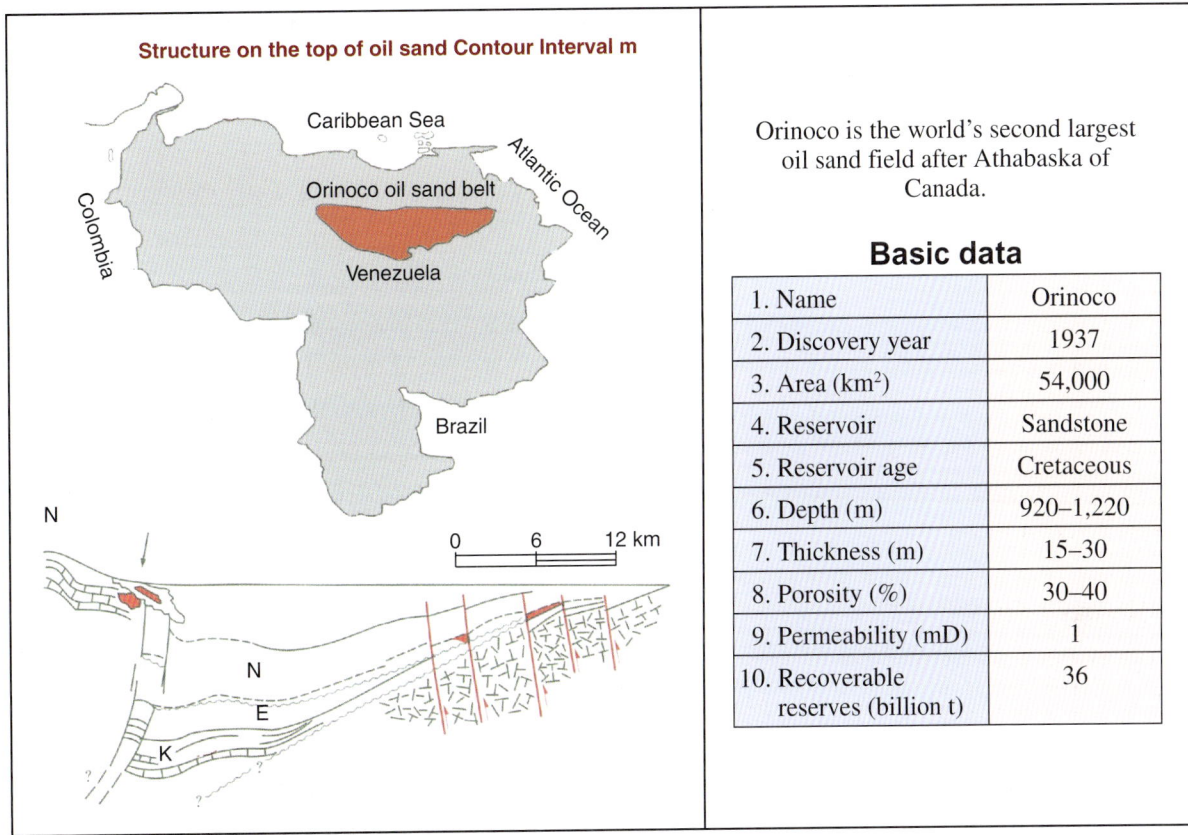

Orinoco is the world's second largest oil sand field after Athabaska of Canada.

Basic data

1. Name	Orinoco
2. Discovery year	1937
3. Area (km²)	54,000
4. Reservoir	Sandstone
5. Reservoir age	Cretaceous
6. Depth (m)	920–1,220
7. Thickness (m)	15–30
8. Porosity (%)	30–40
9. Permeability (mD)	1
10. Recoverable reserves (billion t)	36

Fig. 95.1 Details of the Orinoco oil field.

margin of the basin where Oficina Formation sandstones pinchout onto the Guyana Shield, and it contains over 1200 billion barrels of oil in place. Late Cretaceous source rocks have generated oil and gas in the basin, which is trapped in reservoirs ranging in age from Cretaceous to Pleistocene.

PERU, BOLIVIA AND PARAGUAY

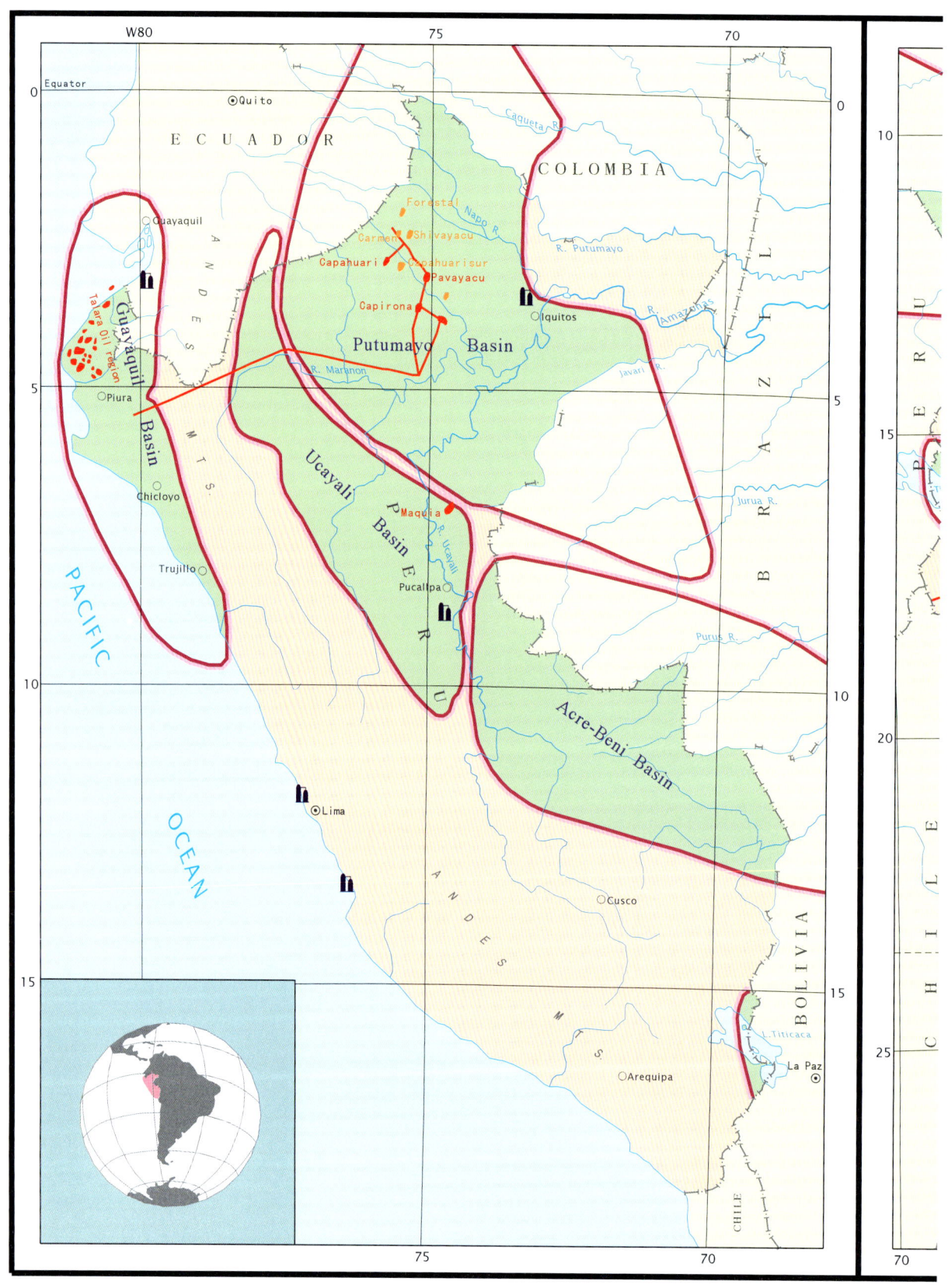

CHAPTER 96

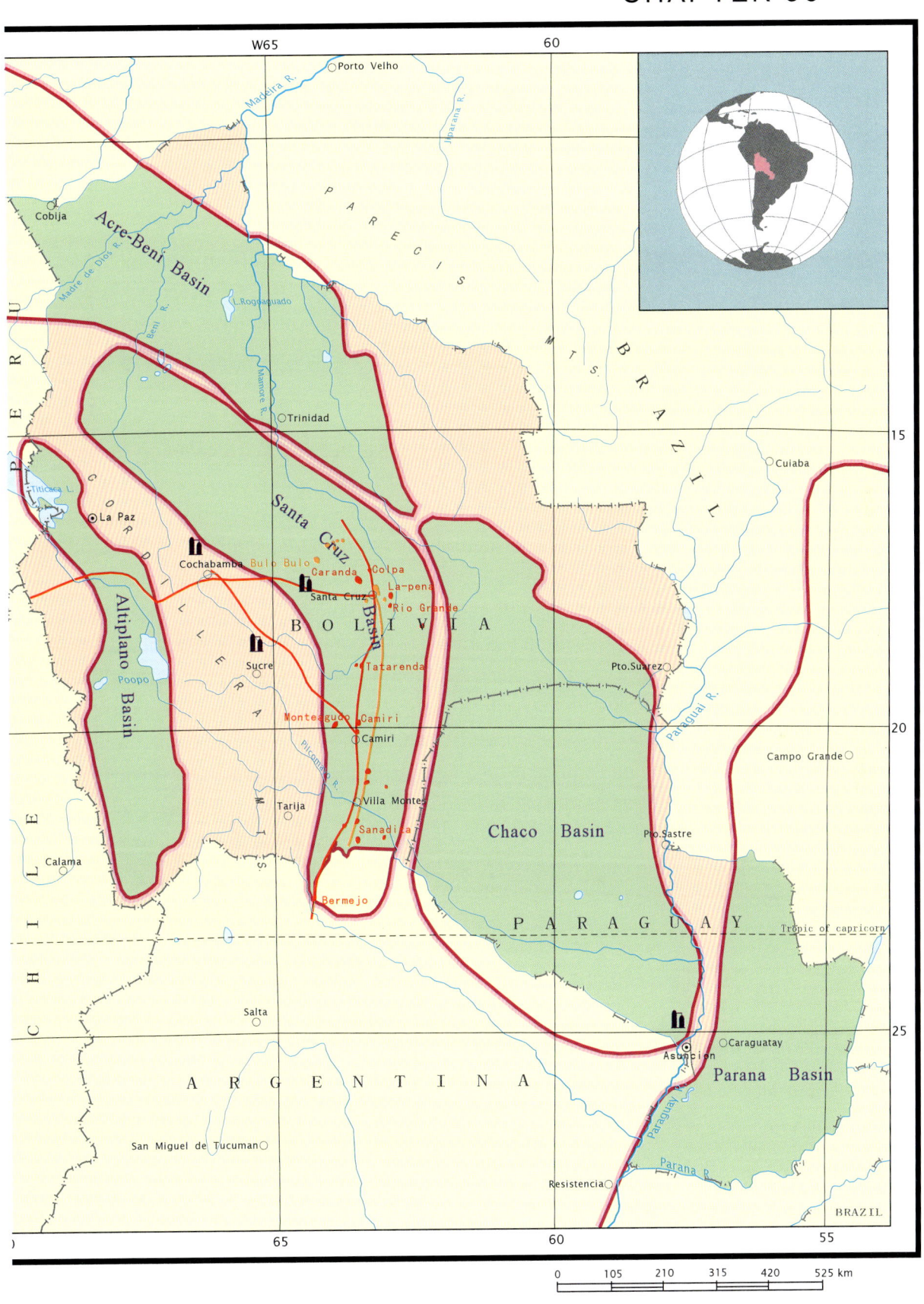

96 Peru, Bolivia and Paraguay

Geography

Peru, Bolivia and Paraguay are located in the western parts of South America, facing the Pacific to the west, and also lying in the foothill belt of the Andes Mountains and the Amazonian areas.

History

Oil exploration commenced over 150 years ago in this area, specifically in Peru in 1864 and Bolivia in 1867. Exploration commenced much later in Paraguay, in 1944. Oil seepages were well known in the Andes foothills in ancient times. Recently, Peru and Bolivia have been classified as oil-producing countries. Details of the major oil and gas basins are provided in Table 96.1.

Regional geology

Seven sedimentary basins lie between the Andes Mountains in the west and the Parecis Mountains in the east. The first wells in Bolivia were drilled in 1918. The sedimentary cover comprises Palaeozoic and Ceno-Mesozoic strata. Source rocks are mudstones, reservoirs are sandstones and limestones, and seals are clays. Trap types are structural, including faulted anticlines. The southern fields of the Chaco Basin that extends into Paraguay and Argentina. These fields are characterized by productive Permo-Pennsylvanian r red conglomeratic sandstone interbedded with red shale.

The most challenging geological problem in this area is to determine the source beds. Most geologists consider the underlying Devonian marine shales to be the source. In many other fields in Bolivia, similar red bed sequences much higher in the section are the producing horizons. This problem of a continental source is common in South America, with similar occurrences in Argentina, Bolivia and Colombia. In many cases, derivation from lacustrine deposits is the more probable explanation.

Most of the producing structures are thrust faulted anticlines. In fields which are deep in the foothills, such as Camiri, the structures are steeply dipping and the faulting has a large displacement. As would be expected, structures removed from the mountain front are gentler. Furthermore, faulting is minor and often normal.

Table 96.1 Major oil and gas basins

Country	Basin	Area (thousand km^2)	Major sedimentary rock Age	Thickness (m)	Reservoir Age	Lithology
Peru	Acre-Beni	250	Mz–Kz			
	Putumayo	470	Pz, Mz, Kz	5000	K	Sandstone
Bolivia	Ucayali	150	C–R	9000	K	Sandstone
	Altiplano	136	Mz–Kz		J,K	Limestone, sandstone
Paraguay	Santa Cruz	250	D–R	6000	D,C,K	Sandstone
	Chaco	450	Pz–Kz	11,000	K	Limestone, sandstone

World Atlas of Oil And Gas Basins, First Edition. Li Guoyu.
© 2011 John Wiley & Sons, Ltd. Published 2011 by John Wiley & Sons, Ltd.

The Mesozoic stratigraphy of Acre-Beni Basin is similar to that of the south, but the sandstones, especially in the Caranda, Colpa and Rio Grande oil fields are cleaner, with a higher porosity and permeability. These sandstones are the principal producing horizons. The Mesozoic strata are all continental. While this limestone may be the source of some hydrocarbon generation, there may be other sources, since in many cases several shale beds in the Tacuru Formation lie between the source horizon and the reservoir bed. Once again, lacustrine clays seem to be the probable source.

North of the Santa Cruz area, in the Bulobulo and Iloilo oil fields, the Cretaceous sands appear to be gas prone. Structurally they are more faulted than the Caranda type oil field, but they are parallel with the northwest–southeast trend of the Andes, which is attributed to the La Paz–Arica megashear. Because of the logistics of the Amazon jungle and the continental environment of the sediments, exploration has been more limited. This is true also for the small Madre de Dios Basin of southern Peru.

Peru Peru is located in western South America, bordering the South Pacific Ocean and lying between Chile and Ecuador. It has an area of $12,852,216\,km^2$ and a population of 27.22 million. There are four sedimentary basins distributed here: Acre-Beni, Putumayo, Ucayali and Guayaquil, with areas of between $150,000\,km^2$ and $250,000\,km^2$, filled by Palaeozoic and Mesozoic sediments with thicknesses in the range of 5000–9000 m. Oil exploration was initiated in the Guayaquil Basin in 1864. The first oil field was discovered in 1869. The offshore oil exploration started in 1955. By the end of 2008 the proven oil reserves were 58.9 million t with annual oil production standing at 3.7 million t. Proven gas reserves were 335 billion m^3 and gas production was 2.6 billion m^3 annually.

Bolivia Landlocked Bolivia sits astride the Andes in the west-central part of South America, occupying an area of $1,098,581\,km^2$ with a population of 9.4 million. There are four main basins in the country, including Altiplano, Acre-Beni, Chaco and Santa Cruz. These sedimentary basins have a combined area of $580,000\,km^2$, occupying about 53 per cent of the land area of the whole country. Huge oil and gas resources are estimated to lie in these basins, which are characterized by older reservoirs rocks (Palaeozoic). Oil exploration started in 1867 in areas nearby oil seepages in the foothills of the Andes. The first oil field was discovered in Devonian sediments in 1924. By the end of 2008 the proven oil reserves stood at 63.69 million t, with annual oil production of 2 million t. Comparable figures for gas were: proven reserves of 750 billion m^3 and annual production of 14.1 billion m^3.

Paraguay Paraguay is located in central South America, northeast of Argentina. It is also landlocked, occupying an area of $406,752\,km^2$, with a population of 5.89 million. At present, Paraguay is classified as a non oil-producing country. There are three tectonic elements: Central Uplift, Chaco Basin and Parana Basin. The Chaco Basin has an area of $450,000\,km^2$, filled by Silurian, Devonian and Mesozoic deposits with thicknesses of up to 11,000 m.

PUTUMAYO BASIN

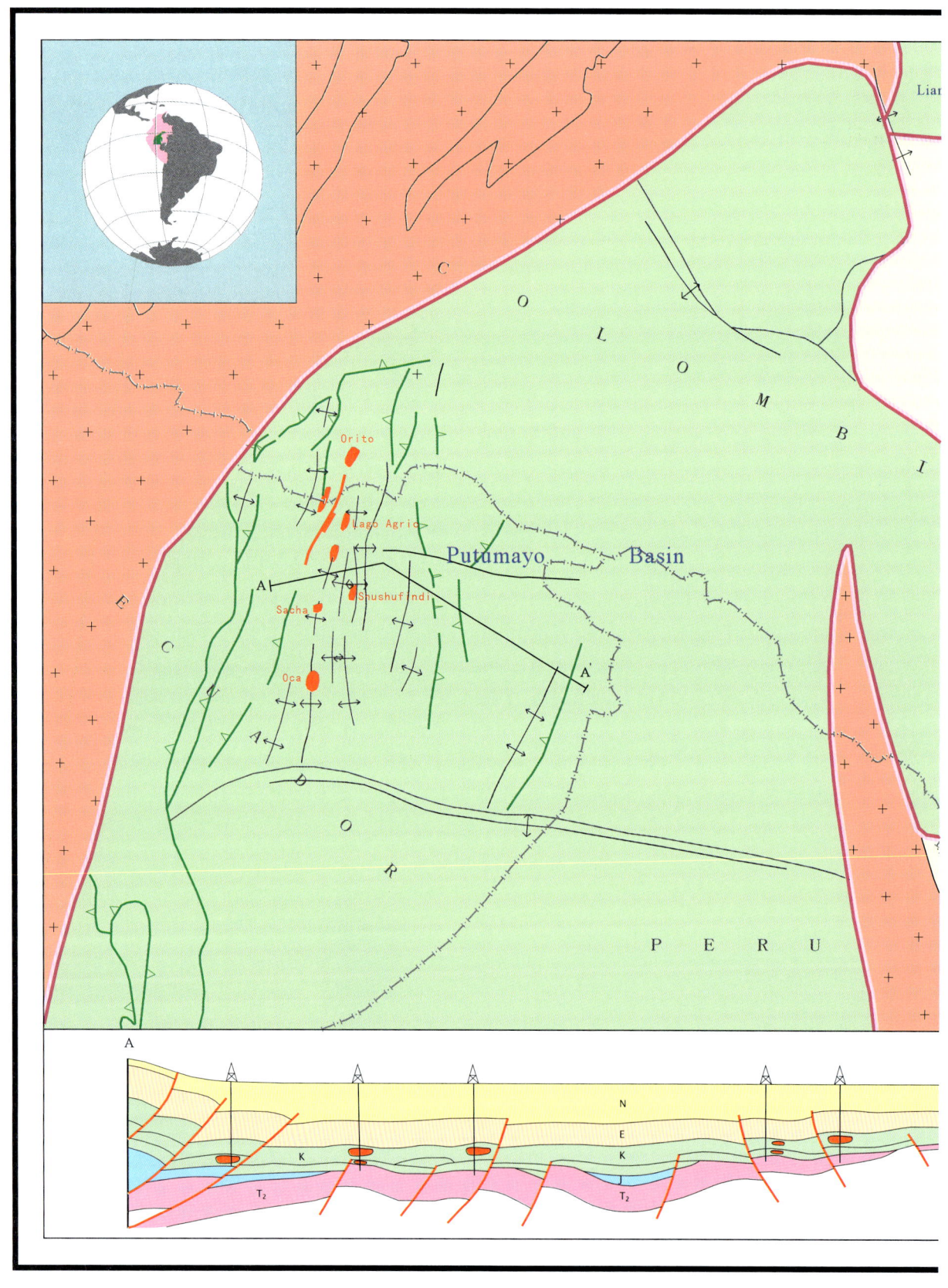

CHAPTER 97

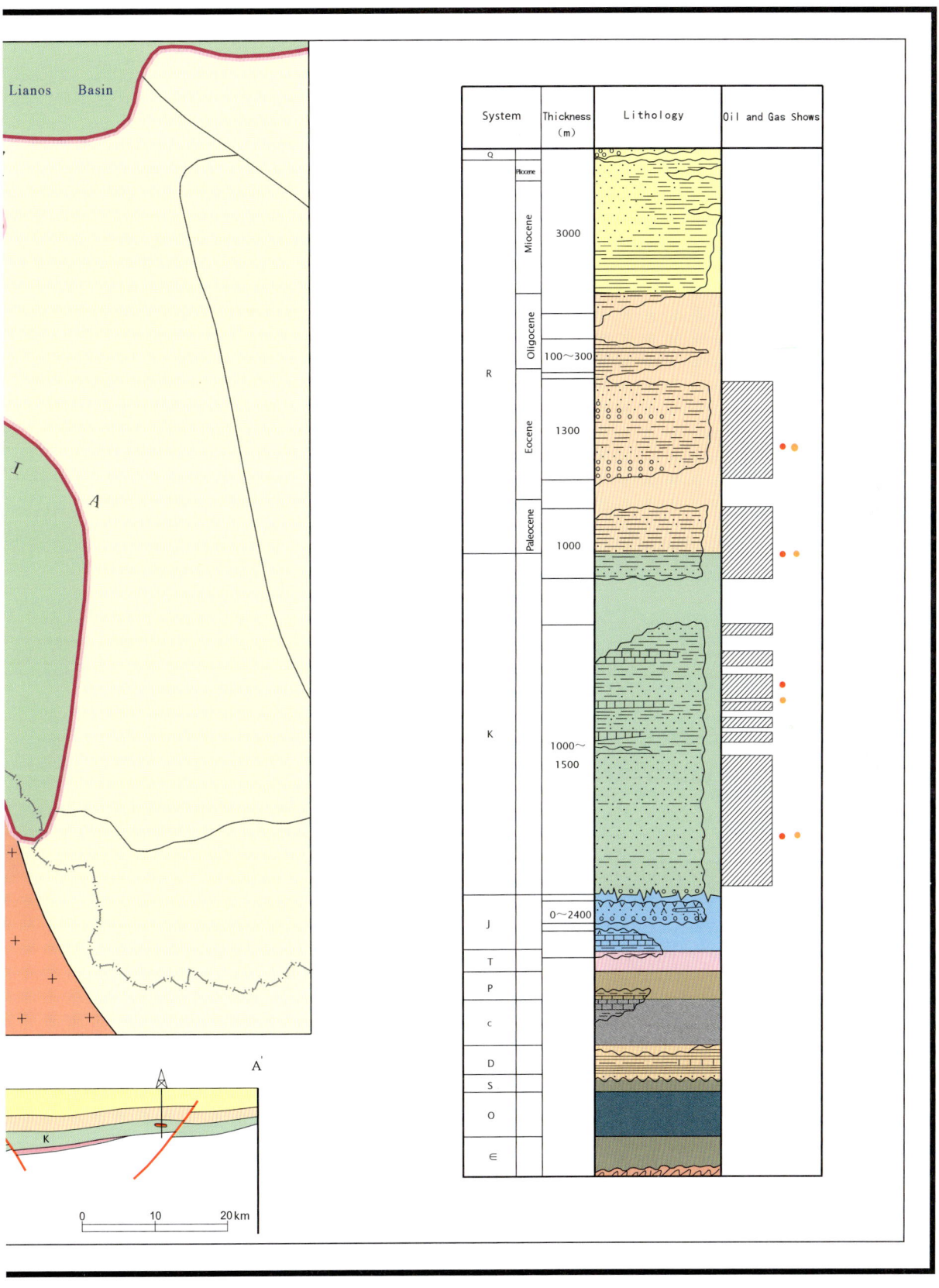

97 Putumayo Basin

Geography

Putumayo Basin is entirely onshore and has an area of 427,000 km². About 68 per cent (300,000 km²) of the basin lies in northeast Peru, and 12 per cent (47,000 km²) in southeast Ecuador, 20 per cent (80,000 km²) in Colombia. The basin is elongate and parallel to the Andean mountain front. The main depocentre is in the west, and the sedimentary succession thins to the east by onlap on to the Guyana Shield.

History

Before 1960 only a few exploration wells were completed. In the late 1960s the giant oil fields of Shushufindi and Sacha were discovered. As a result of these discoveries, this basin became a major oil-producing region. Although development drilling began in the Peruvian sector of the basin in 1972, the main development phase averaging approximately 15 completions a year was undertaken between 1976 and 1983. In the Peruvian sector, large-scale production began in 1978 with the commissioning of the trans-Andean pipeline. Peak production of 128,000 barrels/day was achieved in 1979. Details of the major oil and gas fields are listed in Table 97.1.

Regional geology

The basin developed as one of a chain of sub-Andean foreland basins. A polycyclic basin, with an early history as part of a Palaeozoic passive margin was interrupted by rifting events. Early Andean deformation began in latest Cretaceous to early Tertiary time. Deposition was marine during Cretaceous times, i.e. shallow-marine and deltaic, but changed to continental sandstones and carbonates at the Cretaceous–Tertiary boundary. The Chonta/Napo Formation shale

Table 97.1 Major oil and gas fields in part of Putumayo Basin of Ecuador

Name	Discovery year	Depth (m)	Production of 1994 (thousand t)	Cumulative production (thousand t)	Oil and gas total recoverable reserves (thousand t of oil equivalent)
Auca	1970	3226	1107	14,172	20,913
Cononaco	1972	3426	826	7242	15,091
Lago Agrio	1967	3103	304	17,331	27,105
Sacha	1969	3099	2978	65,448	114,041
Shushufindi/ Aguarico	1959	2980	4927	104,193	216,895
Bermejo	1967	1525	248	1397	17,781
Cuyabeno	1972	2452	280	3059	5639
Libertador	1980	2837	2410	19,624	46,644
Amo	1987		302	559	16,438

World Atlas of Oil And Gas Basins, First Edition. Li Guoyu.
© 2011 John Wiley & Sons, Ltd. Published 2011 by John Wiley & Sons, Ltd.

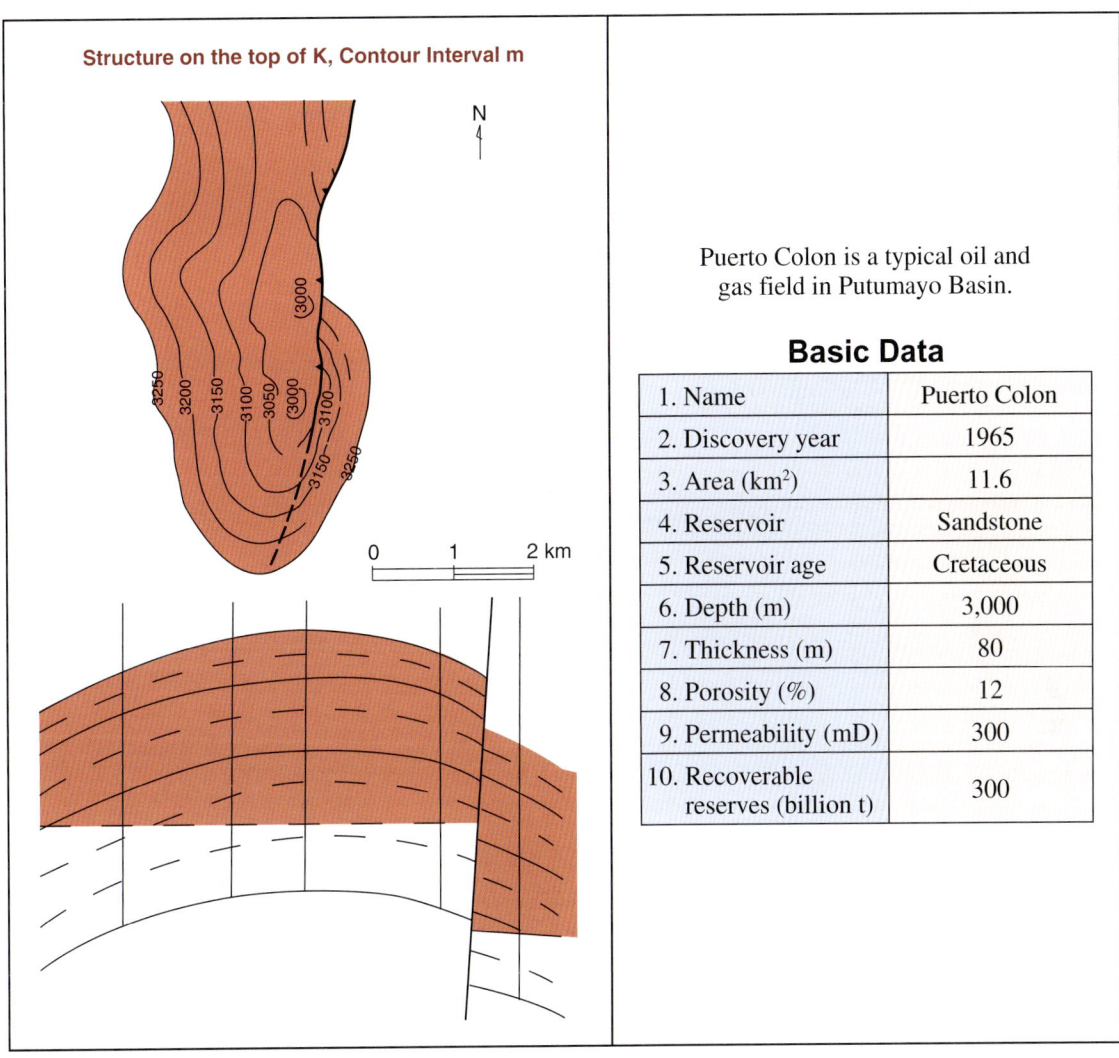

Fig. 97.1 Details of the Puerto Colon oil and gas field.

and bituminous carbonate are the main known oil and gas source beds in the basin. Cretaceous (e.g. Fig. 97.1) and Tertiary reservoirs are sealed mainly by intraformational shale and tightly bound carbonate.

At least three episodes of rifting affected the basin during the Mesozoic, followed by flexural subsidence due to lithospheric loading of the Andes in the Tertiary. They comprised episodes of stretching for Triassic and Jurassic basin evolution accompanied by volcanism and thermal uplift associated with continental sedimentation. The stretching phases were followed by thermal subsidence accompanied by shallow or restricted marine sedimentation. This pattern changed during the Late Cretaceous, when rifting was accompanied by subsidence allowing the accumulation of the main source rocks of the region. This event was followed by flexural foreland basin subsidence during the Tertiary, related to lithospheric loading in the Andean Mountain Belt.

BRAZIL AND URUGUAY

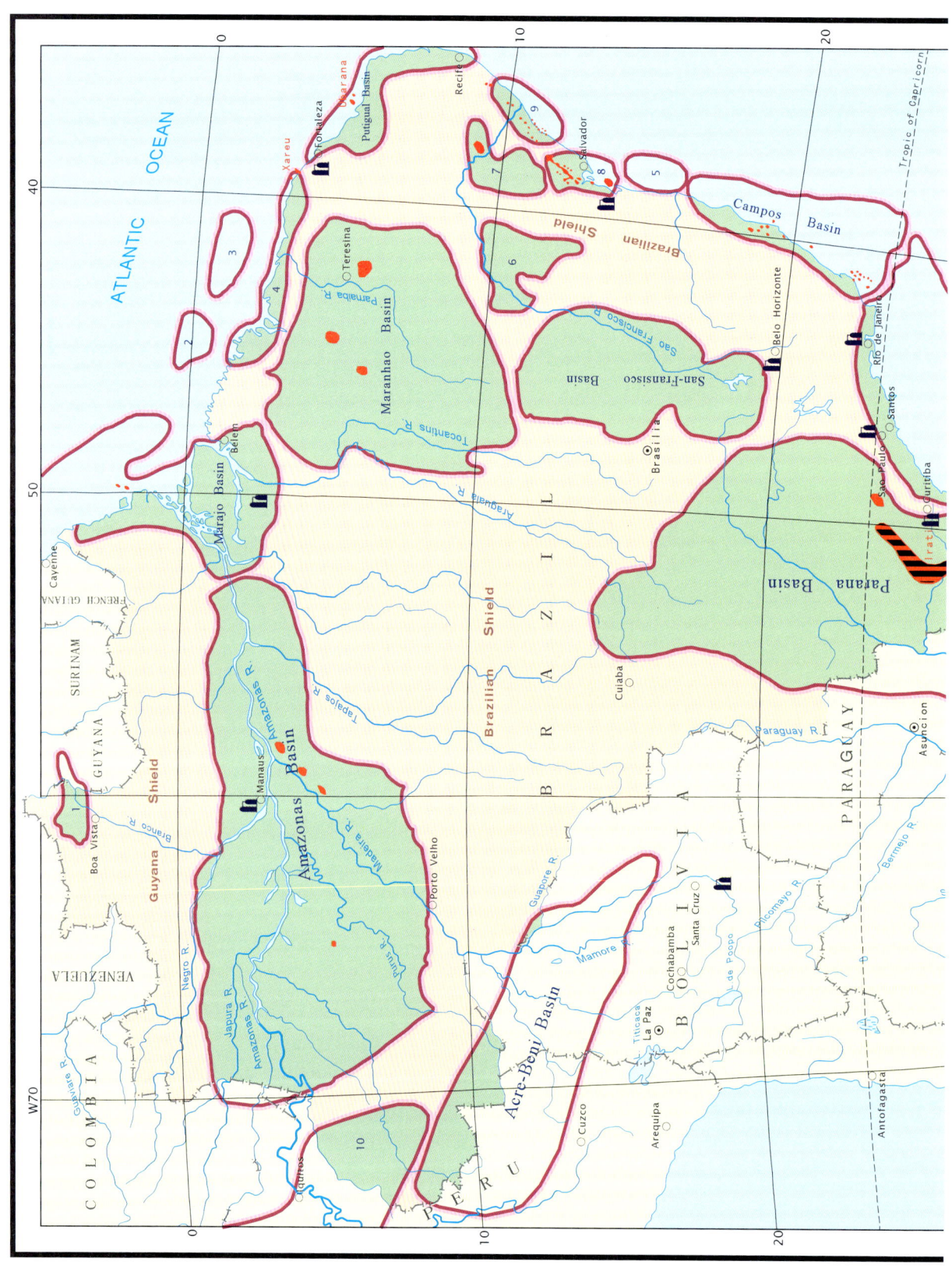

CHAPTER 98

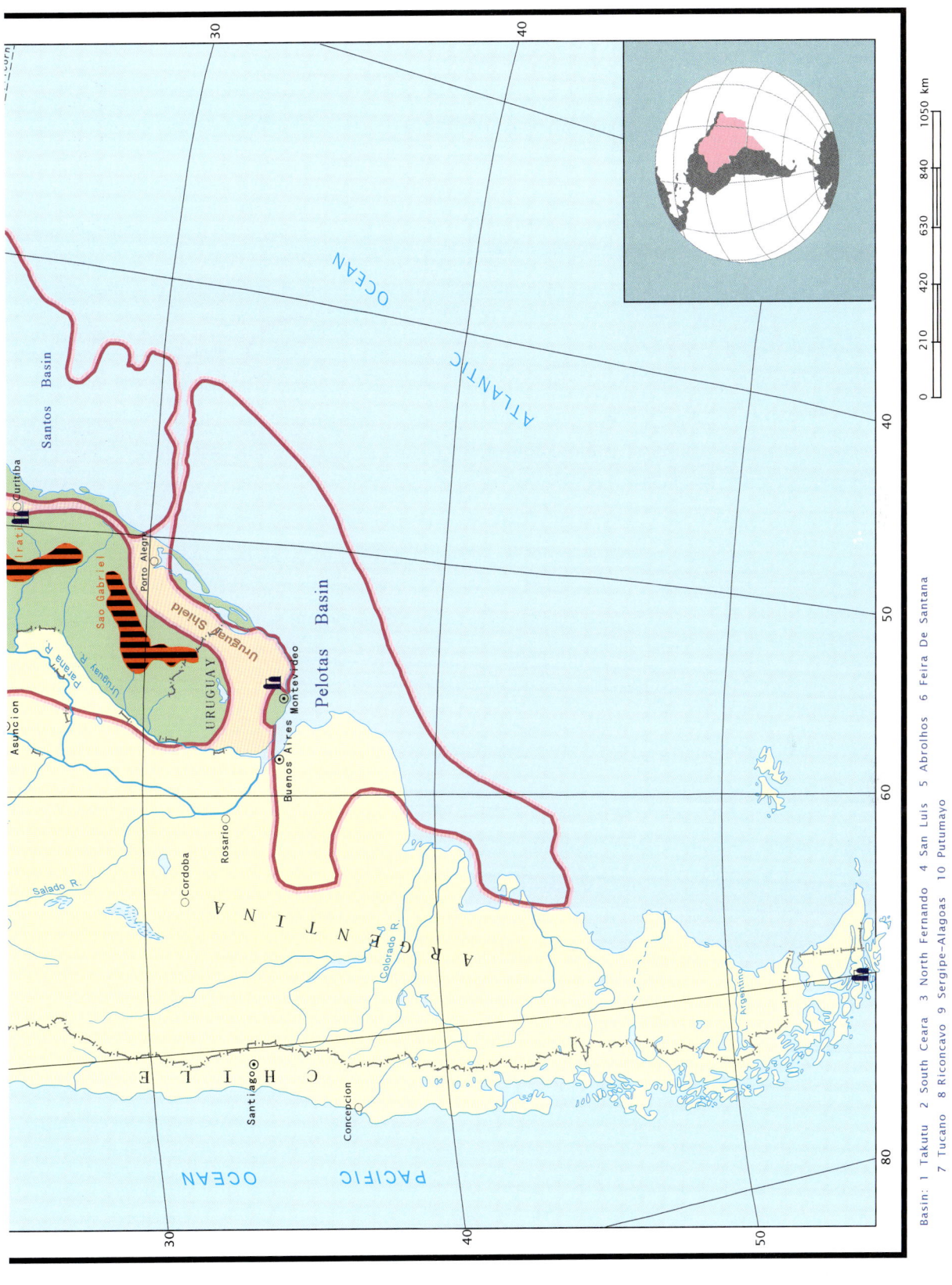

433

98 Brazil and Uruguay

Geography

Brazil and Uruguay are located in the east of South America, facing the Atlantic Ocean to the east and the Pacific Ocean to the west. The River Amazon lies in this region.

History

Oil exploration started in Brazil in 1865. At present, Brazil is an oil-producing country of increasing significance as a result of offshore oil production. Uruguay is a non-producing country.

Regional geology

Twenty basins have been identified within or bordering the region. Commercial oil and gas accumulations were identified the Amazonas and Reconcavo Basins. Tectonically, the basins are associated with syneclises, troughs, grabens and pericratonic margins of the platform.

The sedimentary cover of the basins is confined to platform syneclises (Amazonas, Parana, Maranhao) and is composed of Palaeozoic deposits which account for about 80 per cent of the total thickness of the succession, being some 5–7 km. Trappean magmatism manifestations are typical of the basins. The basins have been very poorly studied. Commercial oil and gas accumulations have been identified in the Amazonas Basin (Table 98.1), where seven oil fields and 12 gas fields have been discovered. Traps are represented by anticlines and zones of sandstone pinching out. Main source rocks are Silurian and Devonian shales, and also Carboniferous limestones and shales. Reservoirs are Middle Devonian and Carboniferous sandstones and, more rarely, limestones. The fields are associated with faulted anticlines, and the potential of petroleum is high in the platform syneclises and troughs.

The basins located on the shields (Reconcavo, Tacutu) are confined to grabens and filled mainly with Mesozoic rocks. Commercial oil and gas accumulations have been identified in Reconcavo Basin. In this basin, 81 oil fields and 16 gas fields have been discovered. Upper Jurassic–Lower Cretaceous deposits are regionally oil- and gas-bearing. Reservoirs are sandstones, more rarely, limestones. Source rocks are considered to be the clayey portions of the succession of the same age. The fields are confined to faulted anticlines. As a rule, they contain more than one reservoir. Petroleum potential of the basins is medium.

Brazil

Brazil is the fifth largest country in the world in terms of area. It occupies almost half the South American continent, stretching from the River Amazon basin in the north and west to the Brazilian Highlands in the southeast with close to 7500 km of Atlantic coastline. Brazil is a major oil-producing country, as well as being a large oil consuming country. Domestic oil production meets 50 per cent of oil consumption. To reduce its reliance on imported oil, Brazil, like the USA, has pursued an active biofuel development programme, based primarily on the country's huge sugar cane production. Oil exploration started in 1865 and the first oil field was discovered in 1939. Annual oil production reached 10 million t in 1981, based on the discovery

World Atlas of Oil And Gas Basins, First Edition. Li Guoyu.
© 2011 John Wiley & Sons, Ltd. Published 2011 by John Wiley & Sons, Ltd.

Table 98.1 Major oil and gas basins of Brazil

Basin	Area (thousand km²)	Major sedimentary rock		Reservoir		Remark
		Age	Thickness (m)	Age	Lithology	
Amazonas	1300	Pz–Kz	4000	D, C	Sandstone	
Campos	260	Mz, Kz	8000	K	Sandstone, conglomerate	
Feira de Santana	80					
Marajo	320	K, R	4000	R	Sandstone, Limestone	Extends into Guyana, Surinam, French Guiana, Venezuela
Maranhao	700	Pt–Mz				
North Fernando	30					
Parana	1270	Mz–Kz				Extends into Uruguay
Putigual	110	K, Kz		K	Sandstone	
Reconcavo	35	Mz	6500	J, K	Sandstone	
San-Francisco	350	Pz–Mz				
San Luis	80	Mz, Kz				
Santos	570	Mz, Kz	8000	K	Sandstone	

of giant offshore oil fields. Following the discovery of supergiant oil fields in deep-water areas of Campos Basin, annual oil production rose steadily to 56.4 million t in 2000. By the end of 2008, the proven oil reserves were 1.72 billion t with annual oil production standing at 90 million t. Proven gas reserves were 365 billion m³ with annual gas production at 9.8 billion m³. Twenty sedimentary basins are located in, or partly within, Brazil with areas ranging from 30,000 km² to 1.3 million km² (the largest being the Amazonas Basin). Brazil has very high petroleum potential. The offshore Campos Basin, north of Rio de Janeiro, is the country's most prolific oil production area and holds the most of its natural gas reserves.

Uruguay Uruguay is located in southeast South America and has an area of 176,215 km² with a population of 3.4 million. This country is located on the Brazil–Uruguay Shield, composed of Precambrian metamorphosed rocks and granite. The Pelotas Basin occupies a small part of south Uruguay onshore and extends to offshore regions, with thicknesses of Tertiary sedimentary rocks reaching 7000 m. The offshore areas of the country are considered to have petroleum potential. In 2008 licensing of offshore blocks started.

CAMPOS BASIN

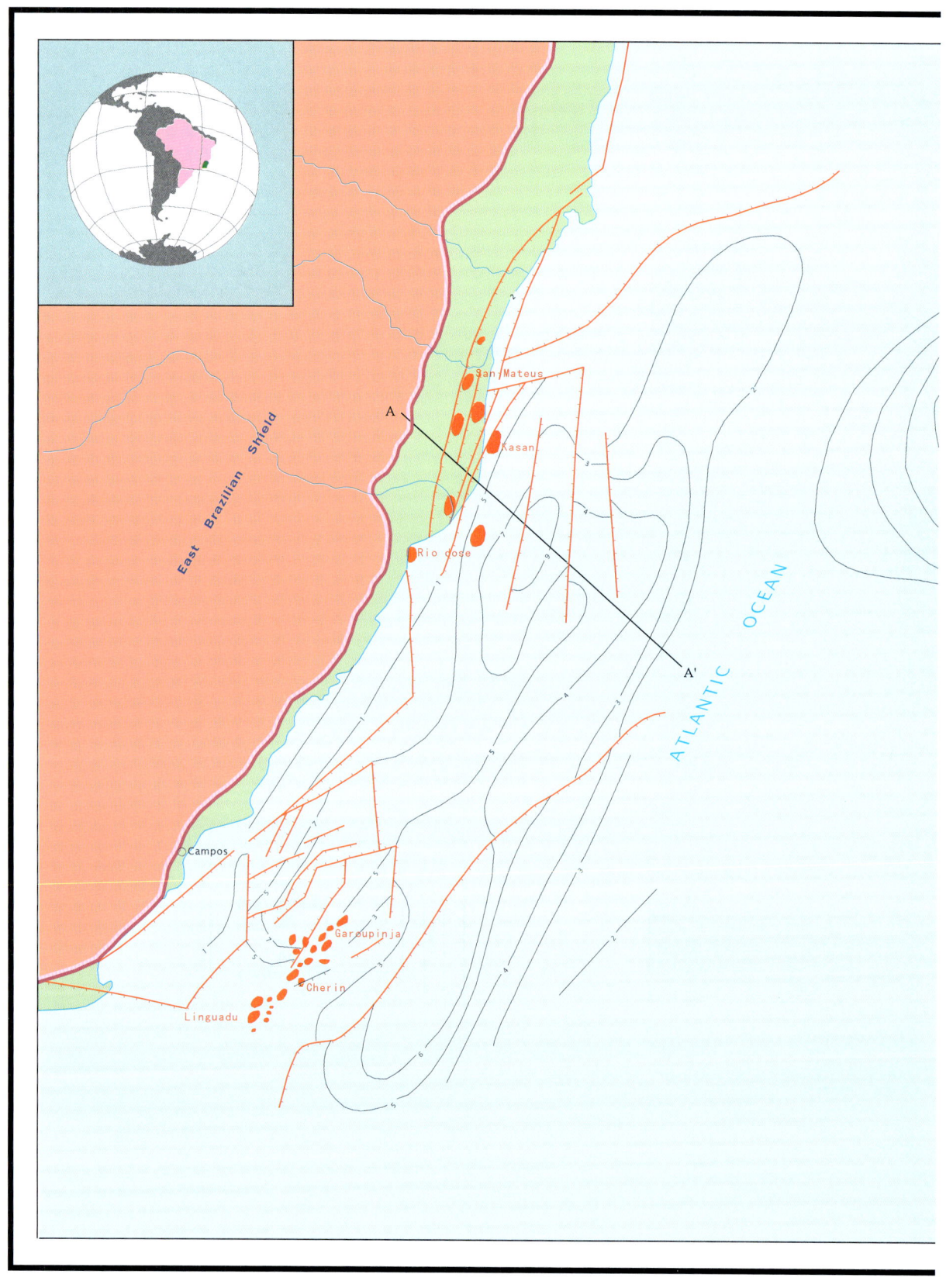

CHAPTER 99

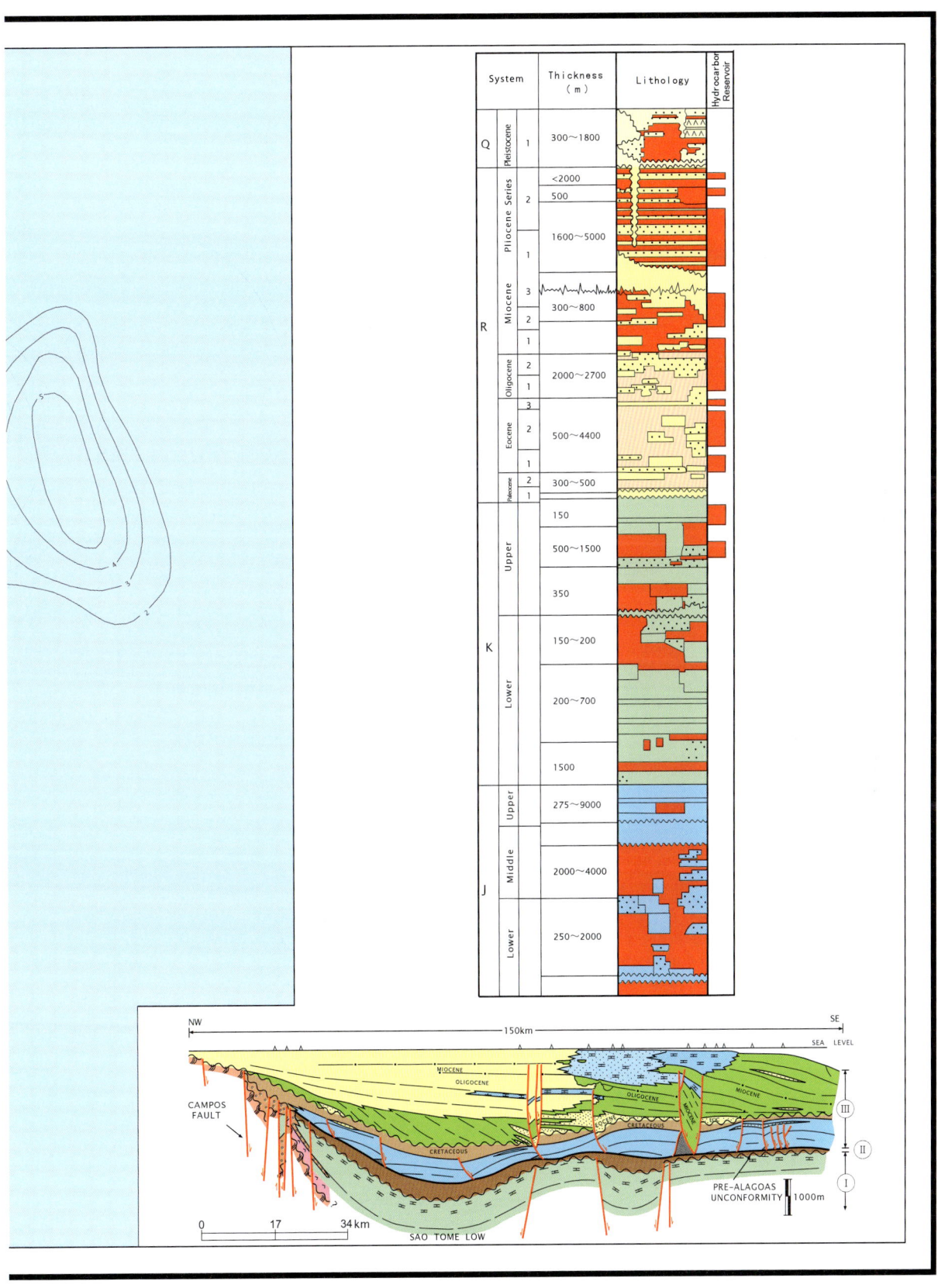

99 Campos Basin

Geography

Campos Basin lies almost entirely offshore the state of Rio de Janeiro in Brazil. Only 3 per cent of the basin lies onshore.

History

Campos is the main producing basin in Brazil, accounting for 65.7 per cent and 38.8 per cent of the country's total oil and gas output, respectively, as at 1993. Onshore, limited exploration activity has been conducted in the small sector of the basin. Offshore exploration started at the beginning of the 1970s and the first discovery, Garoupa, was made in 1974, followed by six finds in 1975. Campos Basin is one of the most important petroleum provinces in Latin America. The first development well was drilled in the Garoupa oil field in 1977, and 377 wells had been drilled by the end of 1988. Recent development drilling is concentrated in the Albacora, Marlin (Fig. 99.1), Pirauna, Marimba, Bijupira, Salema and other fields (Table 99.1), according to their various field

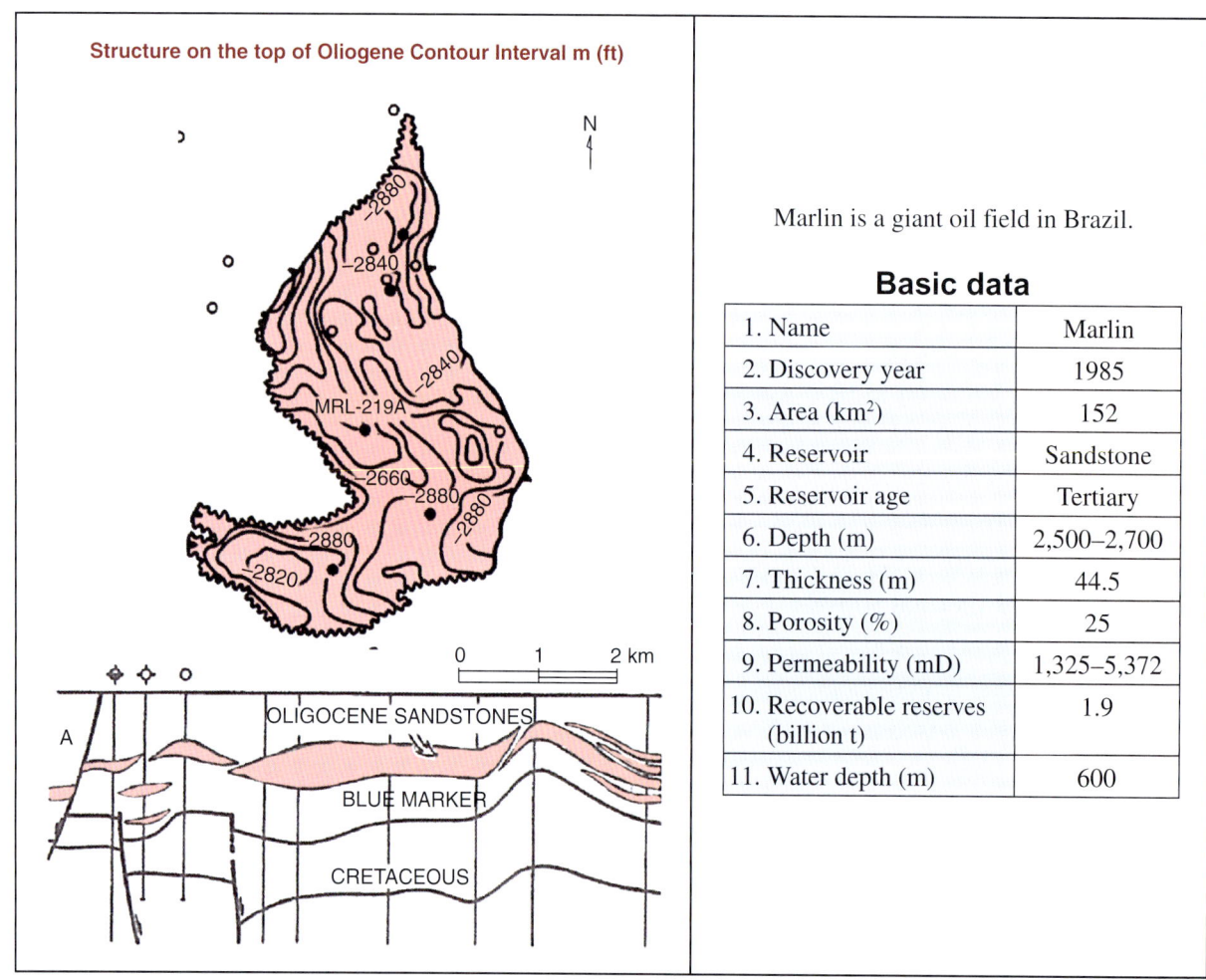

Marlin is a giant oil field in Brazil.

Basic data

1. Name	Marlin
2. Discovery year	1985
3. Area (km^2)	152
4. Reservoir	Sandstone
5. Reservoir age	Tertiary
6. Depth (m)	2,500–2,700
7. Thickness (m)	44.5
8. Porosity (%)	25
9. Permeability (mD)	1,325–5,372
10. Recoverable reserves (billion t)	1.9
11. Water depth (m)	600

Fig. 99.1 Details of the Marlin oil field.

World Atlas of Oil And Gas Basins, First Edition. Li Guoyu.
© 2011 John Wiley & Sons, Ltd. Published 2011 by John Wiley & Sons, Ltd.

Table 99.1 Major oil and gas fields of the Campos Basin

Name	Discovery year	Lithology	Depth (m)	Oil and gas recoverable reserves (million t)	Natural gas Annual production (billion m³)	Year	Oil Annual production (thousand t)	Year
Albacora	1984	Sandstone	2567	100	3	1995	2418	1995
Barracuda	1989	Sandstone	2861	160				
Carapeba	1982	Sandstone	2990	20	0.3	1995	1767	1995
Caratinga	1990	Sandstone	2510	40				
Cherne	1976	Sandstone	3077	30	1.3	1995	1227	1995
Enchova	1976	Sandstone	2126	20	2.5	1995	628	1995
Linguado	1978	Limestone	2703	20	1.1	1995	571	1995
Marimba	1984	Sandstone	2710	40	2.5	1995	2858	1995
Marlin	1985	Sandstone	2713	380	4.3	1995	4214	1995
Namorado	1975	Sandstone	2980	40	2.4	1995	1642	1995
Pampo	1977	Limestone	1917	20	1.4	1995	1108	1995

development programmes. Oil production started in 1977, increasing steadily in the following years. From 1985, production continued to rise steadily, reaching its peak (to date) in 1993. Gas production has been underway since 1978 (Halbouty, 1982).

Regional geology

The Campos Basin is a classic example of an Atlantic-type passive continental margin basin. The post-salt Middle–Upper Cretaceous sequence began with the deposition of Albian shallow-water limestones on a rapidly subsiding platform (Macae Formation). The main source rocks are Neocomian lacustrine calcareous black shales of the Lagoa Feia Formation. There is a wide distribution of reservoirs throughout the stratigraphic column. Production horizons include fractured basalts in the basement and Neocomian coquina (shelly clastic) sediments.

During accumulation of the Aptian salt there was very little tectonic activity. Following the Aptian there was thermal subsidence of the basin and gentle tilting oceanwards, causing gravity gliding of the post-salt carapace. Thin-skinned extensional fault systems were developed, with listric faults that detach on the salt. The salt accumulated in pillows and reactivated grabens in zones of faulting between rafted blocks. The rate of subsidence decreased with time throughout the Late Cretaceous and Tertiary. Gravity glide tectonics and halo-kinetic salt movements decreased so that the Tertiary turbidite sheets are far more extensive and planar than earlier sheets, and are less disrupted by normal faults.

Eocene and Middle–Upper Cretaceous plays contain large quantities of oil and gas. The Lower and Middle Cretaceous plays are less prolific. Most of the current exploration is in water depths ranging from 200 m to 1000 m. Future exploration may concentrate both in deeper waters and also in the shallow waters of the Cabo Frio platform.

CHILE AND ARGENTINA

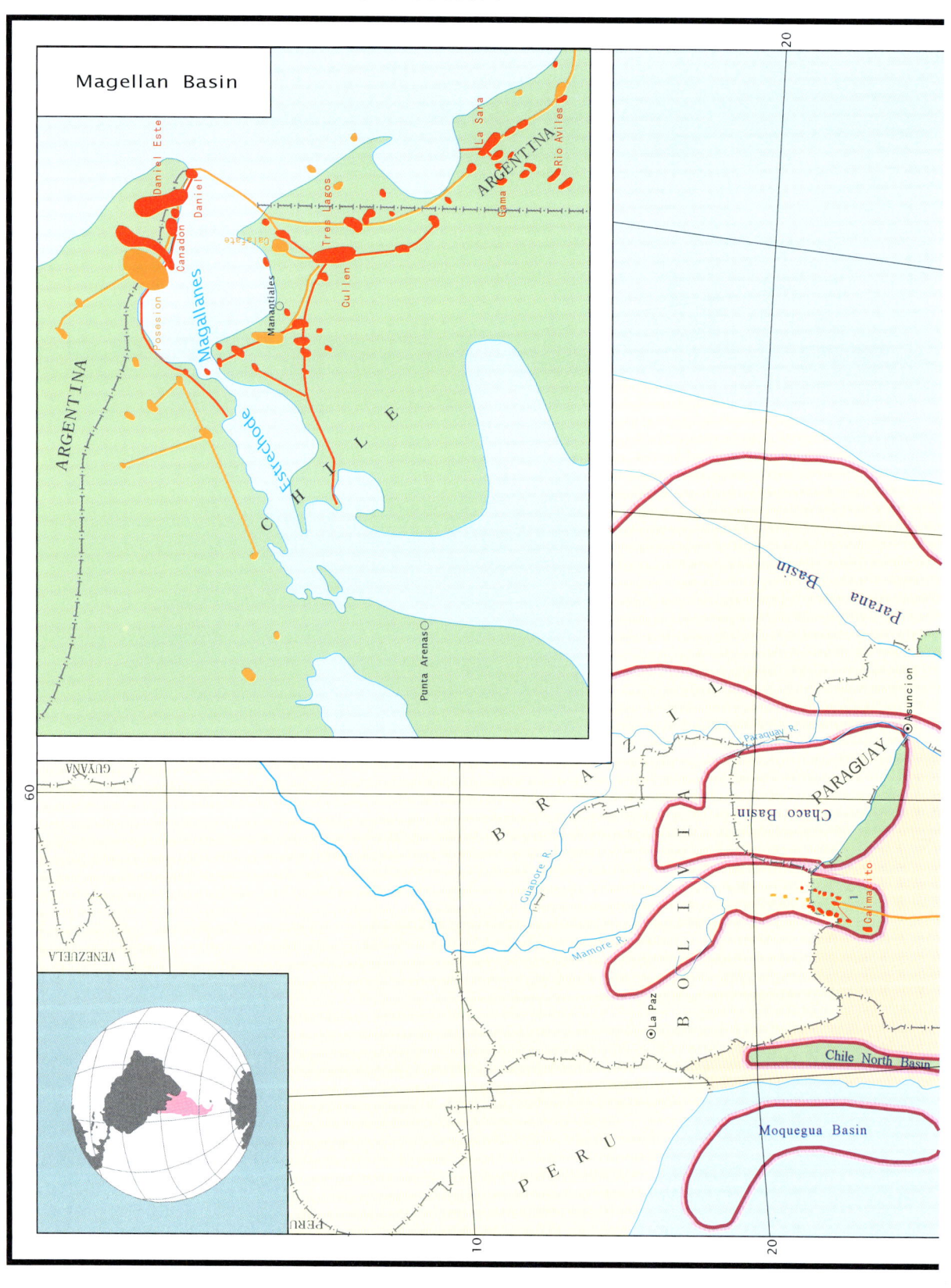

CHAPTER 100

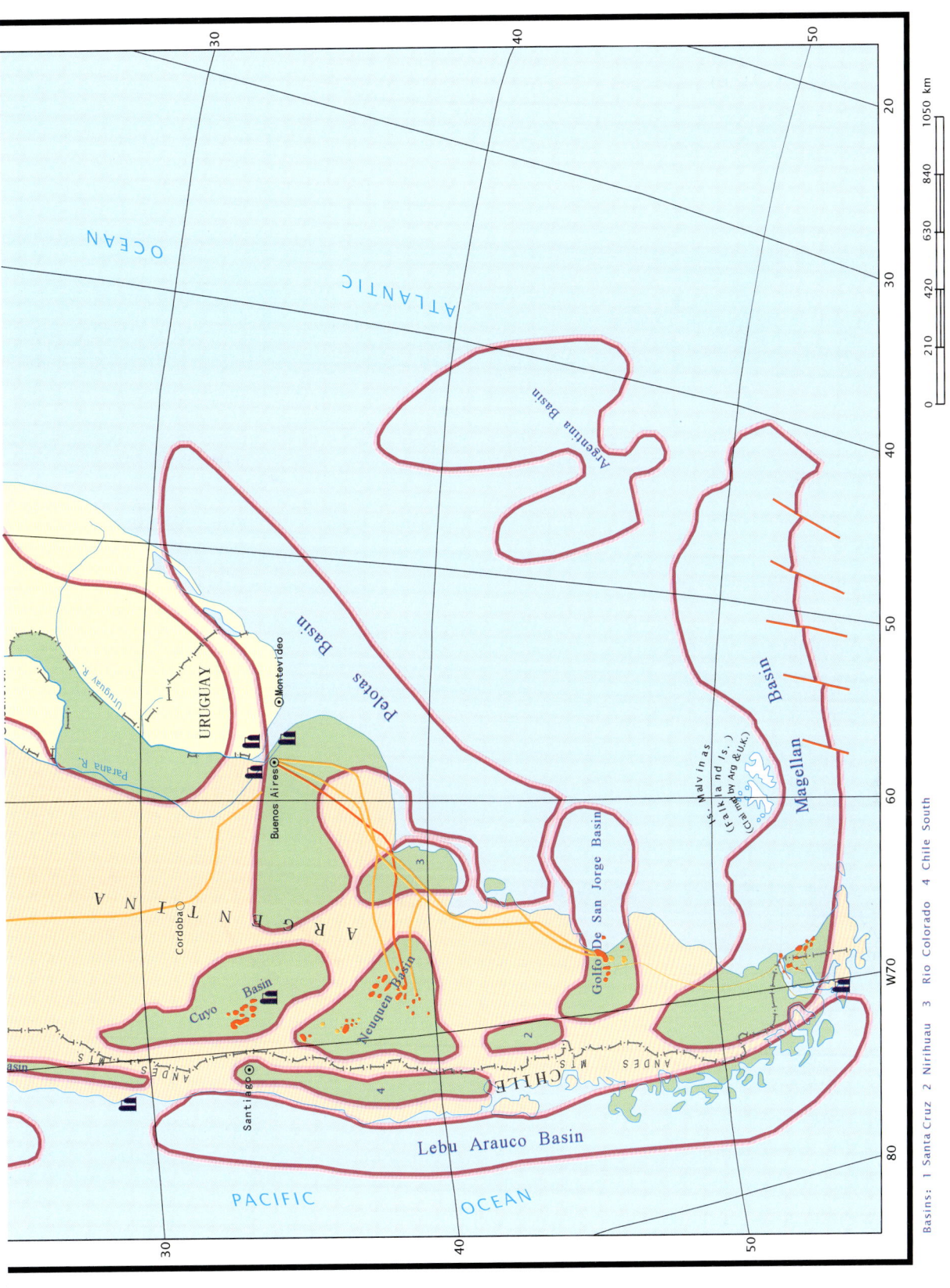

100   Chile and Argentina

Geography

Chile and Argentina are located in the south of South America, bounded by the Atlantic Ocean in the east (Argentina) and the Pacific Ocean in the west (Chile). The narrow coastal Andes Mountain Range runs along the Pacific Coast.

History

Oil exploration started in Argentina earlier than it did in Chile. Surface oil seepages were known to occur widely in Argentina. Oil exploration started in Argentina in 1865 and in Chile in 1909. The first oil field was discovered in Argentina in 1907, in Chile in 1912. Oil exploration continues in both countries on land and offshore. Argentina is one of the major oil-producing countries in South America.

Regional geology

Fifteen sedimentary basins are known to exist in, or partly within, both countries ranging in area from $30,000\,km^2$ to 1.03 million km^2, filled by mainly Mesozoic sediments. The Golfo De San Jorge Basin was originally considered the best prospect for offshore production. This conclusion was based on the fact that the Comodoro Rivadavia oil fields, which are known to extend offshore, are the largest producers in Argentina.

Magellan Basin is bounded to the north by the Deseado Arch, which extends offshore and may link up with the Malvinas Islands, and on the west and south by the Andes Mountains. No wells have been drilled offshore in the Atlantic. Production was found in this basin in 1950, and later oil was discovered on the Argentine side. This primarily marine sandstone is unusual in its source in that it is derived directly from the underlying Tobifera Formation, a Jurassic tuff with a high content of medium grained quartz and almost no mafic minerals.

The last of the three offshore Argentine basins to be mentioned is the Malvinas or Falkland Islands Basin. Extrapolation of the sedimentary conditions of Tierra del Fuego into this basement area would be incorrect and the succession is apparently entirely marine, made up of marls and calcareous ooze interbedded with clays and chalks. The Malvinas Basin is the most favourable for hydrocarbon accumulation. Anticlinal structures may be present and the depth of burial (3.5 km) would favour oil generation.

The onshore Cuyo and Neuquen Basins lie in central Argentina. Both of them have been actively exploited for 60 years. The first tectonic event in the formation of these basins is the separation of the two massifs, the Pampian to the north and the Patagonian to the south. These basically granitic masses have been variously dated from Precambrian to Carboniferous. The massif is overlain in part by Devonian rocks and therefore is at least early Palaeozoic in age.

Argentina   The country occupies most of the southern part of the South American continent. It is bounded by the Atlantic Ocean in the east, has an area of 2.78 million km^2 with a population of

World Atlas of Oil And Gas Basins, First Edition. Li Guoyu.
© 2011 John Wiley & Sons, Ltd. Published 2011 by John Wiley & Sons, Ltd.

Table 100.1 Major oil and gas basins of Argentina and Chile

Country	Basin	Area (thousand km²)	Major sedimentary rock Age	Major sedimentary rock Thickness(m)	Reservoir Age	Reservoir Lithology	Remark
Chile	Pelotas	1180	Mz, Kz				Extends into Brazil, Argentina
	Chile North	66	J–R				
	Chile South	75	Mz–Kz				
	Lebu Arauco	450	Mz, Kz	4000	K	Sandstone	
	Moquegua	220	Kz				
Argentina	Argentina	440					
	Cuyo	150	T–R	3500	T, R	Sandstone	
	Golfo de San Jorge	180	J–Kz	7000	J, K, R	Sandstone	
	Magellan	1030	T–Kz	5500	J, K	Sandstone	Extends into Chile
	Neuquen	200	Mz–Kz	4000	J, K	Sandstone	Extends into Chile
	Nirihuau	830	Mz–Kz				
	Rio Colorado	70	Mz–Kz				

38.59 million. Seven sedimentary basins are located in the country: Argentina, Cuyo, Golfo de San Jorge, Magellan, Neuquen, Nirihuau and Rio Colorado, filled by Mesozoic and Cenozoic sediments with thicknesses ranging between 4000 and 7000 m. Argentina is the country where the earliest oil discoveries were made in South America in 1886, following several surface oil seepages. The wide distribution of sedimentary basins shows high petroleum potential. By the end of 2008 the proven oil reserves were 358 million t with annual oil production of 30.5 million t. Proven gas reserves were 441 billion m³ and annual gas production was 42 billion m³.

Chile This narrow coastal country is located in the southwest of South America, bordering the South Atlantic Ocean and the South Pacific Ocean. The country has an area of 756,628 km² with a population of 16.2 million. Five sedimentary basins are known to lie in the country: Pelotas, Chile North, Chile South, Lebu Arauco and Moquegua, all filled by Mesozoic rocks with thickness reaching more than 4000 m. The offshore basins have high petroleum potential. Oil exploration started in 1909 and the first oil field was discovered in 1912. By the end of 2008 the proven oil reserves were 20.5 million t and oil production annually was 115,000 t. Proven gas reserves stood at 97 billion m³.

Part VII
Australasia and the Poles

AUSTRALIA AND PAPUA NEW GUINEA

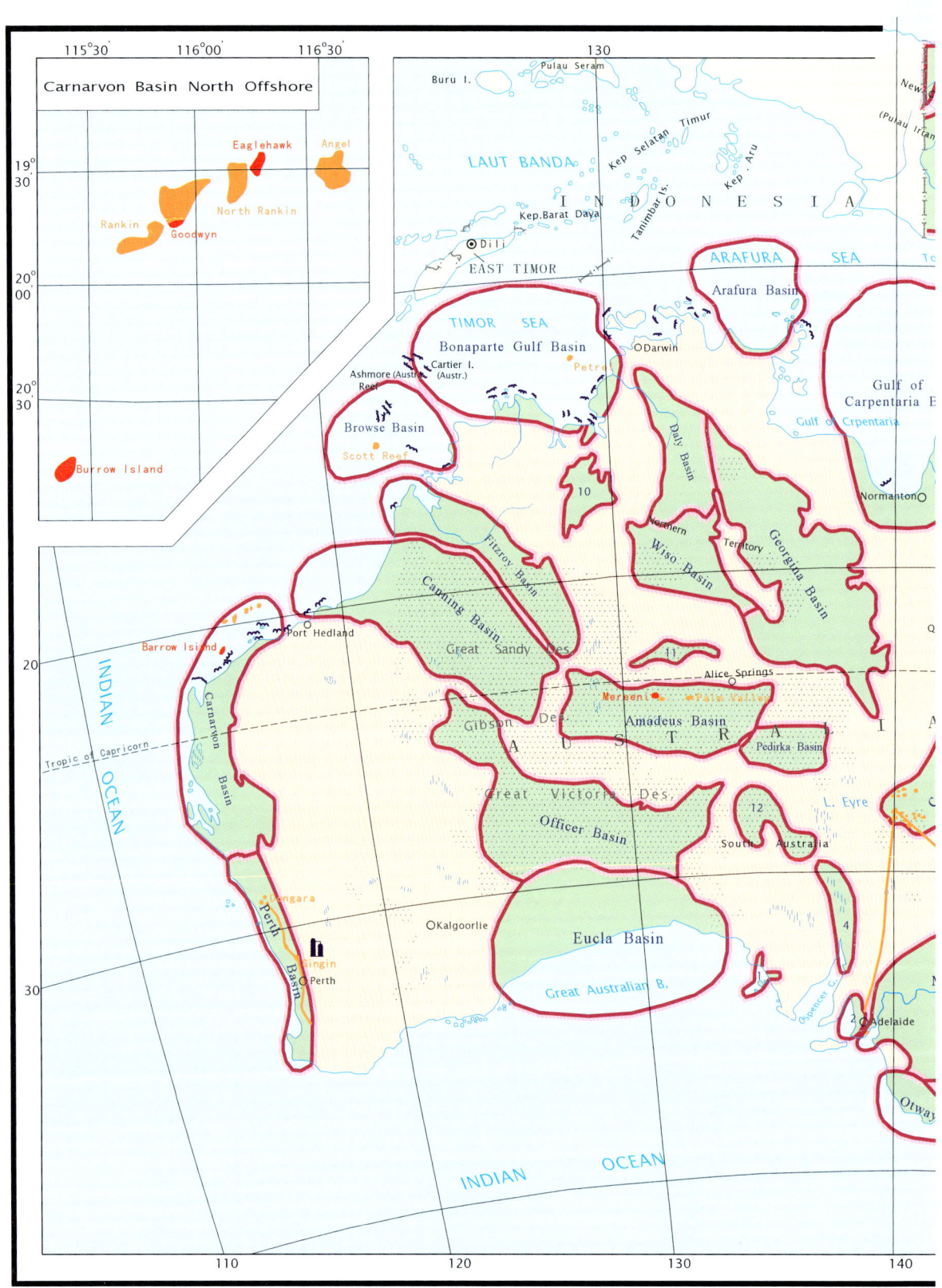

Basin: 1 Polda 2 St.Vincent 3 Tasmania 4 Pirie Torrens 5 Oxley 6 Clarence 7 Maryborough 8 Yareol 9 Laura 10 Ord 11 Ngalia 12 Arckaringa

CHAPTER 101

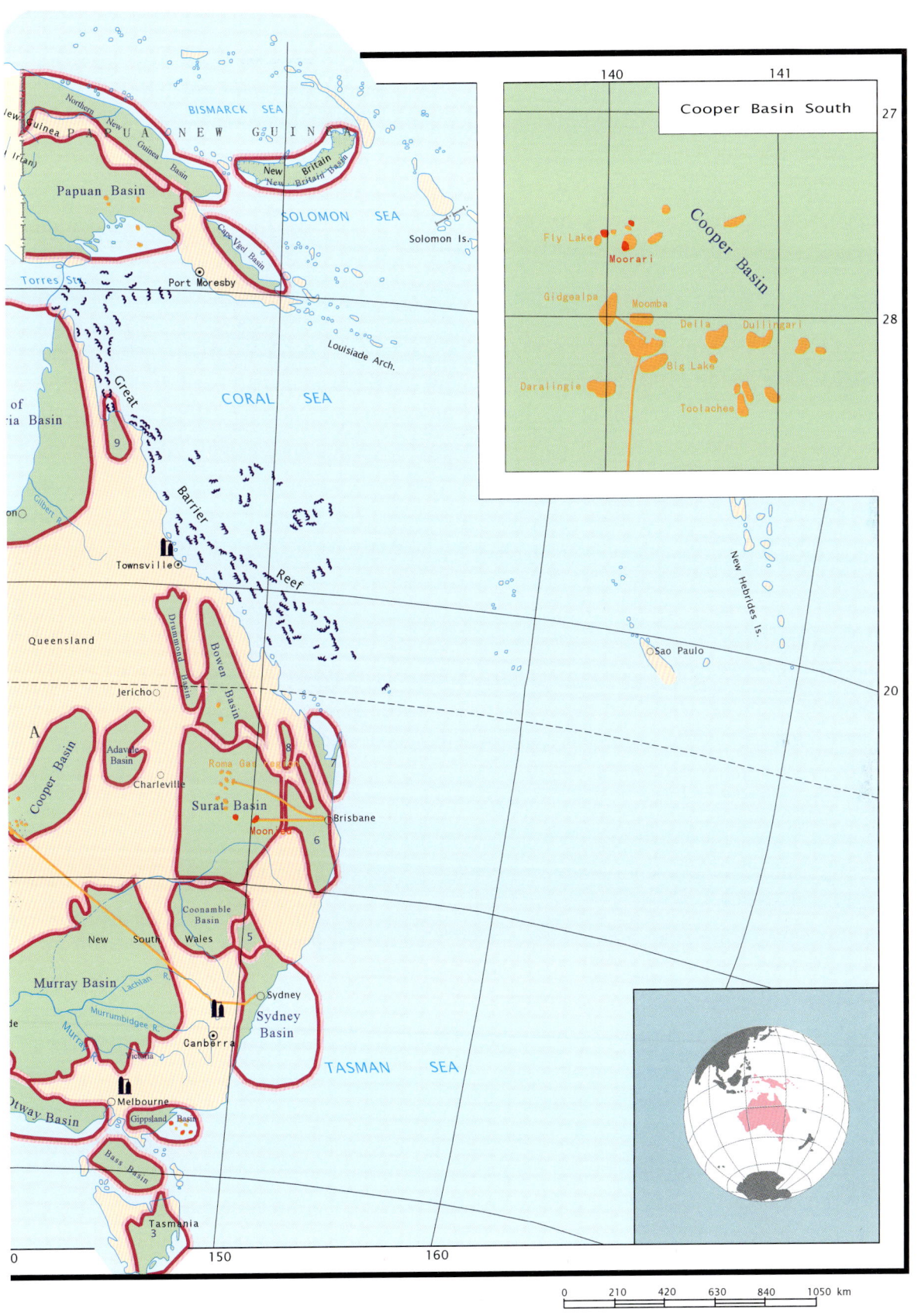

101 Australia and Papua New Guinea

Geography

The continent of Australia is located in the bottom half of the southern hemisphere, between the Pacific Ocean and the Indian Ocean, facing Antarctica to the south. Combined, the deserts of central Australia form one of the largest desert in the world, comparable to the Sahara Desert in Africa. Papua New Guinea is part of the Indonesian islands to the north of Australia, in the South Pacific Ocean.

History

The start of oil and gas exploration was very late compared with other parts of the world. A gas-in-water well was drilled in 1900, and up to 1960 more than 500 wells had been drilled without any large-scale discovery. As a result of this limited success, impetus was provided for Australia to develop its oil shale industry at a faster pace than would have happened had the conventional oil drilling been successful. Just after this strategic move, oil exploration shifted to Bass Strait with the discovery of the Kingfish oil field in the Gippsland Basin. This marked the commencement of the development of the petroleum industry in Australia when a new programme of oil exploration started in 1967.

Regional geology

The region of Australia and Oceania principally represents a part of the Indo-Australian Plate. The Pacific Plate is being subducted under the Indo-Australian Plate throughout their contact. Only the north and east of the continent represent an active margin. The Australian continent is formed from two major tectonic elements. These are the ancient Australian platform and Tasmanian Orogenic Belt. The Australian platform (craton) occupies western and central parts of the continent, the southern half of New Guinea and the area of ocean between them (Arafura Sea). Early Precambrian basement outcrops are observed in one half of the area of the craton. The sedimentary cover succession begins with Upper Proterozoic deposits. Palaeozoic rocks are represented by clayey-carbonate strata, Mesozoic and Cenozoic by sandy–clayey strata with frequent carbonaceous intervals. The Tasmanian Orogenic Belt contains several depressions and a parallel foredeep. Forty-three sedimentary basins have been identified with possible hydrocarbon potential within the region of Australia and Oceania, among which 14 oil and gas fields have been discovered. Details of 18 of the Australian sedimentary basins are listed in Table 101.1.

Australia With an area of 7.6 million km^2, Australia is the sixth largest country in the world. Total population is approximately 20 million. Between 1967 and 1971 quite a few large oil and gas fields were discovered, the most notable being Malin, Kingfish, Halibut (in Gippsland Basin) and North Rankin (in Carnarvon Basin North Offshore). Oil production reached 35 million t in 2000 and 22 million t in 2007. At year end 2008 oil reserves stood at 205 million t and gas

World Atlas of Oil And Gas Basins, First Edition. Li Guoyu.
© 2011 John Wiley & Sons, Ltd. Published 2011 by John Wiley & Sons, Ltd.

Table 101.1 Major sedimentary basins of Australia

Basin	Area (thousand km²)	Major sedimentary rock		Reservoir	
		Age	Thickness (m)	Age	Lithology
Amadeus	155	Pt–Kz	9000	O, D, C	Sandstone, siltstone, limestone
Arafura	130	Mz, Kz			
Bonaparte Gulf	350	Pz, Mz, Kz	9000	C–K, R	Sandstone, limestone, reef
Browse	130	Pz, Mz, Kz	7000	P, J	Sandstone
Carnarvon	220	O–Kz	12,000	T, J, K, D	Sandstone
Canning	360	Mz, Kz		D, C	Reef, sandstone
Cooper	127	P-K	4500	P	Sandstone
Eucla	440	Mz, Kz			
Fitzroy	160	Pz, Mz			
Georgina	300	Pz, Mz			
Gippsland	66	K–Kz	7000	K–E	Sandstone
Gulf of Carpentaria	480	Mz, Kz	2400		
Murray	390	Pz, Kz	2500		
Officer	280	Pz			
Otway	90	K, Kz	6000	K, R	Sandstone
Perth	85	O–Kz	11,000	P, T, J, K	Sandstone
Surat	220	J, K	2400	J	Sandstone
Sydney	176	Mz		T	Sandstone

Table 101.2 Major sedimentary basins of Papua New Guinea

Basin	Area (thousand km²)	Major sedimentary rock		Reservoir	
		Age	Thickness (m)	Age	Lithology
Cape Vgel	41	R	4000	N_1	Clastic rock
Northern New Guinea	110	K, R	9000	N_1	Breccia
Pupan	260	Mz, Kz	6500	J, K, N_1	Sandstone, limestone
New Britain	70	Mz, Kz	7000	R, K	Sandstone, limestone

reserve at 849 billion m³. Annual oil production at the same time was 22.6 million t and the corresponding volume of gas production was 2.9 billion m³.

Papua New Guinea The country is located in the South Pacific and comprises about 600 islands, with a total area of 462,840 km² and a population of 5.9 million. Four sedimentary basins are known to occupy the country (Table 101.2), filled by Mesozoic and Cenozoic sediments with thicknesses of 4000 m to 9000 m. Oil exploration was initiated in 1967. The first gas condensate field was discovered in Pasca reef in 1968. By the end of 2008 the proven oil reserves were 12 million t, annual oil production 2.5 million t, proven gas reserves 226.5 billion m³ and annual gas production 168 million m³.

GIPPSLAND BASIN

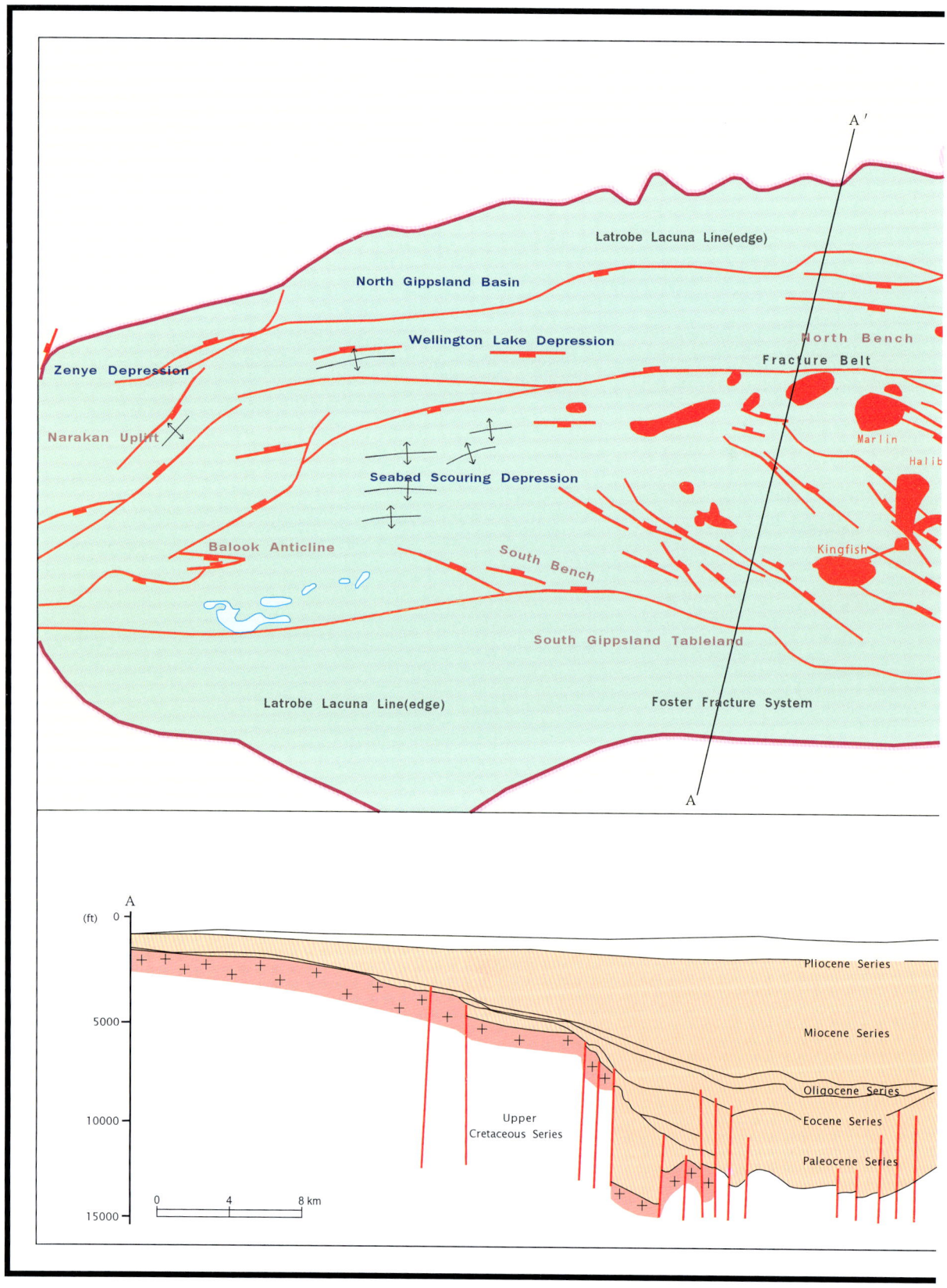

CHAPTER 102

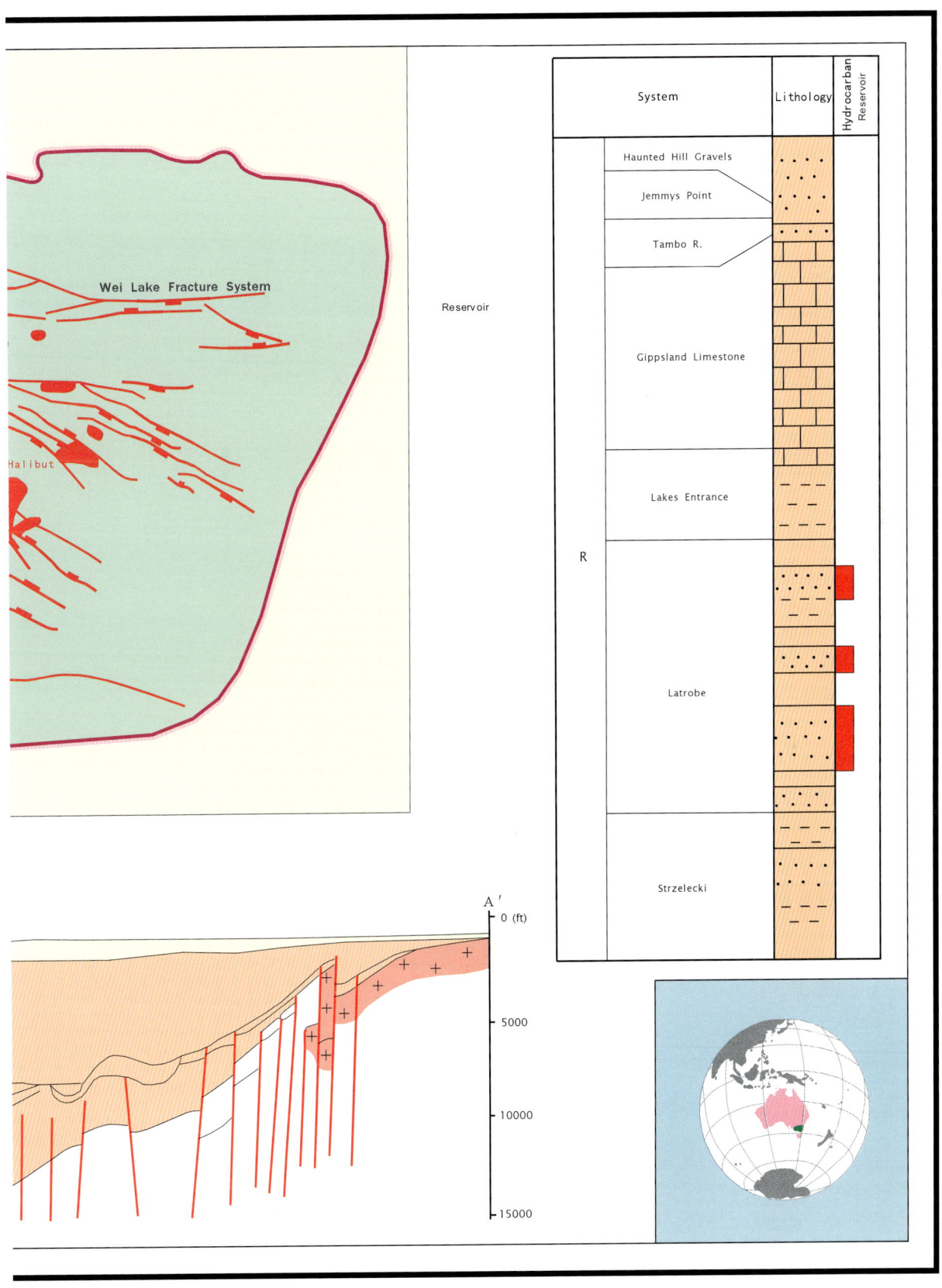

102 Gippsland Basin

Geography

Gippsland Basin occupies part of Victoria and the northeastern part of the Bass Strait, offshore Victoria. It has an area of 644,000 km^2, including 50,000 km^2 offshore, and is a major oil producing area in Australia.

History

In 1924 oil was found in the Lakes Entrance area of Victoria, which provided the impetus for much exploration activity onshore, but without results. During 1962–1963 geophysical survey commenced offshore and several large structures were mapped. Marlin gas field was the first discovered but continuing exploration discovered the Kingfish and Halibut oil fields (Fig. 102.1). The giant Halibut oil field is a stratigraphic trap with an area of 26.8 km^2 in a Tertiary sandstone reservoir at a depth of 2300 m, with a thickness of 62 m, porosity of 18–22 per cent, permeability of 1600 mD and recoverable oil reserves of 160 million t.

Regional geology

The geological development of Gippsland Basin can be divided into five stages:
1. initiation of basement rifting;
2. lower Cretaceous infilling of the rift;
3. rift failure during break up of Gondwanaland;
4. development of Gippsland Basin;
5. Tertiary infilling of the basin.

The sedimentary cover comprises Meso-Cenozoic strata with thicknesses in excess of 7000 m of mainly terrigenous deposits.

The principal source rocks are carbonaceous shales and coals of the Latrobe and Golden Beach Groups. Most of the oil was generated from terrestrial source rocks. Most oil generation took place in the upper part of the oil window, in rocks of Campanian age or older and at a present-day depth of 4 km to 5 km. Gases were generated from an overmature source, with the main phase of generation at depths of 5 km to 6 km. Hydrocarbons have been generated from latest Cretaceous times to the present day. A large section of the Golden Beach Group and the lower part of the Latrobe Group are mature for oil over much of the central basin. The Latrobe Group contains abundant sandstone reservoirs of both marginal marine and fluvial origin with porosities of 18–20 per cent and permeabilities of 1000–1600 mD. The non-marine reservoir sandstones of the Golden Beach Group have moderate reservoir potential.

Both Latrobe Group and Golden Beach sandstone porosities show a marked decrease with burial depth. Diagenesis causes local, significant degradation of Latrobe sandstone porosity and permeability. Authigenic kaolin has also been shown to degrade the reservoir quality of sandstones, particularly in the Golden Beach Group. The Lakes Entrance Formation provides a basin-wide seal for the Top Latrobe reservoirs. The best seals for Intra-Latrobe and Golden Beach Group

World Atlas of Oil And Gas Basins, First Edition. Li Guoyu.
© 2011 John Wiley & Sons, Ltd. Published 2011 by John Wiley & Sons, Ltd.

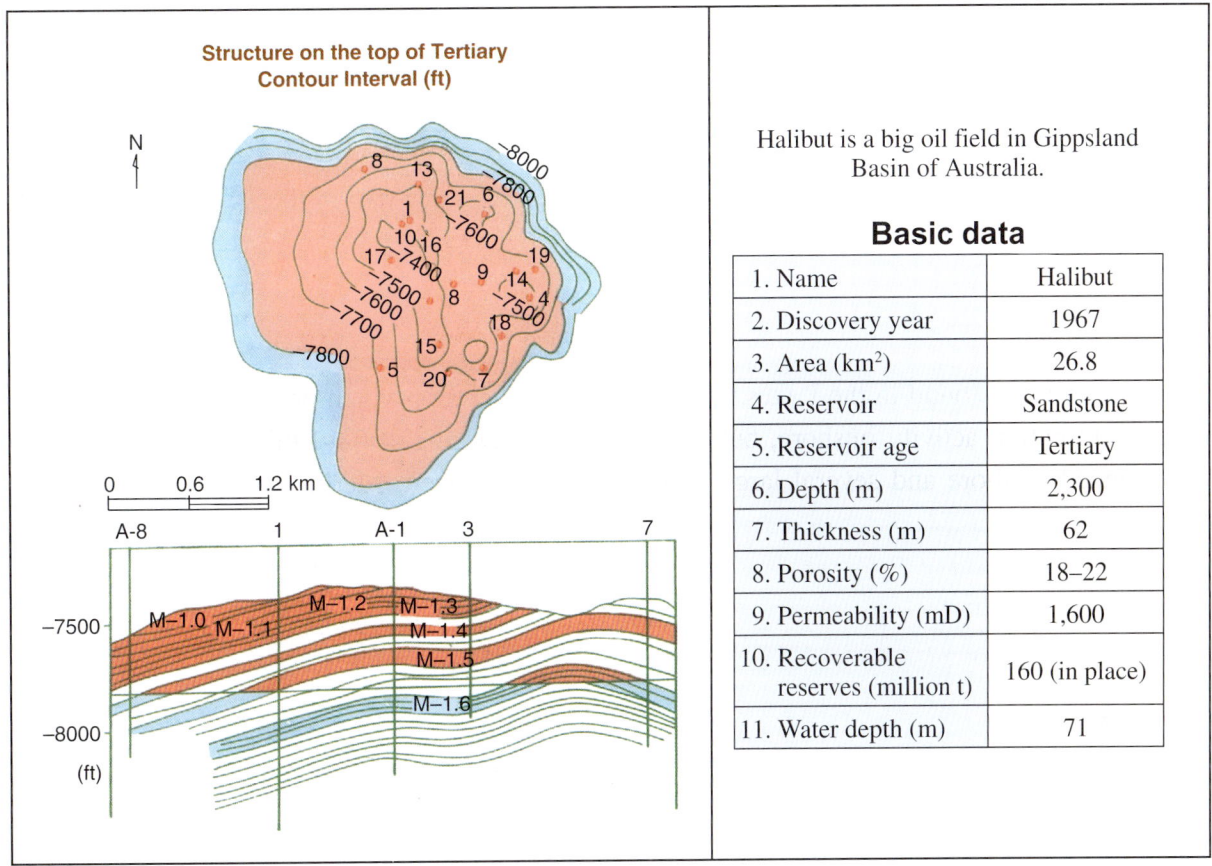

Fig. 102.1 Details of the Halibut oil field.

accumulations are semi-regional marine shale/mudstone and lacustrine shale. Coastal plain shale, coal and volcanics are also proven seals. Fault seals are provided by juxtaposed shale and also by fault plane gouge, especially in the smaller reservoirs.

Top Latrobe plays account for over 88 per cent of all known hydrocarbons. The plays are hence predominantly structural and fault seal-dependent, comprising faulted anticlines and fault-dependent closures. The source and reservoir rocks of the Golden Beach and Latrobe Groups, sealed by the overlying permeability barrier of the Seaspray Group comprise the sole petroleum system present within the basin (Halbouty, 1970).

Exploration of the Gippsland Basin has, by Australian standards, reached a mature stage for all three of its major plays (Top Latrobe, Intra-Latrobe and Golden Beach). However, all three plays have undrilled prospects and further discoveries remain to be made; the Intra-Latrobe is probably the least risky. Significant accumulations remain to be found in stratigraphic traps.

NEW ZEALAND, SAMOA, FIJI AND TONGA

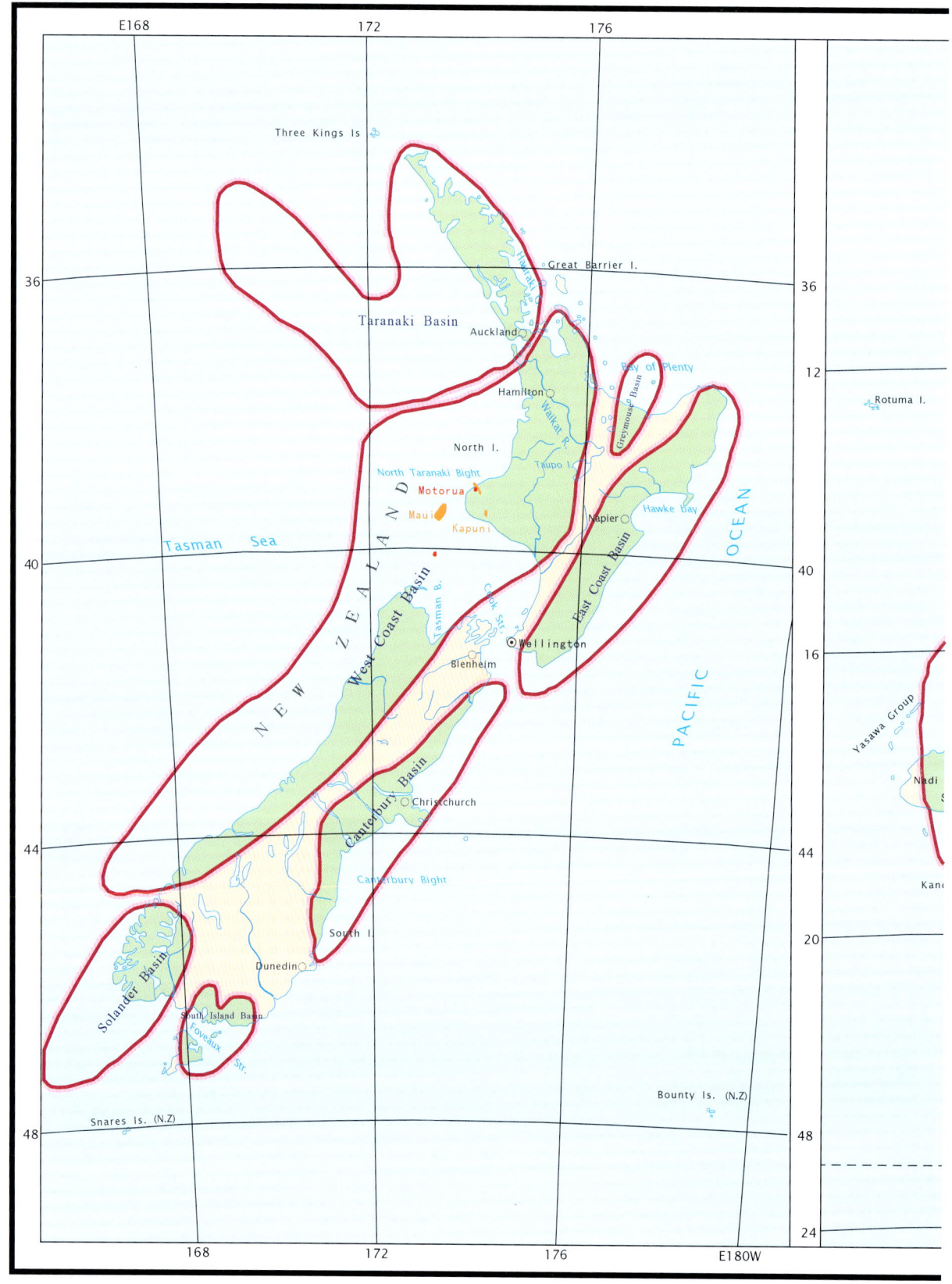

CHAPTER 103

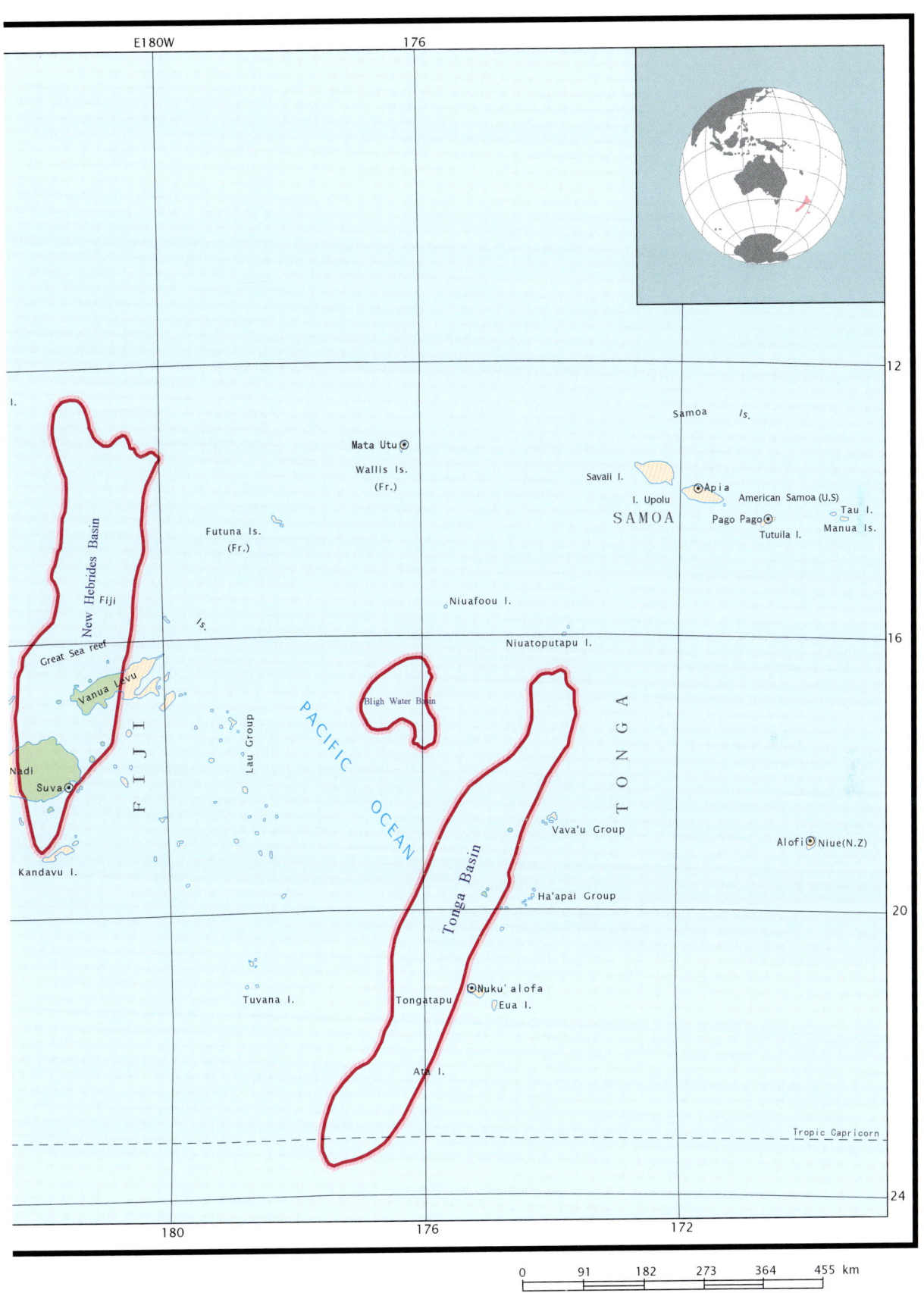

103 New Zealand, Samoa, Fiji and Tonga

Geography

New Zealand, Samoa, Fiji and Tonga are located in the southwest region of the Pacific Ocean.

History

In 1839 oil seepages were extensively recorded as having occurred in New Zealand. Oil and gas exploration was undertaken mainly in New Zealand. At present, oil and gas production activities are ongoing only in New Zealand.

Regional geology

Tectonically Oceania comprises the Pacific Orogenic Belt and young platforms in the form of micro-continents. The Pacific Orogenic Belt extends to the Coral, Fiji, Tasmanian and New Guinea Seas and it forms the eastern half of the Indo-Australian Plate. Within the limits of the region, the belt comprises two island arc systems – older inner and younger outer. The eastern part of the New Guinea Island together with the neighbouring islands form the Cenozoic New Guinea Island Arc. New Britain and New Ireland are recent island arcs along which a Cenozoic foredeep filled with Oligocene–Quaternary molasses is located on the onshore and shelf portion of New Guinea. The North and South islands of New Zealand and a number of other small islands form the Cenozoic New Zealand Arc. Tectonically, it is represented by the Central Anticlinorium, northwest of which the Taranaki (North Island) and West Coast (South Island) Troughs are located, whereas in the southeast there are the East Coast and Canterbury Troughs. The Central Anticlinorium of New Zealand includes a number of depressions such as Murchison, Fairfax and Bayley, which are filled with Mesozoic and Cenozoic rocks. A series of depressions have also been distinguished along the New Caledonian island Arc of Cenozoic. Along the outer side of recent island arcs such as Solomon, Fiji, Tonga-Kermadec and Macquarie, a deep-water trench is traced.

The sedimentary basins have an area of 15,000–30,000 km^2 and their sedimentary cover comprises Cretaceous and Tertiary strata. Source rocks are Cretaceous and Palaeocene coals and possibly marine shales. Reservoirs are Paleocene and Oligocene marine sandstones and seals are Oligocene and Miocene mudrocks. Traps are anticlines and faulted blocks.

New Zealand

New Zealand is located in the southwest Pacific Ocean and comprises more than 1000 islands, many of which are uninhabited. It has a total area of 270,000 km^2 and a population of 4.18 million. The Alps Mountains lie on the west side of South Island. Seven sedimentary basins exist all filled by Cretaceous and Tertiary rocks with thicknesses ranging between 5500 and 9000 m (Table 103.1). The areas of these basins are between 150,000 km^2 and 300,000 km^2. The largest is the West Coast Basin with an area of 300 km^2. Seepages were discovered in 1839. During 1866–1936, 130 exploration wells were completed, with oil and gas occurrence noted in only a few of them. In 1959, the Kapuni gas condensate field was discovered in the vicinity of a volcanic cone. Ten years later in 1969, a significant gas field was discovered in the

World Atlas of Oil And Gas Basins, First Edition. Li Guoyu.
© 2011 John Wiley & Sons, Ltd. Published 2011 by John Wiley & Sons, Ltd.

Table 103.1 Major sedimentary basins of New Zealand

Basin	Area (thousand km²)	Major sedimentary rock		Reservoir	
		Age	Thickness (m)	Age	Lithology
Canterbury	50	K–R	6000		
East Coast	80	K–R	9000	E_2	Sandstone
Greymouse	10	K–R	9000	E_2	
Solander	50	Kz			
South Island	15	K–R	9000	K, R	
Taranaki	123	Mz, Kz	6000	E_2, N_2	Sandstone
West Coast	300	K–R	5500	P–K	Sandstone

West Coast Basin. At present, a total of 13 oil and gas fields exist. By the end of 2008 the proven oil reserves were 8.2 million t with annual oil production standing at 2800,000 t, while the proven gas reserves were 33.9 billion m³ and annual gas production was 4.3 billion m³.

Samoa Samoa is located in the southwest Pacific Ocean and consists of eight islands with numerous volcanic cones. It has an area of 2934 km² and a population of 177,000.

Fiji Fiji is located in the southwest of the Pacific Ocean and comprises 330 islands. It has an area of 18,272 km² and a population of 840,000. Fiji lies within the New Hebrides Basin with an area of 120,000 km² filled with Cenozoic rocks to a thickness of 2700 m.

Tonga Tonga is located in the southwest Pacific Ocean and consists of 172 islands with an area of 747 km² and a population of 110,000. Within the curtilage of Tonga lie two sedimentary basins: Bligh Water Basin with an area of 16,000 km² and the Tonga Basin with an area of 100,000 km². Both basins are filled by Cenozoic sediments. On the Tongatapu Island, oil seepages occurred in 1968 with high sulphur content of 3.7 per cent. Later in 1971 and 1978, some exploratory drilling was carried out without any notable results.

ANTARCTICA

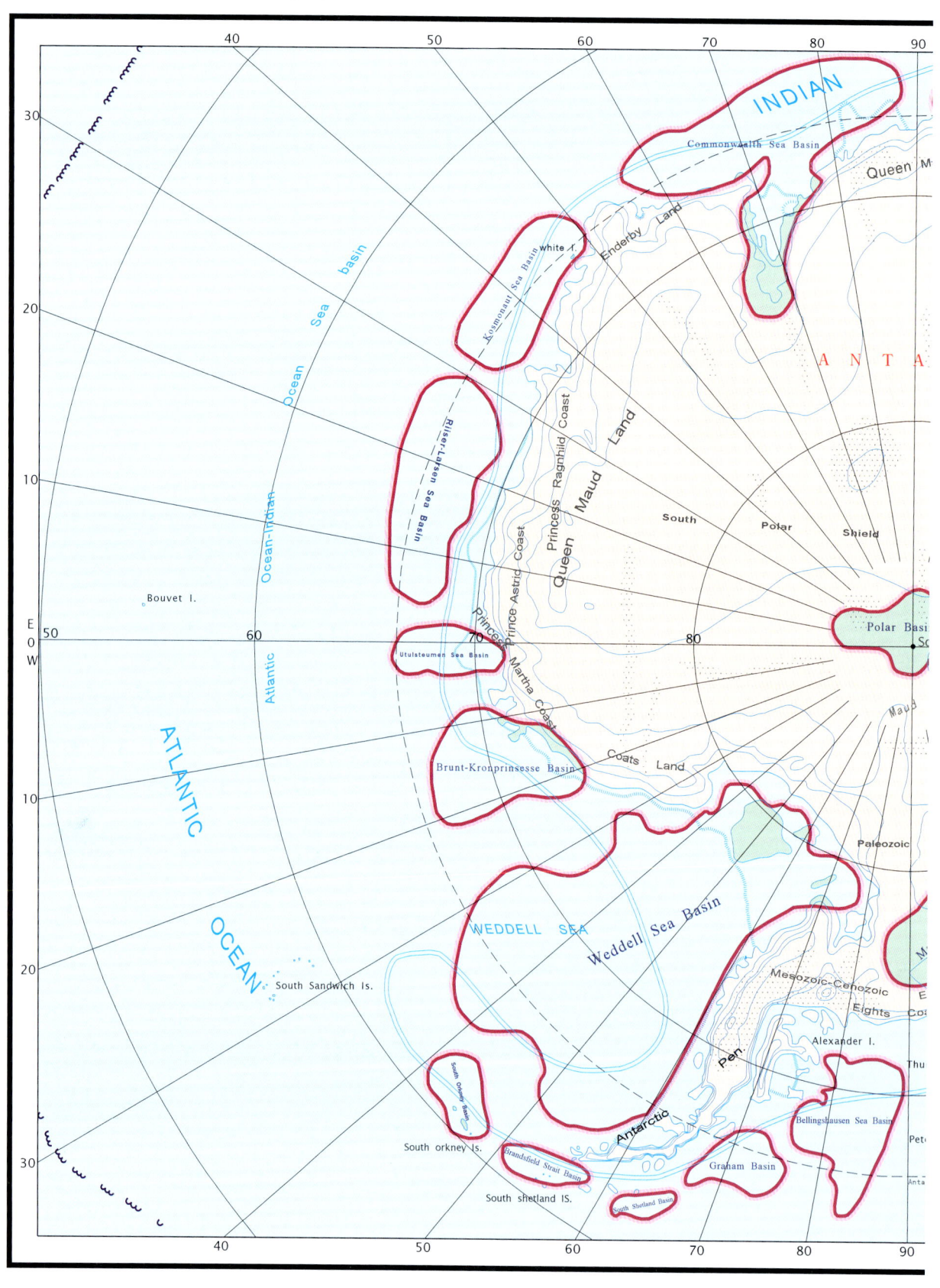

CHAPTER 104

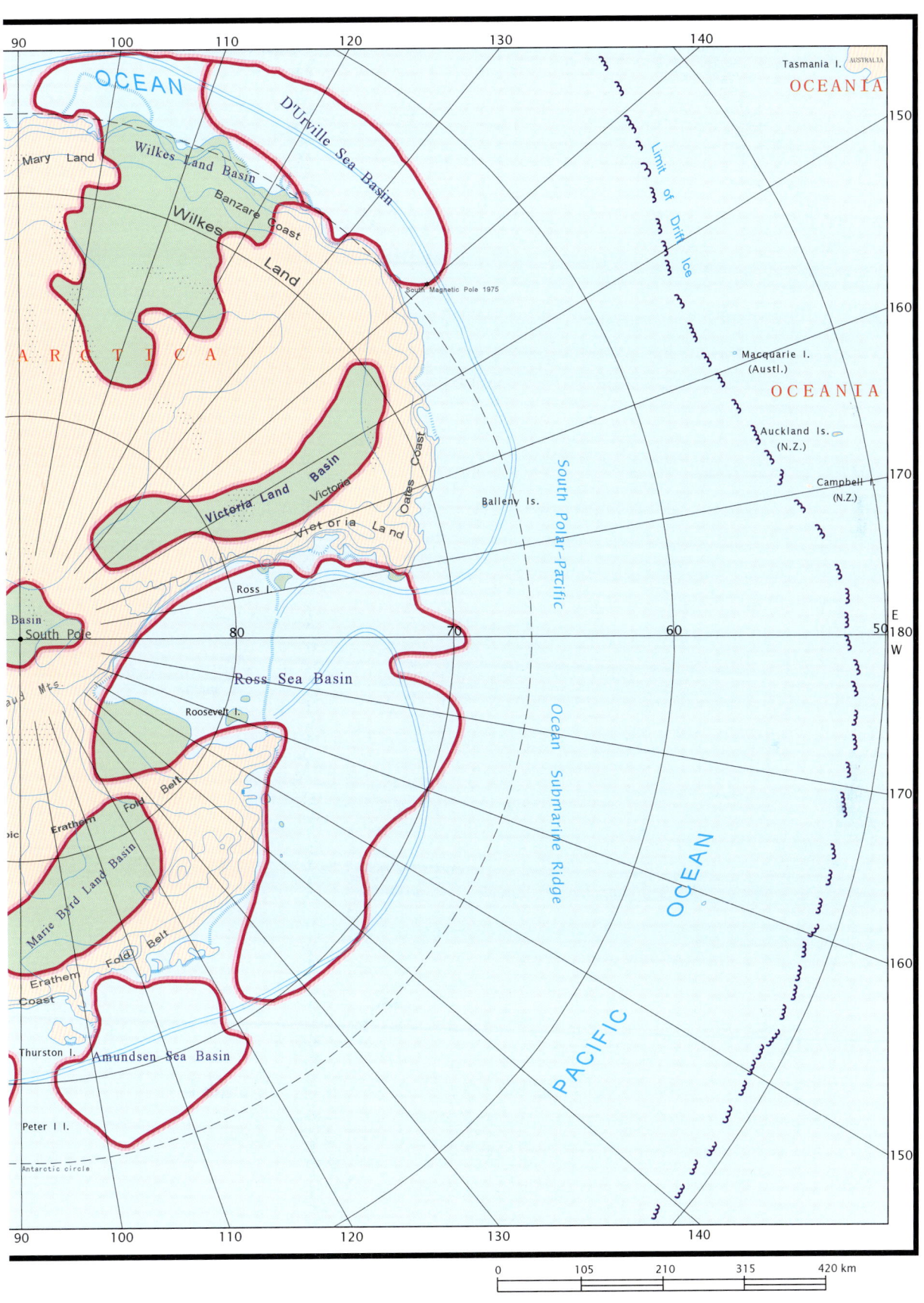

104 Antarctica

Geography

The continent is located at the South Pole. Due to the permanent deep ice cover, limited oil and gas exploration activity has taken place in this region. There are also widespread international concerns regarding the environmental implications of any type of commercial oil and gas activity in this area. The continent has an area of 12 million km^2 from its South Pole to Latitude 70°S around half its perimeter. More than one third of the coastline is fringed by ice shelves or floating ice sheets, which cover another 1.4 million km^2 of area, and this does not include the annual winter pack ice. The continent has a diameter of some 4500 km. The roughly circular shape of the continent is broken by the Ross and Weddell Seas and the Antarctic Peninsula. It is also traversed by the Transantarctic Mountains, a fault-block system extending more than 300 km from West Ross Ice Shelf to Filchner Ice Shelf.

Regional geology

Antarctica is a part of the lithospheric plate of the same name and consists of land, shelf and continental scarp with continental type of crust, and the oceanic crust. The continental margin is passive practically throughout the entire periphery of the continent; it is active only in the northwest.

The Antarctic Plate is separated from the Indo-Australian, African and Pacific Plates mid-ocean spreading ridges. In the northwestern part it is in collisional contact with the Scotia Plate. The continent is built up from two major tectonic elements: old platform (Antarctic) and orogenic belts of different ages (Trans-Antarctic and West Antarctic). The Antarctic platform occupies the eastern half of the continent. The platform basement is ancient, being predominantly Archaean–Early Proterozoic. The sedimentary cover comprises Upper Proterozoic and Lower Palaeozoic sedimentary–volcanic rocks up to 6–7 km in thickness. The succession also includes Permian terrigenous rocks Jurassic basalts, and Mesozoic and Cenozoic deposits. The eastern part of the platform is composed essentially of Mesozoic–Cenozoic rocks. In the west, the platform complexes have been eroded. Vast syneclises are developed within the platform.

There are three orogenic belts in Antarctic: the Trans-Antarctic, the West Antarctic and the Pacific. The Trans-Antarctic Orogenic Belt occurs in the west and was folded during Early Palaeozoic. Depressions and grabens are widely developed. The orogenic belt is overlain by Middle Palaeozoic–Early Mesozoic cover. The West Antarctic Orogenic Belt fringes the Pacific coast of the continent and represents a segment of the Pacific Orogenic Belt. The belt was folded during Middle Palaeozoic–Mesozoic times. Mesozoic–Cenozoic troughs and fore-arcs have been identified within the belt. Within the zone of transition from continent to ocean, pericontinental depressions are widely developed along the passive continental margin. They occupy the land, shelf, continental slope and base of slope portions (Vysotsky, 1995).

World Atlas of Oil And Gas Basins, First Edition. Li Guoyu.
© 2011 John Wiley & Sons, Ltd. Published 2011 by John Wiley & Sons, Ltd.

Table 104.1 Major sedimentary basins of Antarctica

Basin	Area (thousand km²)	Major sedimentary rock	
		Age	Thickness (m)
Amundsen Sea	90	Kz	
Bellingshausen Sea	50	Kz	
Brandsfield Strait	15		
Brunt-Kronprinsesse	40		
Commonwealth Sea	90		
D'urville Sea	80	Kz	
Graham	20		
Kosmonaut Sea	50		
Marie Byrd Land	90	Mz, KZ	5000
Polar	25	Pz, Mz	
Riiser-Larsen Sea	70		
Ross Sea	330	Kz	3000–14,000
South Orkney	17	Kz	2000
South Shetland	15		
Utulsteumen Sea	20		
Victoria Land	70		
Weddell Sea	270	Kz, Mz, Pz	14,000–15,000
Wilkes Land	160	Pt, Mz	

Sedimentary basins

Antarctica has been very poorly studied. The continental portion is covered by ice. The territory and water areas of Antarctica have been investigated mainly by geophysical methods. About nine wells have been drilled there. Within Antarctica and the adjacent seas, 18 sedimentary basins have been identified (Table 104.1), possibly with hydrocarbon potential. There are some large basins in the Ross Sea (330,000 km²), Weddell Sea (270,000 km²) and Wilkes Land (160,000 km²). The thickness of Palaeozoic–Mesozoic–Cenozoic deposits reaches 5–8 km, and in some places is up to 14 km. The 18 sedimentary basins show significant potential for finding oil and gas within the continent. This continent could be a new large oil- and gas-producing region for humankind, although the aforementioned environmental concerns must be borne in mind.

ARCTIC OCEAN

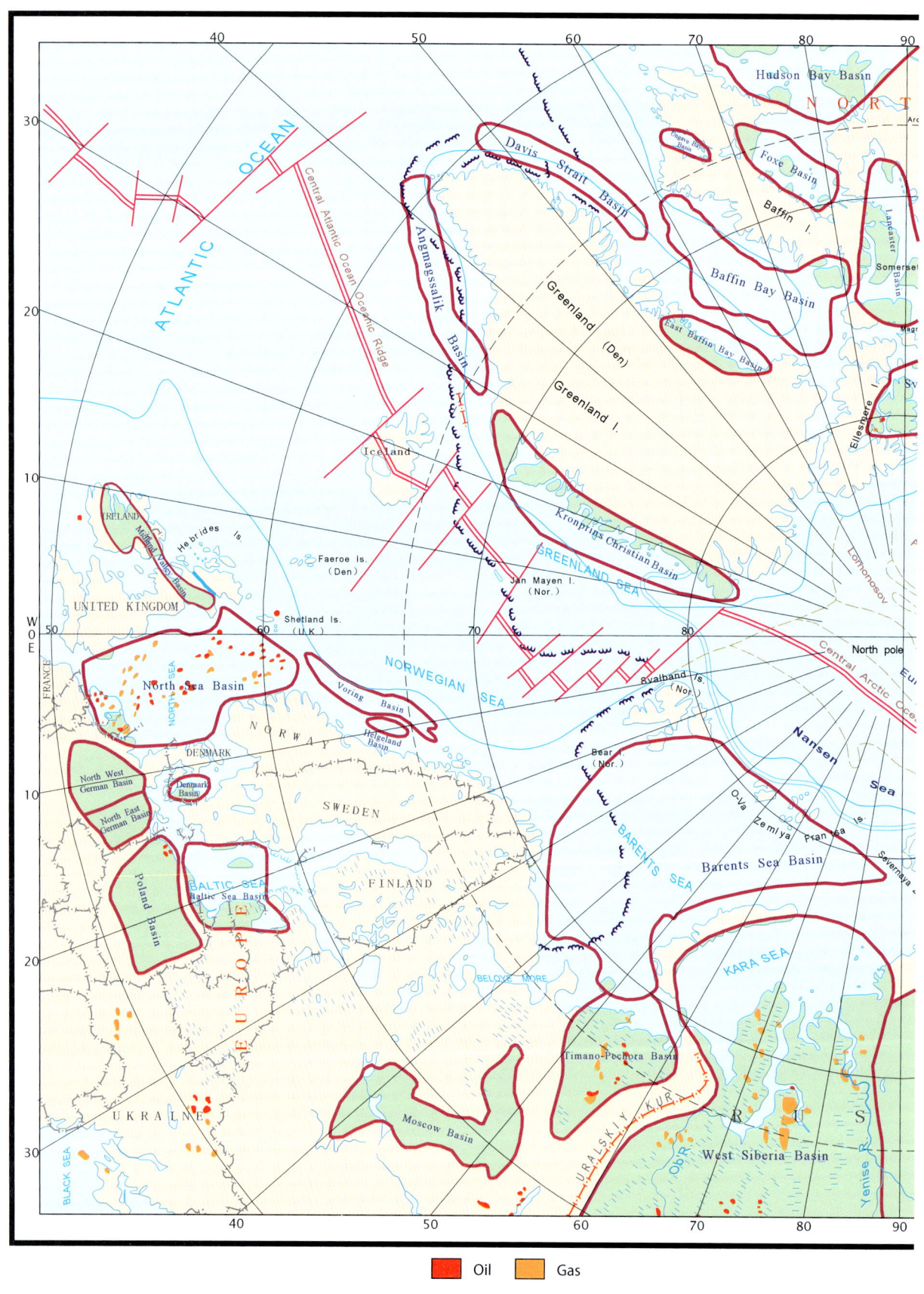

CHAPTER 105

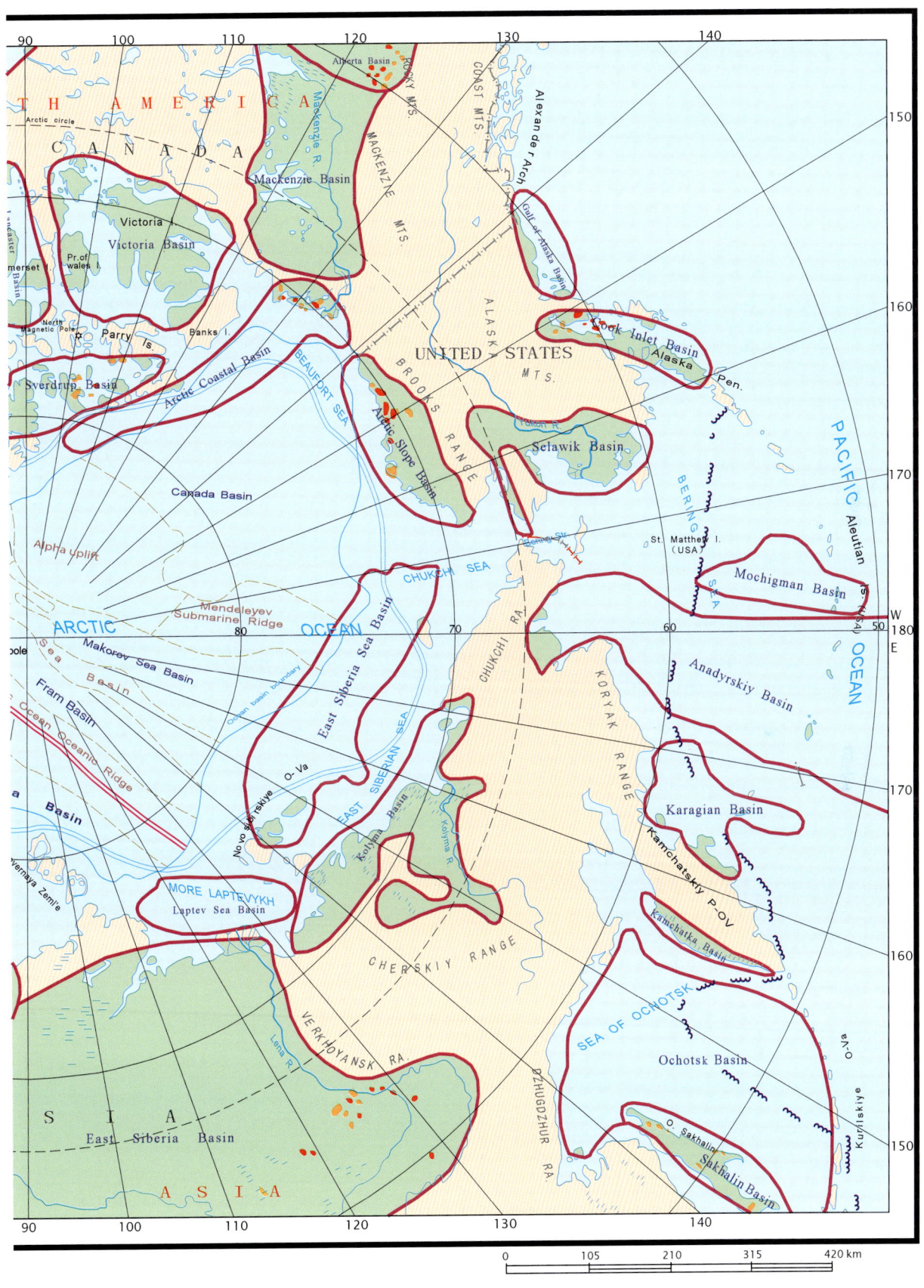

105   Arctic Ocean

Geography

The Arctic Ocean is a body of water that lies mostly north of the Arctic Circle. It is the smallest of the world's four oceans (after the Pacific Ocean, the Atlantic Ocean, and the Indian Ocean), and includes Baffin Bay, Barents Sea, Beaufort Sea, Chukchi Sea, East Siberian Sea, Greenland Sea, Hudson Bay, Hudson Straits, Kara Sea, Laptev Sea, the Northwest Passage and other associated water bodies. Total area is 13.1 million km² with average water depth of 1200 m (maximum water depth of, 449 m). The central surface is covered by a perennial drifting polar icepack that averages about 3 m in thickness.

History

Around the Arctic Ocean there are many sedimentary basins in Russia, the USA, Canada, Norway and Greenland. At present, the oil- and gas-producing basins include the West Siberia Basin of Russia, the Alaska Slope Basin of the USA, the Sverdrup and Arctic Coastal Basins of Canada and the North Sea Basin of Norway. The most notable giant oil field is Prudhoe Bay in Alaska, while the prominent gas field is Yamburg. Details of the sedimentary basins of Greenland with possible hydrocarbon potential are listed in Table 105.1.

Regional geology

The known hydrocarbon-bearing basins bordering the Arctic Ocean are associated with a passive margin setting and occupy the mainland, the vast shelf and part of the continental slope. They are considerable in size and are composed of Proterozoic to recent deposits. Productive horizons are Neogene, Paleogene, Cretaceous, Jurassic and Upper Triassic sandstones and Middle–Lower Triassic, Mississippian and Devonian limestones. Oil and gas traps are anticlinal folds. Source rocks are Cretaceous, Jurassic, Permian, Mississippian, Silurian and Ordovician shales. The region is potentially rich in hydrocarbon resources.

Within the Barents–Kara heterogeneous epi-Archaean–Proterozoic platform, only the Barents and Kara Seas Basin has been identified. The basement of the basin occurs at depths of a few kilometres to 6–10 km in the south. Within the South Barents Depression, the basement is subsided to

Table 105.1   Major sedimentary basins of Greenland

Basin	Area (thousand km²)	Major sedimentary rock		Reservoir	
		Age	Thickness (m)	Age	Lithology
Angmagssalik	25		4500		
Davis Strait	20				
East Baffin Bay	40	Kz		J	Sandstone
Kronprins Christian	30	Mz, Kz	5000–6000	P	Dolomite

World Atlas of Oil And Gas Basins, First Edition. Li Guoyu.
© 2011 John Wiley & Sons, Ltd. Published 2011 by John Wiley & Sons, Ltd.

a depth of 15–20 km. The sedimentary cover has been studied in the southwestern part of the basin. It consists of Upper Palaeozoic terrigenous–marine carbonate and evaporite sequences; Triassic and Jurassic sandstones, clays and coals-; and Cretaceous and Paleogene terrigenous–marine carbonate and volcanic strata. Weakly compacted Neogene rocks complete the succession. In the east, in the Kara Sea portion of the basin, the sedimentary infill is represented by Lower and Middle Palaeozoic carbonate rocks and Triassic and Jurassic marine and continental deposits. Main productive rocks are Jurassic and Triassic deposits. In the eastern portion of the basin, Palaeozoic carbonates are considered to be petroleum reservoirs, and Triassic clays may serve as a seal.

Twelve gas fields have been discovered in the Norwegian and Russian sectors of the Barents Sea Basin. Among them is the unique Shtokmanovskoye gas field (discovered in 1988). The reservoirs are composed of Triassic sandstones and the trap is associated with a large uplift at the side of the South Barents Depression.

Three large basins were formed in Arctic Canada during the mid-Palaeozoic Ellesmerian Orogeny: namely the Sverdrup, Beaufort-Mackenzie and Northern Baffin Basins. The Sverdrup, a successor basin to the Franklinian Geosyncline, subsided into three depocentres, the Sabine, Ellef-Edinburgh and Norwegian, which are separated by transverse arches. The Beaufort-Mackenzie Basin is located in a rift zone between the Canadian Shield and the Brooks Orogen. The intersection of the northeast-trending Kaltag fault system and northwest-trending Brookian structural front localized upper Mesozoic and Tertiary depocentres in the Mackenzie Delta and Beaufort Sea. A Jurassic to Upper Cretaceous marine transgression was followed by Tertiary regressions (Vysotsky, 1995).

References

Beebe, B. W. (1968) *Natural Gases of North America*. American Association of Petroleum Geologists, Tulsa, OK.

Campbell, C. J. (1997) *The Coming Oil Crisis*. Petroconsultants, in association with Multi-Science Publishing.

Cowan, D. S. and Potter, C. J. (1986) *Continent – Ocean Transect B-3: Juan de Fuca Spreading Ridge to Montana Thrust Belt*. Geological Society of America, Boulder, CO.

Cram, I. H. (1971) *Future Petroleum Provinces of the United States – Their Geology and Potential*. American Association of Petroleum Geologists, Tulsa, OK.

Dietz, R. S. and Holden, J. C. (1970) Reconstruction of Pangaea: Breakup and dispersion of continents, Permian to present. *Journal of Geophysical Research* **75**: 4939–4956.

Frisch, R., Brendow, K. and Saunders, R. (1989) *World Energy Horizons, 2000–2020*. Fourteenth World Energy Conference. Éditions Technip, Paris.

Glennie, K. W. (ed.) (1990) *Introduction to the Petroleum Geology of the North Sea*. Blackwell Scientific Publications, Oxford.

Gluyas, J. G. and Hichens, H. M. (2003) *United Kingdom Oil and Gas Fields*. Geological Society Publishing House, Bath.

Grant, N. and Middleton, N. (1987) *Atlas of the World Today*. Harper and Row, New York.

Halbouty, M. T. (1970). *Geology of Giant Petroleum Fields*. American Association of Petroleum Geologists, Tulsa, Oklahoma. AAPG.

Halbouty, M. T. (1982) *The Deliberate Search for the Subtle Trap*. Memoir 32, American Association of Petroleum Geologists, Tulsa, OK.

Halbouty. M. T. (1990) *Giant Oil and Gas Fields of the Decade, 1958–1968*. American Association of Petroleum Geologists, Tulsa, OK.

Halbouty. M. T. (1991) *Giant Oil and Gas Fields of the Decade: 1968–1978*. American Association of Petroleum Geologists, Tulsa, OK.

Halbouty. M. T. (1992) *Giant Oil and Gas Fields of the Decade: 1978–1988*. American Association of Petroleum Geologists, Tulsa, OK.

Hammond. (1988) Discovery World Atlas. Mapcewood, New Jersey.

Hills, L. V. (1974) *Oil Sands: Fuel of the Future*. Canadian Society of Petroleum Geologists, Calgary, Alberta.

Howell, D. G. (ed.) (1993) *The Future of Energy Gases. US Geological Survey Professional Paper*, 1570.

Institute of Microeconomics (2004) *Petroleum Industry of Russia, Middle Asia and China*. Russian Mapping Company, Moscow.

Kewen Gan, Guoyu Li and Liangcheng Zhang. (1990) *China Gas Fields Atlas*. Petroleum Industry Press, Beijing.

King, R. E. (1972) *Stratigraphic Oil and Gas Fields – Classification, Exploration, Methods and Case Histories*. American Association of Petroleum Geologists, Tulsa, OK.

Lapedas, D. N. (1976) *Encyclopedia of Energy*. McGraw-Hill, New York.

Levorsen, A. I. (1954) *Geology of Petroleum*. W. H. Freeman and Company, San Francisco, CA.

Li Guoyu. (1982a) *World Atlas of Oil and Gas Basins (Part 1)*. Petroleum Industry Press, Beijing.

Li Guoyu. (1982b) *World Atlas of Oil and Gas Basins (Part 2)*. Petroleum Industry Press, Beijing.

World Atlas of Oil And Gas Basins, First Edition. Li Guoyu.
© 2011 John Wiley & Sons, Ltd. Published 2011 by John Wiley & Sons, Ltd.

References

Li Guoyu. (1982c) *World Atlas of Oil and Gas Basins*. Petroleum Industry Press, Beijing.
Li Guoyu. (1988a) *China Oil and Gas Basins Atlas*, 2nd edn. Petroleum Industry Press, Beijing.
Li Guoyu. (1988b) *China Oil and Gas Basins Atlas*. Petroleum Industry Press, Beijing.
Li Guoyu. (1990a) *China Oil Fields Atlas (Part 1)*. Petroleum Industry Press, Beijing.
Li Guoyu. (1990b) *China Oil Fields Atlas (Part 2)*. Petroleum Industry Press, Beijing.
Li Guoyu. (1991a) *China Petroleum Geology*. Petroleum Industry Press, Beijing.
Li Guoyu. (1991b) *Collection of Inspection Reports on World Oil Regions and Areas*. Petroleum Industry Press, Beijing.
Li Guoyu. (1991c) *World Atlas of Gas Fields (Part 2)*. Petroleum Industry Press, Beijing.
Li Guoyu. (1991d) *World Petroleum Geology*. Petroleum Industry Press, Beijing.
Li Guoyu. (1996) The theory of sedimentary basins. In *Progress in Geology of China (1993–1996), 30th International Geological Conference*. China Ocean Press, pp. 634–637.
Li Guoyu. (1997) *World Atlas of Oil Fields (Part 1)*. Petroleum Industry Press, Beijing.
Li Guoyu. (2000) *World Atlas of Oil Fields (Part 2)*. Petroleum Industry Press, Beijing.
Li Guoyu. (2002) *The Theory of Sedimentary Basins*. Petroleum Industry Press, Beijing.
Li Guoyu. (2009) *World Atlas of Oil and Gas Basins (English version)*. Petroleum Industry Press, Beijing.
Magon, L. B. and Dow, W. C. (1994) *The Petroleum System – from Source to Trap*. Memoir 60, American Association of Petroleum Geologists, Tulsa, OK.
Miall, A. D. (1980) *Facts and Principles of World Petroleum Occurrence*. Canadian Society of Petroleum Geologists, Calgary, Alberta.
Nakicenovic, N. (1998) *Global Energy Perspectives*. Cambridge University Press, Cambridge.
Odell, P. R. and Rosing, K. E. (1983) *The Future of Oil; World Oil Resources and Use*, 2nd edn. Kogan Page, London.
Perry, W. J., Roeder, D. H. and Lageson, D. R. (1984) *North American Thrust-faulted Terranes*. American Association of Petroleum Geologists, Tulsa, OK.
Riva, J. P., Jr. (1983) *World Petroleum Resources and Reserves*. Westview Press, Boulder, CO.
Selley, R. C. (1985) *Elements of Petroleum Geology*. W. H. Freeman, New York.
Selley, R. C. (1997) *African Basins, Sedimentary Basins of the World*, 3rd edn. Series editor K. J. Hsü, Elesvier Science, Amsterdam.
Shannon, P. M. and Naylor, D. (1989) *Petroleum Basin Studies*. Graham and Trotman, London.
Tarbuck, E. J. and Lutgens, F. K. (1991) *Earth Science,* 6th edn. Macmillan Publishing Company, New York.
U. S. Geological Survey. (2000) *World Petroleum Assessment*. Washington, DC.
Vysotsky. V.I. (1995) *Map of World Oil and Gas Potential*. Moscow: Vnllzarubezhgeologia, Moscow.
Vysotsky, V. I., Isaev, E. N., Klestchov, K. A., Militenko, N. V., Namestnikov, Yu G. and Fyodorov, D. L. (1995) *Map of World oil and Gas Potential, Scale 1:15,000,000, Explanatory Note*. Vnllzarubezhgeologia, Moscow.
Yongxin Jiang and Yishan Dou. (2003) *World Atlas*. Xinqiu Topography Press, Beijing.

Index of Countries

Afghanistan 84
Albania 254, 274
Algeria 176, 194
Andorra 278
Angola 177, 222
Arctic 462
Antarctica 458
Antigua and Barbuda 343
Argentina 405, 440
Armenia 144
Australia 446
Austria 254, 274
Azerbaijan 144

Bahamas 343, 398
Bahrain 156
Bangladesh 84
Barbados 343, 398
Belarus 255, 298
Belgium 254, 282
Belize 343, 398
Benin 176, 206
Bhutan 84
Bolivia 405, 424
Bosnia and Herzegovina 254, 258
Botswana 177, 250
Brazil 404–405, 432
Brunei 68
Bulgaria 255, 258
Burkina Faso 176, 198
Burma (Myanmar) 64
Burundi 177, 242

Cambodia 64
Cameroon 176, 210
Canada 342, 346
Central Africa 176, 226
Chad 176, 226
Chile 405, 440
China (Peoples' Republic of China) 46
Comoros 246
Congo 177, 218
Colombia 404, 408
Costa Rica 343, 398
Cote D'ivoire 176, 206
Croatia 254, 258
Cuba 343, 398
Cyprus 172
Czech Republic 254 266

Democratic Republic of Congo 177, 218
Denmark 254, 290
Djibouti 176, 234
Dominica 343
Dominican Republic 343, 398

East Timor 72
Ecuador 404, 408
Egypt 176, 182
El Salvador 343
Equatorial Guinea 176, 210
Eritrea 176, 234
Estonia 255, 298
Ethiopia 176, 234

Falkland Islands (Islas Malvinas) 405
Fiji 454
Finland 255, 290
France 254, 282
French Guyana 404

Gabon 177, 218
Gambia 176, 198
Georgia 144
Germany 254, 270
Ghana 176, 206
Greece 255, 274
Greenland 342
Grenada 343
Guatemala 343, 398
Guinea 176, 202
Guinea Bissau 176, 202
Guyana 404, 412

Haiti 343, 398
Honduras 343, 398
Hungary 254–255, 258

Iceland 254, 290
India 88
Indonesia 72
Iran 148, 152
Iraq 148, 152
Ireland 254, 286
Israel 148, 172
Italy 254, 274

Jamaica 398
Japan 60
Jordan 148, 172

Kazakhstan 96
Kenya 177, 238
Kyrgyzstan 96
Kuwait 149, 156

Laos 64
Latvia 254, 298
Lebanon 148, 172
Lesotho 177, 250
Liberia 176, 206

World Atlas of Oil And Gas Basins, First Edition. Li Guoyu.
© 2011 John Wiley & Sons, Ltd. Published 2011 by John Wiley & Sons, Ltd.

Index of Countries

Libya 176, 186
Liechtenstein 254, 270
Lithuania 255, 298
Luxembourg 254, 270

Macedonia 258
Madagascar 177, 246
Malawi 177, 222
Malaysia 68
Maldives 88
Mali 176, 198
Malta 254, 274
Mauritania 176, 198
Mauritius 246
Mexico 343, 390, 398
Moldova 255, 298
Monaco 254, 282
Mongolia 56
Montenegro 254, 258
Morocco 176, 194
Mozambique 177, 246

Namibia 176, 250
Nepal 84
Netherlands 254, 282
New Zealand 454
Nicaragua 398
Niger 176, 210
Nigeria 176, 210
North Korea 52
Norway 254, 290

Oman 149, 156

Pakistan 84
Palestine 148, 172
Panama 343, 398, 404
Papua New Guinea 446
Paraguay 405, 424
Peru 404, 424
Philippines 80
Poland 254–255, 266
Portugal 254, 278

Qatar 149, 156

Reunion 246
Romania 255, 258
Russia 255, 306
Rwanda 177, 242

Samoa 454
San Marino 274

Sao Tome and Principe 177, 210
Saudi Arabia 148, 156
Senegal 176, 198
Serbia 255, 258
Seychelles 246
Sierra Leone 176, 202
Singapore 68
Slovakia 254–255, 266
Slovenia 255, 258
Somalia 176, 234
South Africa 177, 250
South Georgia Island 405
South Korea 52
South Orkney Island 405
South Shetland Islands 405
Spain 254, 278
Sri Lanka 88
St Lucia 343
Sudan 176, 226
Surinam 404, 412
Swaziland 177, 250
Sweden 254, 290
Switzerland 254, 270
Syria 148, 172

Tajikistan 96
Tanzania 177, 242
Thailand 64
Togo 176, 206
Tonga 454
Trinidad and Tobago 343, 398
Tunisia 176, 194
Turkey 148
Turkmenistan 96

Uganda 177, 238
Ukraine 255, 298
United Arab Emirates 149, 156
UK 254, 286
Uruguay 405, 432
USA 342, 354
Uzbekistan 96

Venezuela 404, 412
Vietnam 64

Western Sahara 176, 198

Yemen 149, 156

Zambia 177, 222
Zimbabwe 177, 250

Index of Basins

Abidjan 176, 206–207
Abrolhos 405, 433
Abukuma-Oki 61
Acre-Beni 404–405, 424–425, 432
Adana 42, 148
Adavale 447
Adriatic Sea 254–255, 258, 274–275
Afars 176, 234
Afghan-Tajik 42, 84, 96–97, 140
Agusan-Davao 81
Ahnet 176, 195
Akimeugah 73
Akita 61
Alberta 342, 346, 350, 354, 463
Alma-Ata 97
Alor 73
Altiplano 405, 425
Amadeus 446
Amazonas 404, 432
Amundsen Sea 459
Amur (Kara-Kum) 84, 96, 100, 122
Anadarko 343, 354–355
Anadyrskiy 43, 307, 463
Andaman-Nicobar 65
Angmagssalik 342, 347, 462
Anju 52
Anzan 176, 239
Appalachian 342, 355, 382
Aquitaine 254, 283
Arafura 446
Arckaringa 446
Arctic Coastal 342, 346, 463
Arctic Slope 342, 354, 386, 463
Argentina 404, 441
Arkoma 343, 355
Assam 85, 88
Avis 399, 404, 412

Baffin Bay 342, 347, 462
Bafra 42, 148
Baise 46
Baja California 343, 390
Balkhash 97
Baltic Sea 254–255, 266–267, 291, 298, 462
Bangkok 64
Barbados-Tobago 343, 399
Barents Sea 255, 290, 306, 462
Barinas 399, 405, 412
Bass 447
Bayano 343, 398
Bechar 195
Beibu Gulf 46
Beihuanghai 46
Bellingham 343, 346

Bellingshausen Sea 458
Bend Uplift 343, 354–355
Bengal 64, 85, 88
Bengal Bay 43, 64, 89
Benue 176, 211
Bien Ho 65
Big Horn 343, 355, 366
Biru 46
Black Mesa 343, 355, 366
Black Warrior 343, 355
Bligh Water 455
Blue Nile River 177, 227, 234
Bohai Bay 42, 46
Boise 343, 354
Bombay 88, 92
Bonaparte Gulf 446
Bowen 447
Bowser 343, 346
Boyang 46
Brabant 254
Brandsfield Strait 458
Browse 446
Brunei-Sabah 69
Brunt-Kronprinsesse 458
Burgas 259

Cagayan Valley 80
Calio 65
Campos 405, 432–433, 436
Canning 446
Canterbury 454
Cape Vgel 447
Cariaco 343, 399
Carnarvon 446
Carpathian 255, 259, 262, 266–267, 299
Caspian Sea 96, 100
Castilia 254, 278–279
Cauca 405, 408–409
Celtic Sea 254, 287
Central (Algeria) 177, 195
Central (Antilles) 343, 399
Central Gobi 56
Central Iran 42, 149, 153
Central Uplift 354
Ceram 73
Cesar 408
Chaco 405, 425, 440
Chapayal 343, 378, 391, 398
Chelif 177, 194, 195
Chiang Mai 64
Chiba 61
Chile North 405, 440–441
Chile South 405, 441
Chindwin 64

World Atlas of Oil And Gas Basins, First Edition. Li Guoyu.
© 2011 John Wiley & Sons, Ltd. Published 2011 by John Wiley & Sons, Ltd.

Index of Basins

Chobalsan 57
Choconarino 405, 408–409
Chu-Sarysu 42, 97, 128
Chuxiong 46
Cincinnati Uplift 343, 355
Clarence 446
Coastal Llano de Caribe 343, 398
Coastal Llano de Pacifica 343, 398
Colombia 399
Commonwealth Sea 458
Conakry 176, 202–203, 206
Congo 177, 218, 222
Cook Inlet 342, 355, 387, 463
Coonamble 447
Cooper 447
Copper 355
Coromandel 89
Corsica 254, 283
Corum 148
Cotabato 81
Cuanza 177, 222, 250
Cuba Central 343, 398
Cuba North 343, 399
Cuba South 343, 398
Cuoqin 46
Cuyo 405, 441
Cyprus 148, 172
Cyrenaica 177, 187

Daly 446
Davis Strait 342, 347, 462
Delaware 343, 354
Denmark 254, 291, 462
Denver 343, 354, 366–367
Dnept-Donets 255, 299, 302
Doba 176, 226
Dongtinghu 46
Drummond 447
Dunhuang 46
D'Urville Sea 459

East Africa 177, 235, 239, 243, 246
East Baffin Bay 343, 347, 462
East China Sea 47
East Coast 454
East Coastal 342–343, 347, 355, 378
East Gobi 57
East Irish Sea 254, 287, 295
East Kalimantan 69
East Siberia 43, 307, 330, 462–463
East Siberia Sea 307, 463
Ebro 254, 278–279
Eel 354
English 287, 295
English Channel 254, 287
Erlian 47
Essaouira 177, 194
Eucla 446
Eyasi 177, 242

Falcon 398, 405, 412
Fang 65

Farafra 177, 182
Feira de Santana 404, 405, 433
Fergana 42, 97, 136
Fitzroy 446
Flores 73
Forest City 343, 355
Forth 286, 294
Foxe 343, 347, 462

Ganga 42, 84–85, 88
Gao 177, 199
Gauvery 89
Gefara 176, 186, 195
Geluong River 65
Georgina 446
Gercif 176, 195
Gippsland 447, 450
Golfo de San Jorge 405, 441
Graham 458
Grand Banks 342, 347
Great Divide 343, 355, 366
Green River 343, 355, 366
Grenada 343, 399
Greymouse 454
Guadalquivir 254, 278–279
Guajira 405, 408
Guanshi Beach 46
Guayaquil 405, 409, 424
Gulf of Alaska 342, 354, 387, 463
Gulf of Carpentaria 446–447
Gulf of Mexico 343, 354–355, 378–379
Gulf of Siam-Malay 65, 68

Hailaer 47
Hakodate 61
Harney 354
Hassi Homer 177, 195
Hauts Plateau 176, 194, 195
Hebrides Sea 254, 286
Heimand 84
Helgeland 254, 290, 462
Hermand 42
Hetao 46
Hokkaido 61
Hongwen 52
Huabei South 47
Huahai 46
Hudson Bay 342, 347, 462
Hysan 52

Illinois 343, 355
Illizi (Hamra) 176, 186, 195
Iloilo 81
Imperial Valley 343, 355
Indus River 42, 84, 88
Irrawaddy 64
Irrawaddy Delta 64
Iullimeden 176, 199, 210

Jamaica 343, 398
Jazmuri 42, 149, 153
Jianghan 46

471

Index of Basins

Jiaolai 46
Jiudong 46
Jiuxi 46
Junggar 42, 46

Kalahari 177, 250–251
Kalparowite 343, 355, 366
Kamchatka 307, 463
Kandik 354, 387
Kansas Uplift 343, 354
Karagian 43, 307, 463
Karakum 42
Karoo 177, 251
Katale 177, 242
Katawaz 42, 84
Kekexili 46
Khorat 64
Khurtum 176, 227
Kichung 65
Kilchu 52
Kolyma 43, 307, 463
Konkan 42, 89
Korce 275
Kosmonaut Sea 458
Kronprins Christian 342, 462
Kufra 176, 187, 226
Kumukuli 46
Kura 42, 144, 145
Kutei 72
Kyzyl-Kym 42, 96–97

Lamon Bay 80
Lampang 65
Lamu 177, 235, 239
Lancaster 342, 346–347, 462–463
Lanping-Simao 46
Laptev Sea 307, 463
Large Lake 56
Lasu 65
Laura 446
Laut Sulawesi 73, 81
Lebu-Arauco 405, 441
Lianos 404, 408–409
Lile-Xianbin 46
Los Angeles 343, 354, 362
Lower Congo 177, 211, 218, 222
Luangua 177, 223
Lusitania 254, 278
Lut 42, 149, 153
Luzon Central 80

Mackenzie 342, 346, 463
Mac Sot 65
Madagascar East 177, 247
Maga 81
Magdalena Inferior 405, 408–409
Magdalena Medio 398, 405, 408
Magdalena Superior 399, 405, 408
Magellan 405, 440, 441
Magyshlak 42
Majunga 177, 247
Makran 42, 84, 149, 153

Malawi 177, 222, 242, 246
Maluccas 73
Mangyshlak 96, 100, 114
Maracaibo 399, 405, 412, 416
Marajo 404, 412, 432
Maranhao 404, 432
Marib-Jawf 42, 149, 157
Marie Byrd Land 459
Maryborough 446
Masuda 61
Mekong 65
Melut 176, 227
Mentawai 72
Michigan 343, 347, 355, 374
Mid-Continent 355
Middle Tanana 355
Midland Valley 254, 286–287, 462
Miyazaki 61
Mobutu-Sese-Seco 177, 218, 238
Mochigman 307, 463
Moesian 255, 259
Moho 46, 307
Molasse 254, 271, 274
Moquegua 405, 440
Moray 286, 294
Morondava 177, 246
Moscow 255, 306, 462
Mouydir 176, 195
Mozambique 177, 246, 251
Muglad 176, 227, 230
Murray 446–447
Murzuk 176, 186, 195, 210

Nam-Gang 52
Nan-Won 53
Nanxiang 46
Naoetsu-Nagan 61
Nergin 57
Nestos 275
Neuquen 405, 441
Nevada 342, 354
New Britain 447
New Hebrides 455
Ngalia 446
Nicaragua 343, 398
Niger Delta 176, 207, 211, 214
Niger East 176, 210, 226
Niigata 61
Nile River 176, 182
Nile River Delta 177, 182
Nirihuau 405, 441
Noreste 343
North 343, 399
North Cantabrian 254, 278
North Coast 73
Northeast 378, 390–391, 398
North East German 254, 270, 462
North Egypt 176, 182
Northern New Guinea 447
North Fernando 404, 433
North Java 42, 72
North Kavkaz 42, 255, 306, 334

Index of Basins

North Minch 254, 286
North Palawan 81
North Park 343, 355, 366
North Sea 254, 294, 462
North Ustyurt 42, 96, 100, 110
North West German 254, 270, 295, 462

Ochotsk 43, 307, 463
Officer 446
Ogaden 176, 235
Okcheon 53
Okinawa 61
Okovango 177, 222, 250–251
Ontario 343, 346
Orange 177, 250
Ord 446
Ordos 42, 46
Otway 446–447
Outer Hebride Sea 254, 286
Ovorhanga 56
Oxley 446
Ozark Uplift 343

Palawan 81
Palo Duro 343, 354
Panama 398–399
Pannonian 254, 258, 266–267
Papuan 447
Paradox 343, 354, 366
Parana 405, 425, 432, 440–441
Paris 254, 282–283
Pasco 354
Pedirka 446
Pedregosa 355
Pelotas 405, 433, 441
Permian 354
Persian Gulf 42, 148, 152, 156–157, 160, 168, 172–173
Perth 446
Phayau 65
Phosphat 177, 194
Piceance 343, 355, 366
Pirie Torrens 446
Plovdiv 254, 259
Po 254, 274
Poland 254–255, 266–267, 462
Polar 458–459
Polda 446
Potiguar 405
Potwar 84, 88
Powder River 343, 354, 366
Prague 254, 266
Preapennines 254, 274
Pre-Black Sea 255
Pre-Caspian Sea 42, 96, 100, 104, 255, 306
Puerto Rico 343, 398
Pusan 53
Putumayo 404, 409, 424, 428, 433

Qaidam 42, 46
Qiangtang 46
Qinshui 46
Quajira 399

Quebec 343, 346
Queenk 354
Quesnel 343, 346
Qunle Bay 43

Red River 64
Red Sea 42, 148, 156, 176, 183, 227
Reggan 176, 195
Rharb 176, 194
Rhine 254, 271, 282
Rhone 254, 283
Riconcavo 405, 433
Riiser-Larsen Sea 458
Rio Colourado 405, 441
Ross Sea 459
Ruhuhu 177, 242
Rukwa 177, 242

Sabi 251
Sacramento 343, 354
Sagaleh 177, 234
Sahna 354, 355
Sakhalin 307, 338
Sakhonnakhon 65
Salawati 73
Salina 343
Salonica 275
Sand Wash 355, 366
San Francisco 405, 432
Sanjiang 46, 307
San Joaguin 343, 354
San Jose 343
San Juan 343, 354, 366
San Luis 405, 433
Sanshui 46
Santos 405, 432–433
Santa Crus 80
Santa Cruz 405, 425, 441
Santanghu 46, 56
Sardegna 254, 274
Scotia Shelf 342, 347
Selawik 342, 355, 387, 463
Selous 177, 243
Senegal 176, 198, 202
Sergipe-Alagoas 405, 433
Shiratsukari 61
Shiwandashan 46
Sichuan 42, 46
Sicily 254, 274
Siempang 65
Sinu 405, 408
Sirte 176, 187, 190
Sivas 148
Snake River 354
Solander 454
Songliao 42, 47
South Aden 177, 234
South Cape 177, 250–251
South Caspian Sea 42, 96, 100, 118, 149, 153
South Ceara 405, 433
Southeast Luzon 80–81
South Gobi 56

Index of Basins

South Island 454
South Java 72
South Levant 172, 177, 182–183
South Orkney 458
South Shetland 458
South Yellow Sea 47
St Lawrence 342, 347
Strumon 275
St Vincent 446
Subei 46
Suez 172, 177, 182
Sulawesi 73
Sulu 69, 81
Sumatra Central 42, 72, 76
Sumatra South 72
Sumatra North 72
Sumbawa 72–73
Sur 343, 378, 390, 391, 398
Surat 447
Sverdrup 342, 346, 462–463
Sweet Grass Uplift 354, 366
Sydney 447

Tabriz 42, 148, 152
Taetong Bay 52–53
Taixi 47
Taixinan 47
Takutu 412, 433
Talakan 72–73
Tampico 343, 378, 391, 394, 398
Tamtsag Hailar 57
Tanganyika 177, 218, 242
Taodenni 176, 198–199, 203
Taranaki 454
Tarfaya 176, 194, 198
Tarim 42, 46
Tasmania 446
Tengiz 42, 97
Teshio 61
Thrace 42, 148, 275
Thuringian 254, 270
Timano-Pechora 255, 306, 322, 462
Timor 73
Tindouf 176, 194, 198
Tolo 73
Tomini Bay 73
Tonga 455
Towada 61
Toyama-Tsushima 61
Transylvanian 254, 259
Triassic 176, 194, 195
Trinidad 399
Tucano 405, 433
Tuha 46
Tuli 251

Turgay 42, 96, 132
Turkana 177, 239
Tuz Golu 148

Ubsa Nor 56
Ucayali 404, 409, 424
Uinta 343, 355, 366
Umatilla 354
Ungava Bay 343, 346, 462
Usangu 177, 242
Utulsteumen Sea 458

Val Verde 343, 354
Venezuela East 399, 405, 412–413, 420
Venezuela Gulf 398, 405, 412
Veracruz 343, 378, 391, 398
Victoria 342, 346, 463
Victoria Land 459
Vienna 254, 266, 274
Visayan 81
Volga-Urals 255, 306, 318
Volta 176, 199, 207
Voring 254, 290, 462

Wanan 46, 68
Weddell Sea 458
Wessex 254, 287, 295
West Coast 454
West Mindoro 80–81
West Netherlands 282, 295
West Norway Basin 291, 294
West Sheltand 254, 286, 294
West Siberia 42, 306, 326, 462
West Washington-Oregon 343, 354
White Horse 343, 346
Wilkes Land 459
Williston 342, 346, 351, 354, 366–367, 370
Wind River 343, 355, 366
Wiso 446

Yamagata 61
Yanghu 46
Yankman 355
Yanqi 46
Yareol 446
Yilan-Yitong 46
Yingen 47
Yingge Sea 46, 47
Yucatan 398

Zambesi 177, 213, 246, 251
Zengmuansha 46, 69
Zhongba-Baori 46
Zhongjianxi 46
Zhujiang Kou 47